COCKPIT ENGINEERING

Cockpit Engineering

D.N. JARRETT
QinetiQ, UK

LONDON AND NEW YORK

First published 2005 by Ashgate Publishing

Published 2016 by Routledge
2 Park Square, Milton Park, Abingdon, Oxfordshire OX14 4RN
711 Third Avenue, New York, NY 10017, USA

First issued in paperback 2016

Routledge is an imprint of the Taylor & Francis Group, an informa business

British Library Cataloguing in Publication Data
Jarrett, D. N.
Cockpit engineering
1.Airplanes - Cockpits - Design and construction
2.Aeronautical instruments - Display systems - Design
3.Aeronautics - Human factors
I.Title
629.1'3445

Library of Congress Cataloging-in-Publication Data
Jarrett, D. N.
Cockpit engineering / by D.N. Jarrett.
p. cm.
Includes index.
ISBN 0-7546-1751-3
1. Airplanes--Cockpits--Design and construction. 2. Aeronautics--Human factors. I. Title.

TL681.C6J37 2005
629.134'45--dc22

2004030374

ISBN 13: 978-1-138-27377-1 (pbk)
ISBN 13: 978-0-7546-1751-8 (hbk)

Contents

List of Figures

List of Tables

Preface

I joined the staff of the Royal Aircraft Establishment at Farnborough in early 1974, intending to continue a career in materials science. Instead I was asked to join a group dealing with the new topic of helmet-mounted display systems. I took the job thinking it would be an interesting diversion until an internal transfer came along to enable me to resume my original interests. However, over the years the name and size of the organisation changed, firstly in a minor way to become the Royal Aerospace Establishment, then in a major way when amalgamated with similar government research organisations to become a division of the Defence Research Agency. With further rationalisation this morphed into the Defence Evaluation and Research Agency, which was split in 2001, the larger part becoming a commercial research company called QinetiQ. Now in 2005 the world is a very different shape, as is my employer, but I am in the same place and still working in the same field.

I guess my main motive for constructing this book is to explain why I have found the work so engrossing. In part, this is because anything to do with aircraft is an engineering challenge, but it is also because the cockpit presents a uniquely complex design problem. It is not just the place where designers find room for a mass of equipment and let the pilot decide how best to use it. The cockpit engineer must make the interaction with the machine as easy and as natural as possible so that when fastening the seat harness, the pilot should genuinely feel as though he is strapping the aircraft onto his back. I realise that there is nothing new in arranging a workplace to suit the worker, and the principles of human-centred design are probably as old as tool-making and therefore implicit in the thoughts of homo habilis long before homo sapiens. But when a pilot straps a modern combat aircraft onto his back, the immense power of the machine and the weapons it carries confers on the pilot a unique combination of responsibility and vulnerability. He needs all the help the designers can give.

The subject spans many scientific and engineering disciplines, and this has made it difficult to balance the content. Because a combat aircraft presents the greatest engineering and ergonomic challenge, the scope and the layout of the chapters are based on the single-seat fighter. I know from my own formal education and training that engineers are unlikely to be familiar with human abilities and limitations, so I have included condensed explanations of some relevant physiological and psychological ideas. I have also tried to look at equipment from the aircrew viewpoint and indicate how requirements are governed by the way the equipment is used. These interwoven human and operational strands lend colour to what could otherwise be a bland review of the technologies, and I hope that the expert will forgive the many instances where complex issues have been simplified for brevity.

I make two apologies. The first arises because I have not converted quantitative units into the scientific standard, but have retained the units that seem most appropriate to the context. This is because aircraft are operated using kilometers, miles and nautical miles as measures of distance, and both meters and feet are common units of altitude. Airspeed is invariably in knots; nautical miles per hour. Engineers in the US tend to use feet, inches and pounds, and it is only scientists and non-US engineers who generally work in SI units. The case for consistent use of scientific measures is however overwhelming, so I do feel guilty in colluding with the present confusing mixture. The second apology arises because I wanted to avoid the clumsiness of peppering the text with "he/she" and "he or she". General references, for instance to aircrew and designers, should not be taken to imply that all such people are male.

Relevant acronyms and abbreviations are listed at the back of the book, ahead of the index. Instead of including a glossary, the list points to the page where the abbreviation is introduced. Several abbreviations are ambiguous, but I hope that the intended interpretation is clear from the context.

I am very grateful for the help given by a large number of colleagues from RAE/DERA/Qinetiq who have made constructive comments on the draft; Alan Brown, David Bigmore, Ian Davies, Sarah Day, Peter Downs, Laurence Durnell, Eric Farmer, Soo James, Peter Longman, Henry Lupa, Graham Rood and James Sadler. I am especially grateful to Judith Ineson for additional help with proof-reading and for suggesting myriad improvements to the text.

I have also enjoyed working with other colleagues over the last thirty years. There are really too many to name, but in particular, Mark Allen, Judy Aplin, John Banbury, Nick Benger, Phil Catling, Simon Dalton, Tony Doyle, Alan Hepper, Craig Hudson, Tony Karavis, Martin Kaye, Charles Parker, Mark Reed, Howard du Ross, Geoff Rowlands, Barry Short, Fiona Smith, Alan South and David Thorndycraft. I have also been privileged to get to know many scientists, engineers and aircrew during collaborative endeavours. Membership of AGARD Working Party 25 was very educational as the group included Tim Anderson, Karen Carr, Pierre Dauchy, Bernard Hudgins, Alain Leger, Grant McMillan and Dominique Pastor. The quietly-spoken expertise of Josh Borah. has been particularly impressive. Again, there have been a large number of people, including Owen Wynn, Mike Edwards, Paul Marshall, Steve Heptinstall, Peter Wilkinson, Nigel Cox, Brian Tsou, and Mike Haas, who have given their knowledge and experience freely.

Finally, my thanks go to my family and most particularly my partner Patricia, for their patient support and subtle encouragement.

Don Jarrett

Chapter 1

Introduction

1.1 Scope

The English word "cockpit" originally meant a gaming enclosure where a pair of domesticated male fowl fought. By the 16th century, the word also denoted the arena or pit of a theatre, and later the area on the orlop deck on a man-of-war where wounded seamen were treated[1]. By the 18th century it had been shifted by naval humour to describe the wheel well for the helmsman, and at the beginning of the 20th century it flew naturally into the aeroplane. Although there are other words for the workplace of the aviator, such as work-station, crewstation or flight-deck, this word has survived, perhaps because it carries the connotations of frenzy, excitement, discomfort and danger so nicely. This is unfortunate because the aim of the designer is actually to provide the antithesis; a place of calm, professional assurance and safety.

The cockpit of the single-seat combat aircraft has probably been most challenging to design because a fast, manoeuvrable aircraft has little airframe space to house the pilot and the systems, and the pilot must be able to perform all tasks while exposed occasionally to violent forces. The environment can also be extremely hot or cold, bright or dark and it is invariably deafening. On the other hand, because superior performance of any kind enhances the military effectiveness of the craft and the chance of the crew surviving, the designers have received a lot of assistance. Aircrew have been studied to a greater degree than any other worker, and systems that can improve their performance have been the subject of intensive research and development effort. Like most of the engineering of a combat aircraft, the cockpit can be regarded as a forcing problem. Systems devised specifically for this context are commonly adapted for use in other types, and in other vehicles, after the initial cost of transforming an idea into a refined and tested embodiment has been absorbed by a military budget.

This chapter gives a very brief overview of the evolution of the ways that designers have accommodated crew in aircraft and provided facilities for them to perform their roles. The main objective is to state the need for equipment as a context for later chapters that describe the engineered form of the equipment.

1.2 Historical overview

Figures 1.1 to 1.6 illustrate how the single-seat combat aircraft cockpit has changed over the course of the 20th century from a biplane of the First World War to the configuration proposed for the Joint Strike Fighter that will enter service in the USA in several years time. The examples are intended to represent the contemporary standard and illustrate recognisable advances.

The first example in Figure 1.1 shows the cockpit of a typical First World War fighter, the SE5A, which was designed by the Royal Aircraft Factory at Farnborough, England. The cockpit was essentially a reinforced opening in the top of the fuselage that had a padded leather edge and a small frontal glass windscreen. The seat, control levers and pedals were placed so that they could be reached by the pilot when he sat with his head and shoulders proud of the opening, and the necessary gauges and indicators were mounted on a plywood panel fastened below the front.

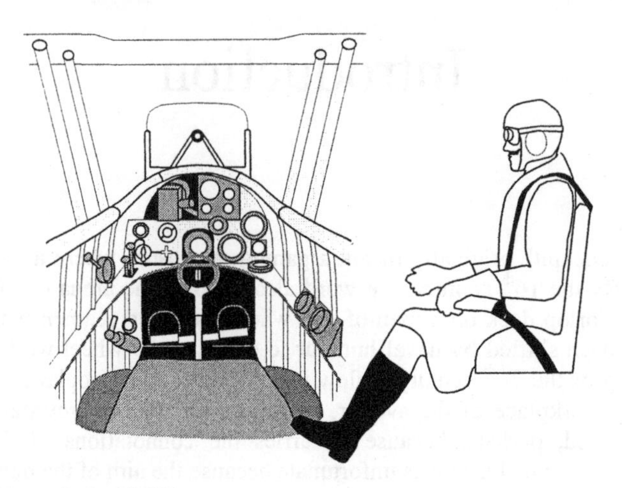

Figure 1.1 SE5A cockpit (1917)

The disposition of the components in this aircraft was as pragmatic as any contemporary automobile or locomotive. The pilot had to manage and fire one machine gun by operating the breech, which intruded into the cockpit through the top left segment of the instrument panel, while looking down the sighting tube to aim the whole aircraft at the target. Although idiosyncratic in many respects, the design was typical of the era in adopting a pair of rudder pedals, a central stick and a left-handed throttle as the main flying controls.

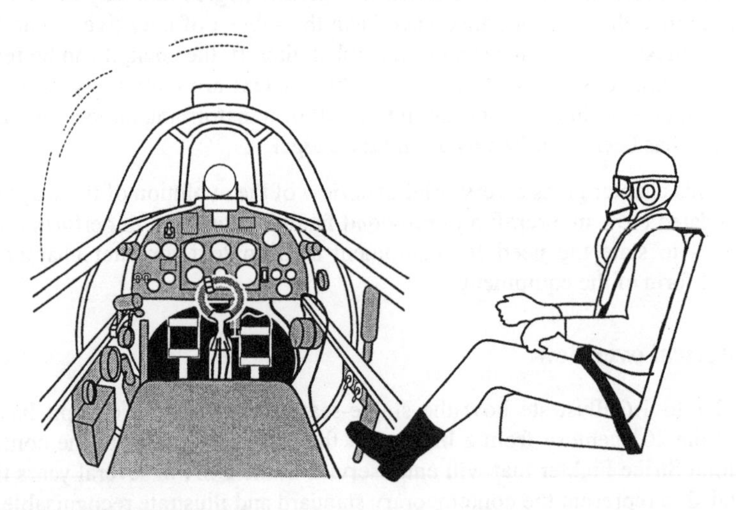

Figure 1.2 Spitfire cockpit (1940)

The second example, the Spitfire made by the Supermarine Aircraft Company shown in Figure 1.2, was a classic of the Second World War. It provided the pilot with vital enclosing protection against the cold windblast when flying at the newly attainable airspeeds, and with pressurised breathing air to counter the lack of oxygen at the newly attainable altitudes. More gauges were installed to show the flying state of the aircraft and the temperatures and pressures of the more complex engine, and there were more ancillary controls for the pumps and valves of the multiple-tank fuel system and for raising and lowering the undercarriage. The primary weapon was still the gun, but the pilot aimed the aircraft at the target using a lead-predicting gyro-stabilised gunsight and he fired multiple guns simultaneously by operating an electrical switch on the top of the stick.

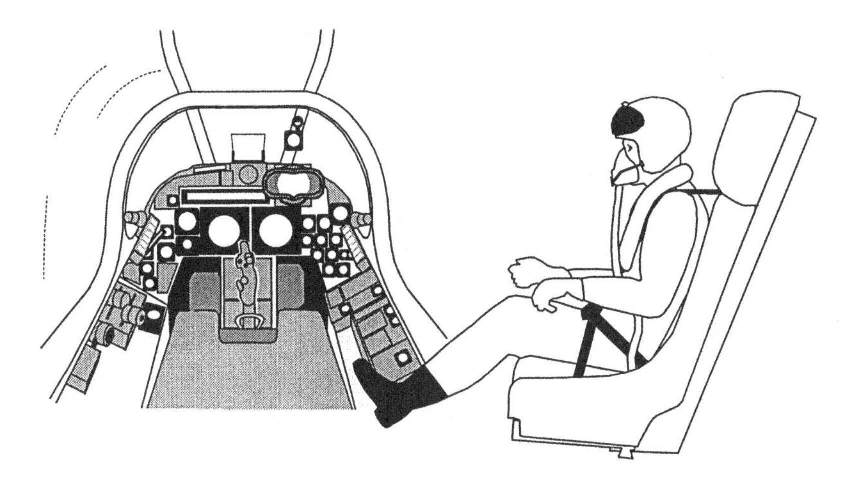

Figure 1.3 Lightning cockpit (1964)

Figure 1.3 shows the cockpit of the English Electric Lightning. This represents the transformation needed to accommodate the extension of flight into the supersonic and stratospheric regime enabled by the jet engine, in particular, a pressurised cabin with a strong stiff canopy frame. It also illustrates the sudden expansion of display and control panels for the complex engines, re-heat boosters, fuel transfer cocks, communication radios, beacon-based navigation systems, central warning system and nose-mounted radar. The last made use of an electronic display device, a cathode ray tube (CRT), which was mounted just beneath the glare shield on the front instrument panel to the right of the gunsight. The CRT produced so little light that the pattern of blips marking the reflections from the objects illuminated by the radar beam could not be discerned unless the screen was shielded from sunlight by a rubber cowl. The panel also housed large, circular attitude and direction indicators and a large horizontal combined airspeed and Mach Number strip gauge. The rear-mounted engines created a relatively quiet ambience and enabled the pilot to see downwards over the aircraft nose. The ejection seat transformed the chance that the pilot would survive a mishap, and brought numerous complexities, including the need for a hard-shelled helmet.

The fourth example, the Panavia Tornado, is perhaps the epitome of apparent complexity. As sketched in Figure 1.4, the glare shield and side consoles are awash with key-pads and switches for the pilot to interact with the newly introduced navigation, communication and weapon aiming computers. The few evident CRT displays are substantially brighter than the Lightning's radar display and incorporate filters to reduce glare, but they are squeezed onto the front panel, along with a large number of electro-mechanical dials and an intrusive head-up display (HUD). The latter can be regarded as an enlarged gunsight, in which the lamp-illuminated aiming reticule has been replaced by a set of symbols projected from a very bright CRT.

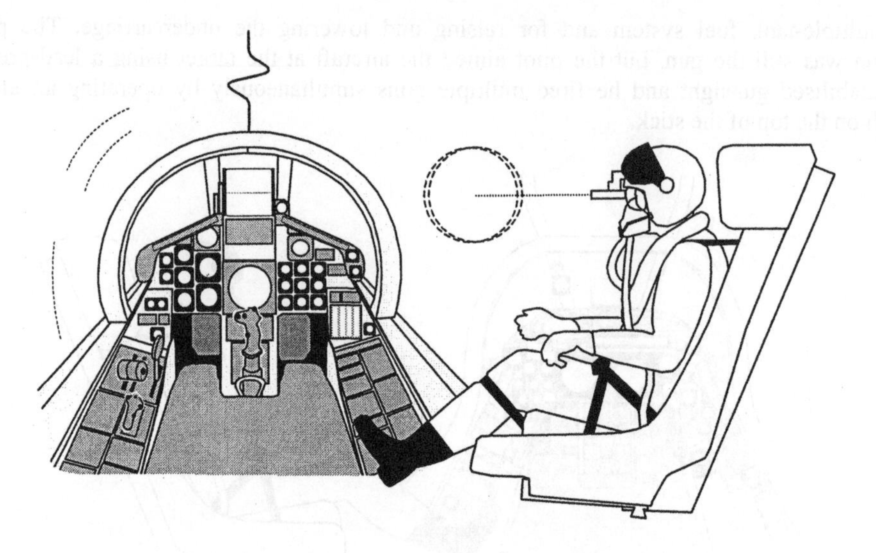

Figure 1.4 Front cockpit of a Tornado (1974)

Another CRT is hidden in the bowels of the Combined Optical Map and Electronic Display (COMED) on the front panel underneath the HUD, and the shell of the helmet provides an anchorage for another kind of electro-optical device, the night vision goggle. Although not apparent in the sketch, the introduction of the latter devices entailed illuminating the cockpit with compatible blue-green light.

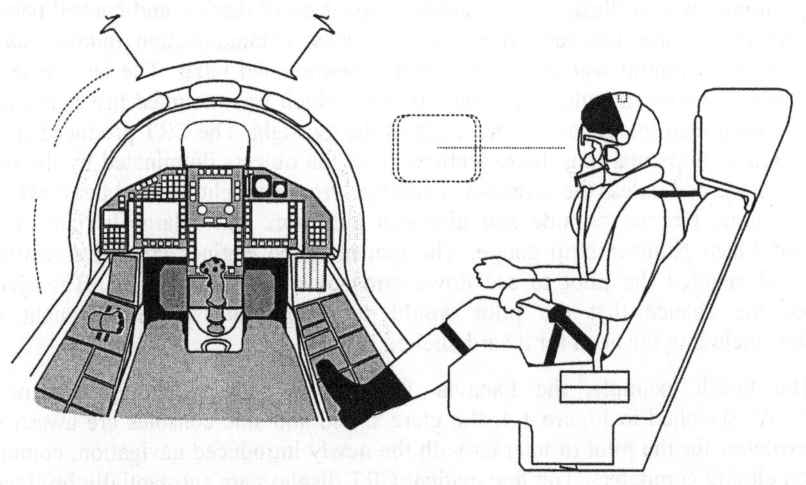

Figure 1.5 Eurofighter cockpit (1986)

The fifth example, the Eurofighter cockpit, represents the consolidated form of "glass cockpit" which arose from the rapid maturation of computing technology during the late 1980s and the adoption of digital electronics for the control all on-board systems. In this case, multi-function electronic display devices are used to show images generated by digital computers, and in comparison with the Tornado the layout seems remarkably well organised and uncluttered. The three full-colour displays are surrounded by keys that enable the pilot to select, for instance, tactical overviews, maps of differing scale or summaries of the, fuel, engine and systems states.

The pilot has a variety of ways to interact with the on-board systems, such as the keypad on the left coaming for entering navigation data, the switches embedded in the hand-grips for fast, routine selections – the HOTAS (hands-on-throttle-and-stick) idea – and he can also make some selections by voice command. So that less time need be spent looking into the office, the HUD imagery is larger and more informative, and symbolic information is also provided by the helmet-mounted display (HMD), at night overlaid onto an intensified view of the world.

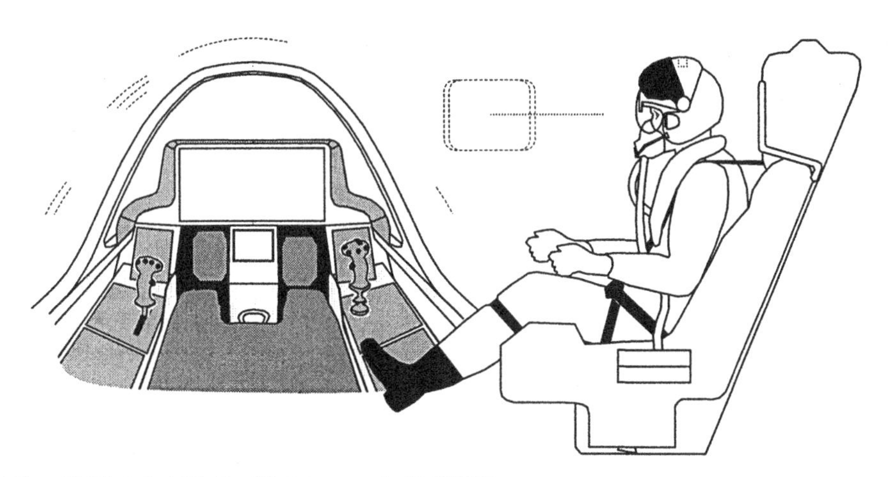

Figure 1.6 F-35 Joint Strike Fighter cockpit (1995)

Figure 1.6 shows a possible layout of the cockpit for the Lockheed-Martin F-35, the Joint Strike Fighter (JSF). The design adopts the philosophy used in the Eurofighter, but exploits recent technological progress. A single large display screen makes significantly better use of the available panel space and provides the versatility to present information in a similar way to the use of "windows" on a personal computer. It is also proposed to rely on the HMD as the sole way of projecting imagery onto the external view of the world.

1.3 Underlying factors

The condensed version of cockpit evolution given in Figures 1.1 to 1.6 fails to bring out a number of underlying threads in the mixture of engineering, human and operational factors that influence the design of cockpit systems.

The most obvious drivers have been the progressive expansion of aeronautical capabilities that have stretched the range of altitudes and airspeeds that aircraft can attain, and the improvement of avionics systems that have enabled aircraft to operate at night and in cloudy skies. There has also been the need to provide protection against modern nuclear, biological, chemical and blinding weapons. Another underlying factor is the accumulation of knowledge about human sensory, muscular and physiological limitations. This has enabled designers to arrange, for instance, that an ejection seat inflicts least injury, that displayed information can be discerned and that switches can be operated reliably even when the pilot is thrown around in a lurching aircraft.

The progressive increase in the automation of avionic systems has however exposed a profound ignorance about human cognitive limitations. The pilot of the SE5A and the pilot of the Lightning were in no doubt that their chief function was to use the throttle, stick and pedals to keep the aircraft safely aloft and on course. In contrast, the pilots of the Eurofighter and F-35 are unlikely to worry about stalling or spinning, and once airborne they will generally hand over control of the flight path to on-board automation. The role of the pilot has changed recently from that of a flier into that of a mission manager who can concentrate on understanding what is happening in the increasingly complex tactical environment generated by modern weapons and their countermeasures. However, the activities involved in flying are overt and the outcomes are measurable, whereas tactical deliberations are largely covert and the choices may not be immediately apparent. The performance of the pilot, i.e. his ability to do his job, is now less easy to assess. This has given those concerned with the design of the cockpit great difficulty in deciding whether one arrangement is better than another, and whether either is manageable. Although the human factors research community has gone to considerable effort to devise ways to measure the workload and situational awareness of the pilot, there are still no generally agreed techniques. Until human cognitive processes are understood better, engineers and regulatory authorities will undoubtedly continue to rely on the judgement of test pilots.

Another practical issue is that aircraft remain in service for a considerable period, typically forty years, often with a gradual transmutation from front line to support roles. Along with many of the mission systems and weapons, the initial configuration of the displays and controls is usually updated. Aircraft such as the Jaguar, the Hercules tactical transporter and the ubiquitous F-5 and F-16 types have received several extensive modernisations that have stretched their periods in service. The avionic architecture of many aircraft types, particularly those conceived within the last few years, has therefore been arranged with such retro-fit programmes in mind.

A final aspect, which may be apparent when looking at Figures 1.1 to 1.6 as a whole, is that the fundamental approach to the design of combat cockpits has not changed. The pilot faces a barrage of information, harnessed to a nearly upright seat, wearing complex clothing and headgear inside a sealed enclosure with a removable transparent lid. Despite the massive and rapid technological changes to airframe structures, systems and weaponry, improvement to cockpit design seems to have been progressive rather than revolutionary. An outsider could conclude either that the technological advances have been exploited in a singular and propitiously optimum line of development or that anything more radical has not succeeded in overcoming some intrinsic conservatism in the piloting community. However, neither of these views is justified because the topic has stimulated considerable inventiveness, and only in hindsight do the successfully adopted novelties seem ordinary and the unadopted seem interestingly unorthodox. For instance, the prone seat (discussed in Chapter 4) and the "see-back-a-scope" (discussed in Chapter 8) both have specific merits, but in the judgement of those faced with the design of a practical combat cockpit these benefits have been outweighed by installation problems and other incompatibilities.

1.4 Aircraft systems

In general, the design of the crewstation is dictated by the needs of the aircrew and the type and the role of the aircraft. Civil types range from microlights, gliders and aircraft for competitive aerobatics to the ubiquitous passenger-carrying airliners with four hundred seats, and include rotary-wing types from the autogyro to the airborne crane. Each is aimed to have an optimum balance of qualities for a niche set of functional and commercial requirements. A similar number of types have been devised to perform specific military rôles, such as transporting equipment, destroying ground-based missile sites, attacking ground targets and destroying aircraft intent on attacking ground targets. Again, each is intended to fulfil specific needs and has been designed to meet a range of cost and performance criteria. The latter cover stealth, survivability and supportability as well as the more obvious aspects of speed, range, load carriage and manoeuvrability.

Despite this rich variety, aircraft of all types have many common elements. Figure 1.7 shows the ingredients of a typical modern combat aircraft separated into their main functions. It is useful to think of these as comprising two broad categories. The airframe, engines, flight control system, electrical and hydraulic power systems, most of the cockpit and the communication and navigation parts of the mission system are a core that constitutes the basic aeroplane. This is the portion that is put together to enable initial flight testing of the prototype. The other elements – the majority of the mission systems and all of the weapon systems – are what transform it into a useful military machine. Specific equipment sets can be chosen in service for performing specific roles. In other types of military and civil aircraft there is a similar distinction between core and role-specific systems. The core systems, albeit in widely varied form, are needed in all types of manned aircraft and are treated by airworthiness authorities as safety-critical. They must perform their intended function under any foreseeable operating conditions, and any single failure condition which could prevent the safe flight and landing of the aircraft must be extremely improbable[2]; failure probability should be less than 10^{-9} per flight hour. Most mission systems, which have become a significant proportion of the cost of an aircraft, can however be built to a lower, mission-critical, standard of reliability and integrity dictated by the customer.

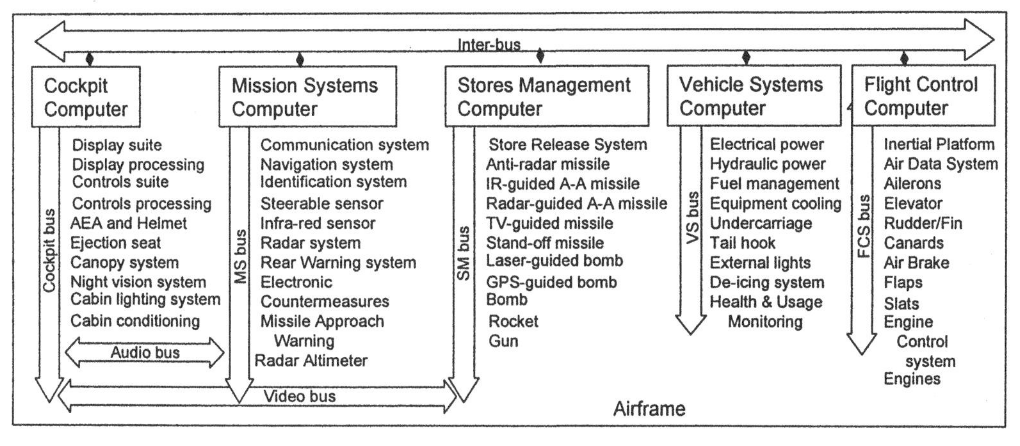

Figure 1.7 Constituent systems of a typical modern combat aircraft

The engineering of the interconnections between systems has undergone significant changes. In older aircraft, both the electrical power and the signals to and from the electronic units were routed from one place to another along parallel tracts of point-to-point wires. These were bunched to form a network of stout cables terminated by complex connectors. The result was weighty, prone to failure and difficult to repair. With the advent of digital computers and the inherently increased digital data flow, the opportunity was taken to send signals through a data-bus. The commonly-adopted 1553B standard[3] used a screened pair of wires to form a daisy-chain linking all electronic units, and digital data was sent in short bursts from a source unit that gained control for an allocated portion of the data transfer cycle and specified the address of the receiver unit. For increased reliability a second daisy-chain passed the same data, and any difference between the two received data sets automatically triggered another transmission. The system was termed "dual-redundant". To cope with the amount of data, it was necessary to split the overall aircraft-carried electronics units into groups that were served by separate buses. It was also necessary to provide special localised links to cater for very high data flow, for instance between the antenna unit and the processors of the radar system.

The approach has been extended in current aircraft, such as Eurofighter, for which Figure 1.7 can be interpreted as a simplified version of the data distribution architecture. Here, five separate fibre-optic data-buses are used within the cockpit, mission, stores, vehicle and flight-control systems. Further data-buses are used to transfer data, video and audio signals across these five systems. The whole arrangement of sensors, effectors, displays, controls, data-bases, computers and software-embedded processes as well as the data transfer media forms a so-called "federated" architecture. Current research is aimed towards an "integrated modular avionic" (IMA) architecture in which most of the computers are general-purpose machines, interconnected so that they can share the processing load in a flexible way and automatically re-configure should individual elements fail. However, the idea must accommodate safety-critical systems, such as the triple-redundant flight control system, in the same network as the mission systems. A hybrid architecture with separate IMA networks for safety-critical and non-safety-critical systems may be an effective compromise for the next generation of aircraft.

1.5 Function of the crewstation

The crewstation is a workplace that must supply the three basic needs of the crew: (a) a view out into the world, (b) protection against the natural environment and (c) means for interacting with the aircraft and its systems. Military aircrew also need protection against weapons. The conceptual linkages between systems supplying these fundamental requirements and the rest of aircraft are summarised schematically in Figure 1.8.

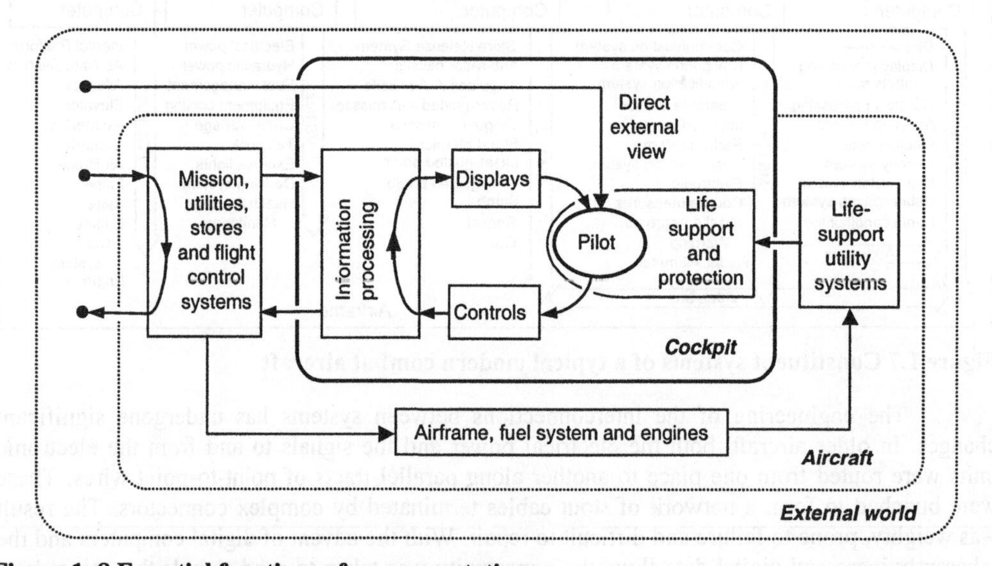

Figure 1. 8 Essential functions of any crewstation

Figure 1.9 keeps the pilot as the central element and expands the broad functions of the cockpit into a detailed delineation of the requirements that the customer could specify and the designer would have to supply. Some of these are only relevant to a few aircraft types, and others such as the need to sense the pilot's physiological state, are as yet only a gleam in the eye of the researcher.

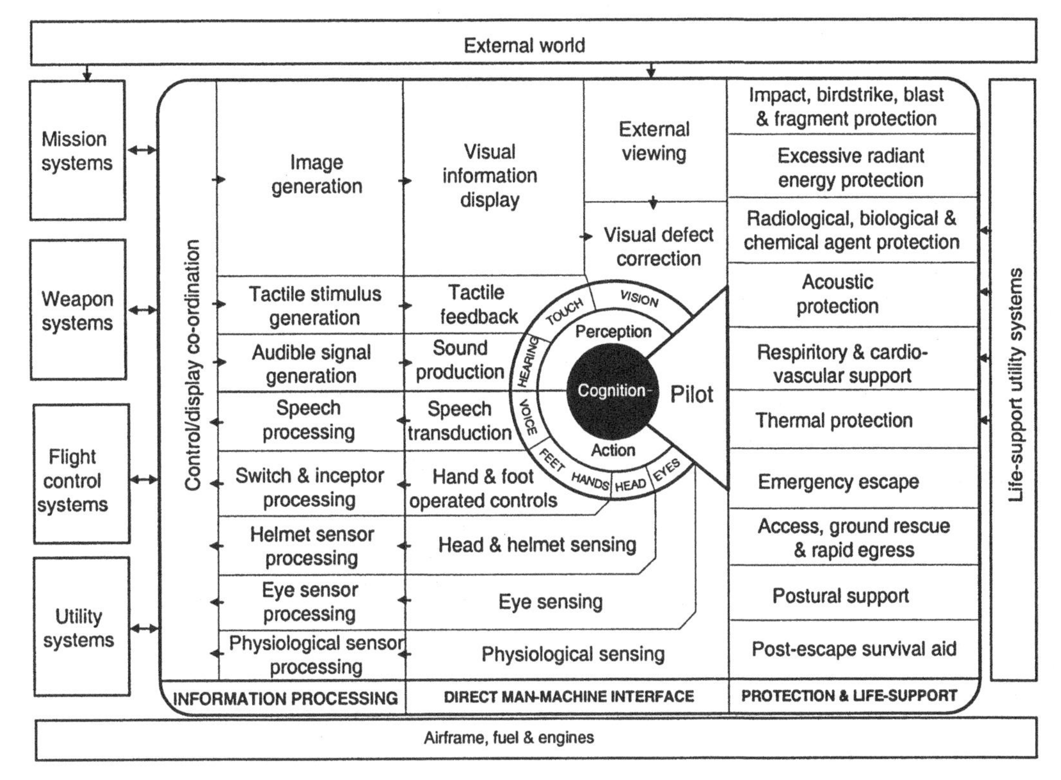

Figure 1.9 A comprehensive delineation of the functions provided by a crewstation

The broad intention of this book is to describe equipment devised to meet these needs. The bulk of the text, in Chapters 5 to 13, covers the rationale, the relevant developmental history and the form and performance of the equipment. This is preceded in Chapters 2 and 3 by a brief review of the human component and the airborne environment. The first treats the pilot as a piece of equipment that takes in information, makes decisions and effects control, and the second describes the hazards that he faces because of his physiological frailty. Chapter 4, which can also be regarded as a pre-amble, describes the considerations involved in coming up with a detailed design for a modern fighter cockpit. The last chapter reviews some current research topics that could have a significant effect on the form of the cockpit and the way the pilot performs his job. It concludes by discussing the relative merits of manned and unmanned combat aircraft.

References

1. Brown L E (ed) *The new shorter Oxford English dictionary*, Clarendon Press, UK, 1993
2. *Advisory circular 25.1309-1 System design analysis*, Federal Aviation Administration, 1982
3. *Aircraft internal time division command/response multiplex data-bus*, MIL-STD-1553B, 1978

Further reading

Coombes L F E, *The Aircraft Cockpit*, Patrick Stevens Ltd, Wellingborough, Northamptonshire, England, 1990

Flight Vehicle Integration Panel Working Group 21 on Glass Cockpit Operational Effectiveness AGARD Report AR-349, 1996

Chapter 2

The Human Component

2.1 Introduction

The design of a cockpit, like the design of anything involving a human operator, needs a broad understanding of relevant human characteristics as well as more detailed knowledge of the aviator's job. This chapter gives a very brief overview of the main human systems and their interaction, concentrating mainly on aspects of speech, hearing, vision and attention that are assumed in later chapters. It ends with a short account of the role, selection and training of the aviator.

2.1.1 The main human systems

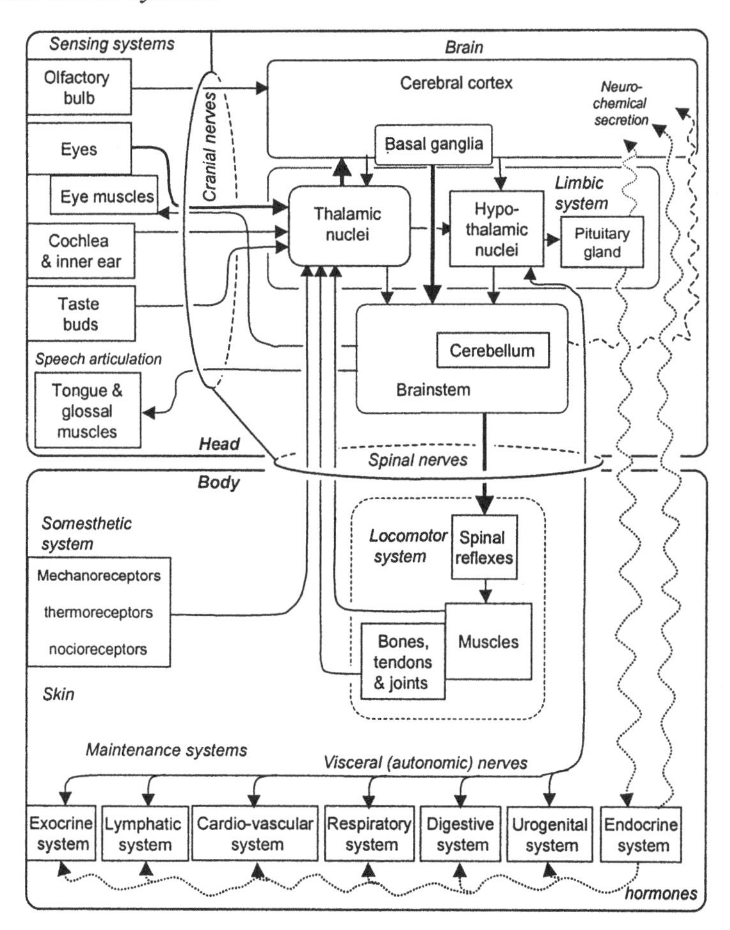

Figure 2.1 Schematic representation of the interconnections between the main human systems

Figure 2.1 is a simplified schematic of the main human systems and their principal linkages. The brain is represented as a set of functional elements that receive information from and exert control over other head-borne units via cranial nerves and over the rest of the body via nerves that run through the spine. In addition to this electrically-mediated signalling, control is also exerted by hormone chemicals secreted into the blood stream and by neuro-chemicals within the brain.

2.1.2 Loco-motor and maintenance systems

The basic chassis is a set of about two hundred bones, with joints and buffer pads, that are held in place and swung around by forces exerted through about five hundred muscle groups acting on tendon attachments. Structural bone tissue is continually under a slow dynamic equilibrium between deposition and dissolution so that it adapts in strength to mechanical stress, a process that also enables self-repair. An analogous process builds muscle tissue.

Most muscle groups are arranged as pairs that apply opposing forces to the bones at a joint. Muscle fibres are energised continuously by local reflex nervous pathways so that the groups are gently but constantly pulling against each other. The resulting antagonistic "tonic readiness" is the normal relaxed state of the loco-motor system. The system is under the control of the brainstem via spinal nerves that make contact with the local reflex loops, and a joint is moved by the turning force generated when, and to a degree, that depends upon the alteration of the balance between the antagonistic muscles. Dynamic actions, such as reaching and grasping, are produced when appropriate muscle groups are activated by higher level processes that have access to sensory feedback.

The products of the organs for food ingestion, digestion, energy storage, energy distribution and waste extraction that are contained in the central torso, along with the sexually-differentiated organs for reproduction, are transported and exchanged through blood vessels that necessarily reach the extremes of the body. A brief overview of the human cardio-vascular and respiratory systems is given in Chapter 3 in the context of the routine hazards of flying. Other systems that supply basic functions are also indicated by boxes in Figure 2.1. The lymphatic system makes the anti-bodies for countering foreign bacteria and also removes some waste products from the blood. The exocrine system is the complex of glands that, for instance, cool the surface by sweating, lubricate the eyes with tears and protect the breathing passages with phlegm; all so vital to a wet organism. The third of these secretary system, the endocrine, is perhaps the most complex and wide ranging because it injects relatively small quantities of hormone chemicals into the bloodstream for distribution throughout the body, and most of these chemicals stimulate or catalyse other processes. For instance, maturation and development over the individual's lifetime are stimulated by estrogens and androgens from ovaries or testes, insulin from the pancreas catalyses energy transfer into cells, and the body is excited to higher levels of effort by epinephrine from the adrenal glands. The endocrine is a remarkably robust and effective control system.

2.2 The nervous system

2.2.1 Neurones

As indicated in Figure 2.1, the electrical signalling paths are significantly more numerous and complex than the purely chemical paths.

Cells that propagate and manipulate the electrical activity have a variety of forms appropriate to their function, but they are all types of neurone. The electrical activity within the neurone is initiated as brief (~1msec) pulses of about 100millivolts in the dendrites (branched inputs) at one end of the cell. The pulse then propagates along the thin membrane wall of the neurone to the axon terminals (branched outputs) at the other end using the differential permeability of sodium and potassium ions. Electrical pulses cross the synapse connecting the output axon of one neurone to the dendrite of another via a chemical neurotransmitter secreted by the neurone. The maximum pulse repetition rate is about 1000 Hz.

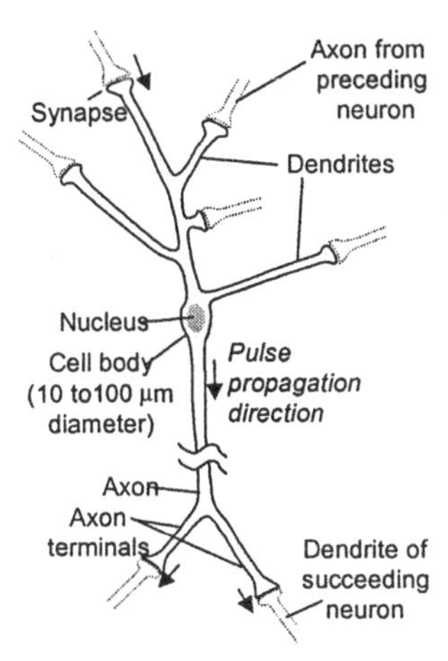

Figure 2.2 Simplified structure of the neurone

Some axons are sheathed in a fatty substance, myelin, which induces faster electromagnetic pulse propagation. In general, the nerve fibre bundles carrying the electrical signals around the body consist of long myelinated neurones, while the short multiply-branched neurones that form interconnected signal-manipulating clusters and nuclei are unsheathed. Ganglion cells, such as the neurones that transmit the output from the eyes and from the base of the brain have a large number of axons but few dendrites.

2.2.2 Brainstem, autonomic nervous system and limbic system

The human brain contains between 10^{10} and 10^{11} neurones, together with glial cells that act partially as a mechanical scaffold and blood vessels that supply glucose and oxygen. It has a volume of about $1100cm^3$, floats in cerebro-spinal fluid within the cranium and, as summarised in Figure 2.3, is composed of symmetrical left/right pairs of separable anatomical regions. In all vertebrates the brain is an elaboration of the interconnections between the head-borne senses and the dense bundle of nerves that are gathered into the spine from where they branch to the internal organs and external muscles. The brainstem, limbic system and cerebral cortex, shown in Figure 2.3, can be regarded as separate organs controlling increasingly complex aspects of behaviour. The brainstem is the most primitive region. It regulates the heart, breathing and the core body temperature through the sympathetic and parasympathetic nervous systems, and it triggers the simple behaviours essential for survival and reproduction that assuage hunger, thirst and sexual arousal. The control of the maintenance systems is termed "autonomic" to distinguish it from higher level processes of the "central nervous system". The brainstem includes the cerebellum, a complex excrescence of multiply-branching neurones, which seems to have the role of controlling the precise timing of muscular activation for co-ordinated movement. In humans this includes fine hand gestures, eye movements and speech as well as posture and gait.

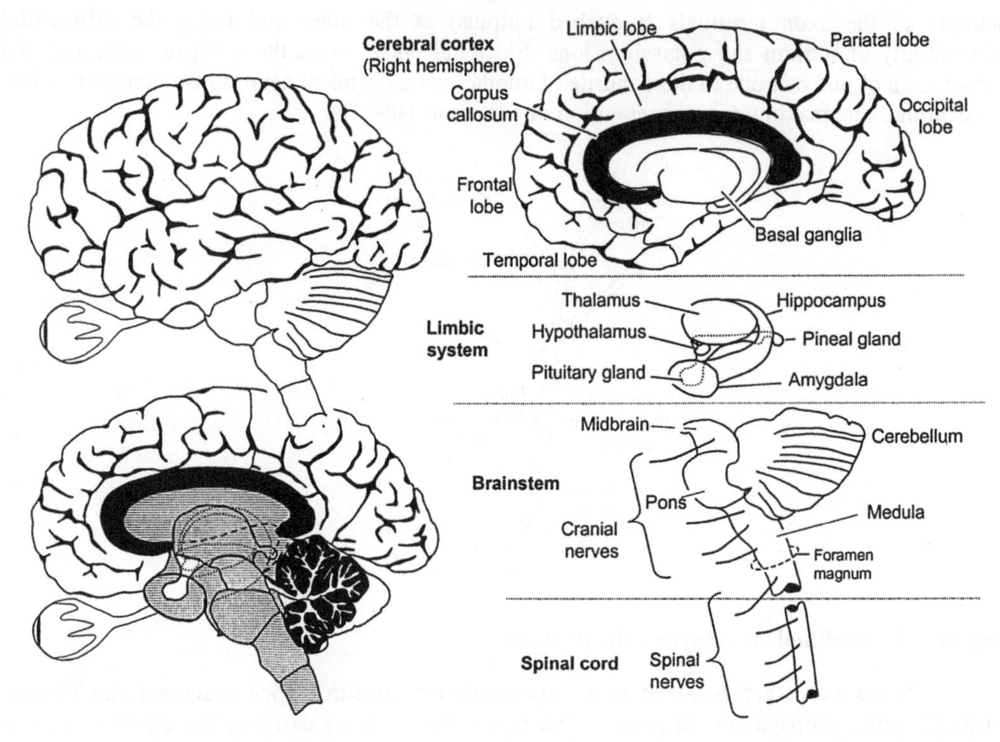

Figure 2.3 Principal anatomical regions of the human brain. The left diagram shows the brain split in half laterally, with the left half above the right. The lighter parts, i.e. the cerebral cortex and the eyes, are intrinsically separate. To split the brain in this fashion it would be necessary to cut through the corpus callosum (the black band), the cerebellum (the large black and white protrusion) and the (hashed) limbic system and brainstem. The right diagram shows the main divisions; the spinal cord, brainstem, limbic system and cerebral cortex. The foramen magnum is the opening for the spinal cord in the base of the skull

The largest parts of the limbic system are the tight collections of neurones that form the nuclei of the thalamus. These receive and distribute signals from all of the senses except the olfactory bulb in the roof of the nose, which is linked directly to the cortex. Although the majority of the sensory pathways are coupled to specialised processing regions in the cerebral cortex, some provide a link to the brainstem for the control of basic behaviour, such as maintaining a sitting posture and stabilising eye orientation during head movements, and some are linked to limbic nuclei such as the amygdala. The function of the latter is as yet unclear, but their close coupling to the thalamus suggests that these regions may trigger automatic attraction or aversion when the senses receive stimulation consistent with, for instance, the proximity of potential mates, predators and steep ground. The hypothalamus and pituitary gland provide the central portal between the electrical and chemical signalling systems that enables higher cortical processes to influence, and be influenced by, the endocrine and maintenance systems.

2.2.3 The cerebral cortex

The comparatively large mass of neurones in the two hemispheres that form the cerebral cortex is peculiar to the human species. This part of the organ is so large that neo-natal humans cannot complete their full development in the uterus. In comparison with other species they are born prematurely and for several years rely on adults for all their physical needs.

The darker outer surface of each hemisphere is a sheet, about 1500cm^2 in area and 3mm thick, containing four or five layers of intensively interconnected neurones. The lighter-coloured inner part is largely tracts of myelinated nerve fibres that either join regions within the hemisphere (association tracts), carry signals to and from lower parts of the organ (projection tracts) or interconnect regions in the left hemisphere with similar regions in the right (commissures). The Corpus Callosum is the major commissure. The irregular folds and bumps (sulci and gyri) in the crumpled sheet surface reflect tract boundaries. As shown in Figure 2.3, each cerebral hemisphere is divided into five regions, the frontal, limbic, temporal, parietal and occipital lobes. The efferent (output) projection tracts connect to the basal ganglia, then through the midbrain and cerebellum to the lower brainstem and into the spinal cord. In the medula, those from the left hemisphere cross to the right and those from the right cross to the left. Afferent (input) pathways rising from the spine to the thalamus undergo a similar cross-over, or decussation, in the same region. The left side of the body is therefore controlled by the right cerebral hemisphere, and vice versa.

Research to understand the cortex has proceeded, much like the exploration of mysterious lands, to map specialised functions to regions of the cerebral surface. Traditionally this has been done by tracing the nerve pathways, by relating abnormal behaviour to local brain damage, and by analysing electrical events picked up by electro-encephalographic (EEG) electrodes either inserted into the cortex or attached externally to the scalp. Newer techniques such as Nuclear Magnetic Resonance Imaging (NMRI) enable finely detailed imaging of the live brain, while others such as functional Magnetic Resonance Imaging (fNMRI), Positron Emission Topography (PET), Near-Infra-Red Spectroscopy (NIRS), and Magneto-EncephaloGraphy (MEG) can also indicate which areas of the cortex are most active when the individual under examination performs specific tasks.

Evidence gathered by these means suggests that specific cortical regions can be associated with specialised functions. For instance, areas in the frontal lobe are associated with general motor control, speech and voluntary eye movements, the parietal lobe has been mapped into regions associated with discrete voluntary movement, touch, pain, audition, language and some mathematical manipulations. The occipital lobes are given over almost entirely to visual processing. Although the two hemispheres to a large extent process information relating to their assigned sides of the body, they are not entirely independent, and some functions are associated with one hemisphere only. There are also large portions of the frontal, limbic and temporal lobes, usually termed the association cortex, that have so far defied easy classification other than the suggestion that they account for such human qualities as imagination, foresight and creativity.

The detailed mechanism of the brain is not understood, but it is certainly characterised by spatio-temporal patterns of electrical activity in the neurone arrays. Individual neurones act like digital integrators in that their axon firing rate, the rate at which pulses are output, is related to the average rate of the pulses received across the synapses by the dendrites. However, some dendrite connections inhibit output. An array containing many layers of neurones with extensive connectivity within and between layers behaves like a selective filter; the output pulse rate of a small group of last-layer axons increases when a particular pattern of activation is received by the first-layer synapses. For example, a small region in the visual cortex is excited when the eyes are stimulated with lines of a particular orientation. The activity of other output axons corresponds to other specific input activation patterns. Neural network filters of this kind have been modelled extensively and have been shown to mimic the sensing, classifying and control processes done by the brain.

As repeated stimulation of a synapse decreases the firing threshold, that synapse becomes progressively more sensitive. Thus layered arrays of neurones, with inter-neurone connections that are initially weak, become progressively more sensitive to consistently repeated patterns of stimulation. Such self-tuning can in principle account for adaptation, learning and long-term memory.

The development of the individual and the acquisition of expected abilities depends upon exposure to a rich environment, particularly during infancy. For instance, the cortical regions that process visual information are connected to receive stimulation from the eyes, and in this sense the regions are predisposed to take on the analysis of visual stimuli. However, they are "plastic" in the sense that they need repeated stimulation to strengthen the synaptic interconnections for the network to act as a set of precise, selective filters. If co-ordinated pointing of the two eyes fails to develop during a crucial period during infancy, the resulting squint is likely to affect the development of the regions of the visual cortex that normally process disparities between the stimulation collected by the two eyes. The enduring "stereo-blindness" that results implies a limited plasticity. However, the connectivity of much of the brain, particularly the higher information processing pathways, changes throughout the individual's lifetime. This has been demonstrated conclusively by Kohler[1] using adults who have worn prismatic goggles that invert the view of the world. Over the course of several weeks, and with constant exposure to forcing conditions that involve walking and eye-hand co-ordination, their perception is "corrected"; they see the ground in the direction in which they feel their feet. After such a lengthy adaptation the individual has the paradoxical sensation of seeing the world upside-down when the goggles are removed, and a similar but slightly more rapid period of re-adaptation is needed before normal spatial perception is regained.

The proper functioning of neurones depends fundamentally on the supply of oxygen and glucose from the bloodstream, as discussed in Chapter 3. Synaptic connections are mediated by minute secretions of neuro-transmitter chemicals, so the vigour of axon pulses and the neurone firing frequency is governed by the availability of appropriate pre-cursor chemicals in the cerebro-spinal fluid. This chemical dependency provides a basis for explaining the activation and inhibition of the cortex by secretions produced by the brainstem. It also accounts for the effect on behaviour of chemicals such as alcohol that can penetrate the blood-brain barrier.

2.3 Hearing and speech

2.3.1 Hearing

As shown in Figure 2.4, the hearing organ is largely embedded in the bony mass of the skull and divided into outer, middle and inner portions. The shell-like shape of the pina (outer ear), gathers sound vibrations into the ear canal and introduces subtle changes that help the brain locate the direction of the source. The middle ear consists of the tympanic membrane (ear-drum) and the ossicles, three tiny bones of the middle ear. The stiff tissue of the ear-drum is stretched across the ear canal and flexes axially in response to the acoustic pressure variations. These movements are transmitted to the oval window of the inner ear by the middle ear ossicles. The inner ear, the cochlea, provides the mechanism for transducing the acoustic vibrations into the electrical signals that are sent along the auditory nerve to the auditory cortex via the brainstem and the thalamus.

The area of the ear-drum is about thirty times greater than the area of the oval window at the start of the cochlea, and the chief function of the small bones of the middle ear is to couple the incident airborne sound waves to the cochlea and the fluid which fills its canals. This acoustic impedance matching is mainly done by the leverage of the hammer on the anvil, which reduces the displacement and magnifies the pressure. The muscles connected to the hammer and stirrup stiffen in response to loud sounds and reduce the efficiency of the energy transfer, so the middle ear also to some extent protects the inner ear against excessive excitation. Unfortunately this oto-acoustic reflex (OAR) cannot provide protection against sudden, intense sounds.

The Eustachian tube connects the middle ear to a vent in the nasal cavity so that the pressure in the middle ear is the same as the ambient pressure in the ear canal. As described in Chapter 3, if this vent is blocked, a change in ambient pressure blows the ear-drum inwards or outwards, which stiffens the drum and produces pain and a loss of hearing,

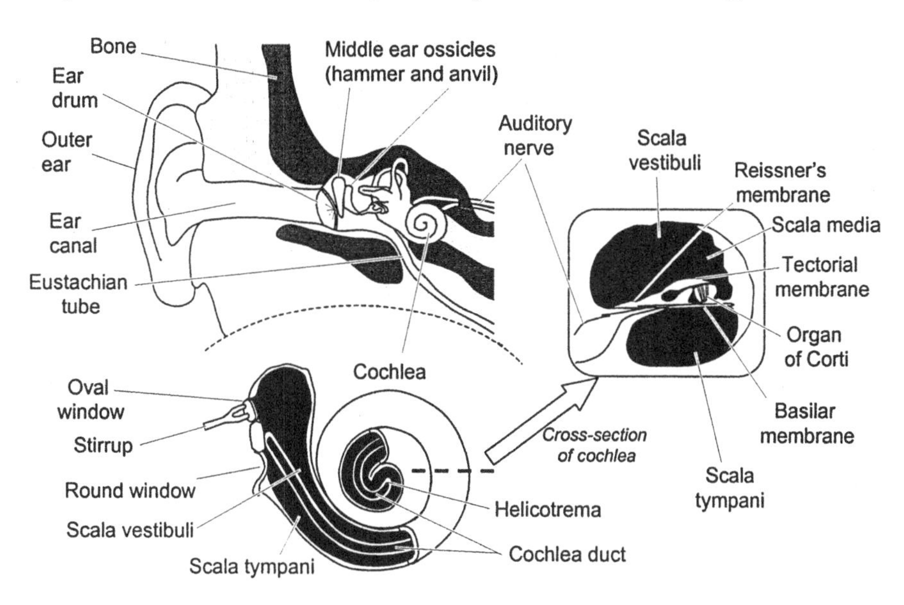

Figure 2.4 Anatomical sketches of the human ear

The cochlea, illustrated in Figure 2.4, consists of a tapered, triple-chambered, fluid-filled tube about 35mm long, wound into about two-and-a-half turns. Active transduction is done by 16,000 or so hair cells on the organ of Corti, which is attached to the basilar membrane. These cells are stimulated by a combination of fluid flow and brushing contact with the tectorial membrane when the basilar membrane vibrates in response to movements of the stirrup on the oval window.

The characteristics of the ear are dictated by the basilar membrane; it is very narrow at the end of the cochlea near the oval window, but at the far end, the helicotrema, it is broader and about 100 times less stiff. As Bekesy[2] has shown schematically in Figure 2.5, longitudinal vibrations of the oval window induce transverse standing waves in the basilar membrane, and the distance along the membrane at which maximum movement occurs depends upon the frequency of the excitation. High frequencies cause most movement at the stiff base and the lower frequencies produce most movement near the relatively flaccid apex. The cochlea therefore provides the brain with information about the strength and the spectral distribution of ambient sounds.

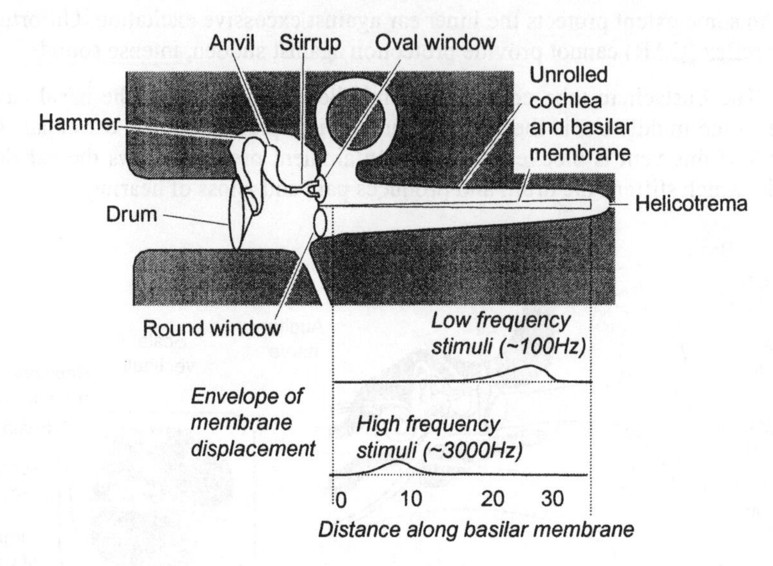

Figure 2.5 Schematic representation of the transfer of acoustic energy from the drum via the linkages in the middle ear to the (unrolled) basilar membrane (adapted from Bekesy G, *Uber der akustiche reizung des vestibular-apparates*, Pflugers Archiv. European Journal of Physiology, 236, pp.59-76, 1935, by kind permission of Springer-Verlag GmbH)

Sounds in gases are small variations in pressure that propagate as longitudinal waves. The human auditory system responds to frequencies between 20Hz and 15kHz. The minimum discernable sound volume occurs when the r.m.s. variation of the pressure is 2.0×10^{-5} Pascal ($=N.m^{-2}$) and the intensity is 10^{-12} $W.m^{-2}$. The amplitude of the pressure variations is very small; about 10^{-9} of the atmospheric pressure. At the other extreme, most people with normal hearing cannot tolerate a sound in which the root mean square variation of the sound pressure is greater than 20 Pa and the intensity is above 1.0 $W.m^{-2}$.

The difference between the intensity of one noise and another is proportional to the square of the fractional change of the pressure variations. A noise level could be related to the minimum value p_0 (2×10^{-5} Pa at 1000Hz) as a ratio $(p / p_0)^2$. However, because it is easier to add and subtract than multiply and divide, the Sound Pressure Level (SPL) has been defined as the logarithm of the ratio:-

$$SPL = \log_{10}(p / p_0)^2 \text{ Bels} = 10.\log_{10}(p / p_0)^2 = 20.\log_{10}(p / p_0) \text{ decibels (dB)}$$

With this formulation, the minimum is 0dB-SPL and the maximum 120dB-SPL. A reduction by -3dB is equivalent to halving the SPL and an increase of +3dB doubles the SPL. A level of 40dB-SPL is quiet, a normal conversation occurs at about 60 to 70dB-SPL and a nearby pneumatic drill generates 90 to 100dB-SPL.

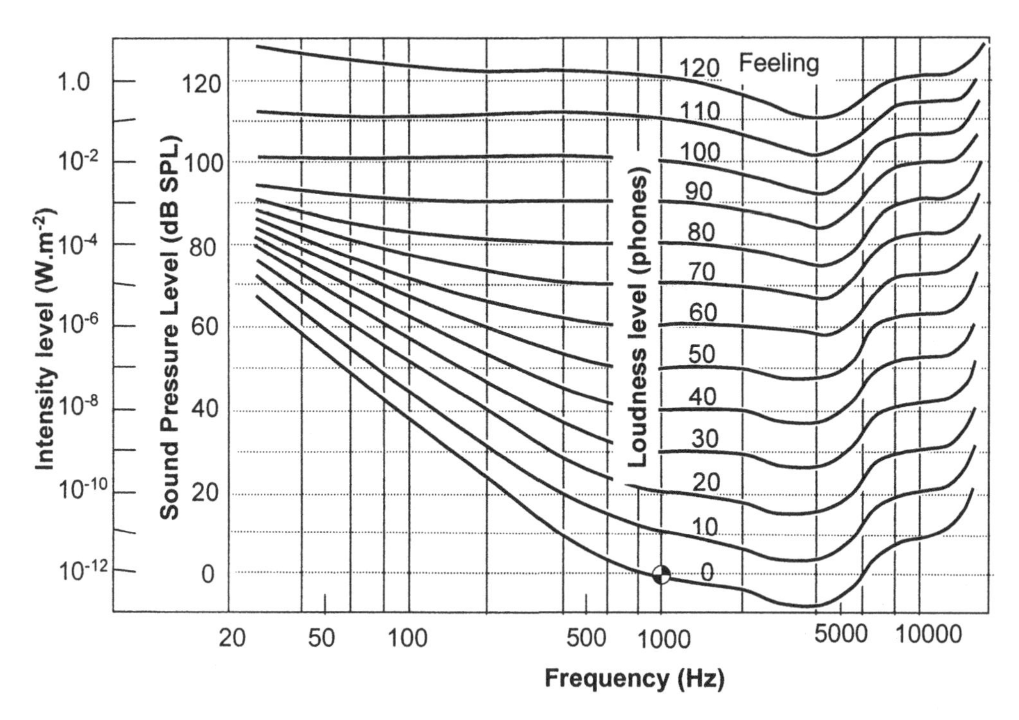

Figure 2.6 Summary of the relationships between objective and subjective qualities of sinusoidal sounds. (From *Normal equal loudness contours for pure tones and normal threshold of hearing under free field conditions.* International Organisation for Standards, ISO R-226, 1962)

Figure 2.6 contains the collated results of psycho-physical experiments that assess the subjective loudness of sinusoidal sounds, expressed in phones[3]. The lowest curve represents the intensity at which a sound source in front of the listener can just be heard when both ears are stimulated: the "zero loudness" characteristic. The other curves show the subjective judgement of loudness level between the zero threshold and the maximum where sounds are felt rather than heard. The figure shows clearly that at low intensity there is considerable variation in the subjective loudness level with frequency. (Some music systems therefore incorporate a loudness filter to boost the low and high frequency sound components at low replay levels.) The sound pressure level is also commonly calculated by weighting the noise spectrum to take account of the variation of human sensitivity to sounds of different frequency. When "A-weighted", the factors are taken from the loudness characteristic at 40 phones, and the resulting sound pressure level is quantified in units of dB SPL(A).

2.3.2 Language

Most animals communicate by making noises, and in humans this has evolved into an ability to convey complex ideas using noises that have become associated with the other qualities of the idea through repeated utterance and hearing. The "speech chain" communicates the abstraction by transforming the internal representation of the sound into activations of vocal muscles, which create sound waves that are received by both the speaker and the listener. The speaker must hear what has been uttered to develop accurate articulation, and the listener's ears and brain must reverse the process to interpret and understand the message. Speech therefore requires a sequence of semantic-to-phonological-to-physiological-to-acoustic transformations, and speech interpretation requires the reverse sequence. The phonological and semantic realms – the relationship with higher-order knowledge representations – are not understood, but it is probable that two regions in the left cerebral hemisphere, called Broca's area and Wernicke's area, generate and interpret sequences of word symbols.

The physiological-to-acoustic processes that deal with words are understood better. Words are made up from syllables, which are combinations of phonemes. Phonemes are message-carrying verbal units having acoustic characteristics that depend upon the age and sex of the speaker as well as the language. Different languages use different sets of phonemes, and during childhood, the channels that process the phonemes of the native language are strengthened while those associated with unused phonemes decay. Phoneticians divide Standard English into 14 vowel phonemes and 24 consonant phonemes. This is a greater number than the 26 letters of the alphabet used conventionally to represent the sounds because some phonemes are represented in written text by the same character and others correspond to diphthongs such as "sh" and "th". Spoken language also carries a great deal of additional information in the form of intonation, pitch and stress that alter the sound of individual phonemes. Such variations are used widely, for instance to convey emphasis or to change a statement into a question.

2.3.3 Speech

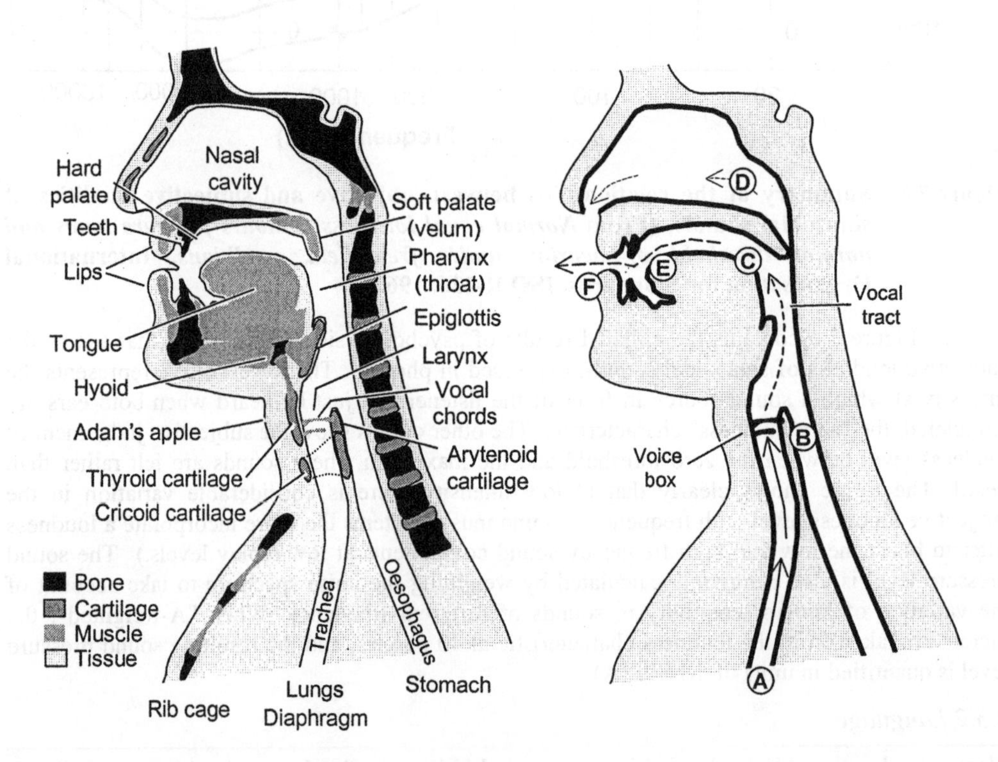

Figure 2.7 (left) Sketch of the vocal anatomy, and (right) an overview of speech production processes (see text for an explanation)

Speech is the process for articulating streams of phonemes, and Figure 2.7 summarises the main effector organs. The vocal cords are essentially a left-right pair of fibrous muscles that cross the larynx between the thyroid cartilage at the front of the neck and the arytenoid cartilages at the back. When relaxed they allow air to pass in and out of the lungs, but when contracted they constrict the airflow and vibrate. Speech production depends upon fine, accurate and timely control of the vocal cords and the other articulation muscles by the motor control regions of the cerebral cortex and cerebellum.

The process of initiating speech is not unlike a musical organ; the chest, stomach and diaphragm muscles compress the lungs to produce exhalation of air (A) through the trachea. The airflow is broken into pulses (B) by constricting the vocal chords in the larynx to produce vibrations of about 100Hz. These sounds, with very strong harmonic overtones at multiples of the basic vibration frequency, are the "formants" of voiced sounds. The shape of the vocal tract - the complex passage between the vocal cords and the lips - is altered by moving the tongue, jaw and lips to create resonances (C) that alter the balance of energy in the formants. Upward movement of the soft palate stops air flow through the nasal cavity, and except for some nasal vowels that excite resonances in the nasal cavity (D), most of the remaining articulation is done by shaping the tongue and lips to form precise constrictions against the hard palate and the teeth. For instance, frictive hissing sounds like "th" and "sh" are produced by placing the tongue against the upper teeth (E) or by pursing the lips (F). Hard sounding consonants such as "b", "d" and "g", called plosives, are produced by the sudden release of the airflow stopped by the tongue or lips.

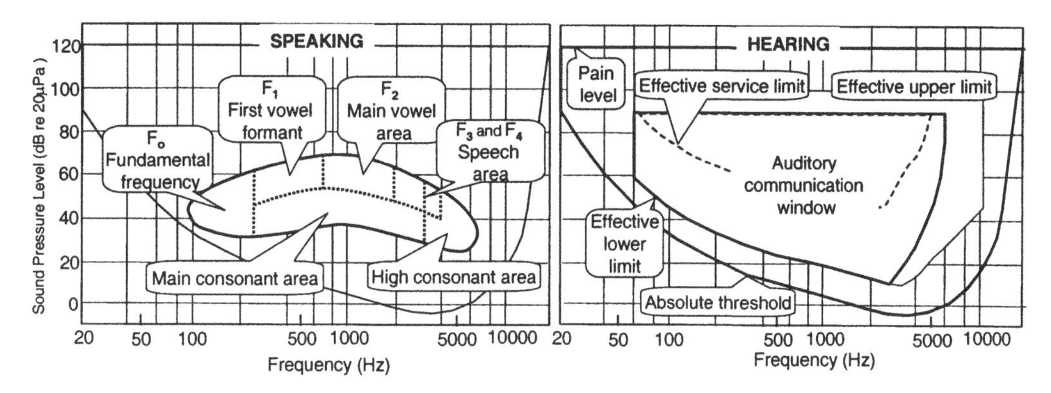

Figure 2.8 The left diagram shows the approximate frequency range and intensity of speech sounds produced at a distance of 1m from the speaker. The right diagram shows the auditory communication window, the range of levels and frequencies of undistorted human hearing. The effective service limit marks the narrower range for personnel with hearing loss (adapted from *Human factors for designers of equipment*, DEF STAN 00-25, UK Ministry of Defence, London, 1991)

Most of the acoustic energy in normal speech is contained in the voiced sounds that carry the vowel formants, as shown in Figure 2.8. These mainly occupy the frequency range between 250Hz and 3000Hz. The subtler sounds produced by consonants are generally less intense and of higher frequency. Alternative modes of articulation produce markedly different signatures. For instance, shouting and singing require the diaphragm and chest muscles to produce stronger, steadier airflow so that maximum effect is extracted from vocal tract resonance. Whispering requires a gentle airflow and avoids voiced sounds entirely. As discussed in Chapter 10, the characteristics of normal speech are important to the design of communication systems and auditory displays. The usage and engineering techniques employed in speech recognition systems are reviewed briefly in Chapter 11.

2.4 Vision

2.4.1 Introduction

The visual system is, as shown in Figure 2.1, one of several sensory pathways that reach the cerebral cortex and other parts of the brain via the thalamus. While supporting diurnal brainstem activation, the main functions are to generate a conscious representation of what is where in the environment and supply cues for the subconscious control of body posture and motion. Although our understanding of its organisation and operation is partial, it is known that most of the sensory mechanisms of the eye are in some way adaptive, that the processing performed along the pathways embodies the individual's experience and that a wide variety of pathological conditions can produce abnormalities.

2.4.2 The eyes

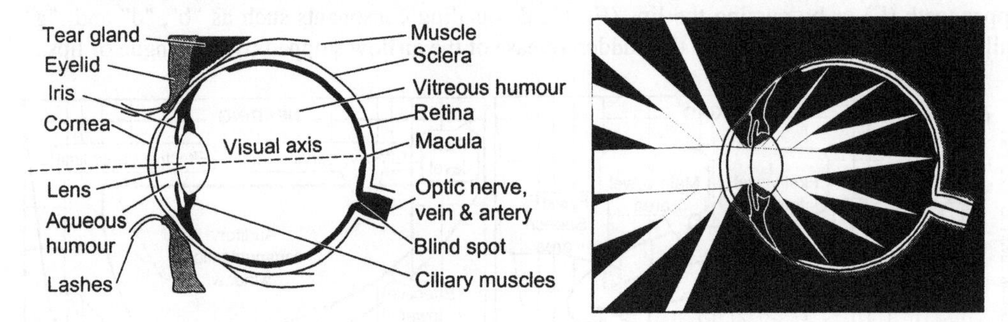

Figure 2.9 Simplified anatomy and optics of the eye

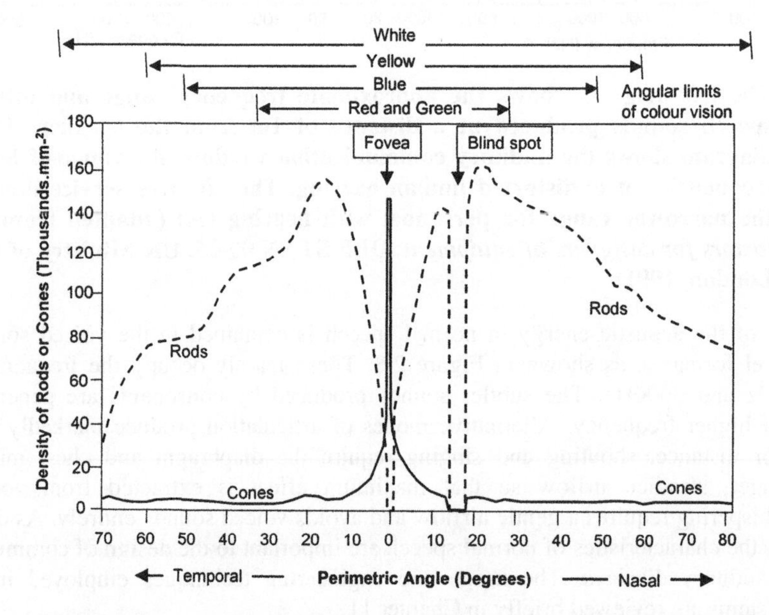

Figure 2.10 Distribution of rods and cones across the retina (from Osterberg G, *Topography of the layers of rods and cones in the human retina*, Acta Ophthal. (suppl.)6, 1935. Reprinted by permission of Blackwell Publishing)

The anatomy of the eye is sketched in Figure 2.9. It is optically analogous to a camera. External light incident on the eye enters through the pupil in the centre of the iris, and is brought into focus on the photosensitive retina by strong refraction at the convex cornea and a weak refraction in the lens. It accommodates variations in light divergence by changing the refractive power of the lens, and adjusts the retinal irradiance by changing the pupil size. In all other respects the eye, and in particular the multi-layered retina, can best be treated as evolved brain tissue. Over most of its area, the retina is inside-out in that light must pass through a fine network of blood vessels, bipolar cells, amacrine cells and ganglionic neurones before reaching the light sensitive rod and cone cells that generate nerve impulses in proportion to the incident light flux.

The distribution of these photo-receptors is summarised in Figure 2.10. Rods, which are sensitive to low levels of irradiance, account for night (scotopic) vision. They adapt slowly to reducing light levels but are desensitised rapidly if exposed, even momentarily, to daylight. The less sensitive cones account for normal daylight (photopic) vision. Cones are less numerous except in the macula, the area where the visual axis intersects the retina.

2.4.3 Sensitivity

Luminous intensity is a measure of the strength of the visual sensation. It is proportional to both the rate at which energy is received and the sensitivity of the receptors to the wavelength of the radiation. Human judgement of luminance is a subjective measure dependant on many factors, in particular the state of the individual eye and the angular extent of the stimulation. The Commission Internationale de L'Eclairage (CIE) has collated data from a large number of studies to create a standard observer with a defined set of spectral sensitivities.

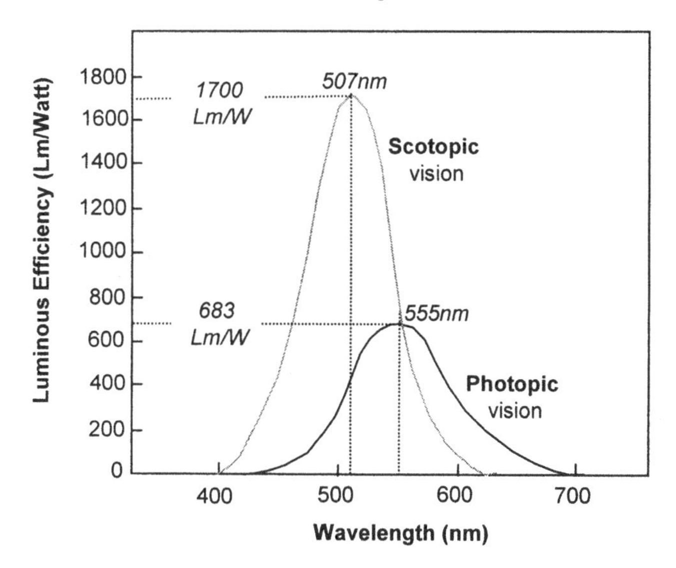

Figure 2.11 The spectral sensitivities of the CIE Standard Observer that are used to transform Radiant Power in Watts into Luminous Flux in Lumens

Figure 2.11 gives the standard luminous efficiency to wavelength characteristics for scotopic and photopic vision[4]. These curves are used to convert the measurable radiometric qualities of electromagnetic radiation, summarised in Table 2.1, into the subjective "photometric" quantities that describe the effect of the radiation on human vision.

When illuminated at 1.0 lux, an ideal Lambertian surface – a surface that scatters all the incident light uniformly – has a luminance of $1/\pi$ candela.m^{-2}. Because this surface luminance is independent of the range to the viewer and the angle at which the surface is viewed, luminance is most commonly used to characterise the strength of the visual stimulation from a surface.

Table 2.1 Relationships between radiometric and photometric quantities

Radiometric quantity		Radiometric unit	Photometric quantity		Photometric unit
Radiant Power	P_e	Watts	Luminous Flux	P_v	lumens
Radiant Intensity	I_e	$Watts.sr^{-1}$	Luminous Intensity	I_v	$lumens.sr^{-1}$ = candela
Irradiance	E_e	$Watts.m^{-2}$	Illuminance	E_v	$lumens.m^{-2}$ = lux
Radiant Exitance	M_e	$Watts.m^{-2}$	Luminous Exitance	M_v	$lumens.m^{-2}$ = lux
Radiance	L_e	$Watts.m^{-2}.sr^{-1}$	Luminance	L_v	$candela.m^{-2}$ = nits

The eye has an exceptional range of sensitivity, from a night vision threshold at about $10^{-6}cd.m^{-2}$ to the damage threshold at about $10^{6}cd.m^{-2}$. When calculating photometric quantities it is essential to use the appropriate spectral sensitivity curve; scotopic for rod vision at low level and photopic for cone vision at high level. The changeover from rod to cone vision occurs gradually as the ambient light level rises through the mesopic range between about $10^{-2}cd.m^{-2}$ and about $10cd.m^{-2}$. The spectral sensitivity in this region should be a proportional blend of the two curves.

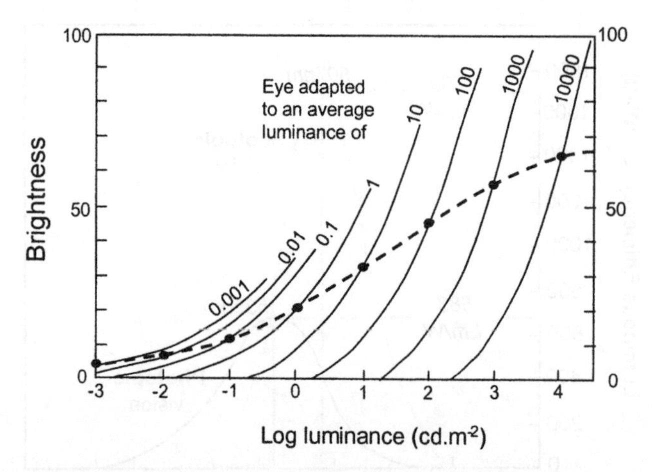

Figure 2.12 The solid lines show the relationships between the subjective judgements of relative brightness of a small uniform area and the logarithm of its luminance when viewed briefly in surrounding fields to which the eye has adapted. The dashed line represents the brightness judgements if several minutes are allowed for direct adaptation to each luminance. (From Higgins G C, *Image Quality Criteria*, Journal of Applied Photographic Engineering, Vol 3, No2, pp53-60, 1977. Reprinted with permission of IS&T: The Society for Imaging Science and Technology sole copyright owners of *The Journal of Applied Photographic Engineering*)

At any instant the eyes respond to only a fraction of the full $10^{12}cd.m^{-2}$ sensitivity range and must adapt to the ambient intensity and spectral composition. Such adaptation is done automatically in a gradually reducing ambient illumination by dilating the pupil to let more light reach the retina and by enhancing the sensitivity of the rods and cones through the regeneration of their photo-pigments. The subjective sensation of relative brightness depends upon adaptation dynamics, because the photo-pigments are bleached more rapidly than they are regenerated, and on the level of the ambient luminance that drives the adaptation. Within the limited effective dynamic range, subjective judgements of the relative brightness of spatially separated regions are markedly non-linear. For instance, a surface having a luminance of only 18% of a reference surface is generally judged to be half the brightness of the reference[4]. The form of the non-linearity is mapped in Figure 2.12. The rightmost curve shows for instance that when the eye is adapted to an ambient luminance of $10000cd.m^{-2}$, the effective dynamic range is between about $300cd.m^{-2}$ and $30000cd.m^{-2}$. Thus when viewing a scene, a photograph or a display at this ambient level, any feature dimmer than this 10^3 range would appear black and anything brighter would appear bright white. As shown by the progressively squashed curves to the left in the figure, the subjective brightness range becomes compressed at low ambient light levels as the eye becomes increasingly reliant on rod excitation.

2.4.4 Colour

Photopic levels of retinal stimulation give rise to the sensation of colour as well as brightness because the photo-pigment (rhodopsin) that mediates energy transduction exists in three forms with different spectral sensitivities. These are used by three types of cone; short wavelength (blue) cones, medium wavelength (green) cones and long wavelength (red) cones. It is thought[8] that the retinal signals from a region in visual space are summed to provide the brightness sensation and differenced to provide two "opponent colours". The red-green opponent dimension is consistent with the difference between green and red signals, and the blue-yellow dimension is consistent with the difference between the blue signal and the aggregated red and green signals[5].

In the same way that "brightness" describes the subjective response to an emission or an illuminated surface, "hue" and "saturation" describe the subjective attributes of colour. Hue describes, for example, *blueness* and *yellowness*, and saturation signifies the degree of purity of the blueness and yellowness. Sources and surfaces with very narrow wavebands, such as laser spectral lines, are fully saturated – the colour sensation is very strong – and those with a broad range of wavelengths, such as sunlit cloud, are unsaturated and give rise to the sensation of bland pastels. Although hue and saturation appear to be constant attributes, the subjective sensations are influenced by the characteristics of neighbouring surfaces, previously fixated surfaces and illumination level. For instance, decreasing illumination exaggerates the sense of redness, while an increase makes red seem orange. Many of these effects are attributed to the differential rate and degree of adaptation of the retinal cones.

In order to ascribe objectively measurable colour attributes to emissions and illuminated surfaces, equivalent in usefulness to the notion of photometric luminance, the CIE formalised the "tri-stimulus" theory of colour vision. Using available psycho-physical data, and with a pragmatic regard for mathematical convenience, the committee defined the relative spectral sensitivities of idealised long-, medium- and short-wavelength receptors. These tri-stimulus sensitivity functions, called $\mathbf{x}'(\lambda)$, $\mathbf{y}'(\lambda)$ and $\mathbf{z}'(\lambda)$ respectively, are sketched in Figure 2.13. As also summarised in this figure, the measured spectral composition of an emission can be transformed into a set of [X, Y, Z] colour matching values by multiplying the source spectrum by $\mathbf{x}'(\lambda)$, $\mathbf{y}'(\lambda)$ and $\mathbf{z}'(\lambda)$ and integrating the three resultant spectra over the visible wavelength range. [X, Y, Z] characterise both the intensity and the colour of the radiation, but because the sensitivity curve $\mathbf{y}'(\lambda)$ is the same as the photopic sensitivity curve shown in Figure 2.11, the colour can be quantified by three normalised chromaticity co-ordinates [x, y, z]. As these sum to unity, one of the values can be computed from the other two, and chromaticity can be characterised by a pair of values [x,y], each having a range between zero and unity.

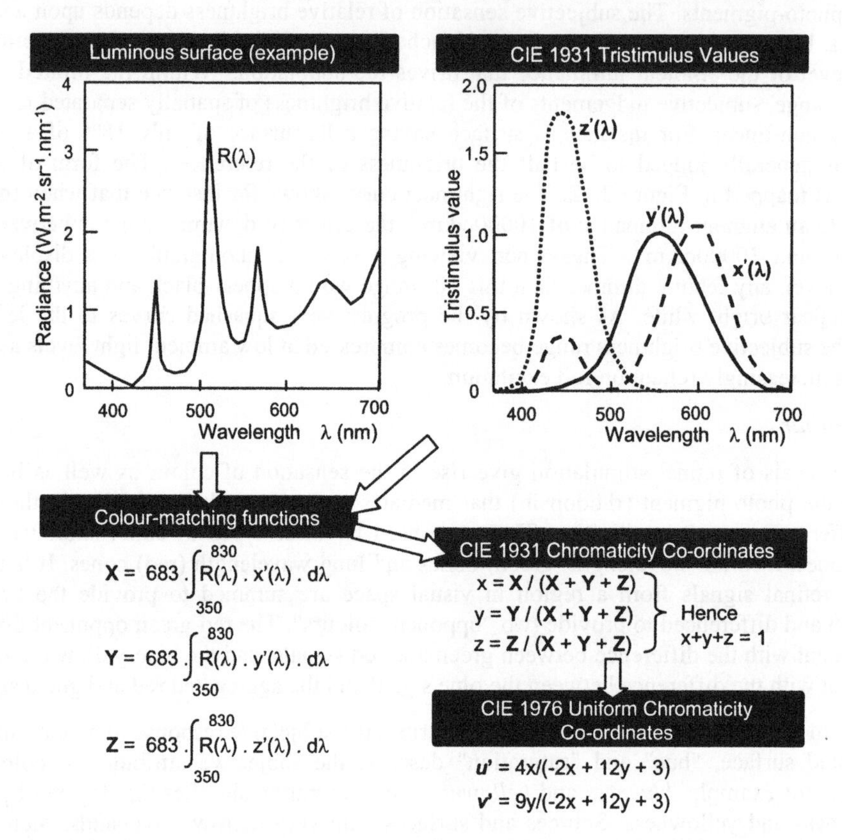

Figure 2.13 Tri-stimulus theory of human colour vision and the CIE representation of colour

The CIE 1976 Uniform Chromaticity co-ordinates [u', v'] are the outcome of further experimental assessments of human sensitivity to small differences in the spectral content of a pair of sources [6]. The co-ordinates and the diagram of colour space shown in Figure 2.14 are currently the most common way to represent the colour of a display or a lamp in aviation.

The spectrum of very narrow emissions from about 700nm (red) through 520nm (green) to 400nm (blue) follows a hook-shaped locus that defines the full colour gamut. The chromaticity values for a light produced by mixing several sources can be predicted by summing the chromaticity values of the components in proportion to their luminance, and it is not necessary to know their detailed spectral composition. For a pair of sources the chromaticity of a mixture lies on the straight line joining the points, the relative distance along the line reflecting the relative luminance of the two sources. Correspondingly, the gamut produced by several sources is the area inside their enclosing polygon. Although purple is not induced by any narrow-band emission, the sensation results from mixing red (~700nm) with blue (~400nm) along the "purple line" joining the co-ordinates for these emissions.

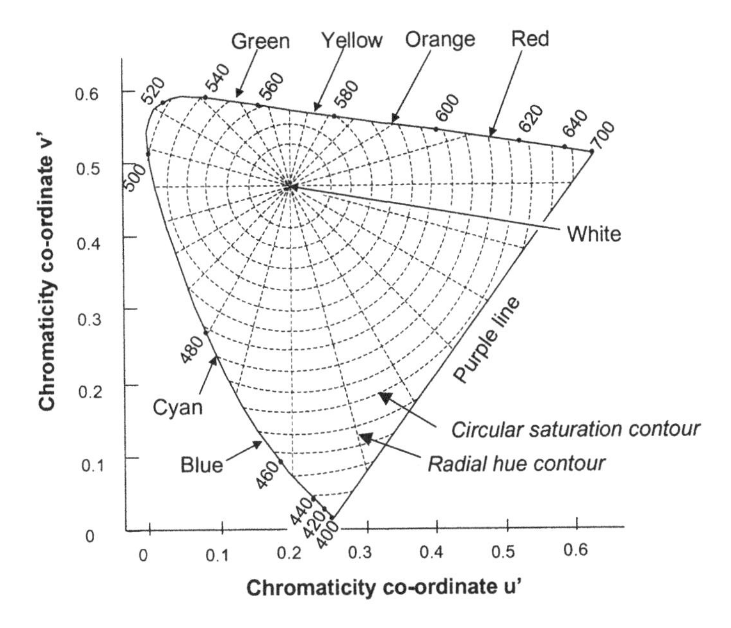

Figure 2.14 Representing colour space using CIE 1976 Uniform Chromaticity co-ordinates

In most individuals the three varieties of cone occur in roughly equal densities, and there is some variation in the periphery as indicated in Figure 2.10. The paucity of blue cones in the centre of the fovea that causes small areas of blue and distant blue lights to appear colourless when fixated is also normal. Colour blindness, the inability to exercise the expected discrimination between stimuli of different spectral composition, arises where individuals lack certain of the cone types, or where there is a marked imbalance in their retinal density. A wide variety of deficiencies can arise, of which the most common is deuteranomalous colour vision. This affects about one in twelve individuals, mainly male, who have difficulty distinguishing what individuals with normal colour discrimination call red and green.

2.4.5 Visual pathways

The two-dimensional array of receptors in the retina is linked to the downstream processes in the brain by the optic nerve by ganglionic neurons. About 100 million rods and 5 million cones converge on only one million ganglions at ratios of 100:1 and 5:1 respectively. The higher spatial resolution of cone excitation is greater in the macula, a small region on the visual axis occupied only by cones and devoid of blood vessels. Acuity is proportional to cone density. As shown in Figure 2.10, cone density reaches a peak of about $150x10^3$ cones.mm^{-2} in the fovea, resulting in a spatial acuity of about one arc-minute. The size of the well-focused diffraction-limited spot image, about 3µm, is roughly the diameter of a foveal cone.

The optics of the eye map the light from the external world to produce an inverted image on the retina, and this spatial correspondence is largely maintained by the downstream processes. Signals from neighbouring receptive fields on the retina are mapped to proximate regions in the thalamus, and thence to the visual cortex, with a spatial compression reflecting the receptor-to-ganglion convergence. The "blind spot", a small region with no photoreceptors, is formed where the optic nerve leaves the eye. The whole system of interconnections is bifurcated into matching left and right structures in which the cross-over (chiasm) in the optic nerves directs stimuli from the right and left fields of view (hemi-fields) to the left and right occipital cortical lobes respectively.

2.4.6 Ambient vision

It has become a convention to regard vision as two almost separate forms of perception[7]. Ambient vision supplies general awareness of the environment, much like the suite of defensive sensors in an aircraft. Stimulation of the relatively wide receptive rod fields in the retina is coupled to the thalamus by the fast, myelinated neurones of the "magnocellular" pathways, and arrives at the cortex 10 to 20msec earlier than signals from the other (parvocellular) set of neurones. Some of the functions of the magnocellular pathways, such as discriminating fine time intervals, detecting moving objects and guiding eye movements, exploit this faster signalling. Others, such as detecting the direction and strength of patterns of stimulation on the retina, make more use of the wide receptive field. These vection patterns, illustrated in Figure 2.15, and their temporal changes contain cues about the direction, speed, angle and likely time to collision with a nearby surface. The perception of such cues, which operate faster than object identification by central vision, contribute to the unconscious control of posture and dynamic movements, for instance during walking and running[8]. It is likely that they play an equally significant part in vehicle control.

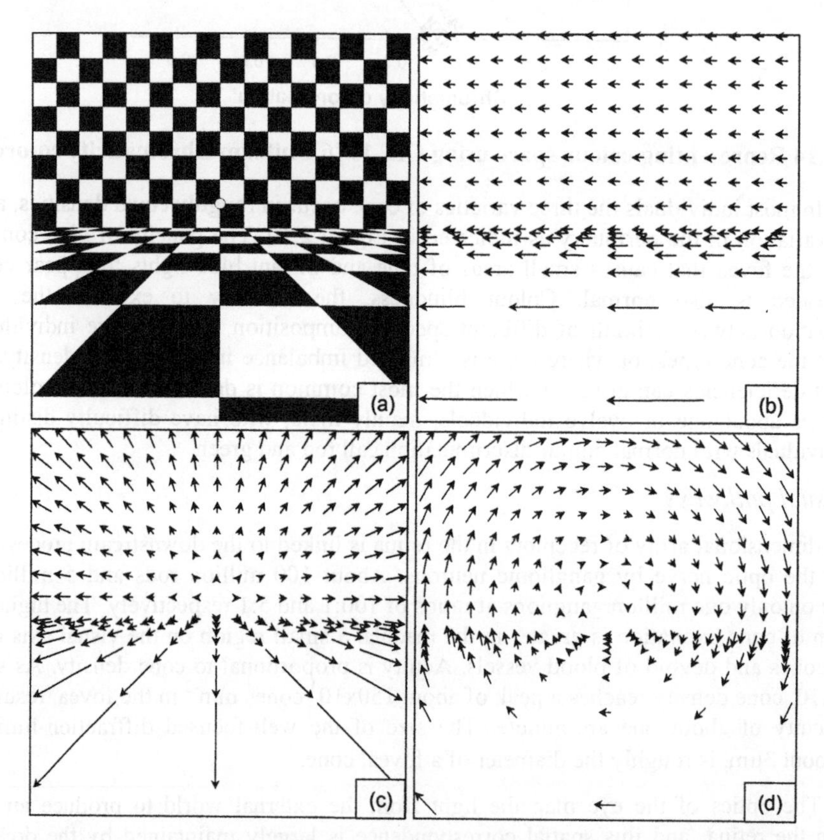

Figure 2.15 Vection cues produced by motion. a) Static view of a horizontal chequered surface and a vertical chequered surface from a point one square above and eight squares back, b) vection field produced by a sideways step of one square, c) a forward step of one square and d) a roll rotation of 10°

2.4.7 Central vision

The second variety of visual perception is termed foveal or central because stimulation is gathered from narrow receptive fields of interconnected cones. The information has high spatial discrimination, and is mediated by the parvocellular system, which differs from the magnocellular system in that the signals are communicated along slower, unmyelinated neurones. This form of vision depends upon the rapid, precise, co-ordinated activation of the extra-ocular muscles to set the orientation of the two eyeballs so that their visual axes intersect the point of interest.

The main visual areas in the occipital lobes break down the parvocellular stimulation into classes of qualities, ranging from depth and distance to colour and orientation. The parietal lobe aggregates these and finds combinations that match significant entities, for instance there is evidence that one parietal region responds solely when presented with familiar faces. The highly manipulated qualities combine with information from the magnocellular stream and terminate in the frontal lobes. Although this large area of associative cortex receives the barrage of information about spatial features and their relative location, there is no generally agreed understanding of how the "what" and the "where" are put together as the conscious perception of the external environment. Appendix D outlines an explanation that links such perceptual integration with attention and the control of eye pointing.

When looking at an object, the lines of sight of the two eyes are driven to converge, and each eye has a slightly different view of the object. These cues are exploited by the visual system to enhance the 3-dimensional representation of the environment. The resulting solidity and depth discrimination is termed "stereopsis".

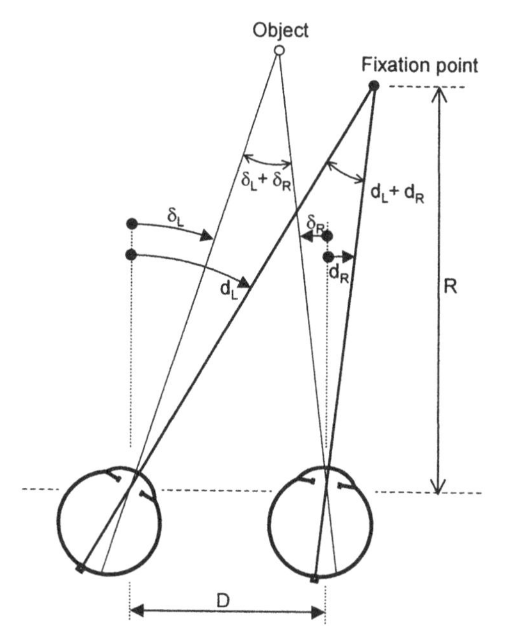

Figure 2.16 The geometry of stereopsis. The vergence angle of the fixation point is the sum of the pointing directions of the two eyes (d_L + d_R) where an inward angle is positive and an outward angle is negative. The retinal disparity of an object from the fixation point is the difference between their vergence angles ($d_L + d_R$) - ($\delta_L + \delta_R$)

As shown in Figure 2.16, the eye-pointing convergence and the disparity between the two retinal images are tied by the viewpoint geometry. Although the eye-pointing convergence provides some depth information, the visual system makes great use of retinal disparities as one of the cues to the relative depths of entities. The minimum resolvable disparity varies between individuals and with conditions. A disparity as small as 5 seconds of arc can be discriminated, and as this is an order of magnitude finer than the one minute of arc spacing between cones in the fovea, stereopsis is regarded as a hyper-acute phenomenon. Stereoscopic cues are most useful for objects relatively near to the viewer. Other cues, such as interposition, parallax, texture and relative size and brightness contribute to the wider perception of the relative distance of objects in the world.

2.4.8 Eye movements

The brainstem controls the muscles within the eyes that adjust the pupil and lens, and also the three antagonistic pairs of extra-ocular muscles that control eye orientation. A very fine muscular tremor keeps the eyes permanently and randomly mobile over an angular amplitude of about a fifth of a minute of arc; a good fraction of the subtense of a foveal cone. As the absence of this Ditchburn tremor causes vision to fade entirely, it is appropriate to think of groups of retinal receptors as detectors that are fundamentally responsive to changes. Most visual performance depends on both the temporal and spatial characteristics of the stimuli.

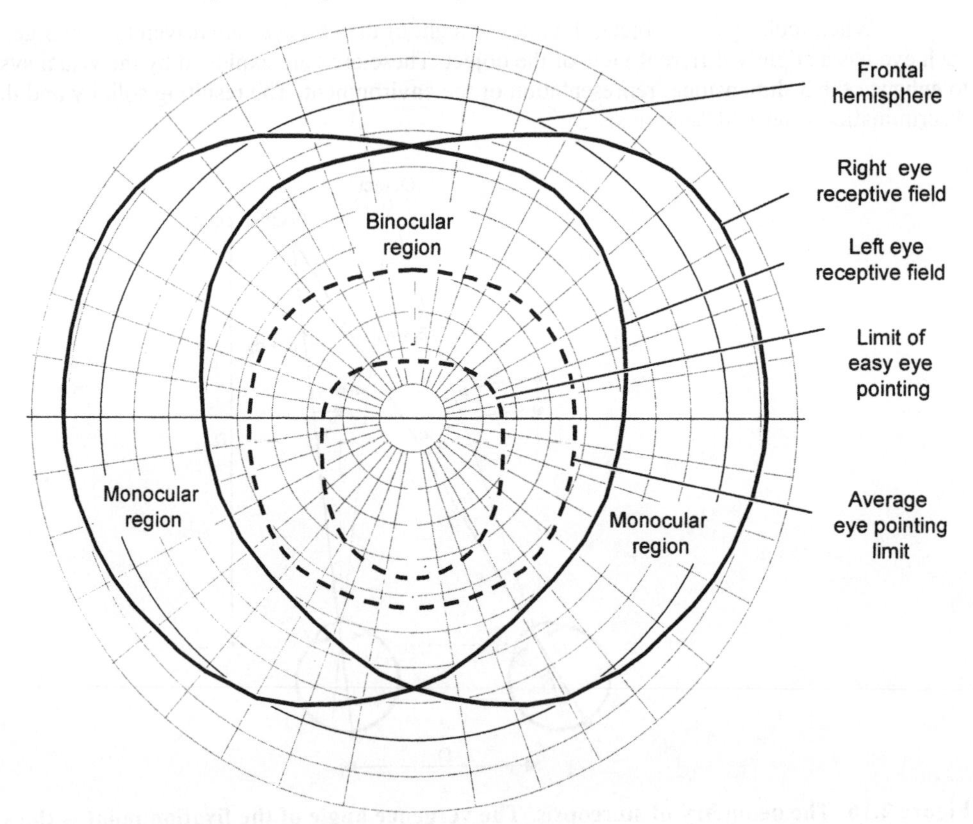

Figure 2.17 Summary of the angular fields of human vision plotted on a 10° orthographic grid (reprinted with permission from QinetiQ)

In the absence of visual stimulation, when for instance the eyelids are shut, the minute muscular tremor is maintained and the lenses relax to a resting focus, usually at about half a metre. The eyes adopt whatever orientation (phoria) results from the balance of forces produced by the relatively relaxed extra-ocular muscles, usually with their lines of sight converging near the resting focus distance. Whether visual stimuli are present or not, rotation of the head causes the eyes to rotate in the opposite direction. This vestibulo-ocular reflex (VOR) tends to stabilise the lines of sight and it is triggered by signals from vestibular sensors in the inner ear; the semi-circular canals that sense the rate of angular rotation of the head. When however a discernable image is present on the retina, eye orientation (and hence pointing direction) falls under the control of higher level feature extraction processes. These direct the eyes to converge on the point of interest, which triggers appropriate lens focus and pupil adjustments. Such fixations or glimpses are maintained for about a third of a second, which is sufficient for feature extraction, whereupon the eyes are directed by attentional processes to jerk (saccade) rapidly to the next point of interest. The eye orientation drifts during fixations, but usually keeps the point of interest within the fovea. During longer fixations there are occasional small jerks (micro-saccades).

Normal eye behaviour is therefore a succession of saccadic jumps and fixating dwells. Figure 2.17 shows that the proportion of the visual field examined in detail is very small in comparison with the total available. The eyes can be turned to point more than 50° from the forward head-pointing direction, but the head is almost always turned in the direction of interest to alleviate the effort needed to sustain eye excursions greater than about 30°.

2.4.9 Binocular vision

Figure 2.17 shows also that the total visual field encompasses almost the full forward hemisphere. A large proportion of this field (about ±60° laterally) is sensed by both eyes, and most of it (about ±40° laterally) can be examined by rotating the eyes to bring foveal vision to bear. Human vision has undoubtedly evolved to make best use of the separated eyes. The co-ordinated eyes have substantially better depth resolution (stereopsis) and about 40% better acuity and brightness sensitivity than a single eye.

When both eyes cannot be used, for instance when viewing through a telescope, microscope or gunsight, about 65% of individuals consistently use their right eye, 32% their left eye and only 3% are ambiocular[9]. The preferred eye is known as the "sighting dominant" eye. A different bias is produced when dissimilar, unfusable, images are presented to the two eyes, for instance if both eyes are kept open when using a telescope or, as described in Chapter 9, the image is presented by a monocular helmet-mounted display. Under these circumstances the viewer experiences "binocular rivalry" in which the two images, or the two eyes, seem to compete for attention. When the two images are of equally brightness, contrast and interest, the visual field is usually seen as an unstructured, changing patchwork of the two. If however, one of the fields appears to be consistently stronger and suppresses the other for longer periods, the bias is known as "sensory dominance". Explanations for rivalry are at the heart of current theories of visual perception. One suggests that objects are perceived as being located unambiguously in visual space by giving precedence to fusible elements over rivalrous elements in the stimuli provided by the two eyes[10]. There may be some correlation between sighting and sensory dominance, and the majority of individuals are right-eyed and right-handed. Handedness arises from an infantile bias for grasping with a particular hand that becomes a defining habit as tools of increased complexity are mastered using the same hand. Eyedness seems to have a similar fixity, but it can also be upturned by voluntary or involuntary transfer of attention. The notion of attention and its interaction with vision is considered in the next section.

2.5 Complex skills

2.5.1 Concepts of attention

Attention generally signifies the selective application of effort or mental resources. When, for instance, an operator does several tasks by applying himself to one then the other in alternation he is said to split his attention between the tasks. This connotation, which implies a versatile but limited and indivisible faculty, underpins some of the workload ideas discussed in Chapter 4. There is, however, a slightly narrower and deeper connotation, which is best discussed using an example. The case of someone reading a book also usefully outlines an explanation for this skill.

The reader faces a book and makes small, rapid eye movements to fixate on some of the marks on the paper. He extracts the meaning from the marks and understands the message. Linking these phenomena is a collection of covert processes that analyse the stimulus, classify the shape of an individual symbol, aggregate this with neighbouring symbols and interpret their combined significance, aggregate this with the preceding words and translate the combination into a significant concept. The sequence is similar to the acoustic, phonological, lexical and semantic decoding used to understand the spoken word, as outlined in section 2.3.2 above, but in this case foveal vision supplies the basic data.

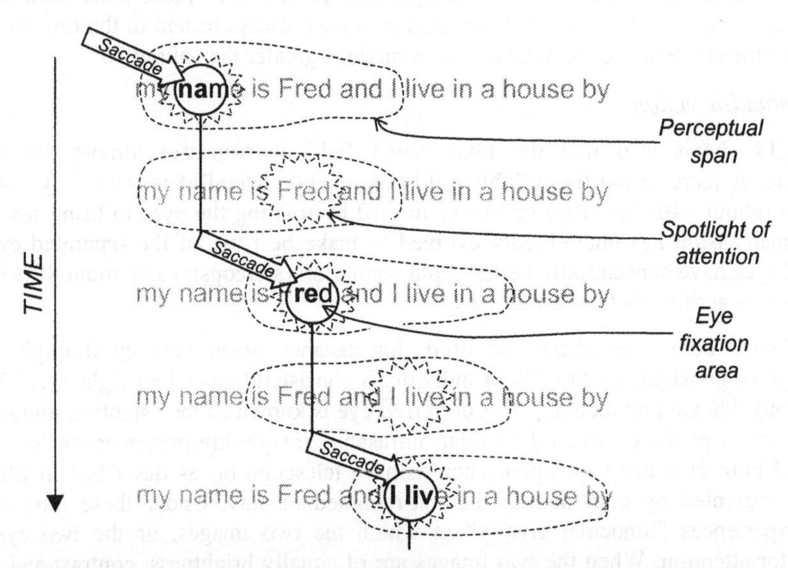

Figure 2.18 The mechanism suggested by Morrison for the control of eye movements during reading. An internal spotlight of interest (attention) shifts within a limited perceptual span to trigger an eye movement (saccade) that brings foveal vision (fixation) in line with the internal spotlight

It is suggested that once foveal vision has been brought to bear on a word and the significance of the symbols has been extracted, the reading "program" decides between three options; to move to the next word, to backtrack up or left, or to skip further along to the right. The second action is chosen if the word or the context is not understood, and the third is chosen if the meaning of the next word or words can be assumed[11]. The outcome of the decision specifies the instruction to the oculo-motor regions in the brainstem to initiate a saccadic jump of the appropriate magnitude and direction. Figure 2.18 illustrates the mechanism postulated by Morrison[12] in which the lexical and semantic decoding processes shift their focus of interest within the available percept, inducing the eyes muscles to produce a saccade that shifts central vision by a roughly equal amount, like a reluctant dog jerked by an elastic lead.

In this model the covert focus of interest is called "attention". The main evidence for the idea of an internal attention shift within a perceptual span comes from intricate experiments in which computer-generated text is manipulated relative to the instantaneous point of fixation[15]. These have shown that a fluent reader of English is usually aware of about fourteen characters to the right of the fixated character and only about four to the left. The converse leftward bias of perceptual span is found for Hebrew readers who scan text from right to left. The bias changes from right to left when bi-lingual individuals changes from reading English to reading Hebrew, so it is likely that achieving reading fluency in any language involves optimising the perceptual span and the eye movement dynamics.

2.5.2 Performance of complex tasks

In the reading example, information is extracted from within a limited perceptual span to enable a decision for an action (saccade size and direction) to achieve a voluntary goal. A similar selectivity is used in other visual tasks such as searching and monitoring, and an analogous form is applicable to auditory tasks such as listening to a particular speaker in a noisy background.

It is obviously important to select and process relevant information when performing a voluntary goal-directed task. However, to maintain a broad awareness of the environment, the brain must simultaneously take in and sift a large amount of irrelevant information, and whatever mechanism is in control of the instantaneous goal of selective attention must be accessible to the pathways that filter the ambient stimuli. The nature of attention-getting signals is important to the design of auditory warnings, as discussed in Chapter 10, but there are deeper issues. These are illustrated when an involuntary glance is made to a peripherally-sensed movement, or one's name is recognised in a previously ignored conversation. The stolen glance reinforces the idea that low level cues are perpetually extracted from sensory stimuli, which is unsurprising, but the overheard name suggests that the phonetic decoding of auditory stimuli is equally irrepressible, which implies that the senses utilise a considerable amount of dedicated, pre-attentive processing resources. Bottle-necks of selective awareness must therefore occur within some more versatile processing resources where higher level goal-directed decisions are made.

Many models have been devised to represent the attentional mechanism that regulates the flow of information through the sequences of sensory, cognitive and neuro-muscular control processes. As outlined in Chapter 4, several have been employed as quantitative tools for predicting the human ability to deal with the competing demands of concurrent tasks. One approach has been offered by Wickens[13] and co-workers as a way of summarising the outcome of experiments that require participants to perform several tasks at the same time. Their "multiple resources theory" considers that auditory and visual information processing are done by independent encoding, central processing and responding resources, and each either takes spatial information to generate a manual output or verbal information to produce a vocal output. This suggests that certain combinations of task can be done without a significant drop in performance if the tasks do not draw on the same postulated resources. For instance, controlling a vehicle makes use of spatially meaningful visual information and produces hand and arm movements, whereas conversation involves listening, logical reasoning, and producing a vocal output. Thus it is possible to converse while driving because the two tasks use different resources. On the other hand, although reading a book requires verbal reasoning and produces no manual output, both reading and vehicle guidance need control over eye movement and visual input, so these tasks can not be done at the same time.

Some experimenters have suggested that this simple description of the cognitive processing streams should be elaborated, for instance to cater for short-term memory conflict[14]. Others have been disinclined to postulate the form of the resources, and have preferred to deal with task interference empirically. For instance, Navon and Gopher[15] suggested that the trade-off between two concurrent tasks could be described using a "performance operating characteristic" (POC), a graph in which the two orthogonal axes represent the level of performance of the two tasks. A point can be plotted on such a graph to show the levels of performance obtained by an individual during an experimental session. A large number of points can be plotted by asking the individual to repeat the tasks but using different priorities, including the two cases where full priority is given to one task and the other is neglected. The shape of the POC curve that best fits the points provides a concise summary of the way the tasks interfere.

Predicting human performance of complex tasks is difficult because tasks that initially require focussed attention can with familiarity become automatic and effortless. It is also difficult to distinguish between the parallel use of separate cognitive resources and an efficient strategy for sharing a central resource. Individuals differ in perceptual and intellectual ability, and in temperament and personality. They also differ greatly in the priorities they attach to the separable tasks, and they perform even a highly defined role in peculiar ways. Although no coherent behaviour would be possible without a mechanism for allocating attention to tasks in order to achieve longer-term intentions, motivation and the business of generating intentions is not understood. Appendix D gives an outline of a conceptual framework that supports the discussion of these factors.

2.6 The aviator

2.6.1 The job

The job of the aircrew is to undertake a flight to achieve a specific purpose. For instance, in a ground-attack fast jet like the RAF Harrier, the job is to fly the machine along a pre-determined route to a location where a target can be found and attacked before returning to a friendly base. Different locations, weather conditions, targets, defences, equipment failures, and interactions with other forces can generate a multitude of specific demands that the pilot must deal with in order to complete the mission satisfactorily. The job can be regarded as the set of safety-critical and mission-critical information-processing tasks summarised in Table 2.2.

A task in Table 2.2 involves interacting with specific aircraft sub-systems and is a responsibility that endures for most of the flight rather than a short, purposeful elementary activity such as represented in Chapter 4, Figure 4.5. The classification corresponds with service practice in that the top section of the table contains the primary skills of controlling an aircraft that are taught during basic training; planning a flight, navigating, communicating and dealing with the aircraft systems. Combat aircrew learn later to deal with tactics, targets, weapons and threats, just as aircrew in other air vehicles learn the special skills, for instance of managing fuel transfer in a refuelling tanker or interacting with passengers in a transport aircraft. Each type of vehicle and mission generates specific tasks.

A number of problems with this classification arise when considering details. Although most activities fall clearly into a task category, some could belong in several. For instance, deploying the flaps before landing could rationally be considered an aspect of either *aircraft control* or *managing aircraft systems* and it is necessary to make a reasonably arbitrary, but consistent, assignment. Other activities need to be categorised carefully. For instance, setting the squawk code on an Identification Friend or Foe (IFF) system is an element of *communicating* even though the receipt and transmission of information is entirely automatic.

These tasks cannot be treated like the separate elements in most multi-task psychology experiments because they are interlinked; failure in one affects another. For instance, flying too high through defended airspace would generate a greater concern with countering threats. The task of *managing the mission* is peculiar in that it is largely covert and that it supplies overall co-ordination. Essentially, if the tasks were allocated to members of a team, one individual would formulate a plan to meet the mission objectives, give suitable operating criteria to the rest of the team, and then monitor their progress. He would only intervene to issue new criteria if, in the light of reports of success and failure, he felt it necessary to revise the plan. The pilot of a single-seat aircraft has, however, to carry out this top level task amongst the others, and ensure that no task is neglected. Effective self-supervision, known as "airmanship" in the services, can be regarded as part of the central co-ordination function of *mission management*.

Table 2.2 A summary of the separable high-level information-processing tasks performed by combat aircrew. S = safety-critical, M = mission-critical. (reprinted by permission of QinetiQ)

Task	Objective		Possible consequences of failure
Aircraft control	Controlling the vehicle speed and flight path	S S M	Crash (stall, spin, too low) Increase exposure to defences (too high) Miss attack window (too fast/slow)
Navigating	Determining location and required course	M	Fail to arrive at target, or base, or diversion
Communicating	Managing radios and sending/receiving information	M	Fail to send or receive crucial information
Aircraft system management	Taking care of fuel, electrics, hydraulics etc	M S S	Abort mission unnecessarily Run out of fuel Allow dangerous condition (e.g. engine fire)
Mission Management	Planning the mission, being aware of the outcome of other tasks and re-planning as necessary	S	Fail to minimise mission risk
Targeting	Locating, identifying and designating targets	M M	Attack wrong target Miss target
Weapon management	Selecting, arming and assigning weapons to targets	M M	Attack wrong target Fail to prosecute attack
Countering threats	Being aware of threat sites and vehicles, and deploying appropriate counter-measures	S	Be shot down
Tactical decision making	Maintaining tactical awareness and making en-route interactive tactical decisions	S	Increase exposure to defences

2.6.2 Human error

Accidents involving manned aircraft can damage property and kill or maim uninvolved personnel as well as the occupants and crew of the aircraft. Civil accidents are investigated by following the procedures in the International Civil Aviation Organisation (ICAO) Manual of Aircraft

Accidents, with the remit of finding the cause, instigating a remedy and minimising the chance of a recurrence. In the UK, except for accidents involving sports aircraft such as hang-gliders and microlights, which are handled by the safety group from the relevant association, the work is done by the Air Accidents Investigation Branch (AAIB). When the accident involves a military aircraft, the AAIB contributes evidence to a Board of Enquiry (BoE) into the circumstances of the accident. The BoE, which can also draw expertise from the RAF Inspectorate of Flight Safety, reports findings to the authorities of the affected service.

In general, about half the accidents and incidents tend to be attributed to equipment failure and about half to human failings. As can be seen in Table 2.3, some human errors can arise from a failure somewhere along the sense-decide-act chain of information processing, perhaps caused by one or more underlying factors. Other types of human error occur, for instance from a slip when executing a well-learned skill, the use of an inappropriate rule or from a lack of knowledge. Some may be due to an intrinsic human fallibility that cannot be corrected. Others can be addressed by better organisational practices or better training, and a few by punitive disincentives. A lot can be corrected by recognising equipment deficiencies and modifying the design.

Table 2.3 A taxonomy of human errors and contributory factors in RAF aviation accidents (From Chappelow J W, *Error and accidents in aviation*, in Ernsting J, Nicholson A N and Rainford D J (eds), *Aviation Medicine 3rd Edition*, Butterworth Heinemann, 1999. Reprinted by permission of QinetiQ)

Error type	Predisposing factors	Disruptive factors	Enabling factors
Perception errors	***Trait***	***Stressors***	***Equipment design***
Visual illusion	Personality	Noise	Ergonomics - displays
Disorientation	Lack of talent	Physiological stresses	Ergonomics - controls
Misinterpreted display	Inexperience	Operational pressure	Ergonomics - layout
Communication failure	Excess zeal	Time pressure	Personal equipment
Threat not detected	Lack of airmanship	Task demand - high	Logic in automatic system
	Sensory limitations	Task demand - low	Aircraft handling characteristics
	Normal feature of group behaviour	Distraction	
	Normal feature of cognitive function	Threat	
Intention errors	***State***		***Process failings***
Deliberate violation	Alcohol/drugs		Training
Inappropriate model	Fatigue		Briefing
	Hypoglycaemia		Administrative support
	Life stress		Selection criteria
Action error	Low morale		
	Under-arousal		
Slip/lapse	Over-arousal		
Slow response	Social factors – crew co-ordination		
Precipitate response	Locally-condoned practice		
Disorganised response	Social context		
Mishandling	Social factors – mutual complacency		

The prevailing assumption, which is almost a dictum of modern management, is that there is a best way to do every task. For each of the tasks delineated above, the desired outcome can be achieved with least risk by determining what information should be selected, what decisions should be made and what actions should be taken. In practice, almost all novices willingly adopt a proven approach and add refinement with familiarity. Dealing with emergencies invariably relies on the ability to recognise the condition and enact a well-rehearsed procedure. In these cases the remedial action is best decided by experienced crew in calm circumstances and tested in a simulator. These, and such safety-critical processes as arming weapons, are endorsed within a squadron as Standard Operating Procedures (SOP). The compulsion to use each SOP varies. They have no legal status, but they represent good practice and are used as benchmarks in the event of an accident.

National aviation authorities realise that accidents represent a small proportion of the hazardous events and circumstances that occur during the many millions of air miles flown by aircrew. Attribution to specific human-related causes is usually uncertain, particularly when the accident is fatal and evidence is gathered under conditions that resemble an adversarial court more than a scientific enquiry. It has been recognised that many potential accidents and incidents go unreported because they were averted or for fear of disciplinary action. In order to gain a broader understanding of such incidents, many authorities have set up a procedure such as the Confidential Human Factors Incident Reporting Programme (CHIRP) in the UK and the Aircraft Safety Reporting System (ASRS) in the USA. These allow aircrew and groundcrew to report hazardous events and poor practice in the confidence that the information will contribute to future safety without jeopardising their relationships with colleagues and employer. Such confessional data is now providing a significantly better understanding of aircrew problems.

2.6.3 Selection

Whether seeking to take control of an aircraft for pleasure or remuneration, the aviator is required by the national regulating authority to have qualities that they regard as necessary for the safe operation of an aircraft. The services and airlines are also keen to ensure that their investment is not wasted and that the candidates they accept for training are those most likely to graduate satisfactorily. In general, potential aircrew are selected on medical, physical, intellectual and temperamental criteria.

Medical standards are assessed on recruitment, and are thereafter checked periodically, usually at least annually, by an authorised examiner. Service standards are usually more stringent than civil standards. The principle concerns are that the individual has the qualities assumed by equipment designers, such as colour discrimination, is fit to take the stresses of the flight environment and has no condition or incipient condition that could lead to a heart attack or other form of incapacitation[16]. A general physical examination, vision and hearing tests, comprehensive blood and urine tests, a chest X-ray and an electrocardiogram are normally done.

Good literacy, numeracy and general academic attainment is needed to show that a candidate will readily understand the aerodynamic and engineering principles and be able to handle navigation computations and spatial reasoning. Non-academic mental and neuro-muscular qualities, such as rapid reactions, accurate short-term memory and the ability to split attention between competing demands, are generally assessed using a computer-controlled battery of tests. Some airforces use sessions in a light aircraft simulator to give candidates an experience very similar to that of controlling an aircraft, and allow the candidate to show their ability to perform manoeuvres. Although lengthier and therefore more expensive to administer than a test battery, this form of aptitude testing has been shown to identify a greater proportion of individuals who will successfully complete their training and reject those who would not.

The personality or temperamental attributes required to become a successful aviator are less easily defined and assessed. Perseverance, acceptance of responsibility and resistance to physical and psychological stress are all needed. As these are also essential for leadership, they are also examined to gain acceptance for officer training, which is a parallel aspect of most aircrew roles. Assessment usually relies on informal judgements made during an interview and observation of the candidate's behaviour and attitude during group problem-solving sessions and when performing the test battery.

2.6.4 Training

Civil aircrew training starts with gaining the minimum, a Private Pilot Licence (PPL), which requires about forty hours flying in a light single-engine aircraft. As well as the ability to control the aircraft on the ground and in the air, the trainee also learns the procedures to recover from a spin and a stall, and to deal with engine failure. Classroom training includes aerodynamic principles, basic airframe engineering, radio communications, navigation, air law and meteorology. Changing to another aircraft type requires conversion instruction by an appropriately qualified instructor for a period that depends upon the type, for instance a dual-engine aircraft requires learning to cope with the asymmetric forces produced by the failure of one engine. The next step is to acquire experience of flying at night using instrument indications, which is done partially in a dual-seat aircraft fitted with a hood that prevents the student from looking outside, and partially in a simulator. The student also receives further instruction in meteorology, general navigation and air law and procedures, and is introduced to flight planning, air-driven and electronic instruments and relevant human factors. After the student passes flight and ground tests, an approved examiner can issue an Instrument Rating. The usual professional qualification, a Commercial Pilot Licence (CPL), requires several hundred hours of command flying, some of which must be at night, with further demonstrations of proficiency in flight and testing of theoretical knowledge.

Successful service pilot candidates undergo elementary flight training using light aircraft powered by a piston-engine. RAF trainees first attend Initial Officer Training and then the Joint Elementary Flying Training Squadron (JEFTS) unless at a university, in which case they are encouraged to enlist in a University Air Squadron (UAS). On completion, they are selected for the next phase to train in either a fast jet, helicopter or a multi-engine transport, for which they are posted to either the Basic Flying Training School or the Defence Helicopter Flying Training School, or attend a short multi-engine lead-in course. The three groups remain separate thereafter, with fast jet pilots going on to the Fast-jet Advanced Flying School and transport pilots going on to the Multi-Engine Advanced Flying Training Squadron. If successful, the pilots graduate and receive their "wings". Candidates selected to train as navigators have to pass a similar selection process and undergo similar basic followed by advanced training phases before receiving their "brevets". Aircrew are then posted to the Operational Conversion Unit (OCU) where they are introduced to the detailed construction, engineering, internal systems and capabilities of the aircraft type they will fly later as a member of an operational squadron. Aircrew training for non-officer roles such as Weapon System Operator in Nimrod, Air Engineer in the ED3 Sentry and Air Loadmaster in Chinook or Hercules have first to pass a twelve week Initial Training Course before undergoing training for about a year.

Service aircrew in the UK generally expect to remain with a squadron for a tour of three years before posting or promotion. If changing aircraft type, it is first necessary to attend the appropriate OCU. The competence and confidence to perform tactical manoeuvres and operate the sensing and weapon systems is acquired under the guidance of Qualified Weapons Instructors (QWI) in the air, and by specialist instructors in a mission simulator. By the end of their first tour aircrew are normally regarded as effective, and equipment such as night vision goggles are introduced when the basic tactical foundation has been laid. Aircrew are also sent regularly to special service schools so that they remain familiar with equipment and procedures for emergency escape, water survival, rescue and evasion; aspects of service flying that most expect not to utilise. Experienced aircrew can, on recommendation, undertake training for special roles such as QWI and Test Pilot.

Although all flying imparts experience, a regular minimum of air time is needed just to maintain skills, and in this sense training never ceases. Training not only imparts the necessary skills, but also enables the aircrew to cope with their physiological and emotional reactions so that they can perform reliably under combat stress. The only distinction between peace-time flying and involvement in conflict is a change of priorities from safety to mission effectiveness. The tasks remain the same.

References

1. Kohler I, *Experiments with goggles*, Scientific American Vol 206, pp. 62-86, 1962
2. Bekesy G, *Uber der akustiche reizung des vestibular-apparates*, Pflugers Archiv. European Journal of Physiology, 236, pp. 59-76, 1935
3. *Normal equal loudness contours for pure tones and normal threshold of hearing under free field conditions*. International Organisation for Standards, ISO R-226, 1962
4. *The basis of physical photometry*, CIE publication 18.2, Commission Internationale de l'Eclairage, Wein, Austria, 1982
5. Walraven P J, *A zone theory of colour vision*, Farbe 15, pp. 17-20 1966
6. MacAdam, *Visual sensitivities to colour differences in daylight*, Journal of the Optical Society of America, Vol 32, 1942
7. Leibowitz A W and Post R B, *The two modes of processing concept and some implications*, in Beck J (ed) *Organisation and representation in perception*, Erlbaum, Hillsdale, NJ, 1982
8. Lee D N and Lishman J R, *Visual proprioceptive control of stance*, Journal of Human Movement Studies, Vol 1, pp. 87-95 1975
9. Porac C and Coren S, *The dominant eye*, Psychological Bulletin, Vol 83 (5), pp. 880-897, 1976
10. Blake R, *A neural theory of binocular rivalry*, Psychological Review, Vol 96(1), pp. 145-167, 1989
11. Rayner K, *Eye movements in reading and information processing: 20 years of research*, Psychological Bulletin Vol 124, pp. 372-422, 1998
12. Morrison R E, *Manipulation of stimulus onset delay in reading: evidence for parallel programming of saccades*, Journal of Experimental Psychology : Human Perception and Performance Vol 10, pp. 667-682, 1984
13. Wickens C, *Processing resources in attention,* in Parasuraman R and Davies R (eds), *Varieties of attention,* Academic Press, NY, 1984
14. Jones D M, *The cognitive psychology of auditory distraction: The 1997 BPS Broadbent lecture*, British Journal of Psychology, Vol 90, pp. 167-187, 1999
15. Navon D and Gopher D, *Task difficulty, resources and dual-task performance*, in Nickerson R S (ed) *Attention and performance VIII*, Erlbaum, Hillsdale, NJ, 1980

16. Evans A D B and Rainford D J, *Medical standards for aircrew,* in Ernsting J, Nicholson A N and Rainford D J (eds), *Aviation Medicine* 3rd Edition, Butterworth Heinemann, 1999

Further reading

Boff K R, Kaufman L and Thomas J P, *Handbook of perception and human performance: Volume I, Sensory processes and perception: Volume II, Cognitive processes and performance,* Wiley Interscience, NY, 1986

Carter R, *Mapping the mind*, Weidenfeld and Nicolson, London, 1998

Denes P B and Pinson E N, *"The Speech Chain - The physics and biology of spoken language,* 2nd edition, Freeman, New York, 1993

Kimble D P, *Biological psychology*, Holt, Rinehart and Winston, NY, 1988

Stevens K N, *Acoustic phonetics,* MIT Press, Cambridge Mass, 1998

Chapter 3

The Need for Protection

3.1 Introduction

Flying exposes its participants to unusual and potentially hazardous conditions that can be parcelled into three categories. The first is the airspeed, altitude and load factor that the aircraft can attain; its flight envelope. This dictates what artificial conditions and protection are needed to keep the aircrew physiologically healthy. The second category encompasses predictable circumstances that arise during a flight that can over-stretch the expectable abilities of the crew and lead to an accident. The third are the deliberate life-threatening actions of an enemy.

This chapter gives an overview of the main concerns and the general approaches taken by aircraft design teams to provide countermeasures against the most prevalent natural hazards. The final sections summarise the additional concerns faced by military aircrew.

3.2 The environment

3.2.1 The atmosphere

The gaseous atmosphere of the Earth is mainly a mixture of nitrogen (~78% by volume) and oxygen (21%), with the remaining 1% made up by small concentrations of carbon dioxide, hydrogen and inert rare gases. It is considered to be composed of layers with different characteristics. The base layer, the troposphere, runs from sea level to 36,000 ft and is characterised by a steady decrease in temperature with altitude (called the "lapse rate") and the inclusion of water vapour that varies in amount between latitudes and seasons. In the next layer, the stratosphere, the temperature remains roughly constant to about 90,000 ft but increases steadily to –3°C at about 160,000 ft. In the layer above, the mesosphere, the temperature again falls to about –113°C at about 300,000 ft. In the final thermosphere or ionosphere it rises again, but the conventional notion of temperature no longer applies as the gases are very dilute and the molecules are ionised by high energy solar particles. The atmosphere ends at about 2400,000 ft (~450 miles). Here molecules rarely collide, and they can escape gravitational attraction to enter deep space.

As the effects of ionising radiation and ultra-violet radiation, either by direct bombardment or by creating ozone from oxygen, have not proven harmful for the crew and passengers of Concorde, this hazard is of concern only to those venturing above 60,000 ft for extended periods. The main dangers for the majority of aircrew are most likely to arise from the low pressure. For those exposed during emergency escape from an unflyable aircraft, the low temperature is also of real concern.

Figure 3.1 indicates how temperature and pressure change with aircraft altitude for the ICAO Standard Atmosphere[1], which was defined to represent the average over the seasons at latitude 45°. In this case the sea level temperature of 15 °C lapses in the troposphere at a rate of 1.98°C per 1000 ft until it reaches –56.5°C at 30,089 ft. Over the same altitude range the pressure falls exponentially from 760mmHg to 170mmHg. The low pressure and low air density at high altitude have many aerodynamic and propulsive consequences that can be exploited to attain higher groundspeed and improved efficiency, but there are also many ramifications for the aircrew.

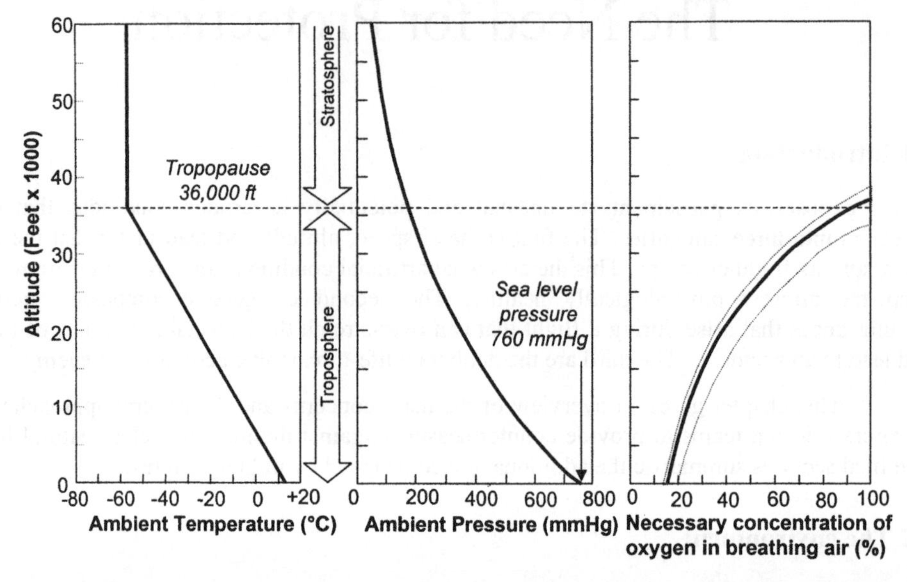

Figure 3.1 The left and central graphs show the variation of temperature and pressure with altitude for the Standard Atmosphere specified by the International Civil Aviation Organisation. The right graph shows that with increasing altitude and decreasing pressure, the air supplied for breathing must contain an increasing proportion of oxygen and above 40,000 ft it is necessary to breath 100% oxygen at a pressure above the ambient level

3.2.2 Weather

Wind, cloud and precipitation are atmospheric states that, although influenced dramatically by local geography, are best considered globally. The heat of the sun evaporates water, largely from the sea, to create a significant partial pressure of water vapour in the troposphere. A combination of coriolis force due to the Earth's spin and convection due to stronger sun heating at the Equator drives this wet air so that it circulates within six bands running parallel to the Equator. These span latitudes from about 0° to 30°, 30° to 60° and 60° to the pole in the Northern and Southern hemispheres. These circulatory movements of the whole troposphere, which vary in strength with sun heating and therefore with season, produce the reliably steady wind used by sailing ships for ocean passage-making; the polar easterlies, the mid-latitude westerlies and the equatorial trade winds.

Complex interactions occur where the air from neighbouring bands come into contact. For instance, around 60°N and 60°S, the cooler, denser air from the polar band pushes beneath warm moist air from the mid-latitudes to form mobile vortexes that circulate in an anti-clockwise sense in the northern hemisphere and clockwise in the southern hemisphere. These cyclones or depressions have lowered central pressure, generally extend for about 300 miles and move in an easterly direction at about 30 knots. Within the mobile vortexes, the temperature of the rising moisture-laden air falls at the lapse rate and on reaching the height at which the temperature-dependent vapour pressure is above the saturation level, the water vapour condenses on airborne dust particles to form clouds of sub-micron droplets. With further cooling, brought about by the up-flow towards the centre of the depression, the droplets grow and precipitate as rain, snow or hail. As shown in Figure 3.1, the air temperature is below freezing at about 10,000 ft, so the cloud layer rarely extends above this altitude. The amount of cloud, the strength of the surface wind and the level of precipitation experienced at any particular location at these latitudes depends largely on the state and proximity of the nearest depression. These have a characteristic life cycle of several days.

Most airline flights cruise at about 35,000 ft and are unaffected by weather conditions at ground level. Aircrew are mainly interested in the wind strength and direction along their intended route and in any restriction to the atmospheric visibility at their destination. Jetstreams are winds formed at the top of the troposphere by the confluence of the polar and mid-latitude circulations. They circulate around the pole in an easterly direction at latitudes of about 50°N and 50°S and extend a distance of about 200 miles laterally and about 12,000 ft vertically. The core air mass moves at a speed of about 180 knots relative to the ground, so aircrew planning east-bound flights attempt to take most benefit from such a tailwind, but stay clear of the strong head-wind when west-bound. However, flying in a jetstream invariably carries the risk of clear air turbulence (CAT) induced by the windshear between layers moving at different speed. Forecasting the phenomenon is difficult.

Condensation forms as cloud, mist or fog wherever wet air is cooled below the saturation vapour pressure. For instance, a thin, low layer of orographic fog condenses when a warm, moist air mass encounters cold ground or water. Hills and mountains force a crossing air mass to rise and cool. Thermal clouds are formed by a rising parcel of air produced by contact with an area of the ground that has absorbed more sun heat, usually because the area is darker than the surroundings or is an island surrounded by cooler water. Although clouds of these varieties are sought by glider pilots to exploit the rising under-current of air to gain altitude, the form and colour of a cloud is crucial. In general a dense nimbus cloud that bears large light-absorbing droplets appears dark in comparison with the other forms of cloud that contain fewer, finer particles. The aviator must be particularly wary of entering a large cumulo-nimbus, thunderstorm, cloud that can extend vertically to the top of the troposphere. The up-draught at the core readily reaches a speed in excess of 80 knots and generates the strong electrical potential that is discharged as lightning. The severe turbulence can break the airframe, and a lightning strike can damage on-board electrical systems. It is now common for passenger-carrying types to be fitted with radar systems that assess the density and movement of precipitated water so that aircrew can assess and avoid the hazard. Extreme manifestations, such as Pacific typhoons and Atlantic hurricanes, are usually anticipated and tracked by meteorological authorities.

When exposed to low temperatures at altitude or in polar regions, and in particular in cloud, aircrew are mindful of the formation of ice. As well as carrying unnecessary mass, the accretion on the aircraft wings can disturb the airfoil lift, and it can constrain air feed to the engines and movement of the control surfaces. Most aircraft that are likely to encounter the phenomenon are equipped with in-flight ice-prevention and de-icing facilities.

3.2.3 The Sun

The Sun[2] acts thermally as a black body with a surface temperature of about 5800°K that bathes the Earth with energy at a rate of 1.37kW.m^{-2}; the "Solar Constant". At the equator about half the thermal radiation reaches the planet surface, the rest being reflected, scattered or absorbed by the gases in the atmosphere. The surface heating decreases progressively towards the poles as more energy is absorbed by the oblique path through the atmosphere and the increased angle of incidence spreads the flux over a larger area.

The effect of Sun heating on aircrew depends on latitude, altitude and the aircraft type. Civil aircrew and passengers are largely shielded by the aircraft structure and when cruising at altitude the solar irradiation combined with the heating effect of high speed airflow usually exceeds the cooling effect of the sub-zero ambient air. The environmental conditioning system (ECS) not only pressurises the air, as outlined in section 3.3.2, but also extracts the excess cabin heat. The imbalance is exaggerated by the large transparency area of the fast jet, which collects a large amount of solar radiation, and without the injection of cooled air from the ECS the temperature would soon become unbearable, especially when operating above cloud. In practice the worst overheating can occur in a combat aircraft when waiting to take-off from a base in a desert. The cockpit temperature can exceed 60°C unless the ECS is powered by an engine or auxiliary power unit, or the aircraft is connected to a ground unit that supplies conditioned air. As reviewed briefly in Chapter 13, one palliative is to include a heat-extracting layer in the ensemble of aircrew clothing.

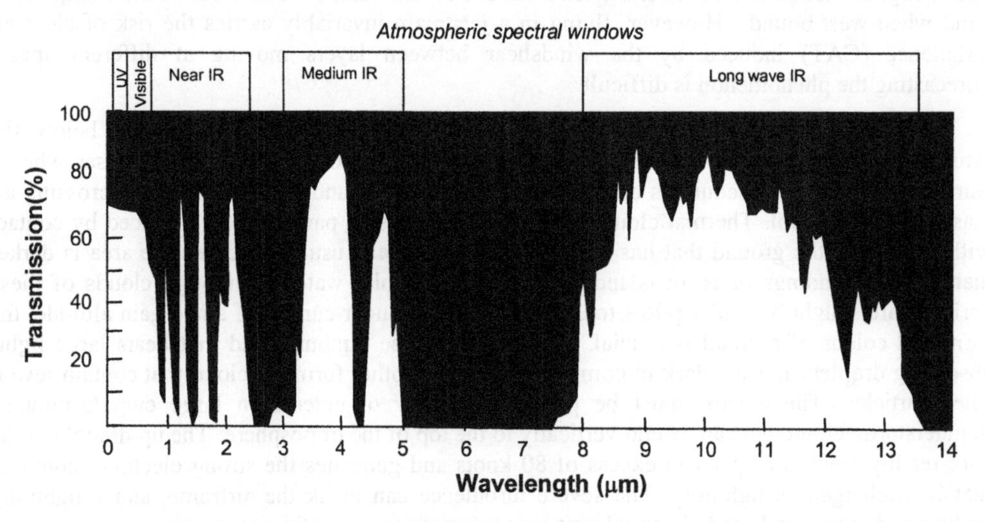

Figure 3.2 Average proportion of radiation transmitted though the atmosphere along a vertical path in daylight in the absence of clouds (Re-printed with permission from QinetiQ)

As well as transmitting visible radiation, as shown in Figure 3.2, there are other spectral bands where absorbing atmospheric gases – mainly water vapour and carbon dioxide – allow electromagnetic waves to propagate with little loss. These windows occur in the ultraviolet (0.25 to 0.4μm), near infra-red (0.7 to 2.3μm), medium infra-red (3 to 5μm) and long-wave infra-red (8 to 13.5μm) spectral bands in addition to the visible band (0.4 to 0.7μm). The global "greenhouse effect", the surprisingly elevated equilibrium temperature of the Earth surface, arises because much of the Sun's energy can penetrate to the planet surface, but the same proportion cannot be re-radiated from the comparatively cool surface because it is blocked by the absorbing regions between spectral windows.

The spectral windows themselves are currently exploited, mainly in the denser atmosphere at low altitude, in a variety of infra-red sensors that can pick up thermal radiation at night, and in cloud and fog that scatter the shorter visible wavelengths. These sensors, as described in Chapter 5, provide pictorial images that can enhance or substitute for the natural view of the world.

In addition to heating the Earth atmosphere and surface, it is notable that the electromagnetic radiation from the Sun is strongest between 0.35 and 0. 7μm, the range sensed by the human visual system. (That these wavelengths are emitted strongly by a black body at 5800°K, as outlined in Appendix A, is less a fortuitous coincidence than a consequence of the evolutionary selection of active chemicals in the sensory systems of Earthly creatures to react most sensitively to the available energy.) However this sensitivity brings vulnerability. The Sun appears as a disc with a surface luminance of about 10^9cd.m^{-2}, and this, as outlined in Chapter 2, exceeds the dynamically sensible range of the eye. If the image of the Sun falls on the same retinal area for more than a normal fixation, the photoreceptors bleach to a desensitised state and a strong, obscuring afterimage is created. Aircrew must be particularly careful to avoid staring at the Sun, and unless flying below cloud or away from the Sun, a dark visor or pair of spectacles must be used to attenuate the glare. The vulnerability of the eye to permanent damage has many repercussions. Aircrew spectacles are discussed in Chapter 5, the design of high brightness displays in Chapter 9 and the necessary qualities of visors in Chapter 13. The need for protection against blinding weapons is discussed briefly in section 3.5.3.

3.2.4 Cloud, fog and light scattering

The need for the aircrew to see out and around defines the conditions in which flight is deemed safe by the responsible authorities[3]. Although detailed aspects of the regulations vary between national bodies, there is a universal distinction between flights undertaken on the presumption that the pilot will or will not be able to see directly what is going on around him. Contact flights are conducted in Visual Meteorological Condition (VMC) under Visual Flight Rules (VFR), whereas flights at night or in poor visibility are conducted in Instrument Meteorological Conditions (IMC) under Instrument Flight Rules (IFR).

Although the details vary between national authorities, VFRs are relatively simple. They specify the way the aircraft must be manoeuvred to avoid the possibility of collision with another aircraft (i.e. always pass on the right), and the way the aircraft should cross controlled or congested airspace. They also specify that that the pilot must keep clear of cloud and for instance, if below 1200 feet above the ground in the UK, be able to see a small aircraft at least one mile ahead, and at 10,000 feet be able to see at least five miles ahead. IFRs are much more complex. They specify the procedures and the equipment that must be fitted to the aircraft as well as the level of pilot training and the number of hours that are needed each year to maintain competence. In general the pilot remains responsible for controlling the flight path and airspeed by interpreting the Primary Flight Display (PFD), described in Chapter 6. To avoid an air-to-air collision in controlled airspace he must follow the flight-path instructions of ground-based controllers who know the position of his and other aircraft from transponding radar systems.

Table 3.1 Certification categories of all-weather landing systems

	Decision Height (DH, feet)	Visibility (VIS, feet)	Runway Visibility Range (RWR, feet)
Category 1	>200	2400	1800
Category 2	200>100	-	1200
Category 3a	100>50	-	600
Category 3b	50>0	-	0

Landing safely at an airport obscured by low cloud or fog is also done routinely by co-operation between ground controllers and aircrew. Although satellite navigation systems, such as GPS, are likely to become standard approach as well as en-route aids, guidance is at present supplied using an Instrument Landing System (ILS) or a Microwave Landing System (MLS).

These use a pair of transmitters to propagate narrow beams out along the approach path from the runway threshold, and the beams are picked up by aircraft-mounted receivers and interpreted automatically to measure the deviation in position of the aircraft from the ideal in-line 3° descending glide-slope. The pilot performs a Precision Instrument Approach (PIA) by following the ILS/MLS guidance cues while monitoring the gradual reduction in the aircraft height above ground from the audible call-out of the radio altimeter. The cues are displayed as movements of vertical and horizontal needles within the attitude indicator. Alternatively, in a suitably equipped aircraft, he can follow a steering bug in a head-up display (HUD) and this assists him to transfer his visual attention between the guidance cues and looking out through the windscreen to discern the runway approach lights. When landing in low cloud, unless he sees the approach lights or the runway threshold before reaching a pre-defined Decision Height (DH) he must abandon the attempt and either go around the circuit again or divert to another airport. In fog the forward visibility is more constraining and two factors are assessed: Visibility (VIS), the actual or forecast distance at which an illuminated object may be seen, and the Runway Visibility Range (RVR), the range at which lights of defined power can be detected objectively by a transmissometer. Again, if the assessed values of these parameters are below prescribed limits it is not safe to attempt a landing and the aircraft must be diverted to another airport.

The performance and integrity of the ground and airborne equipment dictate the values of the crucial parameters. These are summarised in Table 3.1. The combination of ground and airborne equipment conforming to Category 3b enables the output from the guidance system to be connected directly to the flight control system for fully automatic landings, albeit supervised and monitored by the aircrew.

3.2.5 The flight envelope and aircraft manoeuvrability

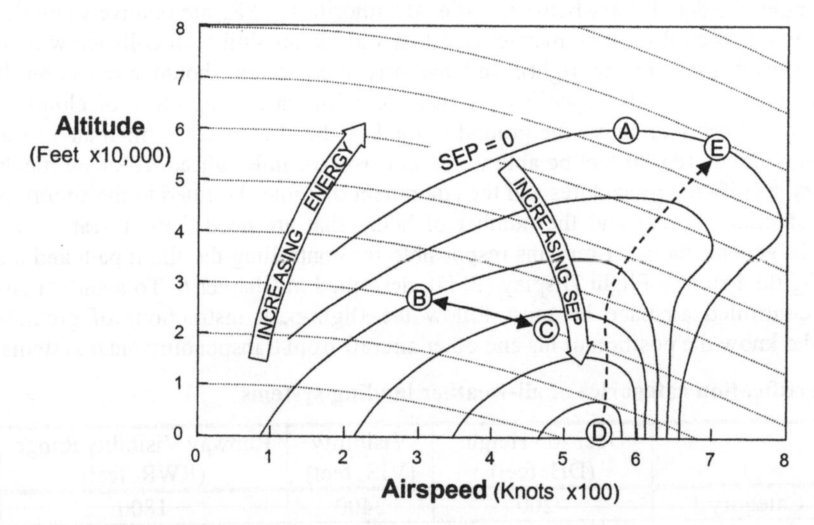

Figure 3.3 Overview of the flight envelope of a typical combat aircraft for a given weight and configuration of stores. See text for definitions of energy and SEP. Point (A) marks the maximum sustainable altitude; the aircraft "ceiling". A diving, acceleration manoeuvre from (B) to (C) maintains energy and gives improved SEP. Maximum energy (E) can be attained most rapidly from take-off by accelerating at low level to airspeed (D), holding this speed while climbing to about 30,000 ft and then gradually accelerating during the remainder of the climb

The performance of an aircraft can be described in many ways, but, as shown in Figure 3.3, energy parameters are particularly useful for combat aircraft. The total energy E of the aircraft is the sum of potential and kinetic components:-

$$E = m.v^2 /2 + m.g.h$$

where m is the mass, v is the true airspeed, h is the altitude and g is the gravitational acceleration. This can be normalised as the Specific Energy (SE) by dividing by the vehicle weight (W = m.g):-

$$SE. = E/W = (0.5/g).v^2 + h$$

In Figure 3.7, the descending parabola-shaped curves are contours of constant SE which have the value of the altitude at zero airspeed. The solid lines describe the aircraft performance. These are contours of constant Specific Excess Power (SEP), which is the rate of change of SE attained with maximum engine thrust:-

$$SEP = d(SE)/dt = v.a/g + dh/dt$$

Here, a is the forward acceleration. Thus SEP has speed units and it is a measure of the ability of the aircraft to gain speed and height. It is also related to the Thrust (T) and Drag (D) forces:-

$$SEP = \text{Excess force x Velocity/Weight} = \text{(Thrust – Drag).Velocity/Weight} = v.(T-D)/W$$

The aircraft can generate energy most rapidly at full throttle when this factor is maximised. For instance, aircrew operating the aircraft shown on Figure 3.3, would know that if the instantaneous airspeed were less than 500 knots it could be worth exchanging altitude for airspeed in order to enter a region of the flight envelope where energy can be gained more rapidly.

In combat the pilot attempts to exploit differences between the performance of his aircraft and the enemy aircraft to gain a position where he can release a weapon, but he must also stay out of the zone in which the enemy can retaliate. The geometrical dynamics have given rise to a wide range of tactical manoeuvres, such as the yo-yo and the split-S in which the interchange between altitude and speed is exploited to get the best airspeed for turning the aircraft towards an evasive adversary. A large part of the skill of a combat pilot involves selecting the most advantageous manoeuvre and executing it efficiently. This is particularly important because rapid turning manoeuvres at high angles of attack (AoA) rob the aircraft of more energy than the engines can generate, and the pilot must not allow the aircraft to become low and slow, where with limited ability to manoeuvre it is very vulnerable to attack.

The amount of energy lost by the aircraft when turning with a particular weight and stores configuration can be shown conveniently by an N-V diagram, as in Figure 3.4, where N is the "load factor" or specific lift force (L/W) and V is the airspeed. The instantaneous combination of load factor and airspeed is important to the crew for three reasons. Firstly, it defines the strength of the centrifugal force to which they are subjected. Secondly, it affects aircraft safety, because it is possible to exceed the maximum AoA and lose control of the aircraft by entering a G-stall at low airspeed, and the airframe can be over-stressed at higher airspeed. Thirdly, it determines the effectiveness of a combat manoeuvre and the rate of loss of energy. The latter is apparent from the N-V diagram which shows that the safe envelope is divided into an inner sustainable region in which SEP is positive (T>D), and an outer unsustainable region in which SEP is negative (T<D). The right hand diagram of Figure 3.4 shows some conditions that result in optimum combat performance. For instance, the maximum turn rate is attained without exceeding the lift and structural limits by aiming for the "corner" combination of N and V where these limits intersect. However, the pilot must use this manoeuvre for a short time, because the

induced drag is much greater than the full engine thrust. To maintain the turn without loss of energy he must relax the demanded load factor and accelerate to the combination of N and V on the zero SEP line that gives maximum sustainable turn rate.

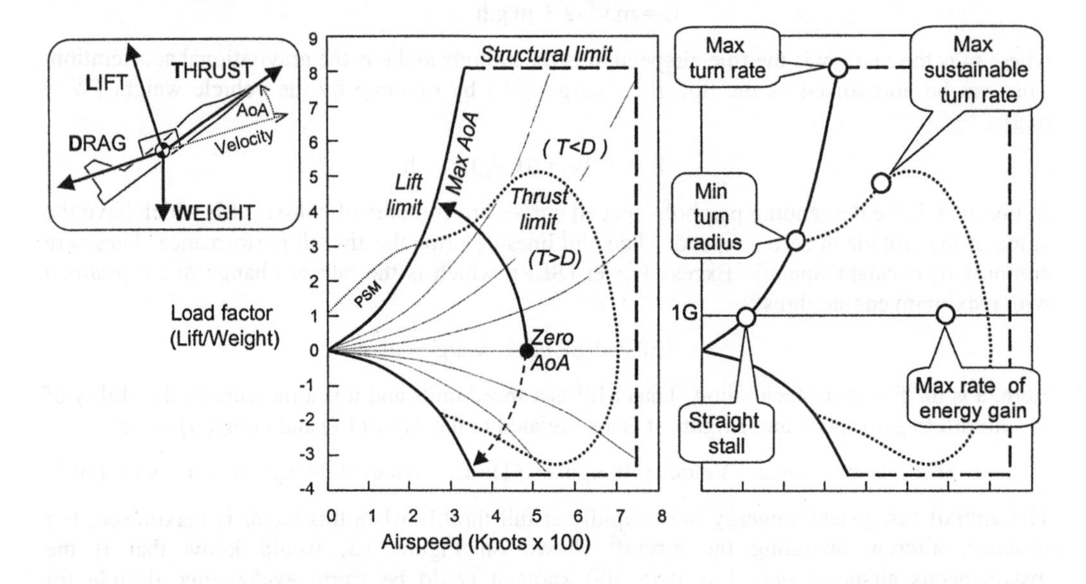

Figure 3.4 N-V diagram for a typical combat aircraft for a particular configuration of under-wing stores, fuel load and altitude. The left graph shows how the specific lift force (L/W) is limited at low speed by the wing stalling at the maximum angle of attack (AoA) and elsewhere by the strength of the structure. Negative lift is produced by negative AoA. The envelope is divided into two regions; a sustainable region in which thrust exceeds drag (T>D) and SEP is positive, and an unsustainable region in which T<D and SEP is negative. The right graph shows the particular combinations of N and V that produce specific optimised manoeuvring states

Aircrew in newer types of combat aircraft can operate in the knowledge that the digital flight control system will automatically prevent G-stall and over-stress conditions. In some, such as the SU-37 and F-22, this automatic control is extended to include the propulsion system and vectoring of the thrust. Taken together with exceptional levels of SEP, these developments not only give carefree handling but generally improve performance and agility. Rapid changes of pitch and yaw are possible, and a Post-Stall Manoeuvrability (PSM) region is opened up in the manoeuvrability envelope[4]. On Figure 3.4 this region would be bounded by the aerodynamic maximum AoA, between the straight stall and minimum turn radius points, and extend to about 90° AoA. Within this regime of very low airspeed and very high AoA, the nose of the aircraft can be pointed by rolling around the velocity vector. Providing the condition is held for a few seconds and followed immediately by a low load factor forward acceleration, the aircraft can change flight path direction significantly faster than a corner point turn and suffer comparable energy loss. However, the practical advantages of this agility are unclear. The forces that act on the crew are complex and the gyrations are potentially disorienting. Also, the tactical advantages may not be so great as gained by carrying a highly agile weapon and a means, such as a helmet-mounted sighting system, that enables the pilot to designate a target at a large angle from the aircraft pointing direction.

3.3 Physiological limitations

3.3.1 Hypoxia

The main concern during flight at high altitude is hypobaric hypoxia; starvation of body tissue, and in particular the brain, of sufficient oxygen to maintain function. The physiological symptoms are complex and variable, typically ranging from an initial light-headedness and nausea to a lack of co-ordination, followed by a dimming of vision and eventually unconsciousness. If exposed to the external air at 25,000 ft, the lack of oxygen would render fit young males unconsciousness in about five minutes. At 36,000 ft this would occur in less than one minute.

An adequate supply of oxygen to the cells depends fundamentally on the partial pressure of the oxygen in the breathing gas. The transport rate is proportional to the rate of flow of the medium containing the oxygen and the concentration of oxygen in that medium. Whether the oxygen is in gaseous form or dissolved in blood and other body fluids, the concentration can be expressed in terms of the oxygen partial pressure.

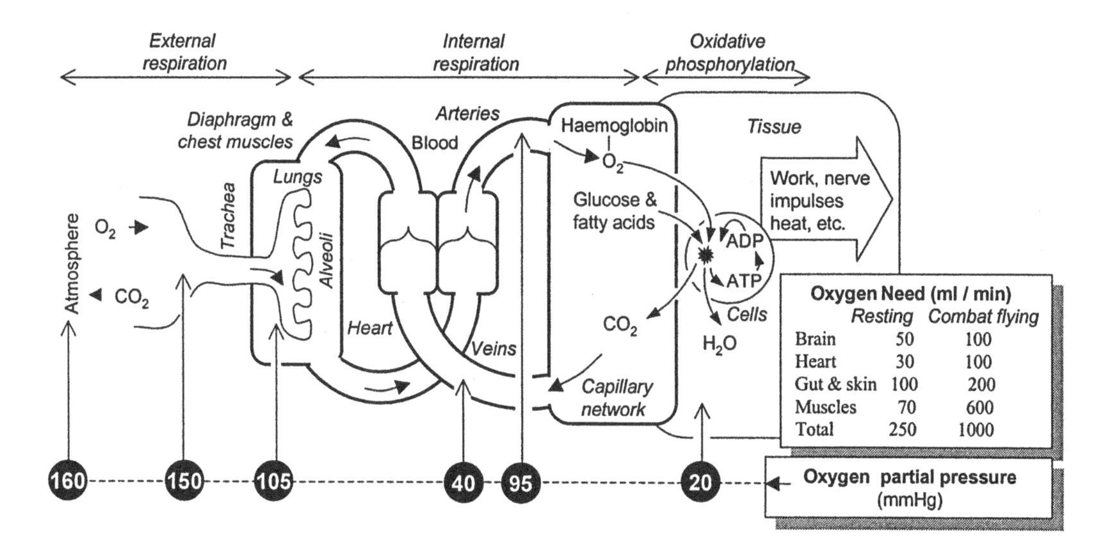

Figure 3.5 A simplified overview of respiratory physiology

The drop in oxygen partial pressure along the feed route is shown in Figure 3.5. As noted in section 3.2.1, at sea level the atmospheric air pressure is about 760mmHg, and as 21% of this is oxygen, the oxygen partial pressure is about 160mmHg. The action of breathing the air into the lungs drops this by about one third to 105mmHg. The oxygen diffuses into the fluid within the air sacks (alveoli) and thence across the walls of the fine capillaries in the lung tissue where it is picked up by the red blood cells in the blood stream to become oxy-haemoglobin. This molecule has a remarkable affinity and the blood is pumped rapidly to the network of fine capillary vessels with only a small drop in partial pressure. The low oxygen content in the intercellular fluid surrounding the feed capillaries induces the oxygen to diffuse through the capillary walls to dissolve in this fluid and then transfer into the cells themselves.

The cells act like engines in which a received nerve impulse triggers adenosine triphosphate (ATP) to decay to adenosine diphosphate (ADP). The result of this decay depends upon the specific nature of the cell, for instance a contraction is produced in a muscle cell, another impulse in a nerve cell and a molecular secretion in a glandular cell. Cells would soon cease to function without a continuous oxydative phosphorylation to regenerate ATP from ADP. This is done by the energy created within the cells by the reaction between the oxygen and fuels, such as glucose and fatty acids, that are also transported via the blood stream and diffusion through the intercellular fluid. The exothermic phosphorylation is also the source of the metabolic energy that maintains the core body temperature at 37°C.

The water and carbon dioxide produced by the oxydative reaction in the cells enter the intercellular fluid as carbonic acid (H_2CO_3) which dissociates into $+H_2$ and $-HCO_3$ ions. These are transported back into the atmosphere by reversing the oxygen delivery route; firstly diffusing into the capillaries, then entering the venous bloodstream, in part by combining with the oxygen-depleted haemoglobin to form carbamino-haemoglobin, and then diffusing through the alveoli in the lungs to be vented during exhalation.

The amount of oxygen needed, and carbon dioxide and water produced, varies greatly with physical workload, ranging from about $250 ml.min^{-1}$ when resting to about $1000 ml.min^{-1}$ during combat. Maximum muscular effort, for instance for running, consumes about 2500 $ml.min^{-1}$. The increased capacity is gained by greater lung ventilation through deeper and more frequent breathing, by increasing the blood pressure and blood flow rate, and by dilating the capillaries. These are controlled autonomically by the brainstem through increased activation of the muscles in the chest wall, diaphragm, heart and capillaries, largely in response to chemo-receptors in the artery walls that sense a small increase in the concentration of dissolved carbon dioxide in the blood.

In an unpressurised crewstation the main hazard during flight at altitudes above about 10,000 ft is the reduction in partial pressure of inhaled oxygen produced by the drop in atmospheric pressure. This hypobaric condition leads to a progressive drop in oxygen partial pressure along the external and internal respiratory routes. If the cellular oxygen pressure falls below a critical value of between 0.5 and 3mmHg, depending on cell type, the oxydative phosphorylation ceases. Although cells can metabolise without oxygen for a short time, this is less efficient and produces lactic acid, which soon stops the cell functioning.

The primary way of preventing hypoxia is to provide an increased proportion of oxygen in the breathing gas using a mask that fits tightly over the mouth and nose. The right graph in Figure 3.1 shows that to receive adequate oxygen from air at the ambient pressure it is necessary to increase the proportion of oxygen in the breathing gas with increasing altitude. Although 100% oxygen would be needed at 33,700 ft to give the ground level partial pressure of 160 mmHg, the breathing mask and regulator described in Chapter 13 are usually arranged to start enriching the oxygen concentration above 10,000 ft and supply 100% oxygen at 40,000 ft. Above this altitude it is still necessary to receive the same amount of oxygen, so with increasing altitude the oxygen pressure in the breathing mask becomes progressively greater than the ambient air pressure.

Flying with oxygen supplied through a mask at a pressure above the pressure acting on the chest walls requires the crewman to "pressure-breathe". Exhalation requires considerably greater effort, and inhalation must be controlled deliberately to prevent sudden over-inflation of the lungs. The excess internal pressure also raises the heart rate and blood pressure, which distends blood vessels and swells tissue, and the pooling temporarily reduces the return flow of blood through the veins. Speaking is difficult and barely understandable, and the eardrums bulge. There is also a good chance of hyperventilation, especially in less experienced individuals when under stress. Rather than producing a slight excess of blood oxygen, this excessive breathing removes carbon dioxide from the lungs and the arterial blood (a condition termed hypocapnia), which causes the brainstem to constrict the supply of blood to the brain. Like hypobaric hypoxia, this produces a general mental and physical impairment and can lead to unconsciousness, but unlike hypoxia it is self-correcting because the victim ceases to hyperventilate when unconscious.

3.3.2 The pressure cabin

Early aircrew gained altitude records in the troposphere in aircraft with open cockpits by breathing bottled oxygen and wearing thick, thermally protective clothing. This became unnecessary with the introduction of strong enclosed cabins and environmental conditioning systems (ECS). At high altitude the internal environment could be kept at a comfortable temperature of about a 20°C and pressurised to be equivalent to that at an altitude of about 8,000 ft.

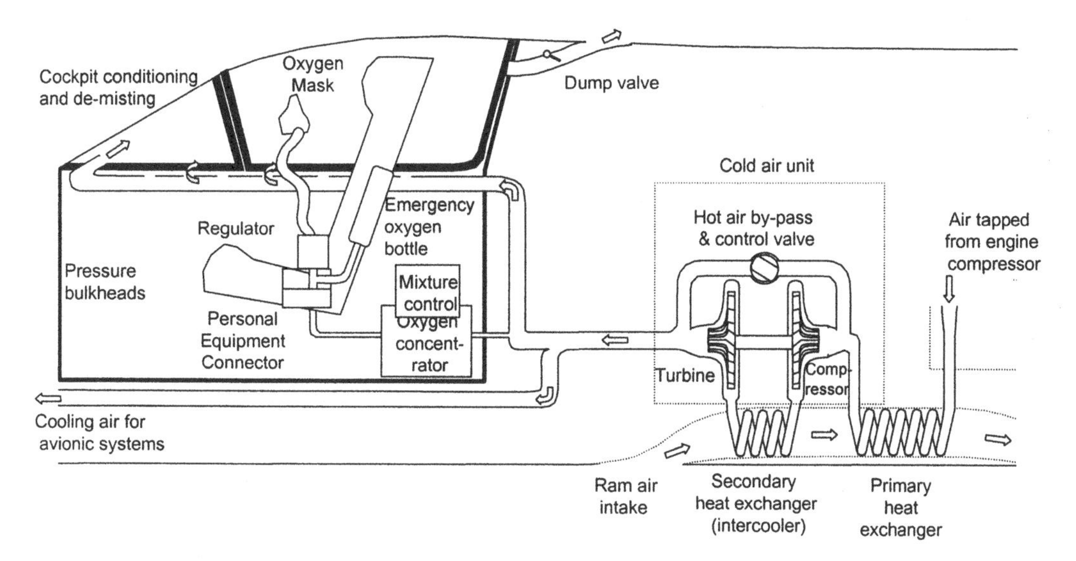

Figure 3.6 A simplified schematic for an environmental conditioning system in a modern fast jet

With a system fitted to a fast jet, as illustrated in Figure 3.6, the intake air is supplied in flight from the final stages of the engine compressor. In larger aircraft it may also be supplied from the Auxiliary Power Unit (APU). In both cases the function of the ECS is to convert the hot pressurised air, at about 250°C and 5 Bar, into a generous flow of clean, temperate, de-humidified air. In most in-service combat aircraft the conditioned air is ducted through channels in the avionic bays to cool the electronics before it is used to cool, pressurise and de-mist the cabin, and the oxygen is supplied from a cryogenic flask that has to be replenished before each flight. However, as shown in Figure 3.6, more recent designs split the output from the cold air unit into separate streams, one of which cools the electronics and the other the cabin, and the oxygen fraction in the breathing gas is enriched by an on-board oxygen generation system (OBOGS). This operates by varying the flow rate through a molecular sieve. The demand regulator that controls the supply of breathing air is, as described in Chapter 13, usually mounted on the ejection seat, as too is the valve that controls the pressure of the air for G-suit inflation.

Aircraft still carry the risk of malfunction or failure of the pressurisation system. Similar problems, exacerbated by the possibility of canopy damage, also arise in combat aircraft, and in this case the small cabin volume in relation to the canopy area can produce a sudden, almost explosive, decompression. All high-flying types therefore carry an emergency means of delivering breathing air with the correct partial pressure of oxygen to the aircrew and passengers. The systems usually feed the breathing gas to simple oro-nasal masks that the users must fit to their headgear or hold over their mouth and nose. Such emergency systems are also deployed if the cabin air becomes contaminated, for instance by smoke or fumes. If the cabin of a fast jet is damaged the aircrew would continue to breathe conditioned air, and should the ECS fail or the normal breathing air become contaminated they would have recourse to the emergency supply of oxygen from the ejection seat.

Transient cabin pressure changes bring the possibility of an imbalance between the pressure on the outside surfaces of the crewman and the pressure of air trapped inside body cavities and the sinus cavities within the head. As noted in Chapter 2, a pressure difference between the middle ear and the outer ear will distort the eardrum, which can be painful. However, because the entrance to the Eustachian tube in the nasal sinus tends to distort and block the passage of in-going air, a barotrauma (permanently damaged hearing) is more likely to occur during the re-compression when descending rapidly.

The low ambient pressure at high altitude may also induce decompression sickness. This condition produces the same symptoms in airmen as in divers who ascend too rapidly underwater, and it is thought to have the same cause and mechanism. The primary factor is the partial pressure of nitrogen dissolved in body fluids, including the blood. This gas is normally in equilibrium with the surrounding atmosphere, but micro-bubbles are formed in the body fluids if the external pressure is reduced at a faster rate than the excess pressure can be diffused through the alveoli in the lungs and vented by exhalation. The mechanism is the same as the release of bubbles in Champagne, or in any fizzy drink that is stored in a pressurised container. The symptoms include cramps and bends when micro-bubbles form in the skeletal joints, and if in cranial tissue, migraine, fits, paralysis and loss of consciousness. The only cure is re-compression, done rapidly enough to stop and reverse the bubble formation. Although the condition is alleviated if the airman breathes oxygen rather than a nitrogen-oxygen mixture before the emergency decompression, it is specifically to combat this hazard that airmen wear pressure suits at high altitude. These garments are described briefly in Chapter 13.

3.3.3 Cold and heat

Protection against heat and cold is a fundamental requirement because the human animal has limited thermal defences. The body is comfortable in temperate air (at about 18 to 25°C) if the air is moving slowly and has moderate humidity. When not producing heat through physical activity, the basic metabolic processes that provide essential maintenance functions and keep the core organs at 37°C dissipate about 40 W of stored energy per unit of skin area. A body with a typical area of $1.5m^2$ therefore generates heat at a rate of about 60W. As indicated in Figure 3.7(b), this is dissipated through conduction with contact surfaces, by radiation, by convective heating of the surrounding air and by evaporation of water. The water is exhaled from the lungs and produced by insensible sweating of intra-cellular fluid through the skin.

In cold air, when more heat is lost than is generated metabolically, as indicated in Figure 3.7(a), the shortfall is drawn from that stored within the thermal mass of the body. To prevent the temperature of the core organs falling below the 37°C homeostatic value, the primary defence is vasoconstriction, an autonomic narrowing of peripheral and surface blood vessels. This reduces the heat supply to the limbs, stems losses by convection and radiation by lowering the surface temperature and incidentally produces the bluish skin hue. The lack of surface blood perfusion also increases the effective thickness and insulating capacity of the outer tissue, and this acts as the body's own thermal shell. Although some heat is generated in the muscles by involuntary shivering, this adds a considerable loss of dexterity to the general mental incapacitation and feeling of fatigue. Continued exposure to the cold produces hypothermia, and at a time dependent upon the rate of extraction of heat, the drop in the core temperature causes loss of consciousness. As water extracts heat about thirty times faster than air, immersion in cold water is particularly dangerous. If unprotected in freezing water, most people would fall unconscious and drown within about 15 minutes. The few who resist would die within the hour it takes the core temperature to fall to about 26°C.

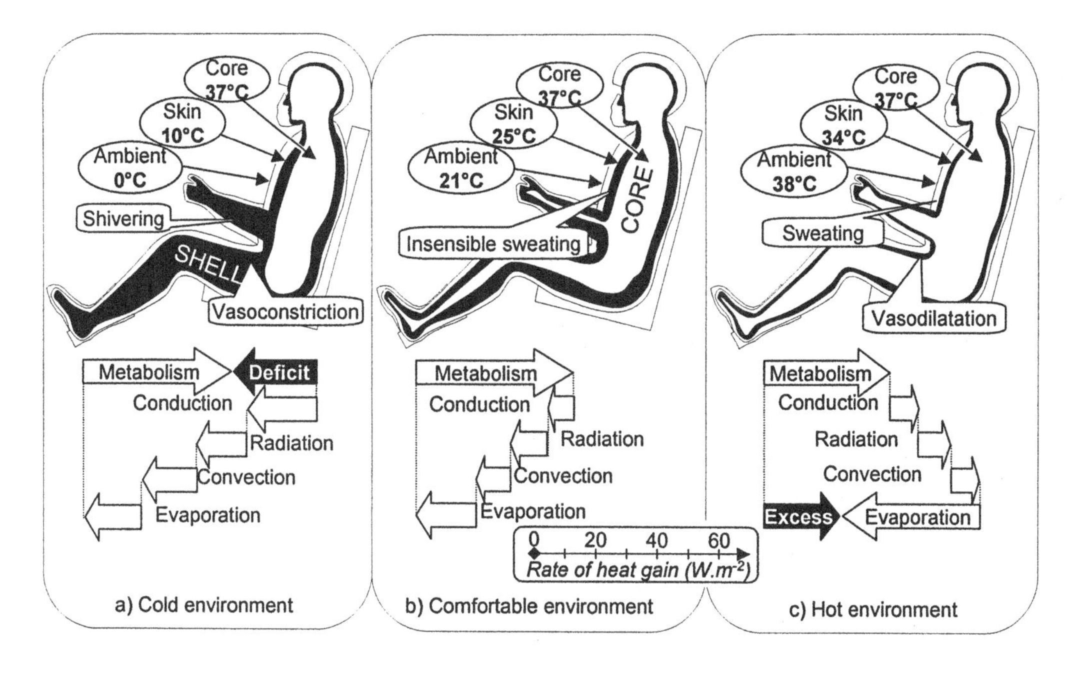

Figure 3.7 Simplified summary of the human reaction to the thermal environment

Vasodilatation and sweating, the opening of surface blood vessels and fluid secretion from surface glands, are the main autonomic reactions to a hot environment, but as shown in Figure 3.7(c), the evaporation of sweat is the only process that can counter heat accumulation when the ambient temperature is greater than 37°C. One or two litres of fluid can be sweated per hour, but this cannot be maintained without intake, and if this or the evaporation of the sweat is impeded, the core temperature must rise. Dehydration and accompanying symptoms such as head-ache and nausea are followed by loss of consciousness and eventually death from thermo-regulatory failure when the core temperature exceeds 42°C.

Although the cockpit is conditioned to provide a temperate environment, it is necessary to equip aircrew with clothing to provide thermal protection from the external environment should the ECS fail, or in the event of emergency escape. These issues are discussed in Chapter 13.

3.3.4 Noise

Despite the use of exhaust diffusers, de-couplers, sound-absorbent shields and baffles, the primary form of cabin noise in most aircraft powered by reciprocating engines is a strong harmonic drone proportional to the crankshaft rotation rate. Jet engines produce a very high pitched whine and a roaring hiss spread over a broad frequency range. The flow of the wind over the canopy also generates a considerable amount of acoustic energy, and in helicopters the impact of the blade downwash on the canopy is, at slow speeds, usually at least as noticeable as the roar of the engines and the whine of the gearbox. The ducting for the cabin conditioning system transmits acoustic vibrations from the engines and other sources outside the cockpit, and this is invariably accompanied by the turbulent efflux from the collection of nozzles and vents that distribute the air. The roar from the cabin conditioning system is the predominant noise in many fast jets, particularly training types, such as the Anglo-French Jaguar, that have comparatively low thrust-to-weight ratio. In most aircraft the crew are also exposed to electrical noise, particularly the 400Hz hum from the engine-driven a.c. generators, picked up by the aircraft audio system and fed to the earphones.

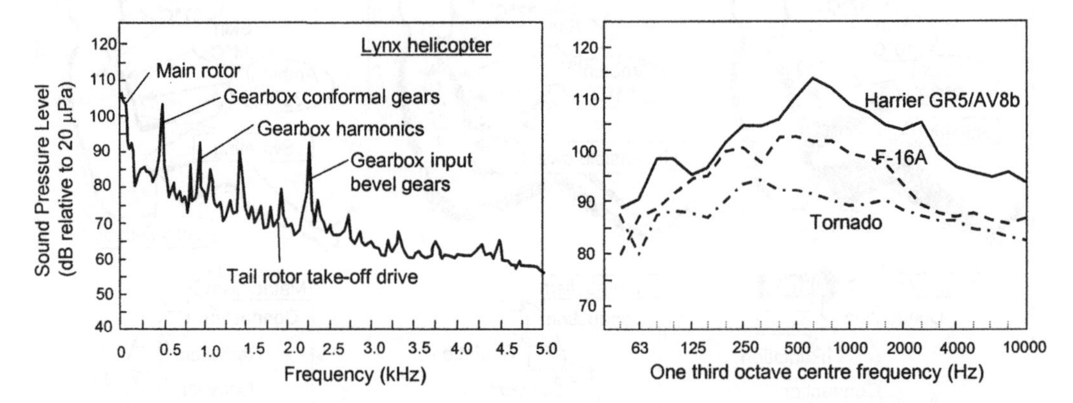

Figure 3.8 Cabin noise spectra in current military aircraft. The left graph shows the spectrum of a sample recorded at pilot head level in a Lynx helicopter during forward flight. The right graph (note the logarithmic frequency scale) shows spectra recorded in the cockpits of fast jet aircraft; in Tornado and F-16 during low-level high-speed flight and in Harrier during a vertical landing. (Reprinted with permission from QinetiQ)

The flight deck of a typical civil transport aircraft is quiet enough, at about 60 or 70dB SPL, to allow normal conversation. As noted in the examples given in Figure 3.8, the crew of a typical multi-engine helicopter can be subjected to a very loud and complex noise field[5]. The crew of a fast jet can suffer almost painful conditions if, as in the Harrier, the cockpit is sited just ahead of the engine and during a vertical landing the whole airframe is engulfed within the reflected noise field produced by the downward jet efflux.

Although noise can induce a variety of short-term effects, ranging from arousal to accelerated fatigue, the prime need for military aircrew is to reduce the SPL at the ear to less than 85dB so that they can think clearly and understand what they hear from their earphones. The effect of noise on the intelligibility of speech is discussed in Chapter 10. The long-term effects are particularly important. The accumulated noise bombardment can causes an irrecoverable loss of hearing sensitivity, and an intense, prolonged dose induces deafness. As most aircrew set the level of speech sound produced by their earphones at least 3dB above the ambient sound level at their ears, even intermittent communication adds significantly to the noise dose.

Accelerated hearing loss has recently become a major concern in most services following the enactment of a formal duty to care for the health and safety of employees[6]. Wearing ear-defenders, usually within the helmet, is the accepted way to protect hearing. The effectiveness of passive techniques and recently developed active forms of noise control is discussed in Chapter 13.

3.3.5 Sustained high acceleration

Although aerial agility is a vital quality for a successful combat aircraft, the cardio-vascular system of the pilot has a limited ability to supply an adequate flow of blood to his brain, and he is very likely to lose consciousness under the strong, sustained centrifugal acceleration generated by combat manoeuvres. Military fast jet aircrew must be selected, trained and equipped to deal with this hazard.

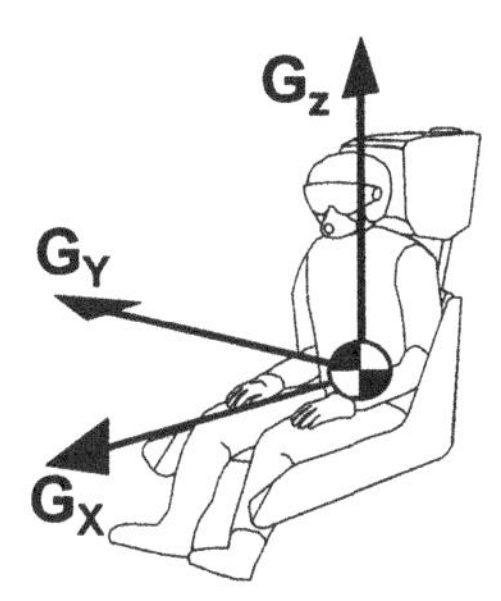

Figure 3.9 The aero-medical convention for the accelerations applied to a seat: Longitudinal (X direction or "shunt"), Lateral (Y direction or "sway") and Vertical (Z direction or "heave") N.B. the occupant experiences the reaction force. This acts in opposition, for instance $+G_Z$ is sometimes known as "eyeballs down" and $-G_X$ as "eyeballs out"

Figure 3.9 summarises the convention for describing the directions of linear forces that act on the cockpit. At a moderate angle of attack the total force acting on the aircraft is the vector combination of the lift along the aircraft +Z direction, the propulsive thrust in the +X direction, the aerodynamic drag in the –X direction and the gravitational attraction of the Earth. The magnitudes of lift, thrust and drag depend on the aircraft manoeuvring state, and the direction of the gravitational attraction depends on the aircraft attitude. During turning manoeuvres the magnitude of the lift force tends to be much stronger than the difference between the thrust and drag, so the acceleration experienced by the pilot is mainly that due to the $+G_z$ lift force. Aircrew invariably quantify the acceleration they experience as a multiple of the gravitational acceleration in "G" units, and this is numerically equal to the load factor. It is also of note that, because the cockpit is usually well forward of the aircraft centre-of-gravity, this acceleration can be exaggerated at the onset of a manoeuvre as the cockpit is heaved by a rapid increase in the angle of attack.

There is no simple threshold level of G at which G–induced loss of consciousness (GLOC) suddenly ensues, because posture, fitness, the rate of onset and numerous conditional factors have an effect on the instantaneous amount of blood reaching the brain. The mechanism is well-understood and arises directly from the way the heart acts as a double pump that circulates a flow of blood through the arteries to all of the internal organs, bones and muscles for return via a network of veins. Each contraction of the heart squeezes a chamber-full of blood into the arteries, and in so doing raises the output arterial pressure from an average of about 100mmHg to a short pulse of about 122mmHg. From a simplified hydrostatic viewpoint the body can be regarded as a flexible envelope containing about 25 litres of fluid in porous cells, about 12 litres around the cells and about 3 litres in a network of flexible pipes. Like any contained fluid subjected to gravity, there is a progressive increase in pressure with depth. Figure 3.10(a) shows the typical vertical gradient of the peak pulse pressure between the top of the head and the soles of the feet of a standing man.

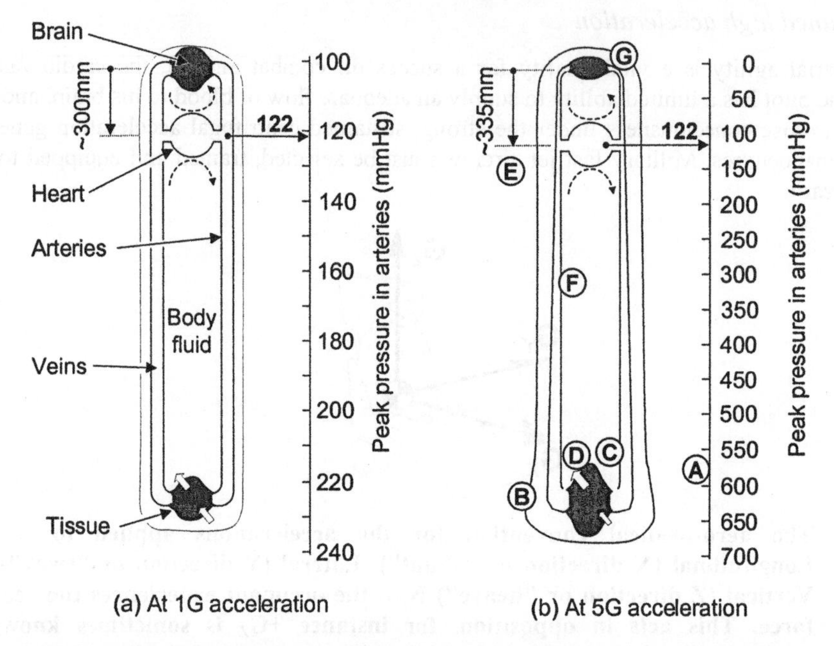

Figure 3.10 Simplified hydrostatic conditions in a standing man at 1G and 5G

Figure 3.10(b) summarises the hydrostatic conditions for a standing man subjected to an acceleration of about 5G. The immediate effects of the high G are an increased hydrostatic pressure gradient (A) that distends the lower limbs (B) and the blood vessels (C), which leads quickly to pooling of blood in the lower limbs. This is accompanied by an imbalance in the cell-capillary fluid exchange and net loss of fluid from capillaries (D), that reduces the amount and the pressure of blood returned through the veins to the heart (F). The comparatively dense heart sinks by about 35 mm (E) increasing the height of the head above the heart. All of these factors contribute to low pressure at the head (G) and a lack of blood flow to the brain. As well as having to exercise strong muscular effort to support the increased weight of his limbs and head, the man would experience a progressive reduction in his visual field before relapsing into unconsciousness. This "grey-out" is a result of the internal pressure of the eyes, of about 20 mmHg, that prevents the low pressure blood from entering the fine tracework of blood vessels feeding the periphery of the retina. He might suffer a complete loss of vision, "black-out", when the excess intra-ocular pressure prevents blood entering the eye. The time course of effects is however complicated by reflex mechanisms that temporarily dilate the blood vessels in the brain, by the store of oxygen in the cells and usually by an anticipatory release of adrenaline that increases the heart rate. These enable aircrew to remain conscious when subjected to a strong, short jolt.

A variety of cockpit designs, crew training procedures and anti-G protective garments have been devised to alleviate the debilitating effects of high-G and protect aircrew against GLOC. In combat and light aerobatic aircraft the pedals and seat base are arranged to raise the knees and feet and the seat is installed at a reclined angle with the intention of minimising the hydrostatic pressure difference in the G_z direction. Although the seat-back angle is reclined by about 30° in the F-16 and Rafale, there is little reduction in the heart-to-head height, partially because a reclined angle of at least 60° is required[7], and partly because the airman needs to hold his head erect so that he can look around. Moderate reclination gives more benefit to comfort than G-tolerance.

In older aircraft combat aircrew trained to raise their tolerance by about +2G using the "Valsalva manoeuvre", which was done by tensing the muscles and breathing out forcibly with the glottis closed. This temporarily reduced the pooling of the blood in the lower limbs and generally raised the pressure in the internal cavities and arteries. An anti-G straining manoeuvre (AGSM) is now preferred. This consists of a sequence of Valsalva manoeuvres in which the muscle tension is maintained but a rapid breath is taken every three seconds to allow venous blood to return to the heart. With practice in a training centrifuge, and under aero-medical supervision, the technique can raise the black-out threshold by up to +4G. The main benefit, a tolerance to +9G, can be gained by using AGSM in conjunction with anti-G garments and pressure breathing. The latter are described in Chapter 13. However, even with such a combination of measures, head movement and speech become almost impossible at high-G and all activity require great effort. As the AGSM itself is very fatiguing, prolonged combat can be exhausting.

3.3.6 Vibration

Vibration is not unusual in aircraft. In most piston-engine aircraft shaking forces are transmitted through the airframe to the cockpit from the engine and propeller at audible frequencies. The blades of a helicopters produce lift impulses proportional to the rotor torque at the blade passage rate, in the region of 5Hz to 15Hz. In fixed-wing aircraft, buffet is induced by turbulent airflow over the wing, but as this provides the pilot with a reliable warning of incipient aerodynamic stall it is more beneficial than disturbing. The most powerfully disturbing vibrations occur when the whole aircraft is thrown around when transiting a turbulent air mass. Aircrew routinely encounter a mild form of the phenomenon when passing through cloud, and they go out of their way to avoid the violent circulating updraft in a thundercloud. As noted in section 3.2.2, large civil airliners are most likely to encounter clear air turbulence during high altitude cruise. Military aircraft and light aircraft, particularly those engaged in low-level flying, are more likely to encounter thermally-induced turbulence produced by vertical air convection induced by differences in the temperature of neighbouring areas of the ground due to differential sun heating. All aircraft are susceptible to the turbulent air left in the wake behind an aircraft, mainly a large, heavy aircraft. This is a particular hazard for lighter types on the approach to a busy airport, and for military aircraft refuelling behind a heavy tanker.

The predominant cockpit vibration is in the Z direction. The heaving is accompanied by pitching, rolling and swaying motions that depend upon the scale of the turbulence, the gust response of the aircraft and the corrective actions of the pilot. Turbulence has most effect on aircraft with low wing-loading, usually agile fighters or high altitude bombers, particularly when operating at high speed and low altitude and carrying little fuel and few stores. In older aircraft the vibration tended to be random in character, with most energy in the 0.1 to 10Hz band and could reach an acceleration level of 0.3G rms, with occasional peaks of ±1G. In designing more recent types, stabilised by flight control systems, the engineers take great pains to sense and counter the main disturbances, transforming what would be a series of low frequency thumps and lurches into a ride over cobble-stones.

The vibrations travel from the seat through the spine and trunk to the head where the main disturbance is a nodding motion, with some rolling and yawing. Enduring, low-frequency vibration can cause hyperventilation, and the higher frequencies upset exhalation and can make speech waver, which affects intelligibility and in newer aircraft the reliability of an automatic speech recogniser. The effects on vision vary with posture, muscle tension, headgear mass and vibration frequency, particularly as the head can suffer a nodding resonance at a frequency of about 5 Hz. However the focus distance is also important because the eye pointing direction is stabilised against both voluntary and involuntary head rotation by the vestibulo-ocular reflex

(VOR), as described in Chapter 2. The VOR helps to maintain stable fixations when looking out into the distant scene, but it is of little benefit to the pilot reading information presented on nearby cockpit displays, as voluntary eye movements are needed to compensate for involuntary vertical head movement. The legibility of information presented on a helmet-mounted display can be affected markedly, as discussed in Chapter 9. The ability to make accurate arm and hand movements also declines, and this affects activities such as writing notes on a knee-pad and operating switches and keys.

The need to be able to operate under severe vibration is tackled by careful design of the seat and harness so that aircrew are restrained comfortably. The use of some equipment, for instance the head-up display, is not affected by vibration because the imagery is projected into the distance. Designers can introduce engineering solutions, for instance, the forces needed to operate switches and controls can be increased over those needed in a static environment to avoid inadvertent mis-selection. However, for the aircrew, the general consequence is that actions require greater deliberation and effort, and they take more time.

3.3.7 Disorientation

It is common in poor visibility conditions for walkers and sailors to be uncertain about their location and direction of travel. Although being lost in the geographical sense can be regarded as a form of disorientation, it is much less of a concern to most pilots who can usually seek help from on-board navigation systems or air traffic control. With their freedom to move in three dimensions, pilots are more concerned about their aircraft attitude in relation to the vertical, and they make a careful distinction between being lost geographically and being disoriented spatially. They also distinguish[8] between "Type 1 Spatial Disorientation" and "Type 2 Spatial Disorientation".

Type 1 SD signifies that the pilot is uncertain about the direction of the Earth's vertical, for instance whether it is necessary to roll the aircraft left or right, or pitch up or down to come to a wings level state. Although the solution is to look at the attitude indicator, mounted centrally in the "primary flight" display panel as described in Chapter 6, because the uncertainty must be resolved in time to avoid a fatal collision with the ground, the experience can be very alarming. Student pilots first encounter this form of uncertainty when they are introduced to stalls and spins when learning about aerodynamic limits and the control actions needed to attain straight, level flight. The trainee also learns how he or she reacts to strongly conflicting, unfamiliar, dynamic cues, and how difficult it is to exercise commanding authority when confused and stressed. Later in their training, when the recovery procedure has been learned thoroughly, the stress reaction abates. Test pilots, aerobatic pilots and most military aircrew regard aircraft gyrations as routine and cease to be discomfited. They are mainly concerned that they could become confused during combat manoeuvres, or through hypoxia and GLOC.

Type 2 SD is a much more subtle and invidious condition where the pilot is unaware of, or deluded about, the orientation of the aircraft. Without a clear view of a highly textured ground carpet separated from a boundless sky by a continuously clear horizon - to stimulate both the ambient and focal visual pathways, described in Chapter 2 - the pilot is susceptible to a variety of spatial illusions. A false sense of the vertical can for instance be induced by a sloping cloud bank, and in the absence of visual references, spatial awareness depends to a greater degree on proprioceptive feedback from muscles, joints and body tissue and from balance sensations. Unfortunately, and contrary to the idea of flying by the seat of the pants, proprioceptive and balance sensors provide an unreliable indication of orientation.

The balance or vestibular system, formed by the left and right inner ear sensors adjoining the hearing organs within the bone structure at the base of the skull, is fast and closely integrated with the visual system. Each inner ear includes a set of three almost orthogonal semi-circular canals and a pair of otoliths. The canals can detect a minimum head rotation rate of about $0.5°.s^{-1}$, and the otoliths a minimum linear head acceleration of about $0.1m.s^{-2}$. Beside this threshold effect, the main limitation of the canals is that signals are not sustained. For example, if the pilot keeps his head still relative to the airframe during a constant rate rolling manoeuvre, after about 5 seconds he will not be aware of the continuing rotation, but when the rotation ceases he will have the sensation of rolling suddenly in the opposite direction. This gives rise to a common illusion, the "leans", when the pilot banks the aircraft gently, below the roll canal threshold, into a nicely co-ordinated turn but stops more suddenly at the required bank angle. The sensation of attaining the banked attitude feels much like flying with wings level, but the roll cessation is perceived as initiating a correcting roll towards the original wings level state. The pilot has the illusion of flying upright while the aircraft is in reality banked. A similar, but opposite, leaning illusion can be induced when recovering to the wings level state.

These quite subtle somato-gyral illusions are usually annulled after a few glances at the attitude indicator, but other effects can be more disturbing. One very disorienting percept is the result of an anomalous cross-coupling induced by coriolis forces when the pilot moves his head while the aircraft is turning. For instance, if in an aircraft banked close to 90° in a rapid horizontal turn, the pilot rotates his head in yaw to look out towards the right wing, he will have a sensation of turning his head in roll as well as yaw. The effect is particularly disturbing during tight turns at very high G, and it is a strong incentive to hold the head still when manoeuvring.

Another effect, usually termed an oculo-gyral disturbance because it affects the eyes, usually follows a rapid rotation, typically a spin. At the start the signals from the canals stimulate the vestibulo-ocular reflex (VOR) to provoke a sequence of jerky slow-fast-slow-fast eye movements known as nystagmus. Here the slower eye movements oppose the spin direction and tend to stabilise the eye pointing direction and the rapid returning saccadic eye movements are made in the spin direction. If the spin persists the signals from the canals decay and this stabilisation breaks down. Fortunately the pilot can still fixate adequately on instruments within the cockpit that are spinning in concert with his head. However, just as the pilot regains control of the aircraft and arrests the spin, the canals register the change in spin rate and stimulate inappropriate, involuntary nystagmic eye movements in the opposite sense. These can persist for a good fraction of a minute. Until the disturbance abates the pilot is unable to read the cockpit instruments or discern small details in the scene.

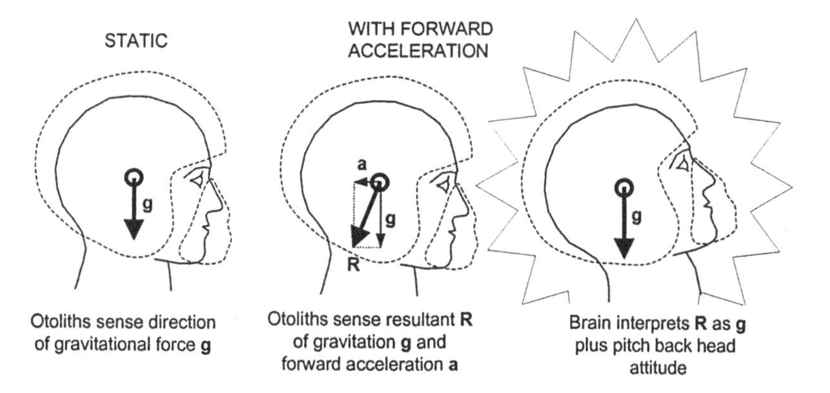
STATIC
WITH FORWARD ACCELERATION
g
a
g
R
g
Otoliths sense direction of gravitational force g
Otoliths sense resultant R of gravitation g and forward acceleration a
Brain interprets R as g plus pitch back head attitude

Figure 3.11 Somatogravic pitch illusion induced by forward acceleration

Otoliths, on the other hand, do maintain an output proportional to the linear acceleration, but as with any mechanical accelerometer they respond to the resultant magnitude and direction of gravitational and inertial forces. For instance, as illustrated in Figure 3.11, despite no confirming signal from the canal that senses pitch rate, a forward acceleration (in the X direction) seems like the whole body has been pitched backwards, a sensation that is reinforced kinaesthetically by the need for extra frontal neck muscle effort. Conversely, a forward deceleration can give rise to the sensation of tumbling forward, head over heels.

The current solution to the problem of disorientation relies on the pilot's self-knowledge that his estimate of aircraft attitude is commonly uncertain or illusory, and for him to develop the habit of checking the attitude by making regular glances at the indicator. The present forms of large, centrally positioned display may be assisted by devices that can supply peripheral awareness and obviate the need for foveal attention, perhaps using a wide angle HMD relaying a synthetic substitute for the external scene. As noted in Chapter 14, it may also become possible for an automatic system to infer, from data sensed by physiological and behavioural sensing systems, that a crewman has become incapacitated or disoriented. The automation could then take control of the aircraft, recover to a safe state and remain in control until the crewman displays a proper level of awareness.

3.3.8 Airsickness

Airsickness starts with a general feeling of unease that becomes a lethargic dizziness, then a headache with a sweaty chill and then the need to swallow back an acrid taste. This is followed, often within a minute, by violent, involuntary vomiting that persists well after the stomach is empty. After emesis the symptoms sometimes abate, but usually the drowsiness and headache endure till the end of the flight. At least half of aircrew trainees suffer the condition to some degree, and those most affected need sympathetic instruction in order to maintain the motivation to learn while incapacitated so that they can accumulate time in the air to adapt to the aircraft motion and become less susceptible. Once adapted, pilots are rarely affected, but other aircrew, particularly navigators and operators deprived of a view of the external world, succumb regularly and just have to perform their job despite the condition.

Like seasickness, space-sickness and carsickness, airsickness seems to be induced by unfamiliar motion, but because some individuals suffer the same symptoms in a static simulator, it is now regarded as a manifestation of a broader condition caused by a mismatch between visual and vestibular sensations[9]. Those with a chronic vestibular insensitivity are immune. A collation of data relating the incidence of sickness to motion conditions suggests that the main causal factors are the duration, frequency and amplitude of vertical (heave) oscillations[10]. Most effects are induced by oscillation the body at a frequency between about 0.1Hz and 0.7Hz. Motion at 0.2Hz is particularly nauseating.

No test has proved to give a sufficiently reliable prediction of an individual's long term susceptibility to airsickness to be useful for selecting trainee aircrew. Once trained, it is undoubtedly best to maintain preventive adaptation through uninterrupted flying duties, but if this is not possible or the aircrewman remains particularly susceptible, the services offer a treatment that is usually effective. This consists of an accelerated, supervised desensitising regime in which the individual experiences complex motions in a variety of ground-based equipment, including a gyrator box, followed by flying with an instructor manoeuvres that would tend to induce the condition. Treatment using available seasickness drugs, such as L-hyoscine hydro-bromide, may be a useful expedient for paratroops or passengers, but is not sanctioned for aircrew; the effective dose causes side-effects, mainly drowsiness, and it prevents the desensitising adaptation.

As noted above, modern flight control systems can alleviate the condition by transforming the gust response of the aircraft from a lurching motion into a jolting rumble, effectively shifting the vibration out of the nausea-inducing low-frequency range. Cockpit designer must however bear in mind that aircrew are not always at a motivational peak, and headgear designers must provide an easy way for aircrew to remove any covering over the mouth and nose so that it is possible to vomit cleanly into a sick bag.

3.4 Survivable incidents

3.4.1 Birdstrike

The crew of a powered aircraft could reasonably expect to experience a variety of mishaps with little probability of serious injury. In addition to the hazards associated with the natural environment discussed above, with emergency escape discussed in Chapter 12 and with the effects of weapons discussed in the next sections, most survivable mishaps involve impact, either with a small object at high speed or a large object at low speed. The main airborne object that an aircraft flying at high speed would be expected to hit is a bird. Most birdstrike incidents occur when flying at low level, typically in the vicinity of the airfield during take-off and approach to landing. Even a small bird can cause severe structural damage, and the crew are commonly presented with the problem of assessing the damage and perhaps dealing with a malfunctioning engine.

In the worst case, when the windshield is struck by a heavy bird, either the frame can be distorted so much that fragments of bird enter the cockpit, or the transparency can shatter and shower the crew with animal, plastic and glassy debris. If unprotected, the face and eyes would be injured severely. Most services therefore stipulate that aircrew, who always wear a helmet, must keep the protective visor in place at all times. With face protection, the crew are immediately aware of the obscuring spatter on the visor and the force and noise produced by the air blast. Their immediate needs are to wipe the visor clear so that they can control the aircraft, then reduce airspeed to lessen the airblast, assess the damage and land as quickly as possible.

Most fast jets designed for naval and ground-attack roles that involve flying at low level much of the time are equipped with a strengthened windshield, and in more recent types the glass combiner plate of the head-up display is constructed to provide a secondary barrier.

3.4.2 Crashing

Collisions while taxying are unlikely to be injurious, and air-to-air collisions or ground impact at normal airspeed are usually fatal. In between are accidents that warrant protection. These usually take the form of a heavy landing, in which case much of the excess energy is absorbed by bending and breaking the undercarriage, or landing with a raised undercarriage, in which case the excess energy is absorbed into the airframe and, if suitably soft, the ground.

If the bent airframe, together with the cockpit and seats, come to an abrupt stop, the occupants travel along their pre-impact trajectory until they too are brought to a halt. What happens to the occupants during the few arresting milliseconds depends on many factors. No amount of muscle effort can hold a braced posture against the force produced by a sudden stop from more than about 25 knots, and given a properly fitted modern harness with leg, lap and shoulder straps, the principal factor is the attitude of the aircraft at impact. This determines whether the body is thrust along the Z direction to collapse into the seat or is thrown sideways or forward against the harness.

Many helicopters are fitted with "crashworthy" or "stroking" seats to absorb Z-directed impact using a support structure that deflects or distorts to dissipate excess vertical impact energy. The principles have also been introduced into several general aviation craft and many transport aircraft[11]. Energy absorption during a head-on X-directed crash places greater reliance on the detailed structure of the harness and associated latching, retraction or tensioning mechanism. As discussed in Chapter 12, in a typical military fast jet the harness is arranged principally for ejection, where it is essential to pinion the torso to the back of the seat and the thighs to the squab. There is also usually much less space around the head and therefore less scope to allow the harness to apply a controlled torso deceleration, or prevent the head being flung into contact with the canopy or front panels.

Injury to the internal organs and the limbs is complicated by dynamic effects; organ resonances, stress concentration from elastic shock waves and the visco-elastic properties of biological tissue that make its strength depend on the strain rate. These can exacerbate damage to blood vessels, tissue and bone. However, the chief problem is to protect the head, or more precisely, the brain, from a blow on contact with the canopy or panels. Data from drop tests using cadavers, and experiments with animals and human volunteers, indicate that the tolerance of the brain to concussion and injury depends on the direction of the blow and the site of the impact. In addition to compaction damage near the area of contact, unless the force is directed through the head centre of mass, the shear stress caused by the violent rotation of the head can rupture blood vessels in the brainstem. As discussed in Chapter 13, headgear must be designed to absorb the impact energy without imposing excessive linear acceleration or a rotational jerk.

3.5 Anti-aircraft weapons

3.5.1 Conventional weapons

No airframe has yet been configured to withstand the blast pressure or deflect the ballistic fragments produced by the nearby detonation of a explosive device such as an anti-aircraft shell or the warhead of an air-to-air missile. Bullets from rapid-fire cannon and small arms are also likely to penetrate the structure and damage every component along their trajectory. In general the airframe and the on-board electronic systems are, and will probably forever remain, vulnerable to conventional weapons of this sort. Some military aircraft, and in particular those that engage ground troops, include partial protection for the aircrew against ballistic weapons. Many military helicopters are fitted with energy-absorbing seats that incorporate composite armoured panels around the sides and rear of the torso and head. Their aircrew also wear some body armour.

3.5.2 Nuclear, biological and chemical weapons

A wide variety of nuclear, biological and chemical (NBC) weapons in the form of gases, vapours, liquids, vesicants, gels, aerosols and dusts have been developed in the twentieth century to kill or incapacitate personnel over a large area. Relatively crude chemical weapons employ toxic liquids and vapours, such as distilled mustard agent (HD) or a derivative that is mixed with arsenical Lewisite (HL). The recently developed G agents such as Soman, Sarin and Tabun, have more complex effects, and modern VX agents interfere with the nervous system. Lethal toxins are produced by naturally occurring species such as anthrax spores and the botulism virus, and these have been selectively developed to be used as biological weapons. As is well known, in addition to an immediate blast, flash and electro-magnetic pulse, nuclear weapons create an enduringly toxic radioactive dust.

These weapons inflict horrific injury indiscriminately. Their products can be spread unpredictable by the wind to endure in the environment and remain lethal for a considerable number of years. Although their use is banned by international convention, many nations hold stocks and have the capacity to deploy the weapons on the assumption that the threat of such retaliation would deter a potential aggressor. The current stance taken by the UK and most western nations is to seek to reduce stocks but to avoid an obvious vulnerability, primarily to discourage a potential aggressor from developing and deploying NBC weapons. The services therefore train to operate under NBC threat, and many military procurement programmes stipulate that personnel and equipment should be resistant to the effects of NBC agents.

The aircraft design team is presented with a complex series of problems in arranging NBC protection for the aircraft and for the aircrew and groundcrew who must operate it. The aircraft can be damaged directly by NBC agents, and the object of "chemical hardening" is to understand the degrading effects, including the dynamics of absorption and desorption, in order to choose materials and to arrange the structure so that the aircraft is less functionally vulnerable. The subject also encompasses the practicalities of de-contamination, the use of sealants and the avoidance of contamination traps, for instance by the use of thread locking compounds. Polymer coatings and structures are particularly vulnerable, and for instance the external surfaces of a polycarbonate or acrylic canopy, and any electrical wiring and connectors within wheel bays and air passages, can be corroded by HD and HL, particularly when contacted by these agents in thickend form. Protection is difficult, but clean surfaces can be protected against contamination using sacrificial layers such as humble cling-film. The better solution is to use resistant materials and design the whole aircraft so that it can be decontaminated by spray washing immediately after landing.

There are two basic approaches to protecting the people. The first is to treat the aircraft as "dirty" and protect personnel using individual protective equipment (IPE), as described in Chapter 13. The second is to provide collective protection (Colpro). The latter term is used to describe on-board toxic-free spaces as well as ground facilities such as the crewroom or decontamination chamber in which IPE is doffed and donned. In practice most combat aircraft operate with IPE because the cockpit of a fast jet is vulnerable to the ingress of airborne particles and vapour when opened to allow the crew to leave and enter. Most rotary-wing aircraft do not have pressurised crewstations, and their interiors are effectively open to any toxic air mass.

Aircrew in transport aircraft and most of the larger, multi-crew types can be given adequate protection without having to wear IPE for long periods of time. One approach involves stowing IPE clothing and headgear on the aircraft for use when required in conjunction with portable, electrically powered filter units. Another approach treats the pressurised cockpit and cabin as a collectively protected space that is always kept above the external pressure so that a net outflow of conditioned air counters inward diffusion. When the aircraft is stationary on the ground the clean internal environment would be maintained by powering the ECS using the APU or an external source. It would be possible to attach temporary air-locks to the doors and have the crew transit between the aircraft and crewroom wearing discardable oversuits.

The more difficult problem is to prevent internal contamination of the aircraft in flight or on the ground, and in particular to prevent toxins entering the air supplied to the crew for breathing and for conditioning the cabin. There are numerous contamination paths. The primary path is to draw contaminated air into the engines or APU that feeds the ECS. Contaminated air can also enter through open doors and canopies, or be passed indirectly into the cabin though pitot tubes, internal equipment spaces and undercarriage bays. Personnel entering the aircraft for servicing or operational flying can also transfer toxic agents adhering to their clothing and equipment.

The problem of countering internal contamination can be understood by considering the ECS, as illustrated in Figure 3.4. Although many chemical and biological agents decompose into relatively harmless products on passing through the engine compressor, those that survive must be trapped by filters inserted along the pipework between the engine compressor and the cabin. The design of the filters is a specialist discipline and practitioners employ specific types for particular threat species. For instance, a centrifugal separator can remove larger particles, but some biological agents require high efficiency particulate air (HEPA) filters made from metals and fibre-glass, while others can be adsorbed by cooled molecular sieves containing activated charcoal and artificial zeolite. The issue is far from straightforward because it is necessary to arrange multiple parallel filters of the same type so that when one is saturated and needs to be regenerated another can be brought into play. The resulting system includes a number of shut-off, by-pass and venting valves, all of which must be leak-tight and made from resistant materials which do not absorb and slowly release contaminants. This precludes the use of conventional silicone seals. In addition, the filters restrict the airflow and place extra load on the engine. It is therefore usual to limit the volume of conditioning air to the cabin during periods of peak demand on the engine so that essential avionic systems can be kept cool and reliable. Such restrictions are most likely on take-off and during combat, when the crew are under greatest stress.

Current research is aimed at improving the effectiveness of filters, introducing detection systems and solving numerous decontamination problems. The overall aim is to deploy a more flexible but adequate level of protection only when under actual attack using personal equipment that is considerably less burdensome than the current protective clothing and headgear. These are outlined in Chapter 13.

3.5.3 Nuclear detonation and directed energy weapons

Aircraft are particularly vulnerable to the broad spectrum of electromagnetic disturbances from a nuclear detonation. The effects endure for several seconds and depend greatly on the intrinsic power of the device and the proximity of the aircraft. Like the blast pressure wave, the E-M field intensity falls with the square of the separation distance, but the sheer power nevertheless creates extensive, immediate damage at considerable distance. The energy at radio frequencies can induce over-voltage pulses in conventional semi-conductor electronic units at several tens of kilometers. In consequence most military avionic systems are shielded against the E-M pulse level that accompanies the blast pressure the aircraft structure is likely to withstand. The emissions in the microwave and far infra-red induce intense heating. This and the energy in the visible, near infra-red and ultra-violet parts of the spectrum can damage on-board sensors, and a variety of protective countermeasures are employed.

The radiation that penetrates the canopy would be experienced by the aircrew as an intense flash. This overloads the retinae to induce temporary blindness, and depending on the proximity of the detonation, the outer surface of the eyes could be heated to produce effects ranging from a searing pain to permanent damage. The most widely deployed form of protection are goggles that detect the onset of the bright flash and switch quickly to an opaque state. These are described in Chapter 13.

The eyes of military aircrew need also to be protected against the intense illumination used by systems such as laser-based range-finders and target designators that are increasingly deployed in battle. These devices are most likely to dazzle and startle the viewer, particularly if the aircraft is manoeuvring close to the ground and the pilot looks directly at the source. In this case the strong foveal after-image acts like a temporary scotoma masking details in the external scene and may render information displayed within the cockpit unreadable. Helicopter crew are particularly vulnerable because they fly relatively slowly and close to the ground. Several dazzling incidents have occurred during recent conflicts, and aircrew in US civil transport aircraft appear to have been deliberately illuminated from the ground by lasers of unusual power[12].

Protection of aircrew vision can be arranged in a number of ways, each having peculiar merits and disadvantages. One requires aircrew to fly with a patch over one eye that can be removed should the exposed eye be damaged; a simple expedient perhaps, but at a cost of considerable visual impairment. Currently, as described briefly in Chapter 13, the approach adopted by most services is to provide spectacles or visors that are coated with optical notch filters to attenuate light in specific wavelength bands. These are chosen to block light from conventional lasers that emit at precisely known wavelengths, mainly the wavelengths summarised in Table 3.2. At night, despite the increased vulnerability brought about by their visual adaptation to the lower ambient light level and their enlarged pupils, most aircrew are reluctant to use such devices because the coatings absorb a large proportion of the available light. Night vision goggles, described in Chapter 5, can provide a barrier providing the goggles are of a type that obscures the direct view of the world. However, the eyes remain vulnerable to off-axis illumination, and the goggles too are vulnerable to overload damage.

Table 3.2 The main laser systems

Laser	Type	Wavelengths (μm)	Typical Power
KrF	Excimer multi-line	0.249 *(note 1)*	100W pulses (~ns)
He-Cd	Metal vapour	0.325	50mW cont
Ar	Gas	0.351, 0.458, 0.488, 0.514, 1.092	10W cont
Kr	Gas	0.413, 0.531, 0.568, 0.647, 0.752	10W cont
Cu	Metal vapour	0.511, 0.578	50W cont
Dye *(note 2)*	Liquid	0.3 to 1.3 tunable	10W cont 2MW pulses(~ps)
Nd:YAG *(note 3)*	Solid state	1.064, 0.532, 1.319	50W cont 1kW pulses (~ns)
He-Ne	Gas	0.612, 0.633, 0.640, 1.153, 3.392	0.1W cont
$Cr:Al_2O_3$ *(note 4)*	Solid state	0.695	10W pulses (~μs)
GaAlAs	Diode	0.84 (0.7 to 0.9)	10mW cont
GaAs	Diode	0.89 to 0.90	10mW cont
COIL *(note 5)*	Gas	1.315	40kW cont 40MW pulses (~μs)
InGaAsP	Diode	1.1 to 1.6	5mW cont
Ho:YAG	Solid state	2.06	50W pulses (~ps)
CO_2 *(note 6)*	Gas	10.6 (9.0 to 11.0)	1kW cont 10MW pulses (~ns)

Note 1 The eye is particularly vulnerable to photochemical degradation by UV irradiation
Note 2 Dye lasers are used widely as "pumps" for other pulsed lasers
Note 3 Niodymium in Ytrium Aluminium Garnet (Nd:YAG) is used in many ranging and target designation systems
Note 4 A ruby crystal was used in the first practical laser
Note 5 The Chemical Oxygen-Iodine Laser (COIL) was developed by the USAF as an airborne weapon
Note 6 The infra-red radiation from carbon dioxide lasers is used for cloud-penetrating LIDAR

Although most nations agreed in 1999 not to develop blinding weapons, it can be anticipated that laser-based, directed-energy weapons (DEW) will eventually be introduced that can deliver pulses at an energy density sufficient to damage the aircraft structure. The military case is clear. Conventional weapons must be carried laboriously to the intended target and on detonation much of their energy is dissipated on harmless entities nearby. In contrast, the novel weapons will send energy at the speed of light in a straight line to produce precisely localised damage. They will undoubtedly be fielded against missiles and aircraft, mainly to defend ships and important installations. Their power to damage sensors and perhaps the airframe at short range will undoubtedly be used to dazzle aircrew at long range and blind them at intermediate ranges.

Visor filters cannot prevent light from frequency agile sources, particularly tuneable dye lasers, from impinging on the eyes. This, and problems associated with the visual degradation imposed by current countermeasures, continue to stimulate developments. One approach would take the form of a goggle or a visor that, like the nuclear flash protection goggle, automatically closes when illuminated strongly. Here the problem is to find a physical phenomenon that can switch to a sufficiently attenuating state within the picosecond onset time of an irradiating pulse. To be useable the effect must be engineered into a lightweight device that does not degrade vision in the open state. Until these challenging requirements are met, the most robust protection against excessive irradiation can be supplied by closing the cockpit, either temporarily or permanently. The form of such cockpits and the issues involved in isolating the crew visually from the external world are discussed in Chapter 14.

References

1. *Manual of the ICAO Standard Atmosphere: extended to 80 kilometres (262,500 feet)*, International Civil Aviation Organization, Montreal, Quebec, Canada, ISBN 92-9194-004-6, 2002
2. Kaye G W C and Laby T H (compilers) *Tables of Physical and Chemical Constants*, Longman Scientific and Technical, 1989
3. *Air Navigation : The Order and the Regulations,* 3rd edition, CAP 393, Civil Aviation Authority, London, 2003
4. *Human consequences of agile aircraft*, RTO-EN-12, AC/323(HFM)TP/32, 2000
5. Rood G M, *The audio environment in aircraft*, in *Audio effectiveness in aviation*, AGARD-CP-596, 1996
6. *Reducing noise at work: Guidance on the Noise at Work regulations*, published by the Health and Safety Executive, London 1998
7. Glaister D H and Lisher B J, *Centrifuge assessment of a reclined seat*. In *On the Pathophysiology of high sustained +Gz acceleration. Limitation to air combat manoeuvring and the use of centrifuges in performance training*, AGARD-CP-189, 1976
8. *Spatial disorientation in military vehicles: causes, consequences and cures*, HFM Symposium, La Coruna, Spain RTO-MP-086, 2002
9. Benson A J, *Motion sickness: Significance in aerospace operations*, AGARD-FS-175, 1991
10. Lawther A and Griffin M J, *Prediction of the incidence of motion sickness from the magnitude, frequency and duration of vertical oscillation*, Journal of the Acoustical Society of America, Vol 82, pp. 957-66, 1987

11. Chandler R F, *Occupant crash protection in military air transport*, AGARD-AG-306, 1990

12. McLin L N, Cheney F, McCracken S, Kosnick W, Reddix M and Ivan D J, *The laser threat*, in *Countering the directed energy threat: Are closed cockpits the ultimate answer?* RTO-MP-30, 1999

Further reading

Gillies J A (ed), *A Textbook of Aviation Physiology*, Pergamon Press, Oxford, 1965

Ernsting J, Nicholson A N, Rainford D J (eds) *Aviation Medicine*, 3rd Edition, Butterworth Heinemann, 1999

Chapter 4

Crewstation Design

4.1 Introduction

Most teams charged with the design of the crewstation of a modern combat aircraft must envy the simplicity of the requirements and facilities in older aircraft types. Now that an embryonic aircraft is regarded as a set of computer-controlled systems with programmable characteristics, there are many more dimensions to the endeavour and the whole gestation is protracted. The aircraft design team may be required to help the customer translate operational aspirations into feasible requirements, explore a wide range of options and choose the optimum before entering into a detailed design that can be built as a prototype for testing. Among the large number of initial considerations is a trade-off between the complexity of automated on-board systems and the necessary number of crewmembers. This, in combination with the variety of possible arrangements of the displays and controls, determines what the aircrew will be asked to do and how well they are likely to do it. These decisions influence the design of the avionics and the airframe, and hence the vehicle performance and cost. They also have repercussions for crew selection, training and numerous logistical issues that affect the cost of operating the aircraft over its service life.

Table 4.1 Stages in the development of a modern crewstation

Stage	Development activity
Define	Initial consideration
	Requirements capture
Study	Mission modelling
	Function analysis
	Conceptual design
	Task analysis
Develop	System architecture
	Mechanical design
	Mock-up
	Rapid prototyping
	Lighting evaluation
	Simulation
Build	Prototype assessment
	Flight test & Qualification
Use	In-service modification

The job of the crewstation design team may therefore be twofold. Firstly the team may be required to examine crewing issues and make a justifiable recommendation so that the whole aircraft design team can consult with the customer to agree on the crew complement. Secondly, in the light of this decision, the team must devise a satisfactory arrangement of the crewstation. This, their time-honoured task, must be demonstrably satisfactory because the later any modification of prototype hardware and software is implemented the greater the expense. This chapter is structured to resemble the sequence of processes that can be followed, as summarised in Table 4.1. It discusses the issues the team must resolve and some of the tools and techniques that they can use.

Several US standards are directly relevant to the organisation and management of the crewstation design process. MIL-HDBK-46855A and MIL-STD-1472[1] give general guidance to any supplier developing man-operated and man-maintained equipment for the US forces. The documents provide forceful justification for including knowledge of human factors and good human engineering practice in equipment development programmes, the handbook doubling as an ergonomic textbook and an introduction to the terminology of the recommended managerial process. The sequence itself – analysis, then design and development, then test and evaluation – is essentially the stages summarised in Table 4.1. It can also be noted that the adoption of a formal process for managing the design of the crewstation is dictated by initiatives such as MANpower and PeRsonnel INTegration[2] (MANPRINT) and Human Factors Integration[3] (HFI). These have been devised by the US Department of Defense and UK Ministry of Defence respectively to ensure that manpower, operator and maintainer selection and training, and the complexities of spares holdings and logistic support receive proper consideration during the development of any new piece of equipment.

4.2 Requirements definition

Few types of combat aircraft have been designed and built speculatively. Most have been conceived in response to a request from the procurement office of a national government. In the UK such customers have started the process by erecting a target that specifies the necessary attributes of the aircraft and, with funding approval, invited prospective manufacturers to tender. The customer selected the bid he judged most likely to fulfil the target and contracted the chosen manufacturer to supply what had been offered. It is notable that most target attributes of the aircraft, and those accepted at the start of the contract and used to assess satisfactory achievements at the end, could be expressed as quantitative performance parameters. Although this made expectations and liabilities clear, technical and contractual difficulties have arisen when the performance of the delivered system has been inadequate. Also, with the rapid technological changes that can occur over a gestation of about fifteen years, many of the component systems of aircraft procured in this way have become obsolete by the time the initial batch of aircraft is delivered.

The recent adoption of "smart" procurement management is intended to avoid this wasteful inflexibility. The idea is to harness the expertise of the customer, the customer's advisors, the contractor and all major sub-contractors, all of whom are represented in an Integrated Product Team (IPT). This top-level team formulates the design solution to the customer's operational needs, instigates crucial research and development, and co-operates to ensure that components are state of the art. The IPT manages the overall development programme through a set of teams, with a similarly eclectic constitution, that manage specialised domains, for instance the airframe structure, propulsion, flight control, utilities, weapon integration, mission systems, support and maintenance, and crew systems.

4.2.1 The design team

The crew systems IPT is the body directly responsible for examining crewing issues and devising a satisfactory arrangement of the crewstation. The crew systems IPT is however a controlling body rather than an engineering team. Cockpit design is analogous to the preparation of a meal in the sense that both are intended to create a satisfying ensemble by combining compatible ingredients. But each new generation of cockpit designers must invent a new recipe, for which knowledge of previous approaches must be boosted by a wider appreciation of current research and a willingness to consider the adoption of novel systems. A dedicated multi-disciplinary team that reports to the crew systems IPT is usually convened to carry out the detailed work. This includes individuals with experience of crewstation design and expertise in mechanical, systems, electrical, electronic, optical, pneumatic, and hydraulic engineering. The other essential skills are contributed by experienced aircrew and scientists with a background in psychology, human factors and physiology.

The team interacts closely with other teams on the aircraft project. A development plan and schedule are drawn up, and members are assigned responsibilities. Ways of controlling and recording decisions and their rationale are set up. The team members are also invited to consider anything that could go awry — technically, managerially or commercially — and affect the satisfactory completion of the development programme. In many organisations the identification of risks is considered wise rather than paranoid, and a formal process of risk management is mandated so that consequences can be anticipated, costed and remedies put in place.

4.2.2 Requirements capture

With the traditional process of managing procurement the customer defined the aircraft role and gave the main requirements as explicit performance parameters that, for instance, covered the operating range, manoeuvre envelope, load carriage, avionics system and the day and night operating conditions. These gave the crew systems team clear guidance about the number of crew, their roles, the seating arrangement and the variety of protective equipment. The customer would also refer to a large number of Standards documents to specify details.

The main Standards relevant to the design of military aircraft crewstations are listed at the end of the chapter. Defence Standards (DEF STAN ... etc.) are produced by the UK Ministry of Defence in London. Military Standards (MIL-.. etc.) are produced by the US Department of Defense in Washington. STANAG documents are produced by the NATO Military Agency for Standardisation in Brussels, with those relevant to crewstation design coming within the remit of the Aircraft Instrument Panel (MAS-AIP). Those listed have been devised almost exclusively at the behest of the government department responsible for procuring aircraft as a means of specifying what they want from a contractor. It must be borne in mind that similar publications are produced in other languages, mainly French and German, and that many refer to others for more detail. The customer can also cite a wide range of non-military national and international standards. In general a standard is treated as a living document under constant review so that the content can be updated to reflect technological developments. Some that have been superseded are not maintained, but they still contain useful information and can be cited.

The UK documents contain many parts, each with many supplements. The main human factors are detailed in fourteen parts of DEF STAN 00-25[4] that cover body size, body strength and stamina, workplace design, stresses and hazards, vision and lighting, auditory information, voice communication, controls, maintainability, systems and human-computer interaction. DEF STAN 00-970[5] is a portmanteau covering design and airworthiness, Part 1 dealing with combat aircraft in nine sections and Part 2 dealing with rotorcraft in three. Section 2 of Part 1 details the manoeuvres and acceptance criteria for flight tests. Most combat aircraft crewstation topics are covered in Section 4 in a series of supplements between No.61 and No.82 that deal with display legends, audio warning signals, anthropometry, panel lighting, the sources, effects, measurement and allowable exposure to noise and vibration, the effects of rain on vision, testing of canopy transparencies, the precautions for ditching and crash landing, and the means of jettisoning and breaking the canopy to enable emergency escape. A few topics, for instance speech communication and protection against laser, nuclear, chemical and biological weapons, are covered in Parts 6 and 9.

Standards have varying specificity. A few describe the detailed form of an entity, for instance so that a plug mates with a connector. Others provide a test for acceptable performance, such as the impact attenuation of a helmet or the required view from an aircraft. The more recent documents, particularly those produced by the US Department of Defence as Handbooks (MIL-HDBK-...), follow the tenets of smart procurement and do not prescribe an engineering solution. They are mainly intended to ensure that operational experience and experimental evaluations are recognised and used in designing equipment. They explain pertinent factors and suggest acceptable parameter values.

The primary indication of the smart customer's need and sense of values, stated in terms that do not indicate how this need should be met, is contained in the User Requirements Document (URD). This is used by the IPT to generate a System Requirements Document (SRD) that also avoids any suggestion of the possible engineering solutions. The SRD translates the high-level aspirations into more concrete parameters, some of which are quantitative, so that the IPT can study the advantages and disadvantages of alternative technological approaches. This document provides a better indication of the nature of the missions, the likely threats, the logistical support and the qualities of the "target user population"; the air and ground crew who will operate the aircraft and who will require training.

4.3 Number of crewmembers and their roles

The crew systems team must consider how the characteristics of the aircraft and the airborne systems, and in particular the degree to which these systems are automated and allowed autonomy, affects the conduct of operational missions and the jobs of the aircrew. At present these issues are handled by conducting a series of separate but inter-related studies that assume the qualities of the on-board systems. The first delineates what happens during the missions and selects the phases that are most likely to overstretch the crew. The second considers how the responsibilities for task execution can be allocated to human and electronic agency. The third considers in detail what the crew has to do and whether they can cope.

4.3.1 Mission modelling

Combat aircraft are designed to perform a specific set of missions. Although fighter types are designed and equipped to engage enemy aircraft, the specific demands of a mission depend greatly on the objective; to destroy enemy surveillance aircraft, escort friendly bombers or provide air policing and surveillance, point air defence or area air defence. In the latter cases the activities also depend upon whether there is enough warning to mount a sentry-like combat air patrol (CAP) or it is necessary to scramble in response to a quick reaction alert (QRA). A similar variety of missions are performed by offensive aircraft types; offensive counter air (OCA) to attack enemy airfields, long-range air interdiction (AI) or deep strike (DS) against militarily important targets, battle-field air interdiction (BAI), suppression of enemy air defences (SEAD) - usually surface-to-air missile (SAM) batteries - and the attack of enemy artillery and armoured vehicles as close air support (CAS) to friendly ground forces. Other types of fixed-wing aircraft also undertake specialised reconnaisance (Recce), electronic warfare (EW), airborne early warning (AEW), air-to-air refuelling (AAR) roles, and many missions involve co-operation between a variety of types. Rotary-wing types tend to have different speed/range/load capabilities and the flexibility to perform a variety of roles, such as troop carriage, coastal patrol and air/sea rescue (ASR). Specialised variants are fitted with the mission systems to perform anti-shipping and other operations with, for example, Special Forces. The most mission-specific types are the attack helicopters that are mainly used to engage armoured vehicles, although they are also likely to be deployed with anti-aircraft missiles to counter other helicopters.

All missions can be divided into a sequence of planned phases that start before take-off with the formulation of a plan and finish after landing with some form of post-flight report. A phase can be regarded as a period with a defined start and finish during which the demands placed on the crew are reasonably constant, for instance the low-level penetration phase starts when the aircraft has descended to a few hundred feet over the ground and accelerated to high speed, and is completed when the aircraft crosses the waypoint that marks the beginning of the run towards the target. Table 4.2 takes the case of a typical ground attack sortie and divides the sequence into two classes of phase; those such as take-off, cruise and landing that are performed during all flights, and those that are specific to the mission. In a similar way, the tasks that the crew perform during each phase - taken from the taxonomy summarised in Chapter 2 in Table 2.2 - are divided into those imposed during any flight and those that result from military concerns.

Table 4.2 A summary of the information-processing tasks performed by combat aircrew over the course of a typical ground attack mission

Mission Phases / Tasks	General							Ground attack mission							
	Mission planning	Pre-flight checks	Take-off & climb	Cruise	Approach & landing	Emergency	Post flight de-brief	Airborne re-fuelling	Medium level ingress	Defensive manoeuvring	Low level penetration	Target acquisition	Weapon delivery	Low level egress	Return to base
Mission Management	3	3	3	3	3	3	3	3	3	3	3	3	3	3	3
Flight path control			3	3	3	3		3	3	3	3	3	3	3	3
Navigating			3	3	3	3		3	3	3	3	3		3	3
Communicating		3	3	3	3	3		3	3	3					3
System management		3	3	3	3	3		3	3	3	3			3	3
Targeting												3	3		
Weapon management		3								3		3	3		
Countering threats									3	3	3	3	3	3	
Tactics	3						3		3	3	3	3	3	3	

Included in Table 4.2 are two columns called "emergency" and "defensive manoeuvring" that, although instigated in response to events outside the control of the crew, can also be regarded as phases. They have definable start and stop conditions and involve an abrupt change in the priorities or required performance of one or more tasks. The first includes handling any unwanted state, such as a fuel leak or an engine failure, and the second covers any diversion from the planned route to evade enemy aircraft or an unexpected SAM site. In general, the tick marks in each column indicates that the task has some priority during the phase, and the analyst's main input is to define a set of representative missions composed of sequences of phases. The phases are characterised by start and end conditions, by geometrical factors such as the distance, ground speed and the nature of the terrain, and by tactical factors such as the disposition of enemy defences. Taken together these govern the allowable duration, the required states of the aircraft and the on-board systems, and what the crew must do to meet the demands of the individual tasks. Because the detailed facilities are not designed at this stage, the description is in fairly broad terms, typically covering periods of several minutes. This is usually summarised as a lengthy table or a time line diagram, such as can be produced by a tool like MicroSAINT, as illustrated in Figure 4.5.

4.3.2 Function analysis

The object of function analysis is to understand how the crew should interact with on-board systems to perform their tasks, as delineated in the previous section, and a significant part consists of understanding the limitations, alternative modes and capabilities of automated systems. This knowledge is needed for the subsequent studies that consider the trade-off between the number of crew and the capabilities of the automation. It is also essential for the design of a satisfactory display and control interface.

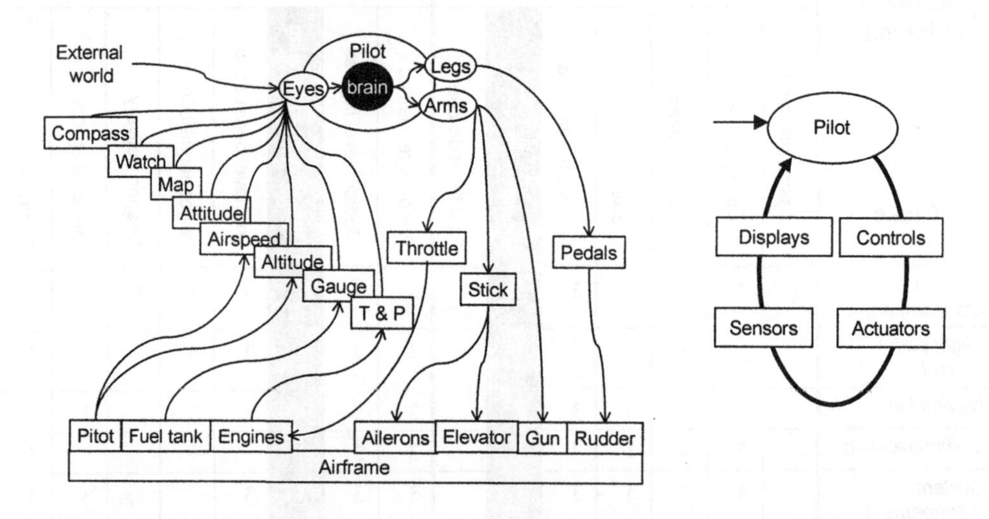

Figure 4.1 Information flow paths in an early fighter aircraft

Early combat aircraft had no automation. The pilot performed all tasks by viewing the world and the meagre sources of information in the cockpit. He executed whatever decisions he made by handling the appropriate device directly; the gun when weapon aiming, the map when navigating and the usual levers for controlling the aircraft. As indicated in Figure 4.1, most tasks involved a loop in which information flowed from the pilot through a control device to an actuator. The outcome was apparent from the pilot's view of the world or through a display device that relayed feedback from a sensor.

The earliest form of automatic aid, the electronic automatic pilot, was a set of closed-loop servo-mechanisms that received analogue signals from the sensors and adjusted the actuators to hold the aircraft on a set height and heading. The pilot had first to attain the desired flying state and engaged the automatic mechanism by depressing a switch on the stick. He resumed control by pressing another *instinctive disengagement* switch, so called because it was positioned under the little finger of the hand that gripped the stick top. Although he was rarely supplied with a positive indication that the facility was engaged, this was usually apparent from the steadiness of the flight path. The back-driven movements of the stick and pedals, which were coupled mechanically to the aerodynamic control surfaces and moved in proportion to the automatically commanded deflection, also gave useful proprioceptive feedback. The authority of the system, the amount of control deflection that it could command, was usually a fraction of that available to the pilot.

With the maturation of digital electronic processing in the 1980s, the analogue computer was replaced by a programmable digital equivalent that enabled more complex control laws and the ability to tolerate larger perturbations. The simple auto-pilot became a flight control system (FCS) that could exercise full authority over the hydraulic jacks that moved the control surfaces. In current aircraft, such as Eurofighter, the flight control system is integrated with the engine control system to confer stability on an unstable aerodynamic configuration and prevent stalls and excessive wing loading. The system can operate in a number of automatic modes, or it can be put into a manual mode where it responds to the commands from the inceptors that the pilot uses like a throttle and stick.

Digital electronics also expanded the range of functions that could be performed automatically. For instance, the navigation computer, weapon-aiming computer and the terrain-following computer joined the flight control computer as core features in ground-attack aircraft. A similar revolution occurred in civil transport aircraft where the flight management computer joined the flight control computer. These systems required the crew to inserted waypoint co-ordinates and other route-following criteria into the appropriate computers, which made the second by second decisions to access the relevant data, calculate the aircraft position, compute the deviation and adjust the aerodynamic control surfaces and engine thrust to maintain the required course. Human involvement in flight-path control and navigation changed from that of performing computations and exercising direct control to that of setting up automated equipment and monitoring progress. This supervision was assisted by concurrent developments in display technology that replaced banks of dials by a few computer-driven multi-function CRT displays (MFD) programmed to show the progress along the route and the state of the on-board systems. The adoption of complementary multi-function controls (MFC) produced the glass cockpit, as described in Chapter 6.

The increased detachment of the pilot and the complexity of managing automated systems through a glass cockpit are illustrated in Figure 4.2. These diagrams extend the notion of control loops used in Figure 4.1 and are simplified versions of diagrams devised by Sheridan[6] to describe the hierarchical nature of supervisory control. Sheridan and other authors also make a useful distinction between the functions performed at different levels. The lowest "dumb" servo-mechanical controllers such as thermostats, gyrocompass and autopilots carry out what would be "skill-based" functions if done by a human operator. "Intelligent" computer-based systems occupy the next level where they receive operator commands, provide task state information and adjust the set-points of the dumb controllers, and this the authors regard as "rule-based" decision making. The highest "knowledge-based" level of activities such as planning, anticipating, teaching, programming, monitoring, intervening and learning are done by the human operator.

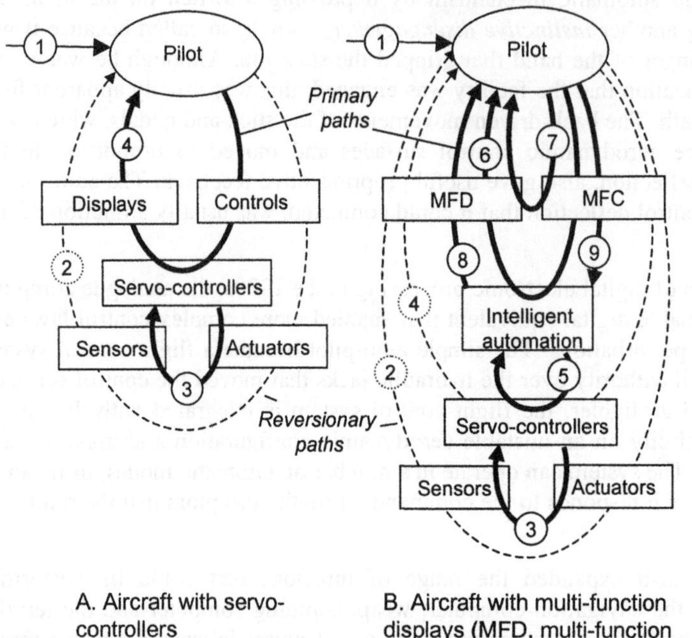

Figure 4.2 The increased number of control loops induced by stratified automation. In (A) the pilot's main way of delegating (4) is through a servo-controller, such as an auto-pilot, that performs fast and fine control (3), and he can revert to direct interaction using the original control loop (2). (B) represents the strata in a glass cockpit that has multi-function displays and controls and puts another layer of automation, such as a flight management system, in direct control of the servo-controllers (5). Although the original control loops (2 & 4) may be available, the pilot operates routinely through the high-level automation and this involves two control loops, one instigated by himself (6) and the other under the control of the automation (8 & 9). Switching between display modes to access appropriate information constitutes another control loop (7)

The development of the glass cockpit has been motivated by the desire to enable flight crew to delegate routine skill-based and rule-based functions to automated systems so that aspects of the job that demand human knowledge and intelligence are given more attention. Much of the crew's work can be done before take-off when devising the flight plan, the details of which are entered into the flight or mission management computer as the co-ordinates of waypoints and the airspeeds for each leg. All being well, the pilot can put the automation into the appropriate mode as the flight unfolds, and check the state of the systems by selecting appropriate modes of the multi-function displays and controls. However, Sheridan *et al* anticipated that the transformation of the pilot into a system manager would not necessarily improve his lot, especially when the flight plan needed revision or it was necessary to bypass some functions performed by the intelligent automation. To devise a remedial plan and re-instruct the automation, or selectively disengage an automated function and use whatever reversionary processes have been put in place by the cockpit design team, the pilot must have a clear understanding of the workings of the whole system. His mental model for each task must accommodate the strata indicated in Figure 4.2 and include numerous subtleties. For instance, to take control over the aircraft attitude, airspeed, height and heading, usually to have the hands-on practice needed to retain this basic skill, the pilot must know whether the mechanism automatically deploys flaps and slats. The other major qualm was that automation would reduce the pilot's routine workload but introduce complications in emergency circumstances.

In practice the glass cockpit has been broadly welcomed by both military and civil aircrew. By optimising the flight plan, minimising fuel usage and enabling the crew to make accurate what-if precautionary plans, including diversions to alternative landing sites, the uncertainties of arriving at the intended destination with adequate fuel reserves are reduced substantially. Safety has also been improved by enabling the crew to attend properly to aspects of the job that were previously neglected or done hurriedly. However, by the mid 1990s the Federal Aviation Administration (FAA) were concerned that some of the difficulties anticipated when operating with the new philosophy may have contributed to the 60% of accidents and incidents in which the actions of the flight crew were cited as primary factors. In aiming to eliminate transport aircraft accidents, the FAA instigated a review of such events that had occurred over the preceding decade.

The review team examined mishaps to commercial transport types made by Boeing, Airbus, McDonnell-Douglas and Fokker, and looked at crew training and selection as well as the flight deck design and the avionic architecture[7]. They concluded that the management of on-board automation by the aircrew had indeed been a significant cause of many accidents. Pilots misunderstood the capabilities, modes and operating principles of the automation, they were surprised and bewildered when the aircraft behaved unexpectedly, and they were not always aware of the current mode of operation of the automation or the proximity of the aircraft flight path to the terrain. Among an extensive list of recommendations the authors suggested that the automation philosophy implemented in an aircraft should be conveyed to flight crew explicitly and in a standard way. Details of the autopilot and autothrottle received particular criticism. It was emphasised that flight crew need clear guidance as to when these facilities should be engaged or used in a mode with lesser authority. The crew should also know when either can or cannot be engaged, and when either will automatically disengage or revert to another mode. The authors also made broader recommendations, for instance that automated systems should be examined in depth to ensure that they could not produce hazardous energy states or other undesirable manoeuvres. They were also keen to avoid the possibility that automation could induce over-reliance or confusion. Over-reliance and hazardous states of awareness such as underload, complacency and absorption should be minimised through better system design and crew training. Confusion could best be removed by unambiguous controls and displays, arranged if necessary in a standardised way.

To a large extent the experience of automation gained by military pilots has paralleled the civil case, but other singular lessons have been learned. One is exemplified by the coupling of a radar-based terrain-following system directly to the flight control computer in several ground attack aircraft, notably the F-111 and the Tornado, so that the flight path automatically followed the undulations in terrain elevation. Pilots soon learned that the forces commanded by the automation were more disturbing than self-instigated manoeuvres. This was partly because machine-imposed manoeuvres tended to come as a surprise and give no time for the pilot to adopt a braced posture and stiffen his neck muscles. It was also because the automatic system was programmed to command a nose-down negative-G bunting manoeuvre to descend after crossing a ridge, whereas pilots generally preferred to roll inverted and apply positive-G. Their urge to anticipate the gyrations, compounded by their incomplete trust in the system, induced most pilots to attend to the display showing the radar-determined terrain profile with as much concentration as they would use to effect direct control over the flight path[8]. Many found that it was preferable to fly the aircraft manually and use the terrain-following system as a source of advice to climb or dive, with this presented conveniently in the direction of flight as a cueing symbol in the head-up display. This technique was later mandated by the need to use the radar sparingly, and only in conditions where the forward visibility was poor, to minimise the chance of detection by an enemy radar receiver.

The second example has come from the task of dealing with enemy defences, particularly surface-to-air missiles guided by ground radar or by infra-red seekers, encountered suddenly during low-level flight. In earlier combat aircraft the pilot relied on on-board systems, such as the radar warning receiver, to detect and classify such threats, but it has been his task, on receipt of visual and auditory alerts, to interpret the warning display and decide on an appropriate countermeasure. Typically the choice is to manoeuvre the aircraft, switch the radar into a selective jamming mode, release radar-scattering chaff or eject infra-red decoy flares. He may also use these in combination. However, it is crucial that the decisions are made quickly and the countermeasure executed at an optimum moment. This classic case of rule-based decision-making and skill-based execution has been the subject of considerable development effort, and fully automatic defensive systems are now available in the majority of modern combat aircraft.

Although many of the on-board mission, utility and weapon systems may themselves be under development at this stage of the aircraft project, to proceed the crewstation team must take pains to understand their working principles and their likely impact on the crew. In the case of the automated systems, this requires understanding the modes of operation, and for each mode the information the system can generate and the parameters that need to be set. The cockpit design team then has the creative task of devising the preferred ways for the crew to interact with each system, for which they have recourse to their own experience plus feedback from operational aircrew and recommendations from reviews such as that instigated by the FAA. The outcome of the exercise is a top-level delineation of the pages presented on the multi-function displays, an indication of the type of device that will be used to make control inputs and the design of the overall structure of the control menu.

It is of note that an increasing proportion of tasks are now supported by on-board automatic systems, and in the not too distant future the collection of more capable, more comprehensive and more co-ordinated systems will take on the character of an electronic crew. As discussed in Chapter 14, this has prompted a need to re-evaluate the role of the human crew and it has stimulated research into alternative approaches to the interface between man and machine that can exploit their complementary intellectual abilities.

4.3.3 Conceptual design

The idea at this stage is for the design team to consider alternative approaches to the physical crewstation configuration. Their experience of arranging an existing design, together with feedback from operational use provides a baseline with advantages and disadvantages that are known. The work is largely done without tools and is essentially a paper exercise, although it may help to create 3-dimensional visualisations of the alternatives using a computer-aided design facility.

The first step is to define the physical constraints of the enclosure. In a combat jet this is the space between the canopy and the pressure-bearing part of the structure; the floor, side panels and the strengthened front and rear bulkheads. If lucky, the team may have some influence on the placement of these bounding surfaces, but decisions are generally made by the airframe design team to meet overriding aerodynamic, stealth and structural criteria. The placement of the cockpit is rarely debated. It is squeezed behind the nose-mounted radar, above the nose-wheel and ahead of the engines. Intrusions, such as those needed to accommodate the stubby canard control surfaces in Eurofighter, are particularly important as they may have a strong influence on the preferred cockpit layout. As the pilot must have an adequate view of the airfield on approach to landing, it is necessary to define the angle of the downward view over the aircraft nose. The *design eye* location and the envelope of the canopy in the vicinity of this datum are therefore defined jointly by the crewstation and airframe teams. Similar considerations are applied in other types, and in aircraft with special requirements, such as observation helicopters, the placement of the transparencies and the design eye locations are dictated as much by the need for a wide visual envelope as the need for a compact airframe.

The position of seats in multi-crew combat types is mainly decided by the available width. Side-by-side seats, as in most transport types, gives each crewmember a comparable view forward and enables them to see what the other is doing, and this helps them to co-ordinate their actions and share the displays and controls along the cockpit mid-line. But it limits the view out to the remote side. Tandem seating is more usual, particularly in the narrow fuselages of modern combat types, although two-seat training variants of older combat aircraft, such as the Hunter and the Lightning in RAF service, had seats set side by side and suffered little loss in performance.

The choice of seat for fixed-wing combat types depends on the mechanism used to effect rapid emergency escape. Unless the team considers the adoption of a form of capsule escape system, or wishes to create a novel mechanism, the choice is limited to the ejection seats developed by specialist manufacturers such as Martin-Baker, Stencil and Zvezda, as discussed in Chapter 12. All of these put the crewmember in a conventional seated posture, and this is likely to remain the most practical way of allowing the pilot to interact with displays and controls.

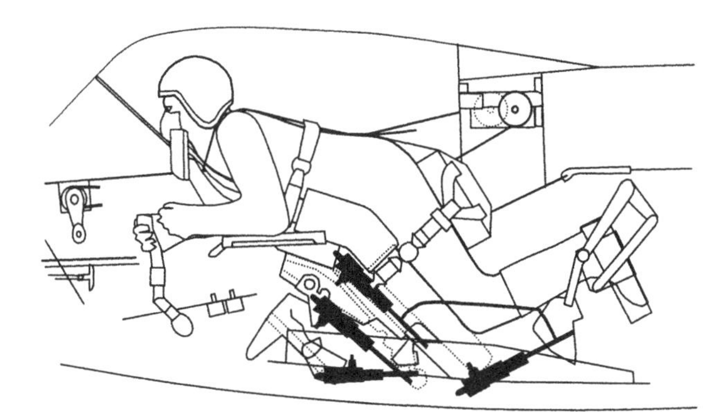

Figure 4.3 Experimental cockpit added to the nose of a Gloster Meteor to investigate the feasibility of prone accommodation (from Green N D C, *The physiological limitations of man in the high G environment*, in *Combat automation for airborne weapon systems; Man/machine interface trends and technologies*, AGARD-CP-520, 1993. Reprinted with permission from QinetiQ)

The departures from the conventional that have been tried have few practical advantages and many disadvantages. For instance, flight trials were conducted by the RAF Institute of Aviation Medicine at Farnborough[9] in the early 1950s to assess the benefit of having the pilot lie on a conformal torso support in a prone posture like a racing motorcyclist, as illustrated in Figure 4.3. The test pilots could look through a small window in the floor during take-off and landing, and the position was not considered unpleasant. However, the pilots could not look around easily or to the rear at all, and there was no enhancement to G-tolerance, only increased chest pain. The layout of the display and control panels was compromised, and even a greatly modified form of ejection seat would not have been practical.

Other postures, such as reclining to an almost supine attitude would only give a useful increase in G-tolerance if the head were laid back so that the vertical heart-to-eye distance was reduced to less than half the normal value. This made it difficult for the pilot to look forward and down into the cockpit, and the necessary head-rest impeded routine head movement. A seat that reclined automatically in proportion to the G-level could obviate these problems, but the articulating mechanism would add mass, complexity and cost. Again, although other types of aircraft have different driving requirements, there is no real advantage in departing from a fairly upright seated posture and fitting seats developed by the specialist manufacturers. All incorporate harness fixings, and some - intended mainly for rotary-wing aircraft - have energy absorbing mounts that can attenuate the jolt from a heavy landing.

The size and shape of the transparent area of the cockpit is a compromise between the ergonomic preference for the largest possible viewing envelope and the practicalities of arranging a structure that is lightweight, strong and which, in combat aircraft, has a low drag. The elongated half-bubble shape adopted for most conventional fast jets has usually been divided into a fixed forward windscreen and a pivoted or sliding rear canopy. Several types, notably the F-16 and the F-22, use a single piece of transparent plastic to avoid the arching framework where the two parts would otherwise join. In the F-16 this removes the visual obstruction of the arch, and in the F-22 it also avoids a discontinuity that could increase the frontal radar signature.

The practicalities of gaining access to the cockpit interior by the aircrew, and of both routine and rapid emergency egress, are largely matters of geometry. In combat aircraft the conventional approach relies on arranging for the canopy to open, close, latch and seal, for which a variety of pneumatic, hydraulic and electrical mechanisms have been successfully employed. Complications arise from the need to remove the canopy rapidly to allow a clear passage for the ejection seat and its occupant. As described in Chapter 12, it is also practical to shatter the canopy material, and this is generally quicker than the use of explosive lifting devices, especially at low airspeed where air blast forces give little help. Clearance is also needed within the cockpit so that the pilot's knees or feet do not contact the front panel, the side panels or the head-up display optical assembly when the pilot is propelled upwards during the initial phase of ejection. Other geometrical factors, such as the accessibility and visibility of the panels containing the displays and controls and the choice of a centre-mounted stick or a side-mounted stick can be examined at this stage. The later detailed geometric design will specify the precise location of the internal components and the adjustments needed to cater for the variations in the size of crewmembers.

The design team must also consider a large number of practical issues, such as the aircrew clothing, the supporting elements of the breathing, anti-G and NBC protection systems. Rotary-wing craft sometimes need armoured panels for ballistic protection and airbags to supplement the seat harnesses as body restraints during a heavy landing. The form of the transparencies can also have an appreciable effect on optical distortions, and these must be assessed. If constructed from multiple layers or coated with materials to reflect radar and selected lasers, the reflectivity of the transparency material must be considered across the electro-magnetic spectrum from UV to RF wavebands. A high reflectivity in the near infra-red could, for instance, be incompatible with night vision goggles or other systems that the pilot would use for night and poor weather operations. Other factors affected by the geometrical configuration, such as the accessibility of the equipment by the groundcrew for maintenance, also need early consideration.

Having delineated alternative configurations for the crewstation, it is necessary to select the approach that is, on balance, most satisfactory. As it is unfeasible to complete the processes of designing, building and testing each option, most teams use a pseudo-quantitative procedure such as "quality functional deployment" (QFD)[10], for comparing options that can in reality only be described qualitatively. The procedure consists of setting out the requirements, either as defined performance parameters or as multi-faceted criteria such as "affordability", "technological risk" and "mission effectiveness", or a mixture of both criterion types. Each of these is then given a weighting factor that indicates the relative value of the requirement to the customer. The bulk of the analysis is done by judging how well each of the options meets each of the requirements, and describing this by a numerical rating, say a number between one and ten. The results are collated into a matrix table. The most satisfactory option is strictly that with the highest sum of weighted ratings, but the process is rarely concluded without a series of iterations. The main value is in forcing the team to address the customer's needs and debate the relative merits of the alternatives. The matrix forms a concise record of the rationale for their decision.

4.3.4 Task analysis

The outcomes of the preparatory studies provide some understanding of the physical form of the crewstation and the way the crew should use the control/display interface to interact with the on-board systems. The object of the task analysis stage of crewstation development is to consider, especially during critical mission phases, the activities that crewmembers must perform and assess their likely ability to cope.

The tools the team can call upon are computer-based simulations that can be set up to represent the time course of the activities performed by man and machine. Some of the simulations are deterministic and others cycle through the activity sequence with small parameter variations to produce probabilistic predictions. All include a human operator model (HOM) that takes account of human sensory, cognitive and neuro-muscular abilities, and this is used within the simulation package to predict the outcome of the activities the operator undertakes and give some measure of the fluctuation of the workload he is asked to manage. All HOMs borrow ideas from one or more engineering disciplines, and it is the relevance of the ideas that largely govern their usefulness.

One quantitative mathematical approach uses the concepts and terminology of control theory to represent the human operator as a mechanism for adjusting the state of a disturbable system[11]. In the case of an aircraft, the system would encompass the aerodynamics, engines and flight control system, all of which could be represented by a series of interacting transformations that contribute to the time-dependent (vector) state variable. A sub-set of this is displayed and used by the human operator to perform a sequence of sensing, deciding and neuro-muscular operations that result in a control output, for instance an arm movement. The input and output processes are represented only in as much as they introduce error and delay. The core element is the "decider" that comprises a Kalman filter and predictor that jointly make the decisions and partially compensate for the limitations of the other stages. The Kalman filter is used to produce a best estimate of the state of the system, from which the predictor — essentially an internal simplified model of the system dynamics — estimates the future state of the controlled system and computes the arm movement to position the flight controls that will bring this to the desired state. Such models are mostly used to represent the human operator performing a continuous control task. Sophisticated variants have been developed to cater for the control of multiple degrees of freedom with serial attention, but they have not yet been used to represent the complex mixture of continuous control and discrete logical tasks that constitute the pilot's job.

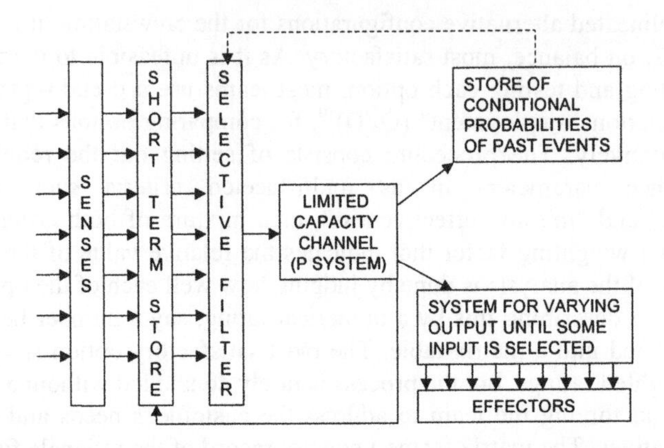

Figure 4.4 Broadbent's diagram of the information flow within the nervous system (Reprinted from Broadbent, D E, *Perception and Communication*, Pergamon Press, London. Copyright (1958), with permission from Elsevier)

The more common approach has been to consider the human as a set of information-processing resources. The prototype was devised by Broadbent[12] who considered information flow using the metrics devised for communication systems. The information contained in a series of (n) alternative stimuli each of which (i) occurred with some probability (p_i) was quantified by analogy with alternative symbols in a signal as:-

$$\text{Information content} = \sum_{i=0}^{i=n} - p_i . \log_2 (1/p_i) \quad \text{bits}$$

The central tenet was that the whole nervous system could be regarded as a single channel that could handle information within a limited capacity, specified in bits/sec, and multiple tasks could be done without interference providing the sum of the information to be handled was less than this limit. Figure 4.4 shows Broadbent's explanatory diagram in which the limited-capacity decision-making system was preceded and protected by a filter that took relevant information from a short-term buffer fed by the sensors, and the results of decisions were the selections of the necessary responses. The notion of filtering the input was later elaborated into the broader concept of categorising or "pigeon-holing". Stress and noise were considered to affect vigilance and attention by influencing the distractibility of the selection mechanism and the capacity of the channel. This theoretical scaffold stimulated numerous experiments and theoretical refinements, in particular to accommodate the salience and redundancy of the available information and predict the required frequency and duration for sampling multiple control dimensions[13]. However, the central notion of a single limited-capacity channel has largely been overtaken by theories that postulate resources that can operate in parallel[14].

The Model Human Processor (MHP) was put together by Card, Moran and Newell[15] to recast the current psychological theories about the broad mechanisms employed by the brain into a form that could be used quantitatively, for instance to predict the time taken to press a key in response to a set of lights or to transcribe text. Like the control theoretic and limited capacity models, the MHP was based on a sense-decide-act architecture, but it employed many ideas from computer engineering. It was composed of a "perceptual system" that transformed visual and auditory stimuli into a symbolically coded image that was used by the "cognitive system", together with information stored in long-term memory, to select responses that were carried out by the "motor system". Unlike a basic electronic computer with a central processor, these sub-systems operated as a set of parallel pipelines that both stored and processed information.

The MHP was like a construction kit containing a mass of psychophysical data accompanied by a diagram and a set of operating principles. The data were reduced to a common form in which each stage in the sequence of operations could be described by a code type, a storage capacity, a decay constant and a cycle time. Cycles had a duration of about 100msec. The operating principles contained the meat of the theory. The first principle stated that the cognitive sub-system performed processing in long-term memory as a discrete number of recognise-act cycles. Here the "recognise" element activated a section of long-term memory to become "working memory" that contained the instructions for carrying out the process, for instance comparing a perceptual code with an item retrieved from memory. The "act" element set the stage by pointing to the next cycle. All complex forms of organised behavior therefore resulted from associatively linked sets of recognise-act cycles, some of which were associated with, and triggered, motor programs. These linked sets constituted "productions", and these also resided in long-term memory. Other principles were succinct statements of ergonomic findings. These included Fitts' Law (relating the time to move the hand to a target of a given size at a given distance), the Power Law of Practice (relating the reduction in time to perform a task with repetition of the task) and the Information Theory Principle (the time needed to make a decision increasing with the uncertainty, i.e. the amount of information to be processed).

The main purpose of the MHP was to synthesise theory and data into a sufficiently good approximation to describe mental activity on a time-scale of about 50msec, and it is used mainly in psychological investigations to dissect phenomena to this fineness of detail. Although elaboration would be possible, for instance to handle multi-tasking, other developers have been more pragmatic in devising models that describe the behaviour of an operator using a complex system to perform a set of interactive tasks within a changeable environment. One such is the Integrated Performance Modelling Environment[16] (IPME), which is now a commercially supported package.

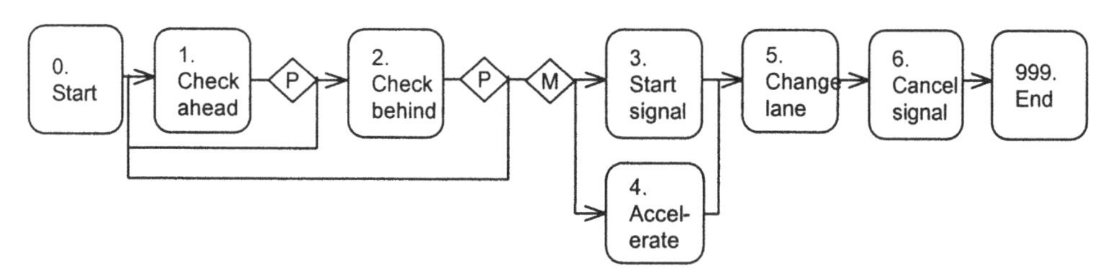

Figure 4.5 Example of a simple task sequence performed by a car driver

To use IPME, the analyst starts by having experts consider the actions and decisions performed by the operator and the system. These are represented as a linked flow diagram, as illustrated in Figure 4.5. The representation can include crucial events and interactions with the environment, and it can be done to an arbitrary level of detail, if necessary by creating a hierarchical structure with nested and linked elements. Whatever the level of detail chosen, the elements are regarded as tasks characterised by a set of parameters: the entity performing the task (the operator or the system), the time taken, the probability of success, the consequence of failure (for instance, repeating the task or jumping to an alternative task), the nature of the task, and task demand ratings. The analyst can select from a library of micro-models, including the MHP, data and conditional equations describing the time needed to do the task. If a task is performed by the operator, the "nature" of the task is classified using a limited pre-defined taxonomy that allows, for instance, overt physical activities like reaching for a switch to be distinguished from covert mental activities such as deciding the severity of a threat. The task demand ratings are based on a "workload" model, for instance the Prediction of Operator Performance[16] (POP) model for which the analyst makes separate judgements about the operator's input, central and output demand and the time pressure.

Having defined what sequence of tasks must be performed it is possible to compute the total time required. This is a rarely a single value because task failures generate recursive branching paths, each with an associated probability. The main outcomes are therefore the time course of the demands placed on the operator and the likelihood of successfully completing the whole sequence within a time span. IPME allows three classes of conditional factor to have a bearing on task duration and error probability. The first are environmental factors, such as time of day, temperature, noise level and humidity, the second are properties of the operator, such as height, weight and intelligence. The third are transient variables, like alertness and the time since waking. The environmental factors can be regarded as stressors that influence the operator state through pre-defined quantitative relationships. In a similar way, specific Performance Shaping Functions define the changes introduced to task duration and error rate by changes of operator state. Although the causal chains span many links, the authors have defined some shaping functions by suitably interpreting experimental data. Also, all tasks of the same nature are affected by the state of the operator to the same degree. This avoids having to define shaping functions for all combinations of state and task.

Human operator models represent the human ability to cope with several simultaneous tasks in different ways. In IPME, the analyst represents concurrent tasks as parallel elements, and the aggregated outcome and workload are calculated from the outcome and the workload associated with each task. Other programmes use different techniques. Two well known programmes, Workload Index (Windex) and Wincrew[17], also require the analyst to decompose the job of the operator into a network of tasks, but they deal with multi-tasking by representing the operator's attentional limitations using a "conflict matrix" based on Wickens' Multiple Resource Theory. In these programmes the analyst enters values between 0 and 1 into the matrix to specify the degree of conflict between each resource. When the simulation is run, a workload value is computed from the pre-defined demands of each task combined with the paired task conflict values in the conflict matrix. Wincrew allows the analyst to specify what happens when the workload value exceeds a threshold. He can for instance specify that the lower priority task is allocated to another crewmember or to automation, or that the individual accommodates the increased workload but incurs time and accuracy penalties.

In the same way that behaviour can be decomposed into a hierarchy of progressively refined actions, the outcome that an operator is striving toward can be decomposed into a set of interdependent subsidiary objectives. The Man-Machine Integrated Design and Analysis System (MIDAS), devised by scientists at NASA Ames Research Centre[18] and shown in Figure 4.6, includes such a mechanism for handling goals. These take the form of "Reactive Action Packets" (RAPs) that are instantiated by a high-level plan and stored in an "agenda structure". The idea is similar to the production rules of the MHP.

The object of the task analysis is to judge whether the crew can be expected to perform the tasks by executing the sense-decide-act routines imposed by the design of the system at a pace dictated either by the plan for the mission or by external events. The HOM should represent the relevant human characteristics with reasonable accuracy, but because so little is known about the workings of small elements and even less is known about the overall architecture, no HOM represents the way the brain actually processes information. Instead, the available tools incorporate models that perform like a human operator in as much as the analyst can imagine how the operator would behave in the hypothesised working conditions. Although the analyst can undoubtedly devise a plausible task network, it is particularly difficult to anticipate the subtleties that influence how an operator would develop a strategy for using novel systems. The analyst must also apply arbitrary criteria, for instance in Wincrew to set the threshold at which the operator unloads tasks. The tools also treat a generalised operator rather than a set of individuals with differing abilities and preferred ways of exploiting idiosyncrasies. These variations are critical in attempting to predict errors and instances of excessive workload.

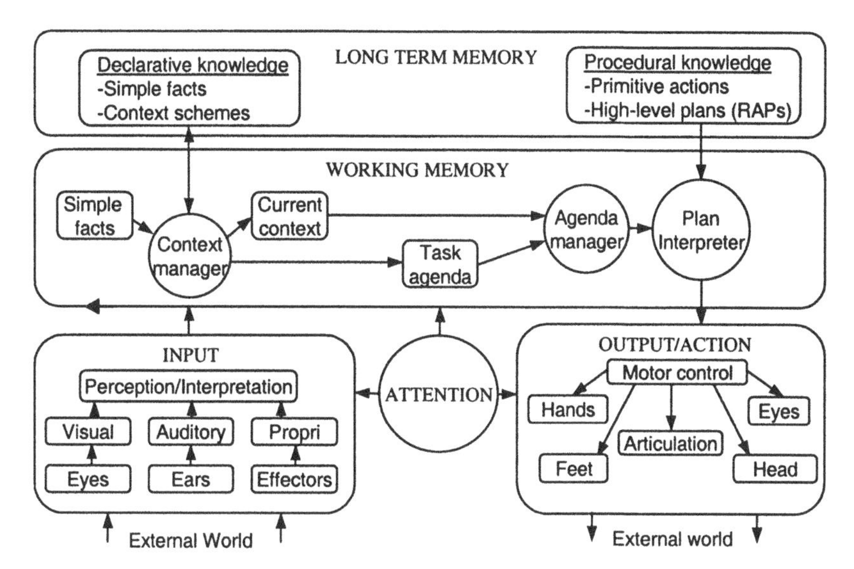

Figure 4.6 Overview of the Human Performance Model used by the MIDAS simulation program

All of these simplifications make it difficult for the developers to validate that their tool gives a reasonable description of the way the human operator would behave and perform in a future scenario. Users generally take the view that although a tool may not give a reliable prediction of the outcome, the analysis is nevertheless useful for exploring and comparing alternative system designs, for instance between single-crew and twin-crew aircraft. These can also be related to a baseline analysis of the mission segment performed in a current aircraft type.

4.4 Crewstation development and testing

By this stage the design team have considered the options and chosen the broad configuration of the crewstation. The next steps of the top-down approach are a series of exercises that consider the elements in more detail.

4.4.1 System design

A typical combat aircraft can be regarded as a set of electronic and electro-mechanical systems that are controlled by computers and interconnected by data-buses. The avionics team will have studied alternative ways of transferring the large amount of data between the systems, and they will have chosen an overall architecture and set the standards and protocols that characterise the aircraft. The crewstation team must devise the detailed architecture of the systems interconnected by the crewstation data-bus and design the layout the electronic units that fit within the cockpit, including the routes for cables and the positions of connectors. The main concerns are to ensure that all of the facilities are provided, that systems can operate reliably without interference and that back-up units can be called upon if safety-critical elements fail. Some thought is also given to the maintenance of the system, for instance the ease of interchanging electronic modules.

An example of the architecture for the display suite is shown in Chapter 6 as Figure 6.7. This uses a pair of computers to generate images and manipulate sensor imagery for presentation on five electronic display devices; three multi-purpose head-down displays, a head-up display and a device that superimposes the "horizontal situation" onto a moving map. The pilot can select back-up sets of information on each display device if one image generator and several display devices fail. The suite also includes a prominent attention-getting warning light, to cue the pilot to the appropriate display, and a set of air-driven and electro-mechanical instruments (powered by an independent battery) that provide essential "get you home" information should the whole electronic suite fail. Combat aircraft designed more recently use a similarly redundant architecture, adding a binocular helmet-mounted display with two independently generated and powered channels, but omit the electro-mechanical instruments. The stick, throttles and pedal levers — no longer linked mechanically to the ailerons, elevators and fin — are now inceptor devices that generate electrical signals for input into the flight control computer. To preserve their safety-critical integrity, each inceptor contains as many transducers as there are multiplex channels in the FCS.

The principal tools available to the engineering team are a rigorous consideration of the ways that individual units can produce an erroneous output, combined with an appraisal of the probability of such unwanted states. Detailed failure mode analysis is then carried out to assess the consequences of each failure taken in isolation. Simultaneous, independent failures are also considered if sufficiently likely. Although it seems logical to consider the errors and omissions attributable to the operator, for instance the probability that he might misread a displayed number or press the wrong switch, it is difficult to predict the consequences and whether he will notice and rectify the mistake. In general, the team only considers system failure states, and endeavours to devise the simplest and cheapest architecture that can give adequate predicted reliability.

4.4.2 Mechanical design

The object of this development phase is to create a detailed mechanical design for the cockpit out of the ideas suggested by the conceptual studies in section 4.3.3 and the preferred architecture of the display and control system devised in section 4.4. The process includes choosing the component systems and defining the internal structures and materials. The outcome should be in the form of computer-aided design (CAD) data files that can be offered to the team designing the airframe so that manufacturing practicalities can be discussed and resolved.

The main elements of the structural envelope, including the design eye position, the transparent areas and the clearance passage for the ejection seat will already be defined. The chief geometrical problem is to accommodate the display and control boxes and ensure that their active surfaces can be seen and reached by the aircrew. The problem therefore centres on the optimal positions for the devices and the range of positional adjustments needed for satisfactory placement of the eyes, hands and feet of individuals with widely differing body dimensions.

Traditionally the draughtsman has used a set of articulating 2-dimensional manikins, usually at 1:5 scale, to represent the side-view dimensions of small (5%ile), median (50%ile) and large (95%ile) men. These could be laid on the drawing so that the buttocks and spine were set into the seat in a harnessed posture and the head was made erect. The eyes were placed in the design eye position, or close to this position and certainly on or above the line defining the required view over the aircraft nose. In later aircraft the eye position also defined the optical axis of the head-up display. The limbs could then be articulated to establish the reach envelopes of the hands and feet. The throttle, stick and switch panels could be fixed where they would be comfortably accessible to crewmen of all sizes, and the pedals could be sited where they allowed easy movement and needed minimum fore-aft adjustment. Inevitably, some compromise was needed to get a balance between over-stretching small individuals and cramping large individuals. The compromise would be eased in the flight deck of a transport type where the seats could be adjusted both fore-aft and vertically. However, in the confines of combat types, where the ejection seat would be fixed to the cockpit floor and the rear bulkhead, and only the height of the seat could be adjustable, an effective compromise was often impossible. Several types, notably the Harrier and the Hawk, were not arranged to accommodate aircrew greater than 70%ile in stature.

The problems are not eased in current combat aircraft by the need to accommodate females, who have shrunk the dimensions of the 5%ile aircrew. The modern approach is to use computer-aiding tools, such as JACK and COMBIMAN[19], that allow the designer to place a manipulable, virtual 3-dimensional human into a manipulable, virtual 3-dimensional cockpit structure. The virtual human, with over sixty body segments and a hundred degrees of postural freedom, is biomechanically more representative than any manikin, and it can be set to have limb dimensions and ranges of articulation taken from several anthropometric surveys. Constraint can be specified, such as imposed by the seat, clothing, tight harness and the need to keep the feet on the pedals, allowing the model to be used for evaluating head mobility and the lines of sight to display devices as well as reaching for controls. The model can also be put through more complex manoeuvres that enable the designer to check that a man of a given size can enter and leave the cockpit, and that the changes in sitting posture during ejection do not bring the head or limbs into contact with fixed structures.

The rest of the mechanical design consists of settling a lot of details. These include the form and extent of the glare shield, the supporting structures for the panels that carry the displays and switches, and the arrangement of the ducts and nozzles for heating, de-misting and cabin conditioning. With the increase of man-mounted equipment it is also necessary to consider the detailed arrangement of the pneumatic, electrical and liquid (for body cooling) connections between the pilot and the cockpit, and decide which are routed through the personal equipment connector (PEC) on the ejection seat. Throughout it is necessary to arrange a modular structure so that items can be accessed for maintenance and testing, and provide adjustments, for instance so that the head-up display and the helmet position sensor can be set in line (harmonised) with the airframe axes. The geometry of the installation can affect electro-magnetic interactions between systems. For instance, the transmitter of an electro-magnetic helmet tracking system must be sited well clear of conductive metal, and the receivers of optical helmet tracking systems must be sited where they can view the helmet but not pick up reflections from the canopy or be flooded with sunlight.

4.4.3 Ambient lighting

The high level of light within the cockpit, particularly the cockpit of a combat aircraft (as described in Chapter 6), has considerable impact on the ability of the crew to do their jobs. The ambient level governs the adaptation state of their eyes, while the strength of glare reflected and scattered from displays and from the canopy can mask displayed information and the outside view. The chance of direct specular reflection of the sun from a glass surface can be avoided by inclining the surface normal below the canopy sill, and most manufacturers take pains to apply anti-glare treatments to displays. However, secondary scattering from within the cockpit, particularly from headgear and clothing, can produce strong glare, and the cockpit design team usually model the effects using computer ray-tracing tools to calculate what aircrew would see for a wide range of sun angles and skylight intensities. Many teams also construct a rig that reproduces the salient light conditions[22] and this allows them to confirm the modelling predictions. The lighting rig is particularly useful for checking the usability of displays at night, and for assessing the effectiveness of the filters that, as described in Chapter 6, are intended to make the internal cockpit lighting compatible with helmet-mounted night vision systems.

4.4.4 Man-vehicle interface design

Having established the roles of the crewmembers, studied what they must do to interact with on-board systems and defined the architecture and physical layout of the suite of display and control devices, it is necessary to decide where and how information should be presented. The job of defining display modes and formats is primarily one of programming the computer suite to present appropriate pages on the head-down, head-up and helmet-mounted displays and defining the appearance and mechanisation of the symbols on each of the pages. It includes decisions about the way input devices such as grip-top switches and keys are used to manage the presentations, the automatic changes of pages triggered by crucial aircraft states, the design of aural messages and warnings, the use of short-cut techniques and the availability of de-cluttered (simplified) formats. If a novel control facility such as a speech recogniser is introduced, decisions must be made about syntax, vocabulary and parallel operation of manual keys. Consideration must also be given to special conditions such as operating at night, in poor visibility and under laser threat, and to degraded states such as failure of one or more display and control systems. The last includes the provision of minimal "get you home" information.

As illustrated schematically by Figure 4.2, combat aircraft are now controlled directly by computers so the interface between the pilot and the aircraft can in principle be as flexible as that between an operator and a computer. The crewstation design team can therefore use conventions such as moveable windows, pull-down menus, soft-keys and selective annotation that were pioneered by computer system designers to facilitate human-computer interaction. The design team must however make a considered judgement about the mind-set of the candidate operators, because whatever operating system they implement will dictate whether the crew regard the whole aircraft as helpful or awkward. Most crew will be familiar with computer-originated conventions and will willingly adopt them in a new aircraft, but they will be wary of too many levels in menu structures and also want to conserve conventions from previous aircraft types.

This stage in the development process tends to be organised as a series of iterative cycles involving design suggestions that are implemented in software, evaluated by the team, refined and then re-evaluated. Computer-aided design tools, such as VAPS[20], can be used initially to visualise and explore a wide range of detailed possibilities for individual pages and formats, but most of the work is done using a simulation rig. This represents the physical layout of the controls and displays, allows the engineers to manipulate the display content and the effects of the controls. It also supplies a representation of the aircraft, the on-board systems and an outside world containing targets and adversaries. Eventually this simulation can be exercised in real-time and evaluated by test pilots.

The business of conducting an evaluation is rarely simple, mainly because much depends upon the chosen measure of effectiveness. Unlike physical work, which can usually be assessed in terms of whatever is produced, the aircrew job mainly requires covert, mental effort. There is an underlying notion that a crewman can cope with decision-making events up to a particular pace but must apply unsustainable effort, relax accuracy or postpone some tasks to handle a greater pace, all the while feeling stressed and over-stretched. Conversely, where decision-making events are infrequent, vigilance and performance also suffer. Much effort has been aimed at devising a quantitative scale of mental workload that spans the gamut from under-loaded boredom to over-loaded frenzy[21].

Subjective workload assessment techniques are based on the idea of asking operators to rate how busy they feel when performing the task. The prototype in aviation research and testing was the Handling Qualities Rating (HQR) scale shown in Figure 4.9. Workload ratings such as the Task Load Index (TLX) and the Defence Research Agency Workload Scale (DRAWS), devised by NASA and by DRA respectively, are superficially similar in structure to the HQR. For instance, the latter asks the operator to assess four aspects (input demand, central processing demand, output demand and time pressure) on a scale running from zero to 100, and to use ratings greater than 100 to indicate overload. The use of continuous ratio scales is intended to allow arithmetic manipulation, for instance within the model of the operator in the IPME simulation described above. The reliability of such techniques depends on the assessor, who must have a remarkable and consistent faculty for self-analysis.

Three classes of objective assessment technique have been devised. The first includes those that measure performance directly, for instance the accuracy with which the pilot controls speed, height and heading or aims a weapon. Although such measures are of prime concern it is rarely possible to quantify all tasks in this way, particularly the covert information-assimilation and decision-making for countering threats and managing the mission. Also, direct measures of performance rarely discriminate between subtle differences in the man-machine interface because the operator usually strives to attain the required performance and differences in compensatory effort may only be apparent subjectively.

The second class of objective methods are intended to overcome these methodological problems by loading the operator to a maximum sustainable level of effort using an additional easily assessed probe or secondary task, such as tracking a continuously moving pointer, pressing a switch at regular intervals to cancel a signal light, or adding three to a number heard over the intercom. The idea is to choose a probe task that cannot be done at the same time as the primary tasks, so that changes in the "spare mental capacity" can be revealed by changes to the secondary task outcome. However, this approach is not without methodological difficulties. Most operators tend to evolve a time-sharing strategy that accommodates the probe, and this perturbs the original way of working and, perversely, gives the probe the attention needed for consistent performance. The most satisfactory solution, devised by Jex *et al*[22], uses as the probe a standardised, adaptive, compensatory tracking task that has dynamics akin to balancing a broom vertically on an upturned hand. In this implementation the time constant that governs the rate of fall of the broom at a given angle from the vertical is automatically adjusted so that the broom can be kept in balance. The inverse of the adaptive time constant λ_x (in radians.s^{-1}) can be divided by λ_C , the value that the individual can attain when giving full attention to the probe task, to get λ_x/λ_C ,the normalised "excess control capacity".

The third class of objective techniques involves assessment of physiological variables such as blink rate, heart rate, blood pressure, heart rate variation (arrythmia), sweat rate, muscle tension and the concentration of adrenal hormone secretions in the blood and urine. These have been shown to correlate with the other forms of workload assessment if calibrated carefully for each individual. Some variables respond quickly, within seconds, but many are only useful for characterising a phase of a job lasting at least several minutes. Because an increase to these measures is generally much more rapid than a decrease, and sometimes the increase occurs in anticipation of the actual need for mental effort, it is more likely that a change reflects systemic physiological arousal than transient mental effort.

The underlying assumption behind workload assessment is a causal chain whereby job demands produce stress that induce arousal that stimulates effort to create performance. Except for direct measures of performance, most of techniques that purport to assess workload can with equal validity be interpreted to signify any of the other concepts. Although much of the research has been aimed at finding a single way to measure mental workload that is sensitive, consistent and easily administered, it seems reasonable to conclude that the concept is multi-dimensional and multi-faceted. It would be unlikely for the manifestations of workload to be captured by one unique, representative measure[23]. A metallurgical analogy may be useful. An engineer would make a choice between materials for the manufacture of a component by considering qualities such as electrical conductivity, density, thermal conductivity, expansion co-efficient and strength. A synthetic concept, for instance "metalness", that aggregates these qualities would have little value.

Recently "situation awareness" (SA) criteria have tended to supplant workload criteria for evaluating alternative ways of presenting information and as goals for designing a cockpit. The idea is that the engineer should organise the man-machine interface to maximise the SA of the operator; SA having been defined[24] as "the perception of the elements in the environment within a volume of time and space, the comprehension of their meaning, and the projection of their status in the near future". This has been put more succinctly by a pilot leading a cockpit design team[25] as "knowing what's going on so you can figure out what to do". From the perspective of a combat pilot, it is suggested that the idea can be regarded as encompassing three levels; "global", "short-term" and "ownship". The first includes the pilot's longer-term knowledge about the tactical conditions in relation to the mission plan, with a scale of hours and several hundred miles. The second is the knowledge of the tactical circumstances within the likely range of defensive weapons, say several tens of miles. The third concerns the immediate flight path of the aircraft and the state of on-board systems.

The idea seems to pinpoint the crucial issue; the pilot's knowledge of dynamically changing conditions. One technique is to measure, or have an expert assess, the operator's performance. If satisfactory it is assumed that the operator is aware of relevant happenings. In man-in-the-loop simulation studies it is possible to freeze the simulation periodically and ask the operator to answer crucial questions. Another technique involves asking the operator after the test to recall his uncertainty about his knowledge of events. Yet another requires the operator to make a regular self-assessment of his awareness, and when prompted call out a set of ratings. However each form of assessment raises methodological problems. Performance measurement offers a very indirect inference about an operator's awareness, the accurate recollection of knowledge during rapidly changing circumstances is difficult, and the time-freezing interruptions can give the operator the respite to collect his thoughts. The self-assessment technique is least satisfactory because the operator must accommodate a regular disturbance. It is also illogical, because he cannot know what he does not know.

4.5 Prototype testing and in-service modification

In general, the final phase of the detailed interface development coincides with the maturation of the simulator into a complete avionics rig. In this, the hardware and software for the prototype crewstation act as the hub of ground-based facility in which all of the utility, mission and weapon systems are connected through their appropriate data transfer buses. This rig is the principal tool for integrating the whole avionics system, assessing data transfer loads and verifying software. Once the design is frozen it is used to test flightworthy components and systems and support the flight tests of the prototype aircraft. The engineering team makes use of the facility thereafter to test software modifications and generally support the aircraft in service.

Thorough flight tests are conducted at the behest of the civil authorities to assess the airworthiness of the aircraft type and identify the safe flight envelope. Although much of the testing can be reported in terms of the objective performance and stability of the aircraft, the interaction between the required manoeuvre, the response of the whole aircraft, the atmospheric conditions and myriad details within the crewstation are only apparent to the aircrew subjectively. In consequence, it is now usual for assessing aircrew to use a standard rating scheme to express opinions about detailed arrangements of the crewstation and other aspects of the aircraft that affect their ability to do their job.

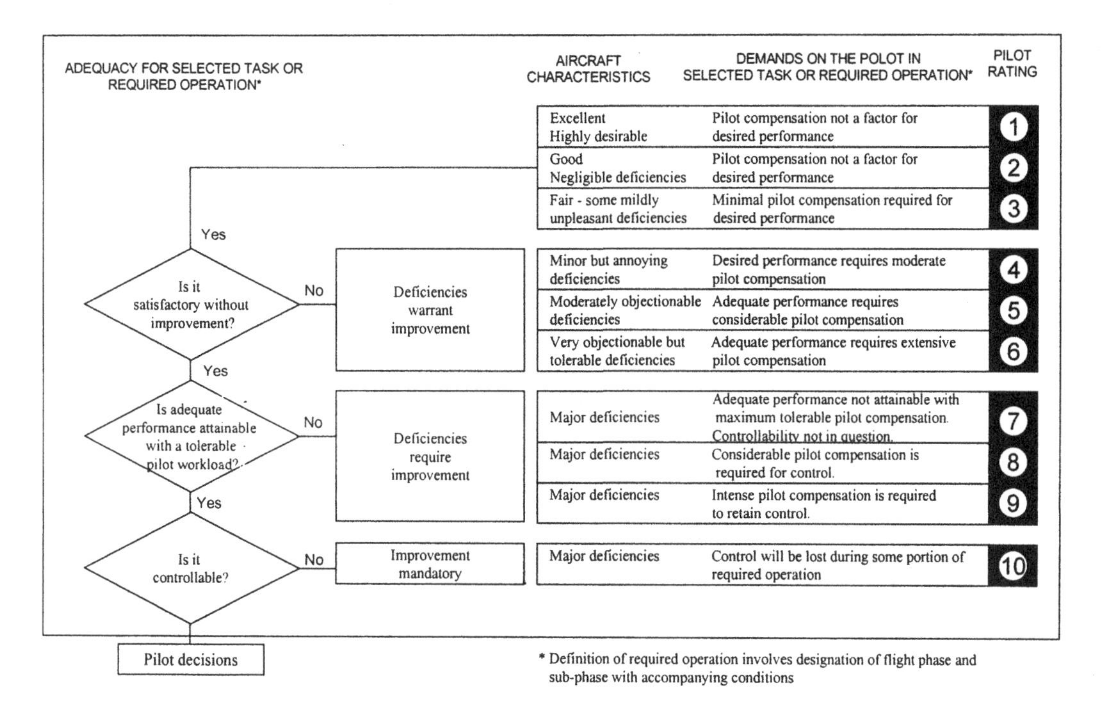

Figure 4.7 Handling Qualities Rating (HQR) Scale (Reprinted from Cooper G E and Harper R P, *The use of pilot ratings in the evaluation of aircraft handling qualities*, NASA TM D-5133, 1969)

In the UK the customer usually specifies the use of the Handling Qualities Rating (HQR) scale, which was originally devised as a simple, pragmatic way for pilots to state their judgements about the controllability of fixed-wing aircraft[26]. After conducting a defined flying manoeuvre, the pilot makes a series of forced-choice decisions, as shown in Figure 4.7, that lead to a rating between 1 and 10. A rating at the low end of the scale signifies that he can achieve good performance with little compensatory intervention, while one at the high end signifies that the aircraft cannot be controlled adequately even with full attention.

In recognition of the complex behaviour of rotorcraft, even those with electronic stability augmentation and full flight control systems, the military rotorcraft community is developing ADS-33E, an Aeronautical Design Standard for Handling Qualities[27]. The work has been led within NATO by the US Army Aviation Engineering Directorate, largely in preparation for the recently cancelled RAH-66 Comanche attack helicopter programme, but it is intended to apply also to scout, utility and cargo types. Here, the ten ratings of the HQR scale are reduced to three Levels in which "Level 1" signifies that the aircraft is satisfactory, "Level 2" that the aircraft is adequate for accomplishing the mission but with increased pilot workload, and "Level 3" that the aircraft can be controlled safely but only with excessive pilot workload. These equate to HQRs between 1 and 3.5, between 3.5 and 6.5, and between 6.5 and 9 respectively. The

Standard considers that handling qualities depend upon the following factors during a specific manoeuvre:-

a) the vehicle agility and performance
b) the state of the aircraft relative to the boundaries of the flight envelope
c) the vehicle configuration and loading
d) wind and turbulence
e) underslung loads
f) whether in a failure state
g) the relationships between movements of the pilot's control levers and the aircraft's dynamic response in all six degrees of motion
h) the relationships between the movements of the control levers and the forces exerted by the pilot
i) the required accuracy of the manoeuvre
j) the required aggressiveness of the manoeuvre
k) the cues available to the pilot from his view of the external scene
l) displayed information
m) any additional attentional demands on the pilot.

Factors (a) to (h) are specified by prescriptive relationships between objective characteristics and HQ Levels that can be used to optimise the helicopter during the early iterative stages of a development programme. Factors (i) to (m) can however only be taken into account during later stages of the development programme when HQ Levels are assigned by pilots conducting simulation and flight tests. To enable consistent pilot-in-the-loop evaluations, prescriptions are given for conducting twenty test manoeuvres, called Mission Task Elements (MTE), that include a hover, pirouette, slalom, side-step and landing. All are done in a Good Visual Environment (GVE), and sub-sets are done in Instrument Meteorological Conditions (IMC) and in Degraded Visual Environments (DVE), typically at night using night vision goggles.

The objective, course layout, required flight path and the criteria for desired and adequate performance are prescribed for each MTE. For instance, the pirouette is intended to test the ability of the pilot to accomplish precise, simultaneous control in the pitch, roll, yaw and heave axes in a moderate wind that is continuously varying with the aircraft heading. In the DVE condition it also provides a check of display symbology for such a multiple axis manoeuvre. Starting from a stable hover at about 10ft altitude, the pilot is required to translate sideways around the 100ft radius circle at a lateral groundspeed of 8kn (desired) or 6kn (adequate) while keeping the nose pointed at the centre. On reaching the start point, the pilot must stop the manoeuvre and settle quickly in a stable hover. The manoeuvre is then done in the opposite direction and finishes with the aircraft in a stable hover back at the start position. The suggested layout of the pirouette course, shown in Figure 4.8, consists of markings on the ground that clearly denote the allowable excursions from the circular pathway that define desired and adequate performance. A central post may also be provided as a vertical cue.

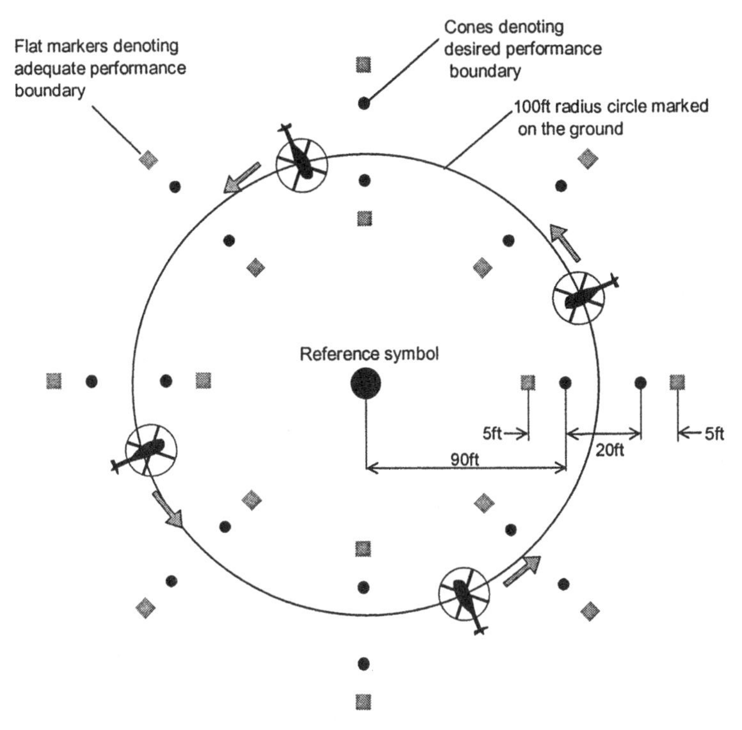

Figure 4.8 Suggested course for a pirouette manoeuvre (adapted from Hoh R H, Mitchell D G, Aponso B L , Key D L and Blanken C L, *Handling qualities requirements for military rotorcraft, Aeronautical Design Standard ADS-33E*, US Army Aviation and Missile Command Aviation Engineering Directorate, Alabama, 2000)

The effects on handling qualities of cues available from the view of the external world, and from cues provided by display devices, are incorporated in the scheme by determining the Usable Cue Environment (UCE) for each flight test. To do this, at least three pilots assess the "stabilisation effectiveness" of pitch, roll and yaw attitude cues, and lateral-longitudinal and vertical translation rate cues, in terms of a Visual Cue Rating (VCR). Here, 1 is "Good" (can make aggressive and precise corrections with confidence and precision is good), 3 is "Fair" (can make limited corrections with confidence and precision is only fair) and 5 is "Poor" (only small and gentle corrections are possible, and consistent precision is not attainable).

The data from all the pilots are pooled for each MTE to get average VCRs for pitch, roll and yaw attitude, and vertical and horizontal translational rate. Figure 4.9 shows how the worst (numerically highest) of the three resulting attitude VCRs and the two resulting translational rate VCRs are combined to determine the UCE for the task element. For instance a UCE of 2 is assigned if the worst VCR for translational rate cues is 4 and the worst VCR for attitude cues is 3. The reckoning is important because the Specification prescribes the Response Type (the vehicle's dynamic response to control movements) needed to gain a particular HQ Level when executing a particular manoeuvre. For instance, when hovering, Level 1 can be gained in UCE1 using rate control, which is the least artificially stabilised form of control coupling. However, attaining Level 1 in the same manoeuvre in UCE2 requires a combination of RCDH (yaw Rate Command, direction Heading Hold) from the cyclic lever and RCHH (vertical Rate Command, Height Hold) from the collective lever.

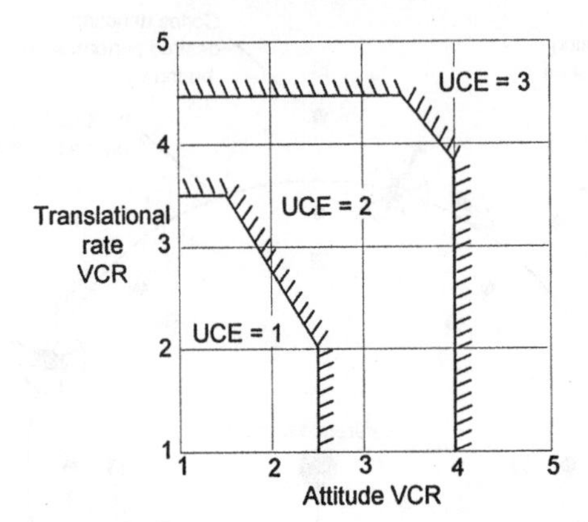

Figure 4.9 Usable Cue Environment (UCE) is determined by the Translational Rate Visual Cue Rating and the Attitude Visual Cue Rating (from Hoh R H, Mitchell D G, Aponso B L , Key D L and Blanken C L, *Handling qualities requirements for military rotorcraft, Aeronautical Design Standard ADS-33E*, US Army Aviation and Missile Command Aviation Engineering Directorate, Alabama, 2000)

The overall objective of HQR and ADS-33E is to ensure that fixed-wing and rotary-wing aircraft are regarded as satisfactory by the aircrew who will operate the type when it enters service. The latter is likely to join the established use of HQRs and become a significant element of the training given at the national test pilot training schools at Istres in France, Boscombe Down in the UK and at Edwards Airforce Base and Patuxent River Naval Air Station in the USA.

The assumption underlying the use of HQRs is that the fundamental issue - the acceptability of the characteristics of a vehicle - can only be defined by test pilots about the ease with which they can accomplish flight manoeuvres. However, such judgements are not simple. One uncertainty arises from the ambiguities associated with terms used in the HQR questionnaire, such as "workload" and "pilot compensation". Each pilot must come to his own understanding as to what is meant, and then consider what is tolerable, for what time period, by whom and with what degree of familiarity with the aircraft. The pilot knows that whatever he decides must be consistent with his previous judgements and with those of other test pilots. Although each rating is meant to be independent, there is a considerable incentive to conform because a judgement is eliminated if it differs by more than a few values from the ratings of other pilots*. Many flight tests allow the pilot to concentrate on flying the prescribed manoeuvre and rarely involve concurrent mission-related tasks such as navigating and maintaining tactical awareness. The pilot must therefore imagine where he must look, what decisions he must make and what controls he must operate to meet these concurrent demands; a further source of inconsistency. The pilot may also consider it to be semantically perverse, especially in a vehicle with a full-authority flight control system, to express a judgement about the acceptability of an automated countermeasures system in terms of vehicle handling qualities.

ADS-33E is also likely to cause debate. It is recognised that being able to see what is happening and having a stable, responsive aircraft are both important, especially for demanding phases of a mission such as landing a helicopter in adverse weather on the rolling, pitching and

* A quibble: Ratings are commonly manipulated, for instance to obtain an average. This is inappropriate for mere ordinal numbers that denote qualities rather than quantities.

heaving deck of a ship. However, the scheme for assessing the UCE seems arbitrary, and users face practical conundrums and must make very fine judgements. For instance, it is not clear whether the cones and ground markers should be visible in the tests. If visible, the pilot must declare a UCE of 1; if not visible, he lacks references for assessing his performance. He must also distinguish between the peripheral sensations of changing attitude and changing position, and between confidence gained from available visual cues and confidence gained through familiarity with the test course and trust in the accompanying safety pilot. It would certainly be wise to have an independent way of assessing performance, for instance from position measurements or from the judgement of the safety pilot. The allusiveness and inappropriateness of handling quality ratings would also be removed if the HQR questionnaire were used specifically to enable the test pilots to express judgements about the stability and responsiveness of the vehicle when flight-path control is not limited by what they can see. Other questionnaires could be devised to enable them to express equally specific judgements about the other aspects of the crewstation that affect their ability to perform their job.

4.6 End note

Table 4.3 Summary of the techniques employed during crewstation development

Development activity	Tools	Input	Output
Requirements capture	MIL-STDs	URD	SRD
Mission modelling	Microsaint	Mission definition	Task time-lines
Function analysis	Automation knowledge	Avionics definition	Control/display modes
Conceptual design	CAD, QFD	Physical requirements	Optimum crewstation configuration
Task analysis	IPME, Windex, Wincrew & MIDAS	Task time-lines	Predicted performance
System architecture	Failure mode analysis	Crewstation architectures	Detailed electronic architecture Reliability estimates
Mechanical design	CAD JACK, COMBIMAN	Preferred crewstation configuration	Detailed mechanical design
Mock-up	CAD	Detailed mechanical design	Acceptability of layout
Rapid prototyping	VAPS	Control/display modes	Detailed display format designs
Lighting evaluation	CAD (ray tracing) Lighting rig	Detailed mechanical design	Refinements to layout & equipment specifications
Simulation	Workload metrics SA metrics	Mission phase definitions	Workload & SA ratings
Flight test & Qualification	HQR(fixed wing) ADS-33E (rotary wing)	Mission phase definitions	Handling qualities ratings Airworthiness certification

The techniques used for the successive stages of the development of a crewstation are summarised in Table 4.2. The tools available for postulating and evaluating mechanical, electronic and optical issues are understandably more mature than those that must handle the complexity and variability of the human component. Fortunately, in a practical sense this does not matter. The design team proceeds much like a research and development team, but with the objective of making a satisfactory product rather than gathering knowledge of how to do so. In pursuing this aim pragmatically, the cockpit design team is aware that all aspects of their creation are unlikely to please everyone. Lurking below all this intellectual refinement is the assumption, which depends on a fine line between exploiting and abusing human adaptability, that operational aircrew will have the self-preserving motivation to work out ways to cope with minor design deficiencies.

To make more fundamental improvements the development team must be aware of the original rationale for design decisions and understand why these have become inappropriate. If the preferred fix is not obvious, in order to evaluate alternatives it may be necessary to revise the earlier off-line modelling to take account of information gained later from the simulation studies, prototype evaluations and in-service flying. What the human operator prediction techniques need most is a detailed understanding of the ways that pilots actually make use of the interface, rather than the ways imagined by the designers, and in particular how pilots split their attention between competing tasks. Perhaps the most useful output from the pilot-in-the-loop studies would be recordings of the speech and overt eye, head, hand, finger and leg movements of the pilots that allow analysts to infer these behavioural strategies.

References

1. *Human engineering program processes and procedures*, MIL-HDBK-46855A and *Human engineering design criteria for military systems equipment and facilities*, MIL-STD-1472
2. Booher H R, *MANPRINT, an approach to systems integration*, US Department of the Army
3. www.ams.mod.uk/ams/content/docs/hfiweb/hfihome.htm
4. *Human factors for designers of equipment*, DEF STAN 00-25
5. *Design and airworthiness requirements for service aircraft*, DEF STAN 00-970
6. Sheridan T B, (ed) *Supervisory control*, National Academy Press, Washington DC 1984
7. Abbott K, Slotte S M and Stimson D K, *The interfaces between flightcrews and modern flight deck systems*, unnumbered FAA Report 1996
8. Koehl F and Seeck D, *Observations of physiological and psychological effects of terrain following flight based on in-flight test experience, in Human factors considerations in high performance aircraft, ADARD-CP-371, 1984*
9. Green N D C, *The physiological limitations of man in the high G environment*, in *Combat automation for airborne weapon systems; Man/machine interface trends and technologies*, AGARD-CP-520, 1993
10. Mizuro S and Akao Y (trans Mazur G) *QFD; the customer-driven approach to quality, planning and deployment*, The QFD Institute ISBN92-833-1122-1
11. Kleinman, D L, Baron S & Levison W H, *An Optimal Control Model of Human Response, Part 1: Theory and Validation,* Automatica, Vol 6, 1970
12. Broadbent, D E, *Perception and Communication*, Pergamon Press, London 1958
13. Senders J W, *Man's capacity to use information from complex displays*, in Quasler H (ed) *Information theory in psychology*, Free Press, Glencoe, Ill, 1955

14. Wickens C, *Processing resources in attention*, in Parasuraman R and Davies R (eds) *Varieties of attention*, Academic Press, NY 1984

15. Card S K, Moran T P and Newell A, *The model human processor*, in Boff K R, Kaufman L and Thomas J P(eds), *Handbook of Perception and Human Performance Vol II - Cognitive Processes and Performance*, Wiley 1986

16. *A designer's guide to human performance modelling*, AGARD–AR-356, 1998

17. op. cit.

18. op. cit.

19. op. cit.

20. www.virtual prototypes.ca

21. Farmer E W, *Stress and workload*, in P F King and J Ernsting (eds), *Aviation Medicine*, 2nd edition, Butterworths, London 1988

22. Jex H R, *The critical instability tracking task: Its background, development and application*, Systems Technology Inc. Paper No 334 1989

23. Gopher D and Donchin E, *Workload - an examination of the concept*, in Boff K R, Kaufman L and Thomas J P(eds), *Handbook of Perception and Human Performance Vol ll- Cognitive Processes and Performance*, Wiley 1986

24. Endsley M R and Bolstad C A, *Human capabilities and limitations in situation awareness*, in *Combat automation for airborne weapon systems: man/machine interface trends and technologies*, AGARD-CP-520 1993

25. *Flight vehicle integration panel working group 21 on glass cockpit operational effectiveness*, AGARD-AR-349 1996

26. Cooper G E and Harper R P, *The use of pilot ratings in the evaluation of aircraft handling qualities*, NASA TM D-5133, 1969

27. Hoh R H, Mitchell D G, Aponso B L, Key D L and Blanken C L, *Handling qualities requirements for military rotorcraft, Aeronautical Design Standard ADS-33E*, US Army Aviation and Missile Command Aviation Engineering Directorate, Alabama, 2000

Other standards documents pertinent to crewstation design and development are listed on the next page.

Standards relating to crewstation design and development

MIL-C-81774A	Control panel aircraft general requirements for
MIL-L-85762	Lighting aircraft interior night vision imaging system (NVIS compatible)
MIL-M-8650C(2)	Mockups aircraft general specifications for
MIL-STD-411F	Aircrew station alerting systems
MIL-STD-850(B)	Aircrew station vision requirements for military aircraft
MIL-STD-1295	Human factor engineering design criteria for helicopter cockpit electro-optical display symbology
MIL-STD-1333B	Aircrew station geometry for military aircraft
MIL-STD-1472B	Human engineering design criteria for military systems, equipment and facilities
MIL-STD-1478	Task performance analysis
MIL-STD-1787C	Military standard aircraft display symbology [R]
MIL-STD-2500B(1)	Aircrew station controls and displays for rotary wing aircraft
MIL-STD-2525b	Standard for symbols in military displays
SAE AS18276	Lighting aircraft interior installation of
DOD-HDBK-743A	Anthropometry of US military personnel (metric)
MIL-HDBK-87213	Electronically/optically generated airborne displays
NASA STD 3000	Man-machine integration standards
STANAG 3216 AI	Layout of flight data in pilots displays
STANAG 3217 AI	Operation of controls and switches at aircrew stations
STANAG 3218 AI	Location, actuation and shape of engine controls and switches in fixed wing aircraft
STANAG 3219 AI	Location and grouping of electrical switches in aircraft
STANAG 3220 AI	Location, actuation and shape of airframe controls for fixed wing aircraft
STANAG 3221 AI	Automatic flight control system (AFCS) in aircraft - design standards and location of controls
STANAG 3224 AI	Aircrew station lighting
STANAG 3225 AI	Location, actuation and shape of airframe controls for rotary wing aircraft
STANAG 3229 AI	Numerals and letters in aircrew stations
STANAG 3341 AI	Emergency control colour schemes
STANAG 3359 AI	Location and arrangement of engine displays in aircraft
STANAG 3370 AI	Aircrew station warning, cautionary and advisory signals
STANAG 3436 AI	Colours and markings used to denote operating ranges in aircraft instruments
STANAG 3593 AI	Numbering of engines and their associated controls and displays in aircraft
STANAG 3622 AI	External vision from aircrew stations
STANAG 3639 AI	Aircrew station dimensional design factors
STANAG 3648 AI	Electronically and/or optically generated aircraft displays for fixed wing aircraft
STANAG 3692 AI	Location and actuation of thrust vector controls for VSTOL aircraft other than rotary wing aircraft
STANAG 3705 AI	Human engineering design criteria for controls and displays in aircrew stations
STANAG 3800 AI	Night vision goggles lighting compatibility design criteria
STANAG 3869 AI	Aircrew stations control panels
STANAG 3870 AI	Emergency escape/evacuation lighting
STANAG 3994 AI	The application of human engineering to advanced aircrew stations

Chapter 5

External Vision

5.1 Requirements

5.1.1 Introduction

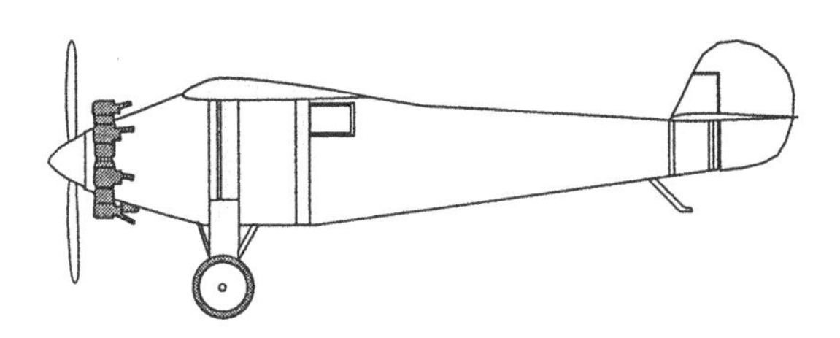

In 1927 Charles Lindbergh flew the "Spirit of St. Louis", a single-engine high wing mono-plane, across 4000 miles of Atlantic Ocean between Roosevelt Field, Long Island, and Le Bourget near Paris. In preparing for the flight, Lindbergh had to take a rational and undoubtedly courageous assessment of the risks of the endeavour, sacrificing every inessential item to maximise the take-off fuel load[1]. He wore lightweight clothing, carried no parachute and sat in a wicker chair facing a bulkhead behind the main fuselage fuel tank. This, he felt, gave more chance of surviving a heavy landing than a seat between the engine and the tank. The bulkhead supported the flight instruments and spanned the fuselage, so although he had some view out through small windows on either side there was no direct view in the direction of flight. He was persuaded to add a pair of mirrors, about three inches by five inches, forming a periscope that afforded a minimal view along the left side of the aircraft nose.

Lindbergh's deliberate sacrifice was exceptional. In general, a pilot's ability to conduct a safe flight, as reviewed in Chapter 2 and mandated in the concept of Visual Meteorological Conditions, depends upon a view of the external world that can supply:-

a) dynamic cues for stabilising the aircraft, avoiding disorientation and judging speed, height, attitude and direction of travel
b) an appreciation of the shape of the surrounding terrain and any obstacles for choosing a safe flight path, or on the ground a taxi path
c) the identity of specific features in the scene that are crucial aim-points, such as runways, and prominent features used to update the navigation system or confirm that the aircraft is where navigation aids say it is
d) the position, track and identity of other air vehicles for optimising tactical manoeuvres and avoiding collisions and wake turbulence
e) cloud and other meteorological phenomena that may affect scene visibility or indicate a hazardous air mass.

The relevance of the information and the required visual envelope vary with context. Cues, such as the texture of tree-tops or the way a field pattern shrinks with distance and is blurred by movement, are of little use to the pilot of an A320 in mid-cruise. However, such stimuli can provide an essential mixture of peripheral and foveal cues for a helicopter pilot to maintain a "nap of the earth" flight path with confidence. Low-level flying in any aircraft is intensely visual, and military flying is particularly so. Flying fast in close proximity to the ground may surprise the enemy defences, but the tactic also compresses the available time for the pilot to respond to a suddenly unscreened target, threat or obstacle. An attacking pilot therefore needs to discern small features at greatest range in his forward view. An air combat pilot would like to be able to discern small airborne objects anywhere in the 4π steradians of his visual space.

5.1.2 Envelope

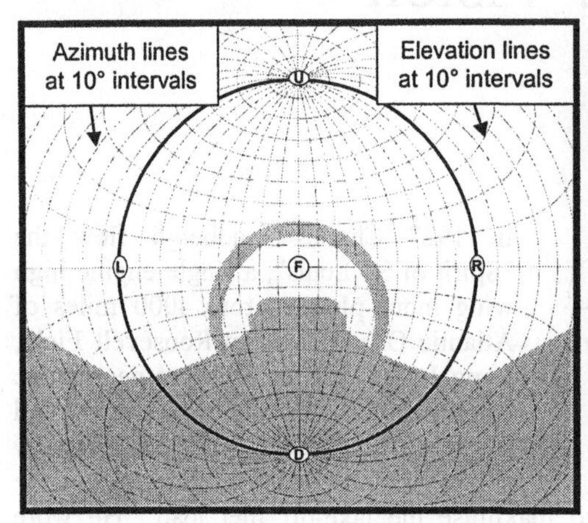

Visual envelope from the datum eye position in Harrier GR5 relative to the forward airframe datum

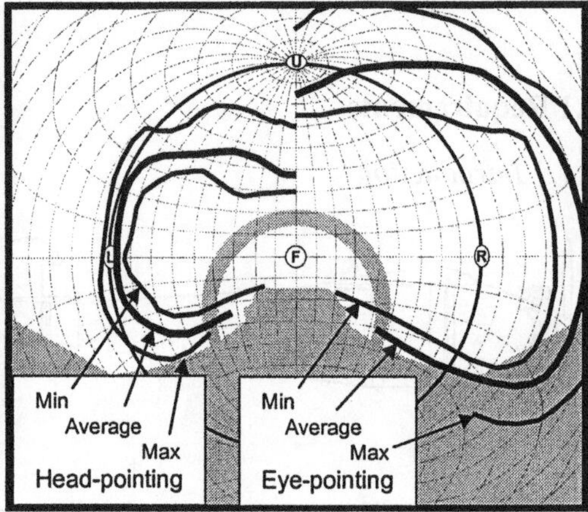

The range of head and eye movement in Tornado GR1 measured with aircrew wearing full AEA and tensioned harness

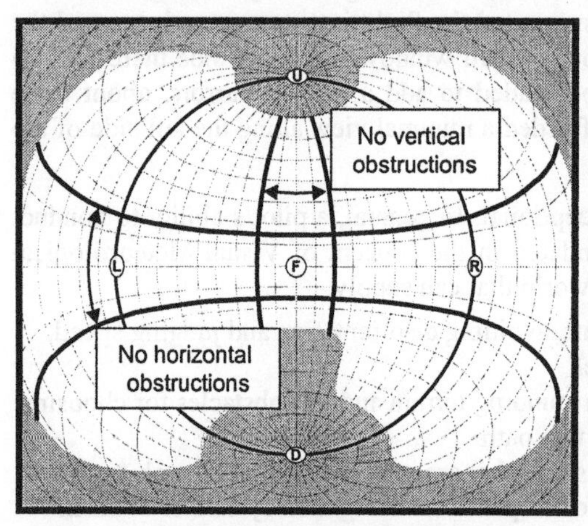

Recommended visual envelope for both crew of a side-by-side observation helicopter

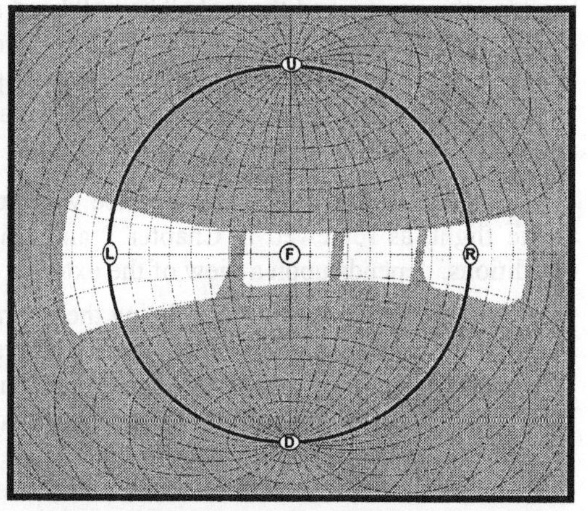

Visual envelope from the left seat of a typical transport aircraft

Figure 5.1 Illustrating the visual envelopes for a variety of roles (Data for the range of head and eye movements in the Tornado are from Manning[2], and the recommendations for the helicopter are from Atkins *et al*[3])

The chief need is to provide a view of the runway threshold on the approach to a landing, and for this the downward view over the aircraft nose must be greater than the angle of attack at the landing airspeed with the appropriate slats and flaps deployed. This angle varies between types and is typically 15° for combat aircraft. As shown in Figure 5.1, the extent of the rest of the envelope depends mainly on the role of the aircraft, with combat aircraft and observation helicopters needing transparencies over more than half the enveloping sphere. Transport aircraft are usually supplied with limited window area and an apparently small viewing field, but this can be extended if the crew can move around within the flight deck.

As indicated in Figure 5.1(b), in military aircraft the angular range of head movement can be restricted severely because the shoulders are almost clamped by the ejection seat harness, and neck articulation is limited by contact between the lower helmet rim and the stiff stole of the life-preserver. The visual envelope is certainly extended by eye movement, as shown, but to look around and search the sky to the rear of the aircraft, any aircrewman must slacken the shoulder harness, lean forward, strain his upper torso through about 90° and force his chin towards his shoulder. In practice this manoeuvre is also constrained by the limited space to swing the bulk of the helmet before it contacts the canopy or the head-rest.

5.1.3 Acuity

One criterion for the ability to pick out detail in the external scene can be derived from the need to avoid an air-to-air collision[4]. Typically, about five seconds elapse between the instant when an aerial hazard is first detected and the aircraft can alter course. This period covers the excitation of the peripheral retina, the triggering of a saccadic eye movement to bring the area containing the potential hazard within the small lobe of highly acute vision, eye focusing, the cognitive processes of object recognition and decision-making, the resulting neuro-muscular action to move the joystick, and then the delay before the aircraft responds. If the aircraft is travelling close to the speed of sound it will cover about 1.5km during this time, so the pilot should be able to discern clearly an aircraft at this range. Depending on the aspect presented to the approaching pilot, a typical light aircraft, helicopter or small military jet will subtend less than 5mrad (about 15 minutes of arc); less than a housefly at two meters, sometimes with little contrast against a cluttered background.

The earliest and most pertinent psycho-physical experiment to assess the human ability to see small targets in different levels of surrounding luminance was set up by Blackwell[5] during the final stages of World War II. The observers were placed at one end of a light-controlled tunnel and targets were presented at the other on a background with luminance L_o that could be set between 10^{-5} and 10^{2} cd.m^{-2}. Observers were given plenty of practice and time to adapt to the ambient level, and they viewed the distant screen with both eyes. Small circular targets were presented in one of eight locations randomly, and the size of the target and the luminance difference between the target and the background (ΔL) were also randomised. A sound cue was given, and the observer had to indicate within six seconds in which location she had seen the target. The large amount of data was collated by the observers themselves and corrected for the 1/8th chance of guessing the correct location. Figure 5.2 shows how the threshold contrast, at which half of the targets were detected, varied with the size of the target and the level of the background to which the observers were adapted. The contrast of a small object is, as noted in Appendix B, the difference between the luminance of target and background expressed as a fraction of the luminance of the background ($\Delta L/L_o$).

The data were scientifically interesting because they showed that for targets smaller than about 30 minutes of arc, the slope of the logarithmic curves had a value of about -2. This suggested that the threshold was determined by the total flux of stimulation from the target, a phenomenon known as Ricco's Law or "spatial summation". Only above a transition size did a curve flatten out and the threshold contrast become independent of target size.

The data were also of practical value. In a bright ambience a target of 30 minutes of arc could be discerned when the contrast was 10^{-2}, i.e. the luminance of the target could be only 1/100th of the ambient level. However at very low light levels the threshold contrast for a similarly sized target was about 10^{1}; the target needed to be 10 times brighter than the ambient level. This represents a substantial reduction in the chance of seeing a small object in the dark, and it also reflects the need to use a different visual strategy. In bright light the observers could fixate directly on the stimulus locations, but when dark-adapted they had to look away slightly because of the absence of rods in the fovea.

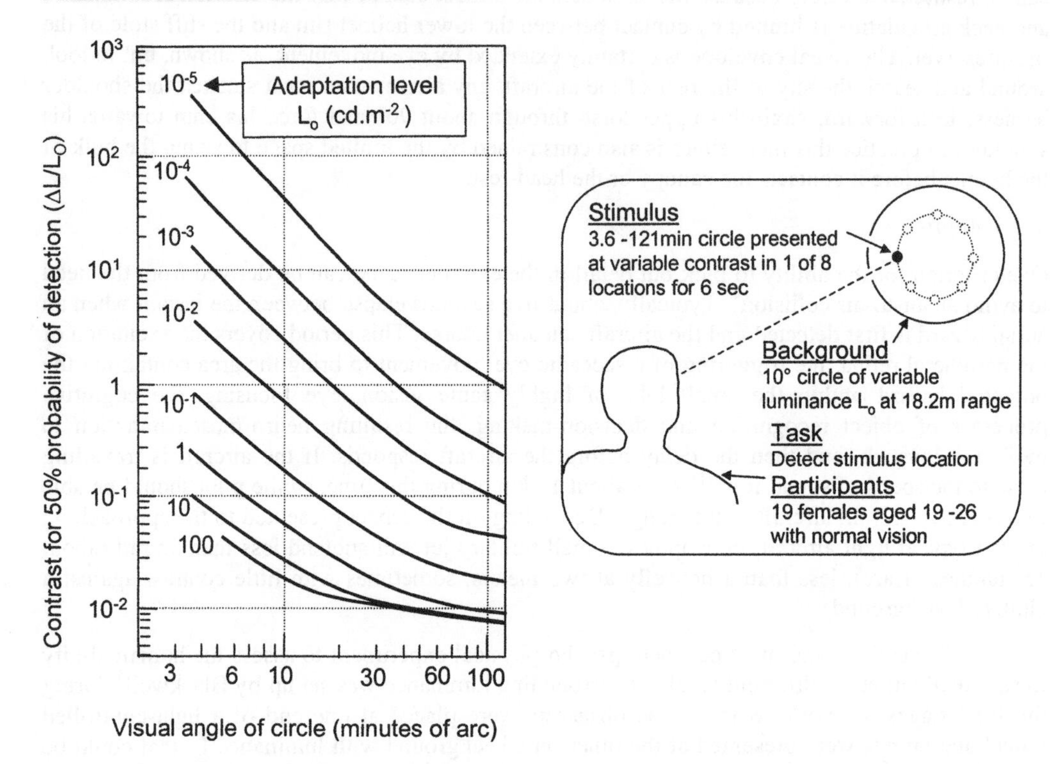

Figure 5.2 Summary results of experiments to assess the probability of detecting a small object in a known location under a wide range of ambient luminance levels (From Blackwell H R, *Contrast thresholds of the human eye*, JOSA 36, pp624-643, 1946. Reprinted with permission from the Journal of the Optical Society of America)

Although photopic vision is intrinsically less sensitive than scotopic vision, which is why stars cannot be discerned in daylight, the generally enhanced detectability of low contrast objects is the fundamental driver for the adoption of night vision aids that raise the luminance of the scene to a photopic level.

5.1.4 Light level

Natural vision is capable of discerning what is where in the external world over a very wide range of light levels. As noted in Chapter 2, the minimum that can be sensed is 10^{12}-fold less than the level causing retinal damage. However, this extraordinary range is covered by two sorts of retinal receptor with markedly different characteristics, and at any instant each sort can respond to luminance differences in the scene over a much narrower range. As shown in Figure 2.11, the sensation of brightness difference within the limited instantaneous dynamic range is:-

a) substantially logarithmic
b) covers about a 1000-fold spatial luminance variation
c) able to distinguish about a hundred shades
d) unable to distinguish features in the scene above or below the instantaneous range

e) adaptive to the mean level of the scene luminance.

The eye behaves somewhat like a video camera that can adjust to temporal variations in the scene radiance by varying the gain applied to the sensed signal. The optimum, where photopic foveal vision is properly stimulated but not overloaded, occurs when the ambient luminance is between about 10 and 500cd.m^{-2}. Most individuals use sunglasses to attenuate higher ambient levels, while some form of aid is essential to raise weak light to these levels when the ambience would otherwise only stimulate relatively impoverished scotopic vision. Figure 5.3 illustrates the notion of a fine but limited spatial brightness discrimination range that slides temporally up and down a wider gamut.

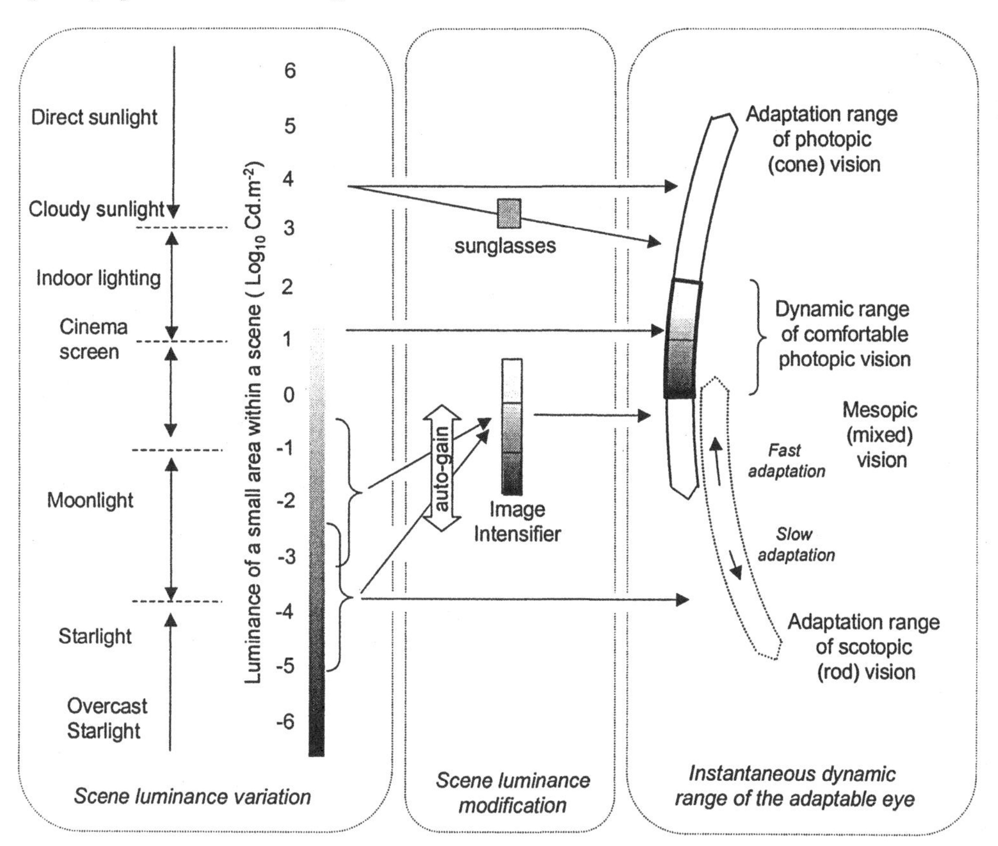

Figure 5.3 Illustrating, on the left scale, the wide range of ambient light levels, and on the right scale, the limited range sensible to human vision. The centre shows some ways of modifying the ambient to match the human preference

Aircrew have flown when adapted to very low levels of light, but this has only proved possible when there is a clear sky and more than a half-moon to take vision into the mesopic region at about 10^{-2}cd.m^{-2} where some central photopic vision is stimulated. For this to be practical, aircrew need to spend some time in a dim red-lit crewroom prior to the mission, and they must then be careful to avoid even momentary exposure to a higher level that would raise their adaptation state and render them insensitive to the low ambience. As discussed in Chapter 6, it has not really been possible to illuminate the cockpit with sufficient light to allow the pilot to read instruments and maps without some disturbance to dark adaptation. It has also been difficult to avoid a vulnerability to external sources when a mission is flown in the vicinity of a town with lit buildings and streets. It is to avoid these vulnerabilities and extend night operations to encompass moonless, overcast conditions that night flying is now done with some form of electro-optical aid. The systems that have been devised are discussed in later sections of this chapter.

5.2 Aircrew vision

5.2.1 Vision standards

The selection standards for military aircrew are particularly exacting. Emmetropy, the ability to discern fine detail over a wide focus range, is usually assessed by asking the individual to read a Snellen chart at a near point of about 0.2m and a far point at about 6m. The latter distance is close enough to infinity. The chart has high contrast black letters of diminishing size on a white background, and an emmetrope can discern a feature, such as the small gap that distinguishes an "O" from a "C" when this gap has an angular subtense of 1arcmin (about 0.3mrad). Such an individual is described as having 6/6 vision in the UK, 20/20 vision in the US and 1.0 vision elsewhere. On the other hand, if an individual can only resolve at 6m what is expected at 18m, his vision is described as 6/18, 20/60 or 0.3. A myopic (short-sighted) individual cannot discern fine distant detail and would have less than 6/6 visual acuity at 6m distance. A hypermetropic (long-sighted) individual would have less than 6/6 visual acuity when asked to read the chart at a short distance.

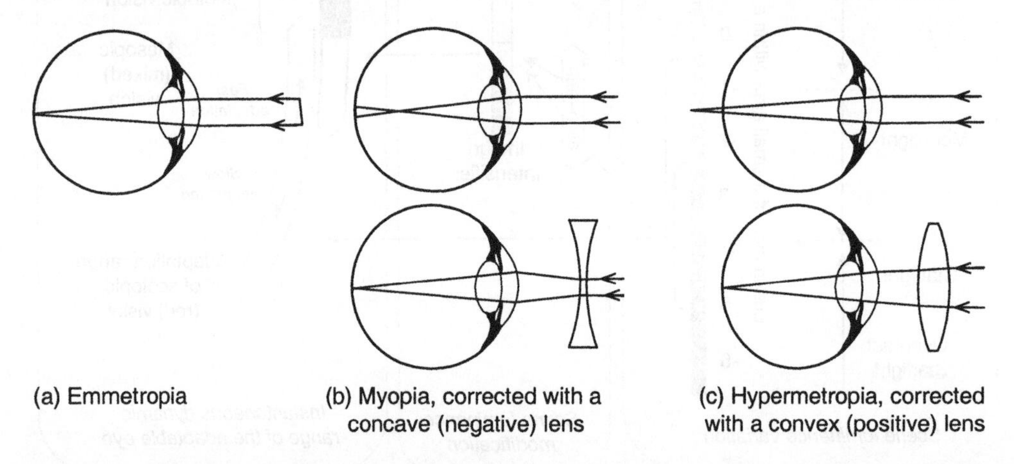

Figure 5.4 Myopia results if parallel light is focused in front of the retina, and hypermetropia if the focus is behind the retina. These are corrected using concave and convex lenses respectively

The refractive error is usually described in terms of the power, or focal length, of the lens needed to correct the error, as illustrated in Figure 5.4. Opticians transform the focal length in metres into a reciprocal unit called the Dioptre (D) such that a convex lens having a focal length of 0.5m is called a +2D lens and a concave lens of 4m focal length is -0.25D. The unit is convenient because the effect of looking through several lenses can be calculated by summing their dioptre values. Thus a combination of a +2D and a -0.25D is 1.75D, or a combined focal length of 0.57m.

In general both pilots and navigators of fast jet aircraft are selected from those with uncorrected 6/6 near and far vision in both eyes. Such individuals have no difficulty transferring focus between the inside of the cockpit and the far outside, they are less likely to be intolerant to helmet-mounted display and vision enhancement equipment, and they are unlikely to need corrective spectacles early in their flying career. Also, if they do need spectacles in visually taxing conditions, such as night flying, these can generally be of low corrective power. This standard may be relaxed for helicopter and transport aircrew. It also depends on a plentiful supply of potential recruits.

Once trained, aircrew are usually retained if their visual acuity deteriorates and can be corrected, because it is recognised that some deterioration is an inevitable consequence of ageing. The most common condition is presbyopia, the progressive hardening of the lens that reduces both the range over which the eye can change focus and the speed with which the change is accomplished. Young children start with about 13D, but this reduces progressively until at 50 years of age most individuals have little effective accommodation[6]. Aircrew are generally considered satisfactory if each eye is within the age-related decline. Although the youthful ability to accommodate over a wide range may confer comfortably clear near and far vision, it may also mask a latent condition that would be manifested fairly early in the individual's career when accommodation declines. Young recruits are therefore usually asked to view the test chart at 6m through a +2.5D lens. If their acuity is unaffected they are hypermetropic.

Other visual standards are also set and appropriate tests conducted. It is exceptional for RAF aircrew to have a family history of myopia, a spherical refractive error outside –0D to +1.75D, astigmatism greater than 0.5D, or a resting imbalance between the eye muscles that affects binocular fusion. They are required to have full colour discrimination, a wide peripheral field and stereoscopic acuity that enables them to resolve a binocular disparity, as defined in Figure 2.16, of 25 seconds of arc. Aircrew, particularly older aircrew, undergo regular ophthalmological examination to ensure they have no incipient retinal damage, such as the formation of retinal dross that can be caused by overexposure to strong blue light, or a scotoma, a loss of visual field, particularly near the fovea.

5.2.2 Spectacles and contact lenses

About 15% of RAF aircrew require refractive correction to achieve 6/6 acuity. Most use contact lenses, but before the advent of soft, high water content, gas-permeable contact lenses that were cheap enough to be discarded at the end of each day, the only solution was a pair of spectacles. Currently aircrew have a choice and can opt for spectacles if they cannot tolerate contact lenses or prefer the flexibility to use different correction for different conditions.

Aviator spectacles are invariably made using non-reflecting black-coated metal frames that fit in the space below the helmet brow inside the inner visor and clear of the oxygen mask. The arms that reach back over the ears are flat and thin so that they do not break the acoustic seal between the earcups and the side of the head. An especially narrow, close-fitting frame has been devised for use within a respirator and the face protection visor necessitated by night vision goggles. The lenses are made from a variety of plastics that, unlike glass, are resistant to shattering under impact. Polycarbonate is preferred because it has a very high refractive index (1.586) combined with exceptional impact resistance and high absorption of ultra-violet wavelengths. The material can be dyed to produce tinted lenses that have a transmission of about 15%, the same as the polycarbonate visor, it can be machined or moulded accurately to produce spherical and cylindrical power and bifocals, and it accommodates a hard anti-scratch coating such as polysiloxane. The principal drawback is its high dispersion (variation of refractive index with wavelength) which makes it unsuitable for thick lenses. Most lenses that provide a correction of more than ±2D are therefore made from an alternative optical grade plastic such as acrylic or CR39.

5.3 Optical transparencies

5.3.1 Windshield and canopy

Although it is desirable for the field of vision to be unrestricted, the aircraft design team can be forgiven for regarding a window as a weak, heavy, protruding discontinuity in an otherwise light and smooth structural envelope. Size, shape and construction are therefore a compromise between the aerodynamic and structural needs on one hand and those of the crew on the other. Supplying freedom for head movement, adequate visual field, minimally intrusive reflections from instruments and displays at night and minimal distortion to the external scene suggests a large, thin, non-reflective, enveloping sphere. However, with available materials that can withstand the excess cabin pressure, aerodynamic temperatures and forces, structural loads and the possibility of bird impact, in a fast jet the compromise is invariably a close, thick, partially reflecting, elongated semi-ellipsoid.

In most combat aircraft the transparency is divided into a frontal windshield and a rear canopy. This allows the former to be fixed to the airframe and made to withstand the high dynamic pressure and temperatures, and provide impact protection. The latter can be significantly lighter and therefore more easily lifted for normal entry and exit, and removed rapidly or shattered for the passage of the ejection seat in an emergency. The most notable exceptions are the F-16 and F-22, where the canopy is formed from a single transparency. This is mainly done to remove the arch that can obscure an opponent in aerial combat, and in the F-22 it also removes a radar-reflective discontinuity that compromises the stealthiness of the airframe.

The windshield is usually constructed from a laminate of glass and a relatively flexible polymer layered to a net thickness of about 18mm. To withstand abrasion the outer layers are always glass, with anti-reflecting coatings on all the surfaces except the sealed innermost surface, which is coated with indium/tin oxide (ITO) and gold so that it can be heated electrically for de-misting. The laminate is bonded into an alloy frame, and this is fastened to the airframe. The canopy is either polycarbonate or acrylic, sometimes a laminate of both, and varies between 8mm and 15mm in thickness for most combat aircraft. The one-piece canopy for the F-22 is unusually thick, about 20mm, and is made from two plies of fusion-bonded polycarbonate.

The manufacture of the canopy, which is in the region of 3m long by about 1m wide and 0.8m high, involves numerous stages. It is firstly necessary to cast acrylic as a slab, anneal this to remove stresses, then grind to form a parallel plate and polish for optical smoothness. This prepared material is then heated slowly and uniformly to a temperature where it is just sufficiently soft that it can be draped over a pre-heated metal former and cooled slowly until set. After trimming and thinning around the edge, the surfaces are coated with an abrasion resistant layer of polysiloxane and then sealed and fastened to the frame. Pipework to carry and distribute the de-misting air from the cabin conditioning system is attached to the frame, as too are any components for a helmet-mounted sighting system and the zig-zag lines of lead-backed micro-detonating cord (MDC) that shatter the canopy to allow a clear passage for the ejection seat. The resulting mass is between 100 and 200kg. A canopy fitted with MDC must be treated with all the care and safeguards afforded any explosive.

5.3.2 Visors

Most aircrew helmets carry at least one visor. These are usually of spherical, ellipsoidal or toroidal curvature and are mounted on a tracks or pivots so that they can be raised and stowed out of view against the helmet brow. When lowered, they protect the eyes and face from fragments of debris should the windshield shatter, for instance by striking a bird. On ejection from a fast jet aircraft, the visor must deflect the violent bombardment of MDC spatter, canopy fragments and wind-blast. As discussed in Chapter 13, the material of choice is polycarbonate, coated with a scratch-resistant layer for durability, and coated and/or dyed to attenuate excess irradiance.

Current aircrew helmets in RAF service carry two visors. The inner is clear and always deployed in flight. The outer is tinted and deployed either with or without the clear visor in bright sunlight. Thus a pilot always has a view of the external world through a clear visor and a windshield or canopy, and he may also wear spectacles and use a dark visor. These layers are rarely optically perfect and generally degrade the view of the external scene. When new they introduce some distortion and unwanted reflections, and when scuffed or dirty they scatter incident light to create veiling glare.

5.3.3 Distortion

External light is generally deflected through a small angle and a small displacement on traversing the canopy or windshield, as shown in Figure 5.5. The effect is an immutable consequence of light waves propagating more slowly in the optically dense canopy material, and is most marked when the light strikes the surfaces at a large angle. In general the small lateral displacement is of no consequence, and only the angular shift is important because it affects the apparent direction of features in the external scene.

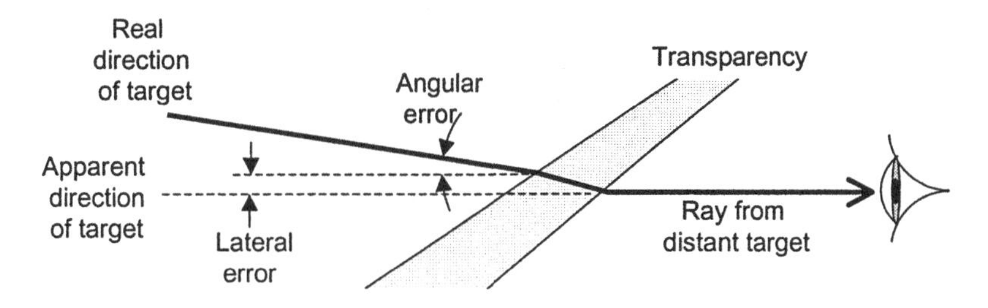

Figure 5.5 Refractive errors introduced by an optical transparency

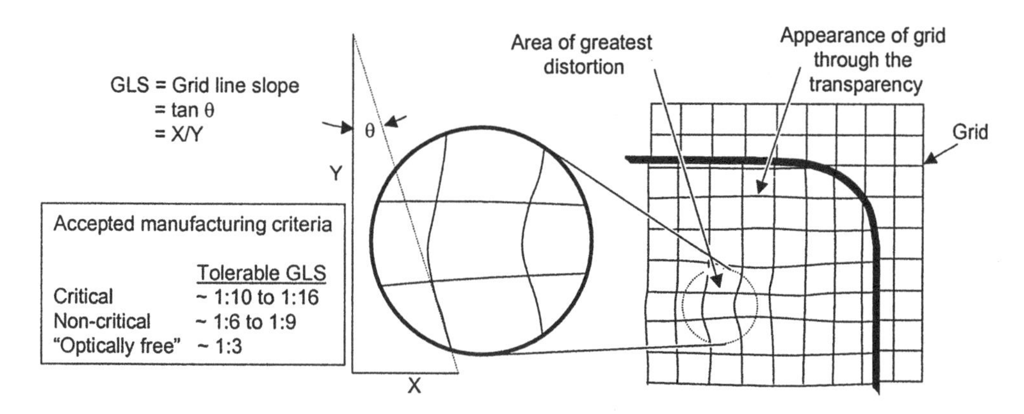

Figure 5.6 The Grid Line Slope method of assessing transparency distortion

Measurement of this distortion to the view through a transparency is done by theodolite, or laser, and involves sighting an array of distant points from the design eye location and assessing the angular differences introduced when the transparency is in place. This is the only effective technique for thick, highly curved structures like canopies and windshields. Angular errors of 3° or 4° are not uncommon. The chief problem is controlling the difference in the angle seen by the two eyes, because even a small inter-ocular difference introduces substantial stereoscopic disparity.

For assessing flatter panels used in transport aircraft, manufacturers tend to use the more convenient technique illustrated in Figure 5.6. Here, the panel is held at a small fixed distance from a high contrast grid. Any lack of parallelism between inner and outer surfaces, or any material inhomogeneity, is seen as a distortion to the grid[7], which is assessed by measuring the maximum grid line slope (GLS) usually from a photograph. The manufacturer calibrates the relationship between the angular error and the GLS and applies appropriate pass/fail criteria to different areas, with least distortion accepted in the forward view. An "optically free" region, where substantially greater distortion is tolerated, is usually declared within about 80mm of the edges.

5.3.4 Scattering

Even when new a transparency will have small defects within the bulk of the material and at the surfaces, especially the surfaces coated with multiple layers and bonding adhesives. When exposed in service to the ravages of ultra-violet radiation, dynamic heating and bombardment by dust, rain and hail, the defects inevitably multiply. Each defect scatters light towards the pilot's eyes, the amount varying with the intensity, angle and spectrum of the source. A field of illuminated defects therefore acts as a luminous surface overlaying the external scene, and the transparency becomes a veil that reduces the contrast of features in the scene. This has numerous consequences, one of which is to reduce the chance of detecting a small airborne object. The quality, cleanliness and opportune replacement of transparencies are of major concern to aircraft operators.

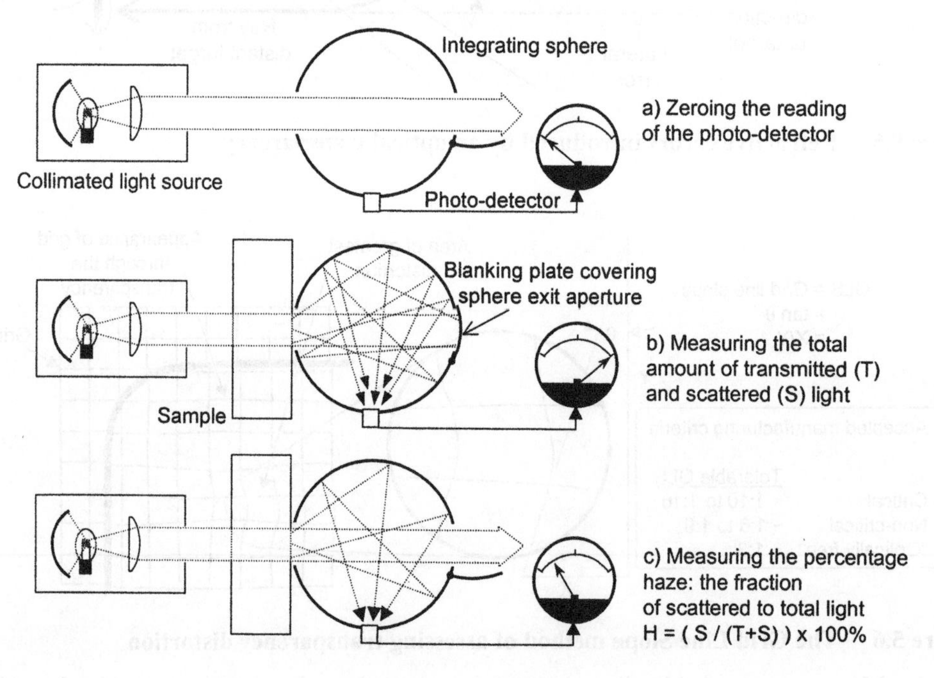

Figure 5.7 The ASTM D1003-61 method of measuring haze

The amount of scatter or haze produced by a transparency can be assessed by several techniques. One approach that is convenient in a manufacturer's laboratory employs a collimated illuminator and an integrating sphere[8]. The latter is a hollow sphere about 150mm diameter, coated internally with matt white paint, containing small entrance and exit holes and a photo-detector filtered to have photopic spectral sensitivity. As summarised in Figure 5.7 (a) the source and sphere are set up in a dark room, (b) the transparency is placed in contact with the entry hole, and the total amount of light entering the sphere is measured with the exit hole closed, and (c) the exit hole is opened to allow the transmitted light through so that the amount of scattered light can be assessed as a proportion of the total. This percentage is called the "haze index", and typically values of 2% to 3% are the maximum accepted in a new transparency. The haze index gives a general indication of diffuseness that takes into account all forward scattered light. However, because the apparent luminance of the transparency depends on the angle at which it is illuminated and the viewpoint angle of the pilot, an alternative approach has been devised that can more readily quantify the amount of contrast lost.

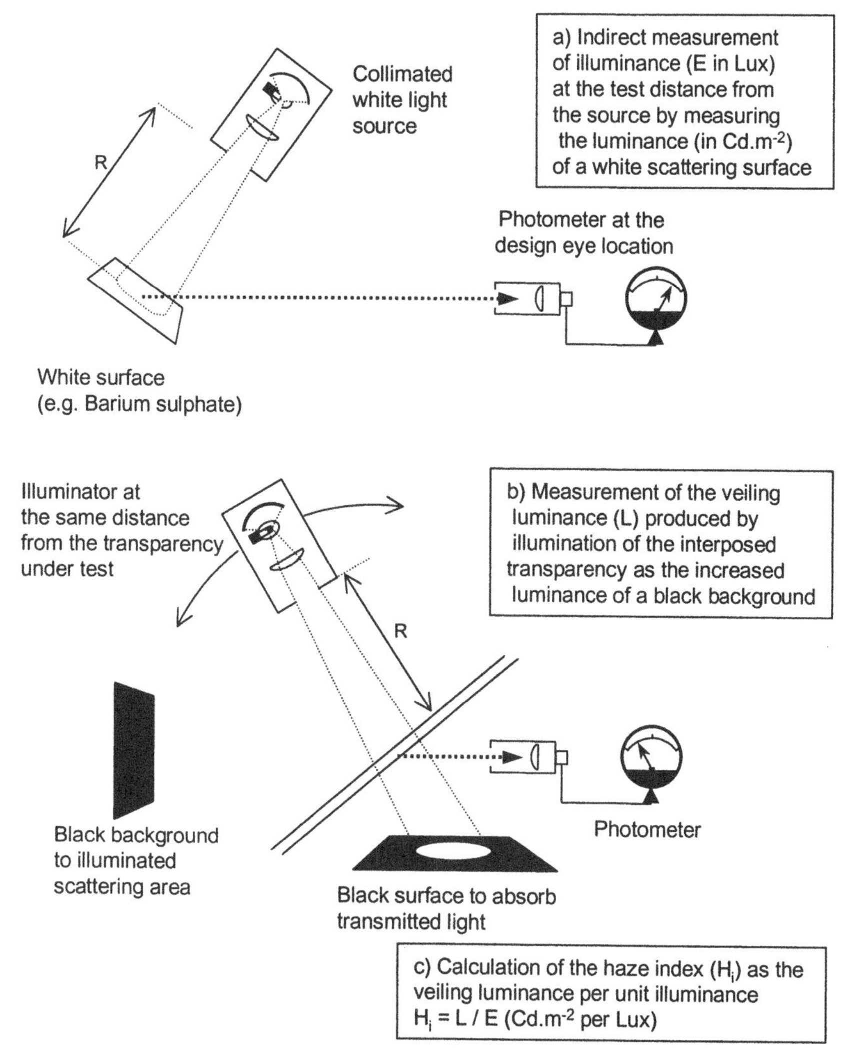

Figure 5.8 Method of measuring haze using ASTM F 773-81

As summarised in Figure 5.8, this alternative measurement technique[9] involves setting up the relevant geometry and characterising the haze index of an area of the transparency in terms of the veiling luminance per unit of illumination. In this case, a transparency that diffused the illumination perfectly would have the same luminance as the white calibration surface shown in the upper figure, i.e. a luminance of $1/\pi$ cd.m^{-2}/Lux. The approach has the advantages of making an assessment from the pilot's viewing location and of allowing measurements with the light source in a wide range of elevation and azimuth angles. This geometrical authenticity is particularly necessary for characterising the directional scatter produced by a holographic layer, such as the HUD combiner glass described in Chapter 8.

5.4 Night vision goggles

5.4.1 Origins

The most ubiquitous aid is the night vision goggle (NVG). The history of development began with the image intensifier tube illustrated in Figure 5.9.

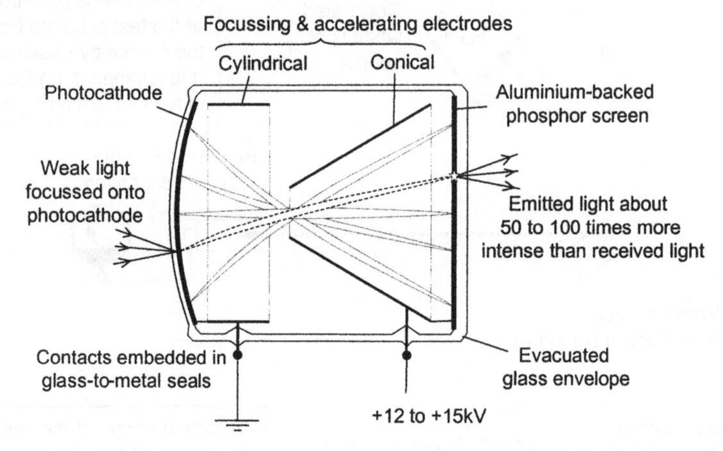

Figure 5.9 Schematic cross-section of a simplified vacuum image intensifier tube

This device grew out of techniques used for the manufacture of cathode ray tubes, as described in Chapter 7, but in this case the vacuum vessel was a glass cylinder of about 50mm diameter and 50mm in length. The inside of the input circular window was coated with a photocathode material that converted light photons into electrons, and the output window was coated internally with a phosphor that converted the electrons back into photons. The intensity of an image formed on the photocathode could be amplified by a factor of about 70 by accelerating the electrons emitted from the photocathode to higher energy so that they produced a larger number of emitted photons on striking the phosphor screen. The configuration of internal cylindrical and conical electrodes needed to create the field that accelerated and focused the emitted electrons, and mapped the output from one face to the other with little distortion, was the product of considerable refinement. As shown, the output image was inverted, so the device was employed with a high quality camera objective that brought external light into focus on the photocathode, and together with a wide angle eyepiece lens this make an upright intensifying telescope. The whole assembly, with battery and power supply unit to generate the 12kV excitation, was a little too heavy to be hand-held and was mounted in military vehicles and tanks for covert operations. Later devices were better integrated and had a wide range of uses, typically as sniper night-sights.

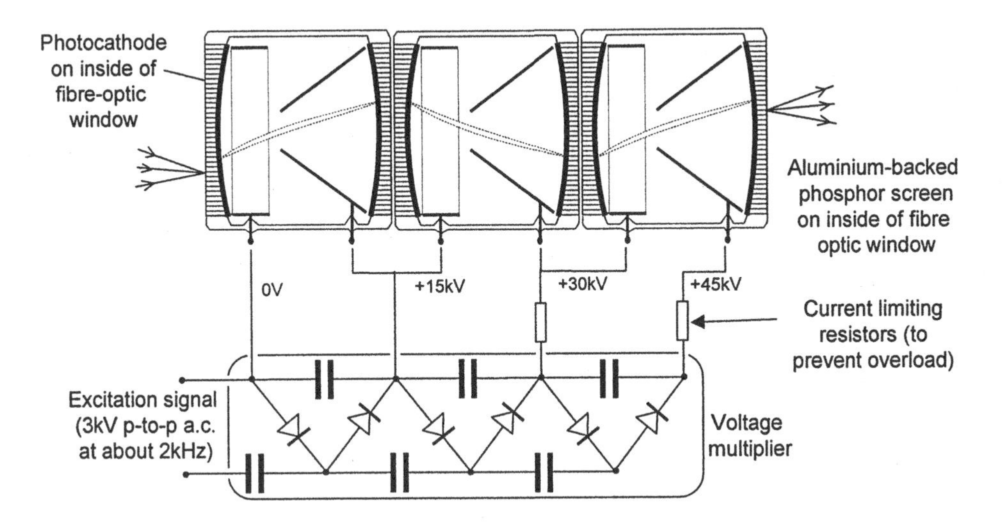

Figure 5.10 Schematic of a simplified three-stage fibre-optically coupled image intensifier tube with voltage multiplier

As shown in Figure 5.10, later versions of the image intensifier tube were made with fibre-optic face-plates. These were stacked so that the emission from one could be coupled directly to another to increase the gain substantially. Intensification of about 10^5 was possible, which gave useful imagery under clear starlight.

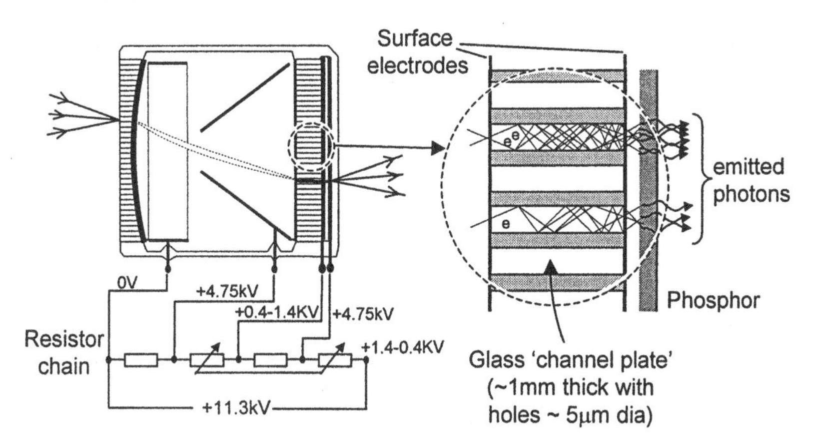

Figure 5.11 Schematic of a channel image intensifier tube and variable voltage supply

The next advance came with the invention of the channel plate multiplier[10] in the 1960s. As shown in Figure 5.11, this consisted of a glass plate containing a matrix of very fine holes that was placed so that it was almost in contact with the phosphor to amplify the number of electrons arriving at the phosphor. Both the method of manufacture and the working principle of the channel plate were ingenious.

The micro-channel plate was made using a process developed for fibre-optic face-plates; fusing together the sheaths of about 10^7 parallel glass fibres to form a cylinder that was heated until soft and drawn axially into a rod about 35mm in diameter. The face-plates were formed by slicing the rod into discs and polishing the faces to have an appropriate curvature. The main difference required to form micro-channel plates was to use hollow tubes of a glass containing lead oxide as the starting material rather than core and sheath glasses of different refractive index. This produced discs that were perforated by a fine array of axial holes, and under a careful heat treatment in a reducing gas such as hydrogen, the surface of the glass became lead-rich. This very thin, semi-metallic surface had two essential properties. The first was a small amount of conductivity, so that a voltage could be applied between the two ends of each hole to establish a uniform electrostatic field along the hole without drawing a significant current. The second was the efficient emission of secondary electrons when bombarded by electrons.

The two flat surfaces of the hole plate were coated with a thin metallic layer, by evaporation in vacuum, and encapsulated within an evacuated glass envelope. When excited by a potential of about 500V, the walls of each hole acted like a photo-multiplier tube, creating an avalanche of electrons at the end of the hole that was proportional to the input flux. Figure 5.11 also shows how the excitation voltage could be reduced to prevent overload. In practice this was done automatically by a circuit that monitored the total current drawn by the tube.

5.4.2 In-service NVG devices

A dramatic reduction in the size of the image intensifying tube occurred when the bulk of the glass envelope was omitted and the photocathode was brought very close to the input face of the micro-channel plate so that electrons could be gathered by the hole adjacent to the emission site.

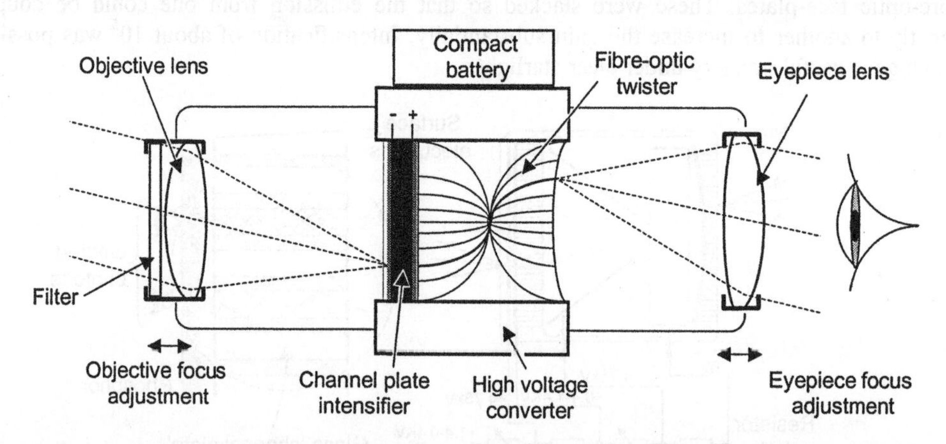

Figure 5.12 Simplified schematic of a modern direct view night vision goggle

The resulting proximity focused[11] channel plate intensifier was a glass sandwich disc about 18mm diameter and 5mm thick, as shown in Figure 5.12, that could be integrated into a "second generation" (Gen-II) sealed unit together with a ring-shaped high voltage conversion module and a fibre-optic twister. The twister was a cylinder about 25mm thick containing fibres, at the same density as the micro-channel plate, that rotated the output image by 180° so that it could be viewed upright through a relatively simple eyepiece lens. With the addition of a matching low f-number objective lens these formed a compact electro-optical unit, weighing only 250g, that could raise the apparent luminance of a quarter-moon scene to a photopic level and gave about 6/12 resolution over a field of view of about 35°. Pairs of these units, mounted on an adjustable bracket to resemble binoculars, were originally used by ground troops. When the mounting was fastened to the helmet brow they became the first practical Aviator Night Vision (ANVIS) goggle.

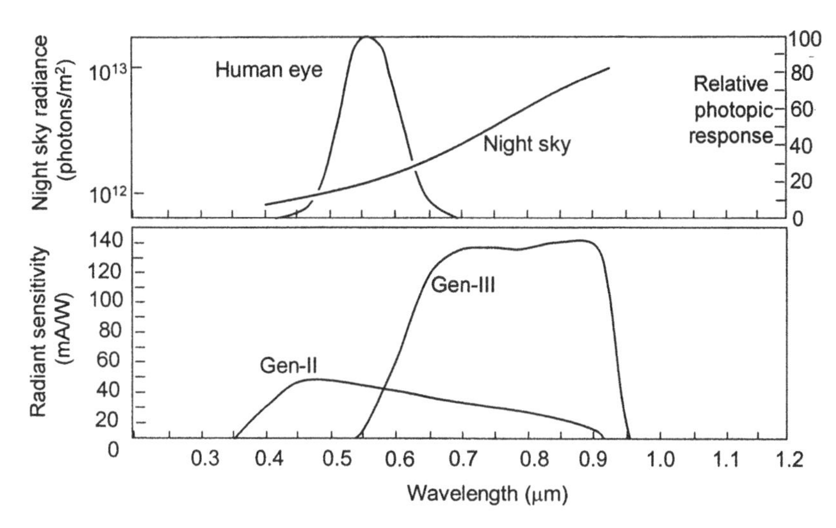

Figure 5.13 Spectral sensitivities of the eye and image intensifier tubes compared with the spectrum of emissions from the night sky (Reprinted with permission from QinetiQ)

Table 5.1 Characteristics of a typical Gen-III image intensifier

Parameter	Unit	Typical value	Condition
Photocathode Sensitivity	µA / lm	1800	Source at 2856 °K (Note 1)
EBI (Note 2)	lux	2.5×10^{-7}	
Signal to Noise	ratio	20	
Luminous Gain	cd.m^{-2} / lux	10^4	At 10^{-5} lux
Output Luminance	cd.m^{-2}	6	
Luminance Uniformity	ratio	3 : 1	Source at 2856 °K
Phosphor	JEDEC code	P-43	
MTF (Note 3)	Output / Input	90% at 2.5 lp / mm 70% at 7.5 lp / mm 45% at 15.0 lp / mm 20% at 25.0 lp / mm 10% at 50.0 lp / mm	

Note 1: The spectral composition of thermal emissions are summarised in Appendix A
Note 2: EBI (Equivalent Background Illumination) is the noise level with no input image
Note 3: MTF (Modulation Transfer Function) is explained in Appendix C

The spectral sensitivity of the photocathode materials used in the Gen-II image intensifiers was far from ideal. As shown in Figure 5.13, the intensifier had maximum sensitivity at about 0.47μm, in the blue-green region of the visible spectrum, while the sky was most radiant at much longer wavelengths in the near infra-red. Also, as described in Chapter 6, the technique for preventing cockpit lights from overloading the image intensifier that used a filter to block radiation below 0.65μm certainly had a marked effect on the amount reaching the photocathode. The developers of the channel plate image intensifier therefore looked for alternative materials for the photocathode that would have greater sensitivity in the near infra-red. The adoption of gallium arsenide produced "third generation" or Gen-III image intensifiers that, as shown in Figure 5.13, are significantly more sensitive above the cut-off wavelength. Typical characteristics are given in Table 5.1.

Gallium arsenide is however prone to damage by ion bombardment produced as a by-product of the electron avalanche, and to prolong the working life an ion-absorbent material is inserted between the photocathode and the hole plate. The major drawback of this barrier layer is to spread the distribution of electrons emitted from the photocathode so that they enter neighbouring holes. This reduces sensitivity, reduces the contrast at the edges of features and enlarges the bloom surrounding any bright point in the scene.

5.4.3 NVG usage

The original ANVIS configuration, with a field of view of about 40° and employing Gen-II and latterly Gen-III image intensifiers, has been adopted by many services and has enabled night missions in helicopters, fast jets and transport aircraft that would otherwise have been restricted to daylight. Their usefulness depends as much on the optics and the mounting as on the central intensifiers. The f-number (the ratio of focal length to diameter) of the objective lens governs the intensity of the image formed on the photocathode. The f-number must be in the region of unity to give a discernible image at the minimum operating light level. The diameter of the eyepiece must also be as large as possible. As explained in Chapter 9, this directly governs the optical exit pupil and indirectly the tolerance to misalignment of the viewer's eyes. Attaining these dimensions while producing well-focused images over the whole range of wavelengths and field angles has presented a difficult optimisation problem. Some arrangements use conventional spherical glass elements, while others incorporate plastics for lightness and aspheric surfaces to enhance the overall image quality.

The mounting bracket, constructed from lightweight materials, contains a self-aligning latch that enables the practiced user to fasten the device single-handedly to the helmet, and remove it equally easily. The bracket provides about 20mm of vertical, fore/aft and inter-axial adjustment so that an individual's eyes can be centred on the optic axes and there is about ±10° of pitch adjustment to cater for variations in head and helmet postures. The mechanism also incorporates a pivot and latch so that the goggles can be held out of view on the helmet brow, plus a battery compartment and an easily operated on/off switch. Some installations compensate for the static pitch-down moment, caused by the large overhung mass of the goggles on the helmet brow, by fitting a counterweight to the nape of the helmet. However, this cannot be used in a fast jet because the protrusion would interfere with the ejection seat head-box. RAF fast jet aircrew, who would normally fly with a clear visor, instead clip a face protection visor (FPV) to the rim of the helmet to provide a barrier between the NVG eyepieces and their eyes. They must also tolerate a small explosive charge in the goggle mounting so that the whole goggle and mounting can be detached automatically in the early stage of an emergency ejection.

Aircrew usually prepare for flight by fitting the helmet tightly to minimise dynamic slippage induced by head movement, and fit new batteries. The quality of the images is checked using a test set; essentially a light-box with two apertures that inject a dim, distant test pattern into the two objective lenses. It is of note that aircrew always set the focus of the objective lenses to get best resolved detail in the distant scene, but the focus setting of the eyepieces is usually more idiosyncratic. Studies suggest that the preferred setting is between -1D and -2D, so that the intensified image of the scene appears to be between 1metre and 0.5metres[12]. This seems to be give comfortable accommodation and eases the transfer of attention between the outside scene and the cockpit displays.

The automatic gain control circuit in each channel adjusts the excitation voltage so that the current drawn by the channel plate is a pre-set value and the output luminance tends to be uniform over time. In good light, for instance when the night scene is illuminated by a full moon and there are few clouds, the intensified scene is clear and rich in cues such as moonlight shadows. When however there is more cloud cover and the moon is absent, the dim star-field illumination needs maximum excitation voltage. Under these conditions there are no high contrast shadows, and the high gain makes thermally excited electron emissions from the photocathode become apparent as an overlaid field of randomly flickering spots, rather like a snowstorm. Fires, street-lights and other luminous emissions can also have a major influence, particularly in goggles fitted with Gen-III image intensifiers. These induce two defects; a local bloom that spreads over neighbouring detail, and a reduction in the gain applied to the whole scene that lowers the contrast of the remaining detail. Image quality therefore depends strongly on the meteorological conditions, the phase of the moon and whether the flight is conducted over town, country or open water.

In some cases in fast jet aircraft equipped with a HUD, it was found to be possible to see the green HUD symbols in the intensified scene despite the cut-off filter. This was a fortunate but unforeseen consequence of the excitation of a small amount of IR emission from some HUD phosphors. However, to control the visibility of the HUD symbols more accurately, a small proportion of the green HUD emission must be allowed, as described in Chapter 6, to enter the goggle objectives through a "green leak" filter.

Many other lessons have been learned using experimental devices that were built to explore alternative configurations. For instance, the use of a single objective lens and image intensifier viewed thorough binocular eyepieces showed that the ANVIS arrangement had two advantages. Firstly, splitting the output light from the single intensifier made the intensified image much dimmer. Secondly, the "snowstorm" at high gain was significantly less intrusive because the random noise pattern seen by the left eye was not correlated with the random noise pattern seen by the right eye.

Other lessons concerned the need to match the attributes of one channel with the other. For instance, it was established that the images could be fused to give a stereoscopic sense of depth when the left and right images differed in brightness by about 40%, departed from unity magnification by about 2% and were rotated differentially by about 3°. On the other hand, a poorly made twister that introduced lines of distortion in the image was not tolerable. It was also found necessary to align the optics of the objective and eyepiece within about 0.25°. Fortunately, with unity magnification the perceived direction of features in the scene was unaffected by misalignment or wobble of the pair of units on their adjustable mount. Most of the other lessons concerned lack of reliable judgements that could be made because of the limited qualities of the imagery[13]. NVGs obviously could not enable the user to discriminate colours, but there were more subtle lessons. Resolution was limited, so elements in the scene could not be discerned until much closer, and judgements about the rate of approach to an obstacle were comparatively coarse. The loss of contrast at high spatial frequencies also tended to destroy vection cues normally conveyed by fine textures in the scene. Thus during low-level flight and, in particular, when approaching a landing area, the pilot would have to rely more on other cues such as the differential motion of identifiable objects in the scene.

In general the use of goggles has certainly brought a substantial expansion to the range of conditions in which operations are possible, but in comparison with normal daylight flying, aircrew have to make do with a colourless, blinkered, fuzzy view of the world, and unsurprisingly they do not fly with the same ease and confidence. Instead they are very careful to check that the intended route avoids obstacles and whenever possible includes easily identified features so that they are less likely to stray. A moving map display is considered vital for such accurate navigation in a single-seat fast jet, and the second crewmember bears this concern in a helicopter. In most phases of flight, even when familiar with the terrain, the pilot compensates for the blinkered field of view by making exaggerated head movements. Transferring attention between the external scene and information displayed in the cockpit requires peering beneath the eyepieces, which is at least wearisome, while the extra head-borne mass strains and fatigues the neck muscles, especially under harsh vibration and high-G. The operational gains have therefore been bought at the expense of considerably greater aircrew effort, and perhaps aircrew health.

5.4.4 NVG development

Alternative configurations have been devised that attempt to overcome the drawbacks of the ANVIS arrangement while maintaining fully independent left and right channels. Several designs have folded the optical paths of both the objective and eyepiece lenses to form a flattened package that fits inside the brow opening of the helmet and obviates the need for a fibre-optic twister. Most of these also use a "see-through" design for the eyepiece. They have the advantage of reducing the overhanging moment and enabling the user to see the intensified images superimposed on the attenuated direct view of the scene, but because the eyepieces must incorporate a beamsplitter, images are dimmer and the exit pupil smaller. The flattened optical path also places some optical elements closer to the face where they mask sideways peripheral vision.

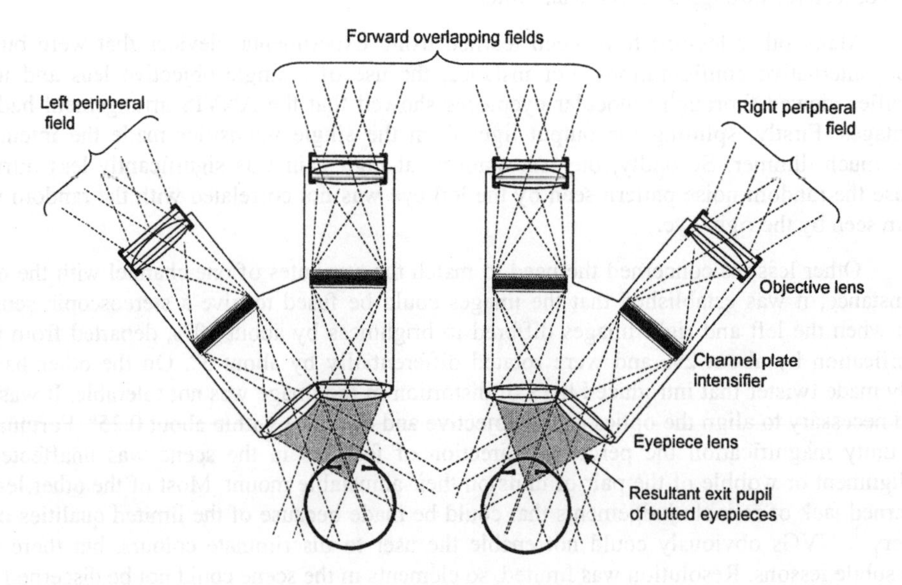

Figure 5.14 Simplified schematic for the panoramic night vision goggle, omitting the folds in the optical paths or the fibre-optic twisters that erect the inverted images

The finite number of channels in the intensifier plate imposes a trade-off between spatial resolution and field of view which is identical to the consideration given to the finite number of pixels in a helmet-mounted display system discussed in Chapter 9. One solution has been to increase the number of channel plates and associated optics to make a "panoramic" goggle, illustrated schematically in Figure 5.14. This extends the sideways field of view from the current value of about 40° to about 100° by supplying each eye with a pair of intensified windows into the world; the original forward view and an additional one on the outside that extends the lateral peripheral field[14]. The forward windows for the left and right eyes overlap, as before, but the left peripheral field is seen only by the left eye, and the right peripheral field only by the right.

The detailed engineering is an example of clever integration, particularly of the folded optical paths and the placement of the high voltage converters. The resultant mass is about 800g; only fractionally more than a conventional ANVIS goggle. The eyepieces are fixed, and different focus settings must be arranged by fitting sets of additional thin lenses over the eyepieces. The most evident difficulties are imposed by the small exit pupil volume. Each eye must be placed close to the butted eyepieces, leaving little room for a face protection visor, and when looking within one of the peripheral fields, the view of the other peripheral field is lost because the eye pupil rotates out of the exit pupil.

The quality of the image and the robustness of the channel plate intensifier will be enduring development drivers. The gain of Gen-III devices has improved steadily, and with a progressive thinning of the barrier layer there is less inter-channel electron transfer and less loss of contrast. The automatic gain control that varies the voltage to limit the current drawn by the plate and avoid excessive luminous output now incorporates a rapid switching or "gating" mechanism[15]. This extends the upper range of light levels in which the goggle can be exposed by altering the on/off time proportion, and it is now possible to fly from dark terrain all the way into a floodlit hanger apron without damaging the intensifiers.

5.4.5 Display night vision goggles

Many goggle designs now accommodate an additional module, usually mounted on the objective lens of one of the channels, that injects symbols into the field of view. These modules are essentially simple helmet-mounted displays, as described in Chapter 9, in which the image need only be very dim because it is intensified by the same factor as the external scene. A wide range of miniature flat panel devices are eligible, including Active Matrix Liquid Crystal Displays (AMLCD) and Active Matrix Electro-Luminescent (AMEL) devices, and the latter is particularly suitable. The filter in the objective lens of the other channel usually has a small opaque area equal to the size of the display injection mirror so that the two channels receive the same excitation and the two intensified images are equally bright.

The information provided by the symbol overlay is similar to that presented in a daylight HMD, and equally useful. It can however be noted that in conditions where the intensified scene contains little useful contrast, such as when flying through cloud, the concentration of attention on the monocular symbols has been reported to induce perceptual anomalies[16]. The main effect is a feeling that the sensitivity of the two eyes is markedly unbalanced, i.e. the eye that is not presented with the symbols has heightened sensitivity. This persists long after the causal condition is removed and makes the individual uneasy about the general reliability of his vision and induces the sensation that the sensitised eye has delayed accommodation when transferring between near and far focus states. The effect may be exaggerated where the symbols are shown to the non-dominant eye.

5.5 Sensor-assisted vision

5.5.1 Origins

Aircrew had used imaging sensors long before NVGs were adopted for night operations. For instance, the navigator in the rear seat of the RAF Buccaneer used a head-down CRT to view the monochromatic image produced by the narrow field-of-view TV camera mounted in the nose of the Martel air-to-surface missile.

Numerous studies had been conducted to assess the ability of trained observers to use a sensor and display, mainly to predict the range at which a target could be found. The tonal range of the system and the rate of movement of the feature within the scene were important, especially in viewing complex scenes, but the factor that had most influence was the size of the feature in relation to the line structure of the TV system[17]. The last finding has been condensed into a set of rules of thumb known as the Johnston Criteria. For target *detection*, the contrast of the feature must be above threshold in only a pair of raster lines in the image. For *recognition*, i.e. that the detected feature has military significance, it must subtend three or four pairs of lines. To *identify* the feature, for instance to distinguish a friendly from an enemy armoured vehicle, more than about seven pairs of lines are needed. Figure 5.15 gives an idea of what can be seen when a tank-like object satisfies these criteria.

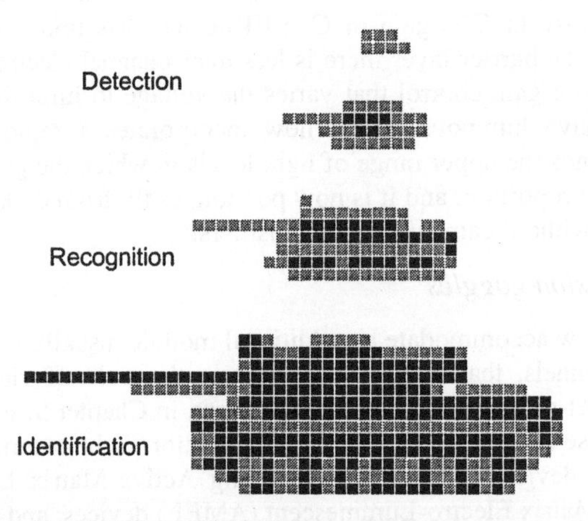

Figure 5.15 Illustrating the Johnston Criteria for detection, recognition and identification of entities in raster imagery

In the late 1960s, flight trials were conducted in fast jets using a variety of low-light TV cameras employing orthicon, plumbicon and vidicon sensors[18]. These indicated that targets could be found at night with a reliability that compared well with the use of a conventional camera during daylight. It was also possible to control the aircraft flight path by displaying on the head-down CRT the image from a wide field-of-view low-light camera overlaid by a set of HUD symbols. The operational benefit was greatest at low altitude, but the pilot was disinclined to spend enduring periods ignoring the view through the transparency no matter how little information it provided in comparison with the picture on the head-down CRT. Fortunately, this qualm was largely met by the contemporary development of the raster head-up display that allowed a suitably collimated image to be superimposed accurately at 1:1 magnification on the direct view of the scene.

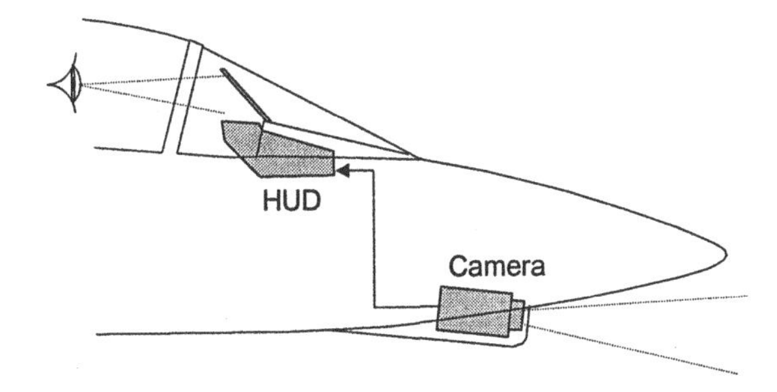

Figure 5.16 The idea of relaying camera imagery on a head-up display

As shown in Figure 5.16, the combination of a fixed forward-looking sensor and a HUD matured into a way of giving the pilot what was effectively an artificial forward window into the outside world that could be used in conjunction with the normal HUD symbols. Such systems have been installed in RAF Tornado GR4 and Harrier GR7 aircraft, and in some aircraft operated by other Airforces.

In most cases the camera was a thermal imager or Forward-Looking Infra-Red (FLIR) sensor employing transducer elements made from a mercury, cadmium and tellurium alloy that sensed radiation in the 8μm to 14μm wavelength range. As summarised in Appendix A, radiation in this band is emitted strongly by objects at the surface temperature of the Earth, about 300°K, and as shown in Figure 3.2, most of this range was transmitted by atmospheric water vapour and carbon dioxide. Although the density of water in a rainstorm would cause too much attenuation, the sensor could give a useful image at night and in some types of fog that would occlude or scatter the shorter wavelengths of visible light. The thermal imager, with a spatial resolution comparable with a TV picture, that could respond to temperature differences as small as 0.2°C was a particularly effective way of detecting small objects a few degrees hotter than their surroundings. Thus, although the window onto the world was monochromatic, fixed to the aircraft boresight and, at about 25° square of very limited field-of-view, it had great tactical value for ground attack aircraft.

It is of note that temperature differences can be represented by the varying shades of grey in a monochromatic display in two ways; the "white hot" mode representing relatively hot features as bright, and the "black hot" mode representing hot features as dark. The former tends to enhance the visibility of man-made entities and is generally preferred for finding and identifying targets and tactically relevant features. The latter tends to make the scene appear more natural, mainly because the cold sky appears brighter than the ground, and this mode is generally better for appreciating the lie of the land. Most pilots prefer the facility to switch between modes and, despite automatic optimising electronics, be able to adjust the way the temperature range is represented by shades of grey on the display. In general, aircrew learn to interpret the thermal imagery, for instance to beware the darkness of water and the flat, texturelessness of snow and rain-sodden ground. The last effect is particularly important because snow and surface water tend to even out temperature differences, and with limited temperature variation to map onto the display dymanic range, texture and texture flow information can lose considerable contrast. The subtle cues that, for instance, stimulate ambient vision subconsciously to detect the differing flows of texture that correspond to rising rather than falling ground, can disappear. Cues also changes strength over time, the classic example being the progressive wash-out of thermal contrast after nightfall as temperature differences settle out.

The navigation pod for the Low Altitude Navigation and Targeting Infra-Red for Night (LANTIRN) system, which can be carried by some F-16C, F15E and F-14 aircraft, provides the pilot with a 28° horizontal by 21° vertical field-of-view thermal image on the HUD. One of the HOTAS grip-top switches can be used to slew the thermal sensor to point 11° to the left or right, which is particularly useful for checking the relevant terrain when entering a turn. To avoid the possibility of flying unintentionally with a misaligned forward view, when released the switch springs back to the central position and the sensor is re-locked to the boresight. Such a facility is not included in the Tornado and Harrier. Instead the system is usually operated in conjunction with NVGs that enable the pilot to look around more flexibly and gives imagery in the near-IR that complements the longer wavelength information picked up by the boresighted thermal sensor. However, as noted above, the pilot views the collimated HUD through the NVG and this generally degrades the dynamic range and resolution of the HUD image.

5.5.2 Visually-coupled system

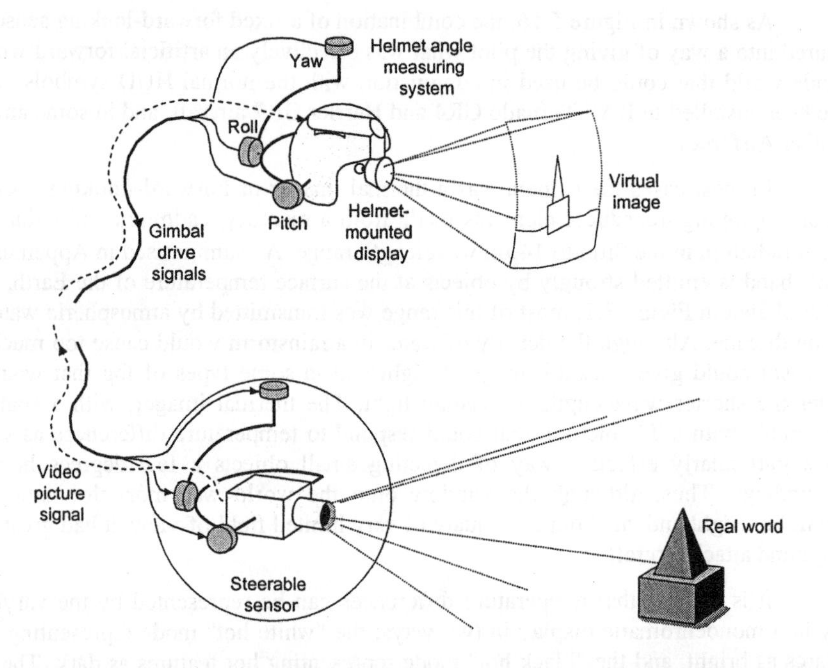

Figure 5.17 The idea of linking a steerable sensor to a helmet-mounted display and a helmet orientation measurement system to create a Visually-Coupled System (Reprinted with permission from QinetiQ)

As shown in figure 5.17, a visually-coupled system (VCS) projects a virtual image, using a helmet-mounted display, from a gimballed sensor that moves in response to measurements of the helmet orientation. With one-to-one correspondence between changes in the helmet and sensor orientation, and between the display field of view and the sensor field of view, the helmet wearer feels that he can look around in a natural "egocentric" way from the sensor location. Although the instantaneous field-of-view of the image can be small, the freedom to move the head extends the overall field of regard to whatever can be provided by the sensor window, or the helmet sensing system and the gimbal articulation. The possibility of providing a human operator with a veridical sense of immersion within a remote, inaccessible or hazardous location has stimulated investigations into a large range of uses, limited only by the engineering difficulties of inserting a suitable steerable camera and arranging the two-way signalling. For instance, a stereoscopic pair of cameras mounted on a steerable gimbal has been incorporated into robots that are sent to examine and disarm unexploded bombs. Similar sensors are used on submersibles to examine the underwater structures of oil-extraction platforms, and miniature endoscopic cameras are routinely inserted into patients for non-invasive surgery. Rather than relay the pictures on fixed display units, these can be made into immersive visually-coupled systems using a head-mounted display with a suitable orientation measurement system.

The best known application of a VCS in aviation is the Integrated Helmet and Display Sighting System (IHADSS) fitted to the Apache AH-64 attack helicopter. IHADSS relays imagery to the pilot via a 40° by 30° monocular helmet display from a thermal sensor in response to a helmet tracker employing scanning infra-red beams. The latter is described in Chapter 11, and the display system and the attendant perceptual problems are described in Chapter 9. It is the perceptual oddities as well as the need to provide some immunity to failure that had encouraged all subsequent projects, such as the US Commanche AH-66, the Franco-German Tiger attack helicopter and combat jets such as Eurofighter, Rafale, F-22 and the Joint Strike Fighter, to adopt binocular helmet-mounted display systems. The latter have benefited from pre-cursor flight test experiments to evaluate the engineering practicality and operational feasibility of the VCS for conducting ground attack missions in fast jets. The main USAF-sponsored activities[19, 20] began in the late 1980s with a well integrated steerable thermal sensor mounted above the nose of the Falcon-Eye F-16 test aircraft, coupled to a monochromatic biocular 30° HMD and pointed by a Honeywell magnetic helmet tracker. These tests showed that the head lag delay of 20msec was tolerable and that most of the problems were due to detailed aspects of the equipment, such as the mass and fit of the headgear, rather than the intrinsic idea. Trials in the UK using a Tornado test aircraft have reinforced these conclusions and emphasised the need for automatic optimisation of the thermal sensor.

The Distributed Aperture System (DAS) under development for the JSF is an array of fixed thermal sensors that are fitted in multiple sites on the airframe and oriented to cover the full 360° azimuthal envelope. The system is linked into the mission avionics to supply information for navigation, warning of missile attack and for searching and tracking ground and airborne targets. The imagery is also intended to provide the pilot with a view of the world, and it is likely that this will be implemented in the form of a VCS. Although the use of a fixed sensor array will remove difficulties associated with the size and siting of apertures needed by gimballed sensors, the system designers are left with a difficult computational requirement. As described in Chapter 9, the pictures from the sensors must be remapped to preserve their spatial correspondence with world when displayed on the pilot's HMD. This will need a significant amount of image processing, because the mapping transformations change with head orientation and the processing must be done with sufficient speed to keep pace with dynamic head movements. A system of this sort is envisaged in the Flying Infra-red for Low-level Operations (FLILO) program undertaken by the Advanced Airlift and Tanker Systems group of the Boeing aircraft company[21]. The configuration has the additional benefit in this case of providing enhanced vision to both of the flight crew from the same array of sensors.

Despite such difficulties, the VCS is an idea with plenty of scope for useful improvement. The use of an auxiliary targeting sensor or suitable zoom optics could enable the pilot to point his head so that an object in the scene is placed at the centre of the HMD field of view, then press a HOTAS button to get a magnified image. This would enable target identification at much greater range. Significant improvement can also be gained using a more versatile system of sensors. The current thermal imager, which senses long wavelength (8 – 15μm) IR emissions, could operate in concert with other passive devices covering the (0.7 – 1.2μm) visible and near IR, the medium wavelength (3 -5μm) IR and short microwave (0.1- 0.5mm) bands. These could provide 2-D pictorial detail in almost any combination of meteorological and luminous conditions. An active sensor, such as a high pulse rate laser radar, could fill in the third dimension, albeit at lower spatial resolution. Knowledge about the shape of the terrain and the distance to objects such as pylons and electrical cables that stand proud of the background is extremely valuable. Although the pilot could be given the flexibility to choose images from the most useful sources as conditions evolve, it would undoubtedly be more satisfactory to incorporate some form of automatic sensor manager. This would spare the pilot the periodic job of scanning through the imagery, and it would open up the possibility of fusing the information from all of the sensors in the suite. Current research is aimed at understanding how to perform such "image fusion" so that the most relevant features and cues are extracted, and how the result should be represented using the available gamut of displayable shades and colours so that it can be assimilated easily by the pilot.

5.6 Challenge

One of the notable paradoxes of modern military operations is that aircrew are selected for excellence of vision but they must fly at night and in poor weather using aids that give them a view of the world as poor as seen by a colour-blind, blinkered myope. The main engineering challenge is to develop the present sensor-aided vision systems so that they provide a view of the world that is at least as useful as optimum natural vision, without the degradation that befalls natural vision in low and high light levels and in cloud and fog.

The relative advantages of natural and artificial vision, and the contingent effects on the form of the crewstation, are considered further in Chapter 14.

References

1. Hall D A, *Technical preparation of the airplane "Spirit of St. Louis"*, Technical Note No 257, National Advisory Committee for Aeronautics, Washington, 1927
2. Manning T K, *Head mobility in the Tornado GR1*, Unpublished MOD report
3. Atkins E R, Dauber R L and Price J W, *Study to analytically derive external vision requirements for US Army helicopters*, Report No 73-1, US Army Aviation Systems Command, St. Louis, 1973
4. Brennan D H, *Vision and visual protection in fast jet aircraft*, in *Visual effects in the high performance aircraft cockpit*, AGARD LS-156, 1988
5. Blackwell H R, *Contrast thresholds of the human eye*, Journal of the Optical Society of America 36, pp624-643, 1946
6. Turner M J, *Observations on the normal subjective amplitude of accommodation*, British Journal of Physiological Optics 15, 70-100, 1958
7. Task H L, *Vision through aircraft transparencies*, in *Visual effects in the high performance aircraft cockpit*, AGARD LS-156, 1988
8. *Standard test method for haze and luminous transmittance of transparent plastics*, ASTM D 1003-00, American National Standards Institute, 1961
9. *Standard practice for optical distortion and deviation of transparent parts using the double exposure method*, ASTM F 1733-00, American National Standards Institute, 2003
10. Goodrich G W and Wiley W C (Bendix Corporation), *Electron multiplier*, US Patent 3,128,408, filed 1960

11. Catchpole C E, *The channel image intensifier*, in Biberman L M and Nudelman S (eds) *Photoelectronic imaging devices* Vol 2, Plenum Press, New York, 1971

12. Kotulac J C and Morse S E, *Focus adjustment effects on visual acuity and oculomotor balance with aviator night vision displays*, Aviation, Space and Environmental Medicine, pp. 348-352, 1994

13. Crowley J S, Rash C E and Stephens R L, *Visual illusions and other effects with night vision devices*, in Lippert T M (ed) *Helmet-Mounted Displays III*, SPIE Proceedings Vol1695, 1992

14. Jackson T W and Craig J L, *Design, development, fabrication, and safety-of-flight testing of a panoramic night vision goggle*, in Lewendowski R L, Haworth L A and Girolamo H J (eds) *Helmet- and Head-Mounted Displays IV*, SPIE Proceedings Vol 3689, 1999

15. Estrera J P and Saldana M R, *High-speed photocathode gating for generation-III image intensifier applications*, in Rash C E and Reese C E (eds) *Helmet- and Head-Mounted Displays VIII*, SPIE Proceedings Vol 5079, 2003

16. Jarrett D N, Ineson J and Cheetham M, *Visual anomalies and display night vision goggles*, in Rash C E and Reese C E (eds) *Helmet- and Head-Mounted Displays VIII*, SPIE Proceedings Vol 5079, 2003

17. Johnston D M, *Target recognition on TV as a function of horizontal resolution and shades of grey*, Human Factors, 10(3) pp. 201-210, 1968

18. Karavis A, *Minions of the moon: Low-light TV*, in *Royal Aircraft Establishment News* Vol 33 No 9, pp. 18-21, 1980

19. Lydick L N, *Head-steered sensor flight test results and implications*, in *Combat Automation for Airborne Weapon Systems: Man/Machine Interface Trends and Technologies*, AGARD-CP-520, 1993

20. Church T O and Bennett W S, *System automation and pilot-vehicle-interface for unconstrained low-altitude night attack*, in *Combat Automation for Airborne Weapon Systems: Man/Machine Interface Trends and Technologies*, AGARD-CP-520, 1993

21. Guell J J, *FLILO (Flying infra-red for low-level operations) an enhanced vision system*, in Lewendowski R L, Haworth L A and Girolamo H J (eds) *Helmet- and Head-Mounted Displays V*, SPIE Volume 4021, 2000

Chapter 6

The Display Suite

6.1 Rationale

It is possible to control the aerodynamic state of an aircraft, particularly a glider or a hang-glider, by interpreting kinaesthetic feedback forces and subtle wind-flow noises, and it is possible to execute manoeuvres using only a clear view of the surrounding terrain. However, in any more complex air vehicle or regime the crew will need additional information to fly safely.

The required information depends fundamentally on the tasks that the aircrew must perform. In a simple powered aircraft with no on-board navigation aids or radios, the pilot needs only the airspeed and altitude. To reach the intended destination the pilot will also need a compass and a map, although some indication of the fuel tank contents, the engine speed, the engine temperature and the oil pressure give comforting reassurance that the powerplant is likely to continue operating satisfactorily. At the other extreme, as discussed in Chapter 2, a military pilot must perform a wide range of tasks by managing sophisticated on-board systems in almost any meteorological and tactical circumstances. The suite of display devices must provide the necessarily comprehensive range of information in a clear, convenient, unambiguous and easily assimilated way.

The intention behind this chapter is to provide an overview of the factors taken into account in the design of the suite of display devices, with a brief explanation of pertinent physical principles to clarify inherent limitations. The engineered form and the perceptual oddities of head-down, head-up, helmet-mounted and auditory display systems are discussed in more detail in separate chapters.

6.2 Background

The history of information presentation in aviation began with the airframe manufacturer supplying an instrument panel containing a cluster of dials showing individual parameters, such as airspeed and compass bearing, together with lights that alerted the pilot to potentially consequential situations such as low engine oil pressure. Toggle switches and rotary switches, which activated systems and selected settings, also acted as display devices in as much as the toggle or marker invariably pointed clearly to a caption alongside the switch to show the setting. It was left to the operator to provide all the other information; maps, emergency procedure flip-cards, slide-rules, lists of radio channel frequencies and a notepad for writing down information, such as way-point co-ordinates received over the radio. Instruments were usually clustered into groups, such as engine indicators and fuel indicators, and systems such as radios were not separated from their control panels. With limited choice of component devices and little flexibility, it was usual for the manufacturing engineers to lay out the cockpit with a mixture of precedent and pragmatism, moderated by advice from the factory test pilots. Until the late 1940s there was considerable variation between the designs produced by different manufacturers.

One of the earliest studies to consider whether it would be advantageous to place instruments in a particular arrangement was carried out by Jones, Milton and Fitts[1]. In this classic human factors study, to understand how pilots made use of instruments during the approach to a landing, the investigators assessed the pilots' eye pointing direction by examining long sequences of film shot by a compact cine camera mounted near the instrument panel. The outcome was the suggestion that a particular arrangement, shown in Figure 6.1, would minimise the number and frequency of glances and the time required to take in the information.

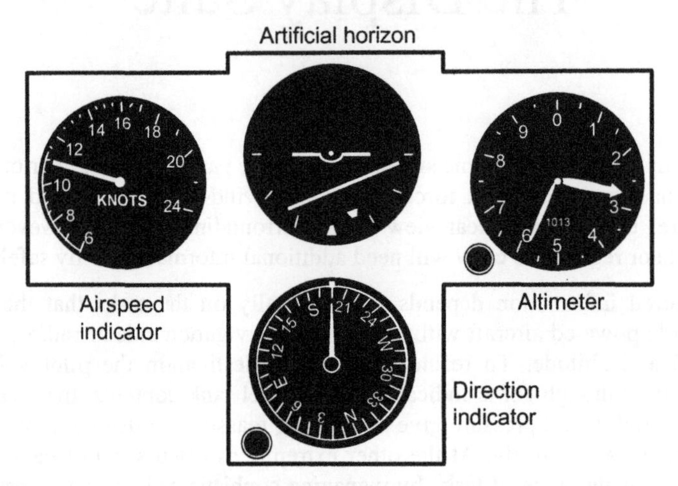

Figure 6.1 Standard T arrangement of the primary flight instruments

In clear conditions the pilot can control aircraft attitude from the general view of the terrain and judge the progress of the aircraft along the glide-slope from the size and aspect of the approaching runway. It is only necessary to glance down periodically to read the altimeter to check the gradually descending height, and at the airspeed indicator to ensure that this is safely above the stall. But when the external view is unclear and the attitude and progress of the aircraft cannot be gleaned from an indistinct horizon and runway, the Standard T comes into its own. The arrangement is easily learned, and it provokes an eye scan pattern that radiates from, and returns to, the central attitude instrument, the instrument that must be read most frequently. Most operators and Air Forces adopted this disposition of the primary flight instruments, and with the inclusion of additional markers in the attitude indicator that guide the pilot onto a prescribed runway approach path, it is known as the "blind flying panel".

As aircraft have evolved to include sophisticated and automated flight control, utility and mission systems, information produced by these systems has become increasingly important. In a modern aircraft, where the systems are controlled directly by computers, the crewstation is really a human-computer interface, albeit one designed to enable the conduct of a very demanding set of tasks under particularly onerous environmental constraints. Information presentation has now a great deal of potential flexibility, and the problem has moved from the optimum layout of panels containing individual gauges to become a consideration of the preferred arrangement of electronic display devices and the symbols they provide.

Information can be classified into categories, most of which are self-explanatory. For instance; safety-critical and non-safety-critical; quantitative and qualitative; continuous and discontinuous; spatial and logical; transient, temporary and enduring, aural and visual. Symbols may be numerical, alphabetic and iconographic. The display devices also have categories; dial gauges provide dedicated information, whereas modern devices are flexibly formatted and usually multi-function. Both can provide quantitative information using analogue and digital symbolic representations. In addition, modern devices present images in shades of grey or binary, in monochrome or colour, to be viewed monocularly or binocularly, as real or virtual. The latter can have narrow or wide field of view, and symbols can be conformal or non-conformal, stabilised to the frame, the ground, to space or to the aircraft, and separated stereoscopically to appear near or far. Thus, despite the wealth of guidance available from the human factors literature to assist the designers, there is much to play with and many uncertainties, mainly because so much depends upon the limitations and peculiarities of different types of display device.

6.3 Mechanical instruments

6.3.1 Cultural factors

The significance of a circle marked with short radial lines and numbers, together with a set of long radial pointers, is taught from infancy. So much so that the notion of time and the clock-face are inseparable, and clockwise and anti-clockwise are fundamental descriptors of rotational direction. Most quantitative information can be displayed with little risk of misinterpretation by exploiting this cultural familiarity, and we have come to regard gauges that use a dial resembling a clock-face as somehow natural. Interpreting the significance of the indication is straightforward; the pointer position is an analogue of the quantity, clockwise rotation of the pointer signifies an increase and the rate of rotation reflects accurately the rate of change of the quantity. The dial size, numerical font, pointer thickness and paint colours - usually a white pointer on a black background to maximise contrast but avoid glare in high ambient light - are all chosen to make the indication clear.

As illustrated in Figure 6.1, the addition of a word such as "knots" can remind the user of the relevance of the information, and as this relevance does not change the instrument is dedicated rather than multi-functional. Resolution is usually limited by the interval between scale makings, the pointer width and friction in the mechanism. The altitude instrument shows that different pointers can be used to extend the range of the indication without sacrificing accuracy by having one pointer for the thousands and another for the hundreds of feet, the latter rotating at ten times the rate of the former.

Another common way of displaying the value of a variable is to show it as a decimal number, represented by sufficient digits to give adequate range and resolution. Thus some modern altimeters may include a six digit counter inset within the dial. The pilot can check the aircraft height unambiguously by reading the numerical value, but monitor height maintenance and the rate of change of height easily by noticing how the pointer moves. This combination exploits the chief virtues of each form of indication. It should be noted that, to read a digital indicator, the pilot must fixate accurately on the individual digits. However, an analogue indication allows a pilot to glance down to fixate in the vicinity of the instrument and he need only partially re-focus his eyes in order to check that the orientation of the pointer is roughly as expected, or roughly in line with pointers on neighbouring instruments. A digital read-out cannot convey differences or temporal changes as readily as an analogue pointer.

Although some instruments are labelled to show what information they convey, e.g. "knots" or "feet", the pilot must for instance remember that the short pointer of an altimeter gives thousands and the long pointer the hundreds of feet, and that these refer to height above sea level rather than above the ground. This process of removing as much fixed information as possible from the display has reached a peak in the design of symbols for the head-up display which must be de-cluttered to the extreme. The general culture in aviation is for designers to assume that pilots will be familiar with the layout, significance and units of any displayed quantity, and any annotations are usually small and deliberately inconspicuous.

6.3.2 Pressure-driven instruments

Airspeed, altitude and vertical speed, the rate of gain or loss of altitude, are measured using a pitot probe which protrudes along the direction of flight and extends into the undisturbed air ahead of the aircraft. The probe contains an open-ended pipe into which the air is compressed by forward vehicle motion, and an outer concentric pipe with a closed forward end and carefully positioned vents in which the air comes to the external ambient pressure. The probe is normally heated to prevent ice accretion. In a simple installation these "dynamic pressure" (P_T) and "static pressure" (P_S) sampling pipes are connected directly to a set of mechanical indicating instruments.

Measurement of aircraft altitude exploits the reduction in atmospheric pressure with ascending altitude, noted in Chapter 3. In the altimeter, the static P_S sampler is connected to a chamber surrounding a partially evacuated aneroid capsule, a squat, thin-walled, metal cylinder with springy, convoluted end faces. The capsule expands in proportion to the difference between the internal and external pressures, and the movement is coupled through a series of linkages, gears and shafts to drive a pointer around a scale.

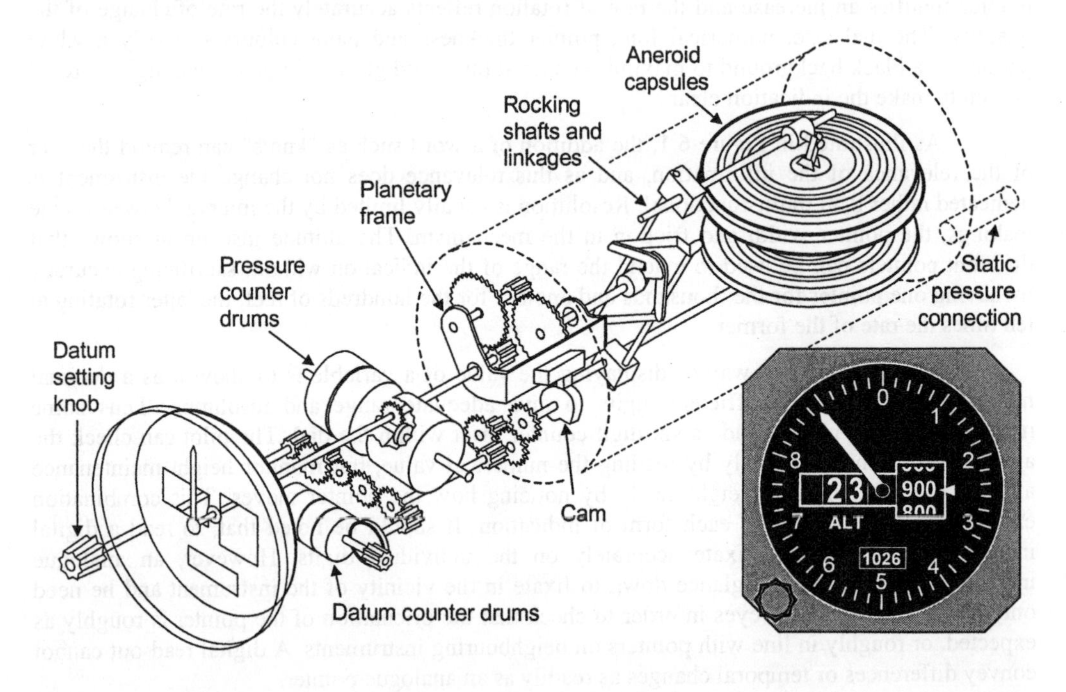

Figure 6.2 Schematic for a baro-mechanical altimeter giving a digital read-out of 1000s and 100s of feet using indexed drum counters. A pointer revolves once for each 1000 feet and can be read to ±10 feet. The datum pressure (here in millibars) is set using a dial-face knob

Figure 6.2 shows the internal arrangement of a typical baro-mechanical altimeter that, to avoid the possibility of confusion between multiple pointers, has a digital read-out with a single pointer. The instrument measures P_S but is calibrated directly in feet to show "barometric altitude" (Baro Alt, or H_P), which is related to P_S by a formula for the ICAO Standard Atmosphere.

For subsonic flight within the troposphere this relationship is:-

$$P_S = 1013.25\ (1 - 2.25577 \times 10^{-5} \times H_P)^{5.255879}$$

where P_S is in millibars (mb), H_P is in metres and 1013.25mb is the Standard pressure at sea level. Other formulae are needed for supersonic and stratospheric flight[2]. As noted in Chapter 3, the actual variation of pressure with altitude on any flight is invariably different from the Standard relationship. To minimise the possibility that the error can cause an accident, the pilot must be aware that there are three distinct, internationally recognised, ways to interpret instrument readings. Each of these requires a different value for the datum pressure, which is set using a dial-face knob shown in Figure 6.2.

One mode of operation, called QNE, sets the datum to the value for the ICAO Standard Atmosphere; 1013.25mb. This is used when the main concern is to avoid air-to-air collisions between aircraft operating well clear of the ground under guidance from an Air Traffic Controller. In this case, to emphasise the distinction between the actual altitude and indicated reading, pilots are asked to attain and hold a "Flight Level", which is the indicated Barometric Altitude expressed in hundreds of feet, so that an indicated Barometric Altitude of 30,000 feet is equivalent to Flight Level 300. Thus a pilot following ATC instructions to fly at Flight Level 310 will know that his aircraft will be safely separated in altitude from another at Flight Level 300, providing both altimeter instruments are calibrated to the same formula and both pilots have set the same 1013.25mb datum.

In the second mode, called QNH, the pilot must know, from the local meteorological report, the existing surface pressure at sea level for the current geographical location of the aircraft. With this value set as the datum the altimeter gives the pressure altitude above mean sea level (Alt AMSL). The third mode, QFE, is useful when operating close to terrain, for instance on approaching an airport when and the pilot is most concerned about the altitude of the aircraft relative to the airport; the height "above ground level" (Alt AGL) at the airport. To operate in this mode the pilot must know the current static pressure at the airport and set this as the datum value in the altimeter.

With careful design and manufacture, and the incorporation of bi-metallic strips to counter temperature changes, barometric altimeters can generally register changes of less than 20 ft and give errors of less than ±30 ft at sea level and ±300 ft at 30,000 ft. Most instruments also have an appreciable lag, in the order of a few seconds. Greater sensitivity, accuracy and speed are gained using electronics to calculate values from pitot pressure values obtained using transducers with a wide dynamic range. This is the rationale for the Air Data Computer (ADC), described below, which is used in most commercial and military aircraft. It is now normal practice to calculate altitude from a variety of sources, including the ADC, the inertial platform and the GPS system, and display the result on an electronic display that has the same appearance as the face of the instrument shown in Figure 6.2. The vertical clearance from the terrain immediately beneath the aircraft is measured using a downward-pointing radar sensor. Such RadAlt systems are fitted to many helicopters and to most fast jets performing a ground-attack role.

In the Air-Speed Indicator (ASI) the inside of the instrument case is connected to the static tube (P_S), as in the altimeter, but the inside of its aneroid capsule is connected to the open-ended tube of the pitot head to receive the air compressed by the forward motion of the aircraft (P_T). The volume of the capsule expands with increasing dynamic pressure and reducing static pressure in proportion to the difference (P_T - P_S). Movements of the capsule are coupled via levers, cams, shafts and gears to move a pointer around a dial annotated in units of airspeed, usually knots, which is known as Indicated Air Speed (IAS). Owners of light aircraft and other types with a limited envelope usually assess errors by calibrating the IAS against a stopwatch or a GPS, and attach a card alongside the instrument to enable easy conversion to Calibrated Air Speed (CAS). However, the drop in air density and temperature with increasing altitude, combined with the changes in angle of attack and the errors introduced by the installation of the pitot head, make both IAS and CAS very unreliable indicators of the True Air-Speed (TAS). Most aircraft that fly above a few thousand feet and a few hundred knots need more accuracy for navigation calculations. The ASI is used mainly as a back-up and for landing and take-off, although pitot measurements of IAS at low airspeed under the disturbed airflow of a helicopter rotor are particularly unreliable.

The insides of the Vertical Speed Indicator (VSI) and the aneroid capsule are both connected to the static pressure line from the pitot head, but the connection to the inside of the instrument is restricted by a small aperture. If P_S is constant, the pressures inside and outside the capsule settle to be equal and the instrument reads zero. During a diving manoeuvre the increase to P_S causes the internal pressure in the aneroid to exceed the restricted pressure inside the casing, so the capsule expands in proportion to the difference, and this drives the pointer to indicate a particular rate of descent. The opposite indication is given when ascending into lower pressure air. The most significant errors occur with altitude, because the rate of change of pressure with altitude decreases with increasing altitude. In most instruments this is partially compensated by a conical valve mounted on a bi-metallic strip that restricts the air flow through the aperture in the cooler air at high level.

6.3.3 Air data computer

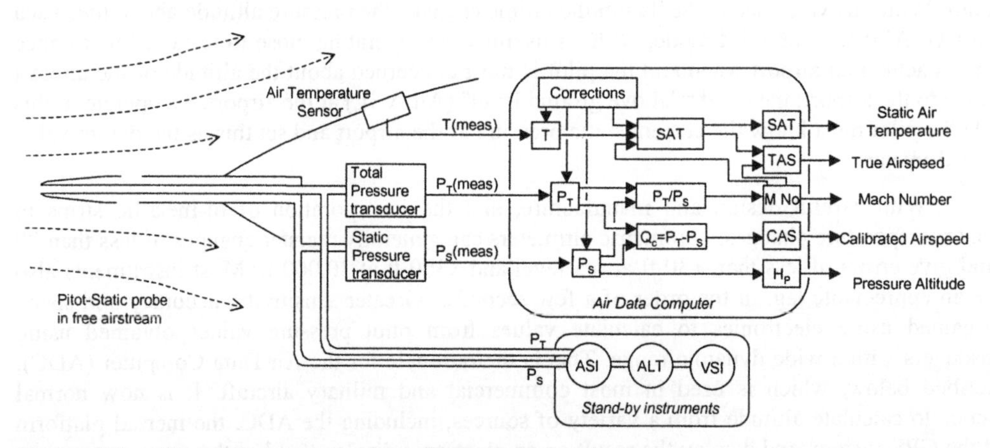

Figure 6.3 Simplified schematic for an air data computer

Aircraft with a wide speed-height envelope, electrical instrumentation and fly-by-wire controls still have a pitot head, with a heater to prevent ice accretion at altitude, and they still rely on accurate sampling of P_T and P_S. However, as shown in Figure 6.3, the dynamic and static pressures, and temperature are transduced into electrical signals and the computations are performed electronically in the Air Data Computer. The pneumatic instruments are usually retained as a stand-by in the event of electrical system failure. Electro-pneumatic transduction and the more complex corrections that can be implemented by digital processing in the ADC enables significantly more accurate measurement of flight parameters.

Within the subsonic tropospheric portion of the flight envelope, the relationships are defined as follows. Calibrated Air-Speed (CAS or V_C) is calculated from the dynamic pressure Q_C ($=P_T - P_S$) using :-

$$Q_C = 1013.25((1 + 0.2(V_C / 340.294)^2)^{3.5} - 1)$$

where Q_C is in millibars, V_C is in $m.s^{-1}$ and the speed of sound at sea level is 340.294 $m.s^{-1}$. Mach Number (M) is obtained from the ratio of the total pressure to the static pressure :-

$$P_T / P_S = (1 + 0.2 \times M^2)^{3.5}$$

The measured air temperature T_M can be corrected for dynamic heating effects and used to calculate the static air temperature T_S :-

$$T_S = T_M (1/(1 + r\, 0.2M^2))$$

where T_S and T_M are in °K. Finally the True Air Speed (TAS or V_T) can be calculated :-

$$V_T = 20.0468M(T_S)^{1/2}$$

As shown in Figure 6.3, the ADC also takes account of a variety of systematic errors. The principal sources are transducer non-linearities and disturbances to the sampled airflow that arise from the siting of the pitot head. These are measured carefully, usually in the early stages of flight testing of a new aircraft type, and the values of the corrections are stored as look-up-tables in the ADC memory. Variations also arise because the probe points away from the impinging airflow as the angle of attack and angle of sideslip increase. Improved accuracy can be attained by coupling the output from the Angle of Attack (AoA) sensor, a miniature horizontal wind-vane mounted close to the pitot head, into the ADC. Current fighter aircraft, such as Eurofighter, employ an integrated sensor unit in which the front of the AoA wind-vane contains a miniature pitot that points directly into the airflow.

The measurements produced by the ADC are the source of most of the primary flight parameters needed by the flight control, weapon aiming and navigation systems in modern aircraft, as well as the systems that display this information to the crew. The aircraft altitude is also encoded into the identifying signal transmitted by the transponder in most civil aircraft in response to an interrogation request from the Air Traffic Control radar.

6.3.4 Inertial instruments

The main inertial instrument gives the pilot an easily understood indication of the aircraft elevation and bank angles. This attitude information is vital if the pilot is to avoid the disorientation that occurs, as noted in Chapter 3, whenever the pilot has no view of the sky, horizon and ground.

Figure 6.4 illustrates the operating principle of an Artificial Horizon attitude indicator (AI), which is mounted on the front panel of the cockpit facing rearwards with the roll axis parallel to the aircraft fore/aft axis. The instrument exploits the inertial properties of a gyro; a small, dense, flywheel that spins rapidly on very low friction bearings within a set of gimbals. Before take-off, the gyro is caged - the gimbals are locked - so that it can be spun up about an axis aligned with the local Earth vertical. The gyro is uncaged before take-off, and throughout the flight it continues to supply a vertical reference while the aircraft pitches, rolls and yaws around it. The innermost (yaw) gimbal, which is free to move in the aircraft pitch plane, is coupled to a pitch arm. The pilot sees the arm as a white-painted horizon against a matt black background relative to an aircraft reference symbol fixed in the centre of the instrument glass. As shown in Figure 6.4, a pair of pins that move along slots in either side of the pitch arm reverse the movement so that when the aircraft is pitched nose up the arm is moved downward and the pilot sees the horizon line descend. A pitch down manoeuvre causes the opposite movement so that the horizon rises above the aircraft symbol. During banking manoeuvres the gyro continues to point vertically and hold the artificial horizon parallel with the real horizon while the rest of the instrument, including the central aircraft symbol and the bank angle reference marks, roll with the airframe.

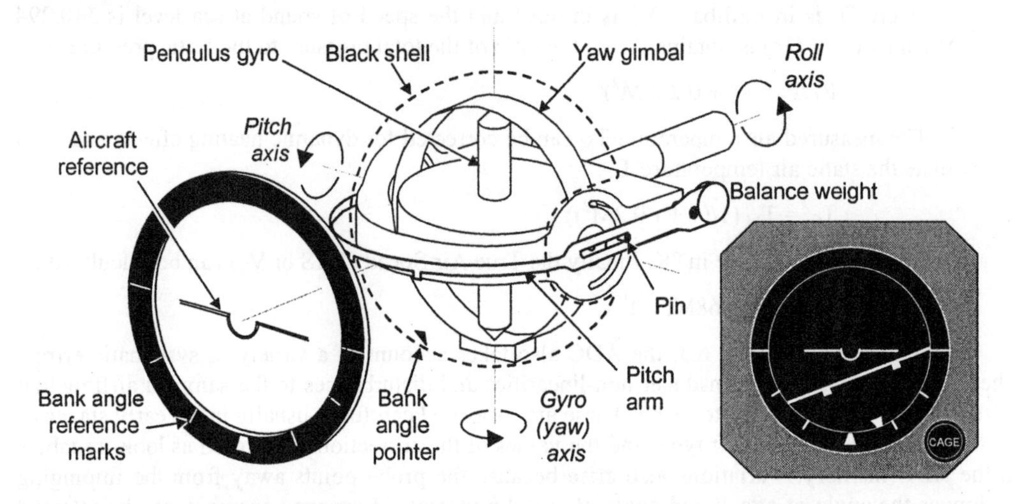

Figure 6.4 Main elements of a typical Artificial Horizon attitude indicator

The principal errors arise from precession; the intrinsic tendency of a spinning mass to turn about an axis perpendicular to the spin axis when perturbed by a torque about the third orthogonal axis; a torque about the roll axis induces a pitch rate, and vice versa. The rate of precession is proportional to the torque and inversely proportional to the gyro spin rate and inertia. Friction in the gimbal bearings and the pitch arm therefore induce the gyro to drift slowly away from the vertical, and manoeuvres induce a cross-coupled transient misreading. Also, to enable the gyro to self-erect, the centre of the spinning mass is set below the pitch axis, and the small torque applied by the pendulous action induces a precession error that is particularly noticeable during take-off acceleration. The most serious consequence arises if a gimbal is impeded, for instance on reaching an end stop, and the resulting torque causes a very rapid precession; the gyro is said to topple or tumble.

Many embodiments have been devised to allow unimpeded gimbal rotation, minimise friction and enable recovery to the vertical. Early versions were air-powered. The bearings were conical and supported by a thin air gap, and gyro rotors were driven at about 15,000 rpm by air jets that blew against cupped slots machined in the outer cylindrical surface. The air flow was arranged by coupling an engine-driven vacuum pump to the instrument to produce a partial vacuum inside the case, and this drew in air from the cabin to the jets. An ingenious mechanism was incorporated that re-erected the rotor during steady wings-level flight by inducing a corrective torque through the action of a set of pendulous vanes that impeded the efflux through slits in a cylinder formed around the rotor.

Most modern instruments are electric. They are less susceptible to degradation through contaminated air, and continue to work under emergency battery power if the engine or engine-powered pump fail. However, in the event of electrical failure a solenoid-actuated flag appears in the corner of the glass face as a positive indication that the indication should be ignored. The gyro is in effect the rotor of an electrical motor, and power is transferred through the bearings via slip rings and low friction brushes. The design is similar to that shown in Figure 6.4, but with unlimited pitch and roll gimbals, and a clear delineation between a dark brown ground and a light blue sky.

A gyro is also used in the Turn and Slip Indicator, which enables the pilot to execute a balanced turn by showing the rate of the turn and whether the aircraft is side-slipping. In this case gyro precession is exploited to sense the rate of turn in the aircraft yaw plane, the rotor spinning on a lateral (pitch) axis in a gymbal that is free to yaw but constrained by springs in roll. When turning about the yaw axis, the torque induces a roll precession that is resisted by the springs. The equilibrium roll angle depends upon the rate of turn, which is indicated by a roll pointer moving across a scale. When the turn ceases, gyro precession stops and the springs restore the gyro to the lateral aircraft axis and the pointer to the midpoint. The scale is calibrated in multiples of $3°.s^{-1}$, or 180° per minute, which is a "Rate 1" horizontal turn for a pilot using instrument flying procedures.

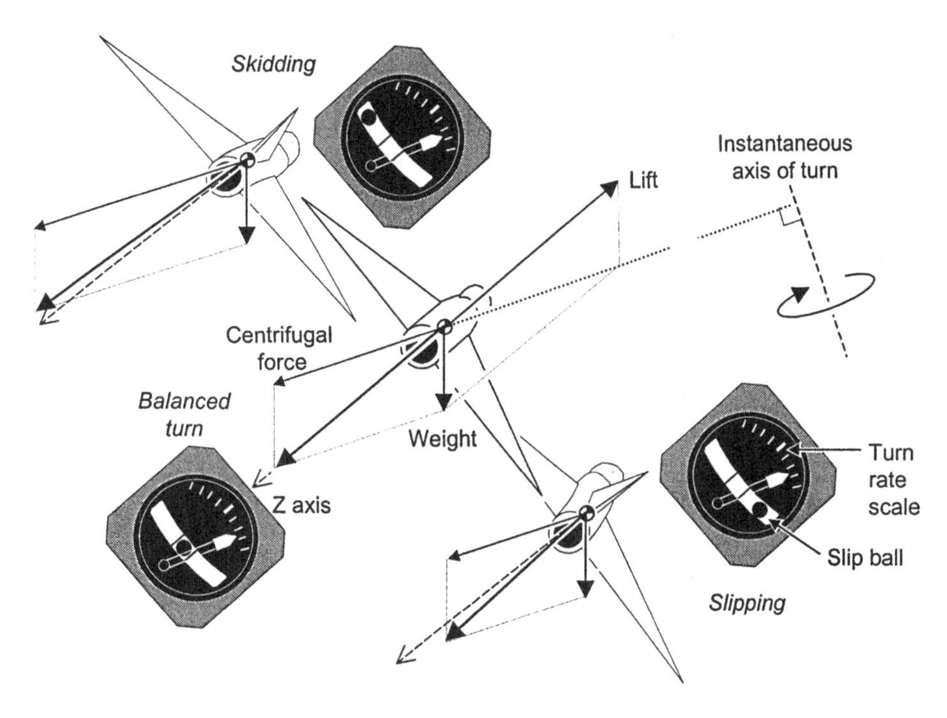

Figure 6.5 Turn and Slip Indicator illustrating cases where an aircraft is turning to the right and the instantaneous velocity vector is directed into the paper

The slip indicator is either a pendulum that can swing about the roll axis or, as in most modern instruments, an inclinometer that takes the form of a ball in a curved glass tube filled with liquid paraffin or oil to damp the ball movement. The displacement of the ball is proportional to the side-slip angle, and it helps the pilot make best use of the rudder. As shown in Figure 6.5, when the pilot is executing a balanced turn, the resultant of the gravitational and centrifugal forces is directed along the aircraft Z-axis and the ball remains centred. Too little bank or too much into-turn rudder causes excessive centrifugal force and a skid; the ball moves to the outside of the turn. Too much bank or too little into-turn rudder results in too little centrifugal force and a slip; the ball moves to the inside of the turn.

The gyro compass or Directional Gyro (DG) operates on the same principle as the Attitude Indicator (AI), but with the rotor spinning on a horizontal axis in a gymbal that allows the aircraft to pitch and roll, within limits, without affecting the gyro pointing direction. The scale takes the form of a shallow black cylindrical ring, marked with white lines at intervals of 5° and labelled every 10° from 0 to 350 with a two-digit identifier. This is attached to the outer gimbal that rotates about the aircraft Z-axis. Before take-off, with the aircraft pointing along the centre-line of the runway, the pilot uses a knob on the instrument face to rotate the outer gymbal so that the heading of the runway on the scale is aligned with the "lubber line", a vertical reference line marked in the front glass of the instrument. Thereafter, the pointing rigidity of the spinning gyro stabilises the scale, and changes of aircraft heading are apparent from the movement of the lubber line around the scale.

Both vacuum and electrical forms of DG are used, and all share the same variety of precessional errors and corrective mechanisms as the AI. They also share a susceptibility to tumbling when a gimbal hits a stop. The pilot then has no alternative but to fly straight and level and reset the instrument using a magnetic compass as a reference. The effective answer has been the Gyrosyn, a system in which the panel-mounted heading indicator contains a gyro-based sensing unit, and this is automatically corrected using a flux-gate magnetometer[3]. The latter is mounted well clear of magnetic interference, for instance in a wing-tip, and the rotational signal is transferred as three-phase a.c.synchro currents amplified in an additional electronic unit. This form of heading reference system has the advantage of providing signals for additional indicators and an auto-pilot.

Servo-mechanical coupling has allowed the sensing element to be separated from the cockpit-mounted indicator, and in doing so provided space in the instrument for designers to incorporate related information from other systems.

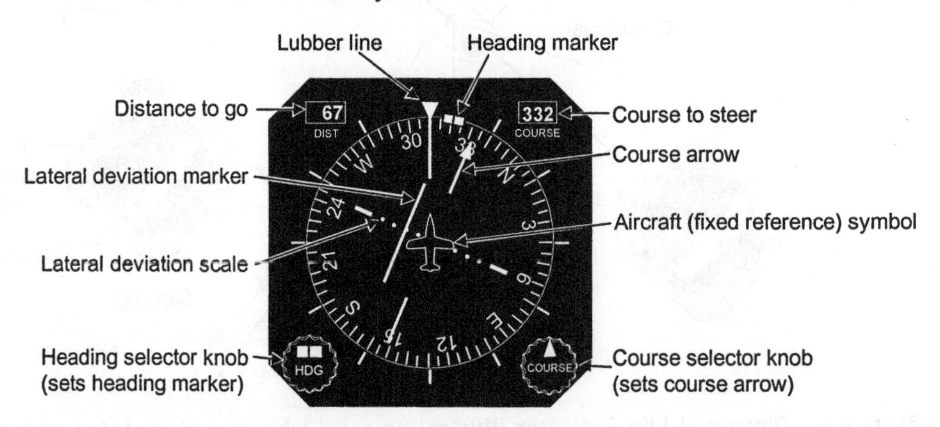

Figure 6.6 Compass and Position Indicator

Figure 6.6 shows how the compass has become the Compass and Position Indicator that supplies information from VHF Omni-Range (VOR) radio and Distance Measuring Equipment (DME) in the form of read-outs and moving markers. The instrument gives the pilot a summary of the "horizontal situation". He can think of the small aircraft fixed in the centre of the instrument glass as representing his vehicle and the circular compass card as representing the ground over which he is travelling. Rotation of the card corresponds to changing magnetic heading, with the lubber line indicating the current value. An arrow can be positioned on the compass card to show the required course, for instance to approach the runway centreline, and a separate marker can show the required magnetic heading corrected for crosswind. Having selected the VOR and DME channel using the appropriate control panel, the distance to go is presented automatically as a digital read-out, and the angular displacement from the selected VOR radial is shown by the lateral deviation marker. In the case shown in the figure, the pilot can fly the current heading until the small aircraft crosses the lateral deviation line, indicating that the aircraft is crossing the desired approach path, and then fly along this path by turning right to bring the heading marker onto the lubber line.

This dissociation of indicator from sensor has also been used in other instruments. The attitude indicator, with the addition of glide-slope and localiser bars that move in response to signals from appropriate beacons, has been transformed into a Vertical Situation Indicator that can guide the pilot to touch-down. Where electronic display systems are used, as in the "glass cockpit", and data is derived from gyros and accelerometers and combined with the data from a wide variety of beacon-derived and satellite-based GPS navigation aids, electro-mechanical indicators are now obsolete. Although such fine pieces of mechanical engineering are now relegated to a back-up role, the way they present information has endured in the design of computer graphical formats.

6.3.5 Other indicating instruments

A magnetic compass is fitted to most civil and military aircraft, mainly as a back-up that the pilot can use to check other heading indications and as a get-you-home reference in the event of electrical power failure. The best known UK device, the E2B compass, differs little from the version fitted during the 1940s. A pair of aligned permanent magnets are attached to the underside of a squat black-anodised aluminium cylinder, the compass card. The card rim is marked every 10° with vertical lines, dots at the intermediate 5°, identifying two-digit numbers every 30° and the four cardinal headings. The card is suspended centrally on the tip of a bearing post that protrudes from the base of the case so that it rotates freely in yaw, and can roll and pitch by about 30°. The front of the casing is a bowl-shaped glass window marked with a vertical lubber line, and the case is filled with clear silicone oil that supports the card and damps the card movement. The oil index of refraction matches that of the glass, which removes glare-inducing reflections at the inner glass surface and enables the outer spherical surface to magnify the card markings. The rear of the case contains a metal bellows to accommodate thermal expansion of the fluid. The unit is usually mounted on the windscreen supports so that it can be seen easily and is separated from magnetic disturbance by the other instruments. Errors due to field distortions can be minimised by adjusting small corrector magnets in the top of the casing.

The G-meter is a relatively simple inertial instrument that consists of a pointer showing the displacement of a mass that is constrained by a spring and a damper to move in one direction between buffered stops. It is calibrated in G-units, usually with a range of -4G to +10G, and installed in the instrument panel or on the glare-shield with the axis of movement in line with the aircraft Z-direction.

Instruments such as those indicating engine temperature, oil pressure, fuel tank contents and rotor torque - parameters that are sensed by electrical transducers - are moving coil gauges that are calibrated specifically to match the relevant transducer. They are constructed with well-damped movements and fine, robust springs and bearings, and the cases are invariably anodised alloy, filled with dry gas. All are expected to function and maintain calibration for more than 40,000 hours.

6.4 Modern cockpit displays

6.4.1 Introduction

Although the cathode ray tube (CRT) has gained prominence as the device for showing television pictures, other specialised forms of CRT have an equally long history, for instance as the "radarscope" and "oscilloscope", and bespoke CRTs have played an interesting part in cockpit development. Early versions were used to mimic the instruments described above, as a display device that was necessarily coupled directly to a specific sensing mechanism, albeit with some form of intermediate electronic processing. Only later were systems designed with a more flexible linkage between sensor, programmable processor and display device that enabled the latter to show different forms of information and for multiple devices to be assembled to form the glass cockpit. A somewhat prosaic nomenclature has arisen. A device that is placed in the main instrument panel in front of the aircrew and produces a directly viewed image is termed a head-down display (HDD). This is in contrast to a device that is placed above the coaming and produces a virtual image superimposed on the forward view through the windscreen, which is called a head-up display (HUD). The latter consists of a display device, such as a CRT, with additional optical elements that transform the real image into a projected virtual image. It is described in Chapter 8. Display systems that are mounted on the helmet and project either one monocular virtual image, like a telescope, or two images, like a pair of binoculars, may also be called head-up displays. However, to avoid confusion here, these systems are always called helmet-mounted displays (HMD). They are described in Chapter 9.

The earliest use of a HDD in fighter aircraft was as a radarscope. This illustrated one of the problems nicely. For instance, in the English Electric Lightning, a small CRT showed the returned energy from the scanned radar beam as a bright blip on a graph that could guide the pilot towards the unknown object. The vertical axis showed the range to the detected object, and the horizontal axis the instantaneous pointing direction of the radar beam as it swept left and right of the aircraft's forward axis. The CRT screen was set in the front panel alongside the primary flight instruments, and in daylight it was necessary to attach a large protruding rubber cowl, like an elongated scuba diver's mask, around the screen to help stabilise the pilot's head and eyes and shield sunlight from the screen. Without the cowl and the blocking action of the head, the sun glare reflected from the grey phosphor coating that formed the screen on the inside surface of the tube would have been much more intense than the sketchy picture emitted by the phosphor. Thus the rubber cowl was a practical, if inconvenient solution to a problem that still remains a challenge to display developers; to decrease the intensity of the unwanted reflected ambient light and increase the intensity of the emissions so that the picture can be discerned even when sunlight falls directly on the screen.

Fortunately, efforts have been sufficiently successful for HDDs to be accepted as the main means of presenting information in both military and civil aircraft, primarily because they are versatile. Unlike dial instruments, which are dedicated in the sense that they can only report specific states, such as the contents of a fuel tank, the nature of the information given by a HDD is whatever the designer chooses, and this can change in flight. In describing such programmable, or multi-function, displays it is therefore necessary to draw a distinction between the *surface* and the *format*. The first term refers to the physical device, and the second to a specific design of information-bearing pictures and symbols. Naturally, the second depends greatly on the first.

The use of cathode ray tube HDDs for radar was soon extended to other needs. Although the first of these were essentially replicas of mechanical indicators, four classes of format evolved: (a)Vertical Situation Display (VSD), (b) Horizontal Situation Display (HSD), (c) System status diagrams and (d) annotated sensor imagery.

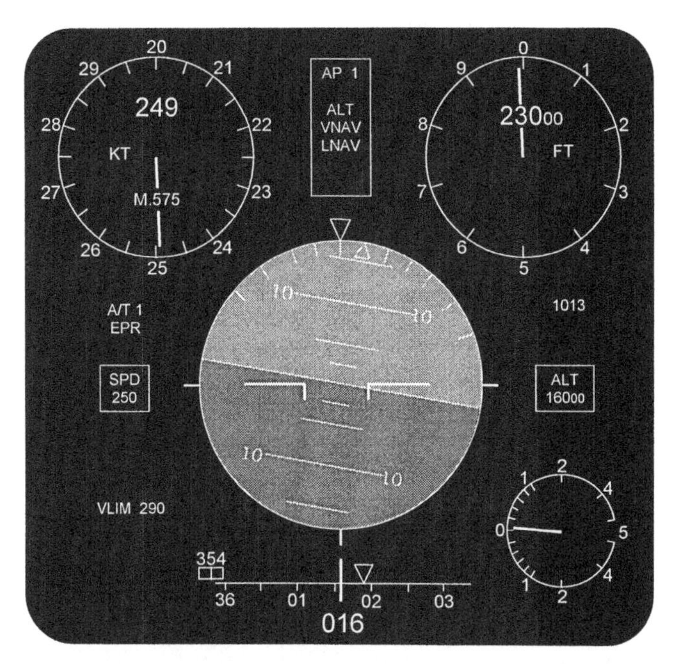

Figure 6.7 Example of Vertical Situation Display

Figure 6.8 Example of a Horizontal Situation Display (This and Figure 6.7 are from Chorley R A, *Electronic flight deck displays for the military transport aircraft*, in The impact of new guidance and control systems on military aircraft cockpit design, AGARD-CP-312 1981. Reprinted with permission from Smiths Aerospace)

As shown in Figure 6.7, the VSD provided the essential information for controlling the aircraft by replicating the primary flight instruments in the Standard T layout. High contrast monochromatic symbols are clearly legible, and colour can be used as a redundant, reinforcing attribute if the format is presented on a colour display, for instance by colouring the ground and sky portions of the attitude ball brown and blue respectively.

The HSD initially took the form of a Compass and Position Indicator, as in Figure 6.6, but this was soon developed to provide all navigation information, as shown on the right of Figure 6.8. Later versions superimposed the planned aircraft route onto a back-projected coloured map as described in Chapter 7. When, in the late 1980s, colour displays became available with sufficient resolution, the symbolic overlay became part of an electronically generated map. In current civil transport aircraft, the HSD can also show a congruent depiction of the strength of reflections sensed by ground-mapping and weather radar systems. In combat aircraft, such as Eurofighter, the HSD has become the main way of summarising the tactical environment. However, information from the main nose-mounted radar is presented separately in a variety of graphical formats, as too is information from the passive radar warning system.

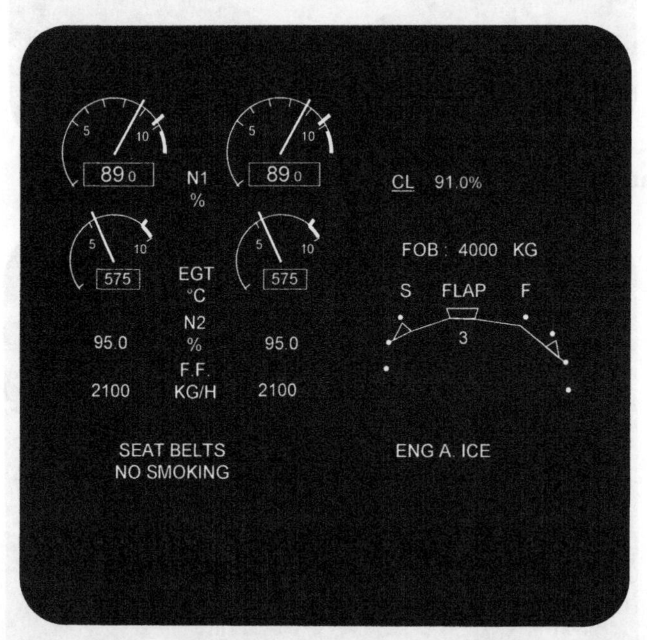

Figure 6.9 Typical arrangement of engine and systems status information in a multi-engine transport aircraft

A wide variety of formats have been designed to summarise the status of the on-board systems. As illustrated in Figures 6.7, 6.8, 6.9 and 6.10, these can use a mixture of text, analogue dials and simple status diagrams. The crew can also select checklists, usually in the form of abbreviated text, when carrying out routine start-up and close-down procedures. These are also available in flight as a reminder of emergency actions. Reising and Emerson[4] were early advocates of the use of simplified, almost cartoon-like diagrams to convey the state of systems. They suggested that such intuitive and immediately understandable depictions could also be used to summarise the consequences of a degraded state and indicate to the crew what actions could be taken. Their ideas are now beginning to influence format design.

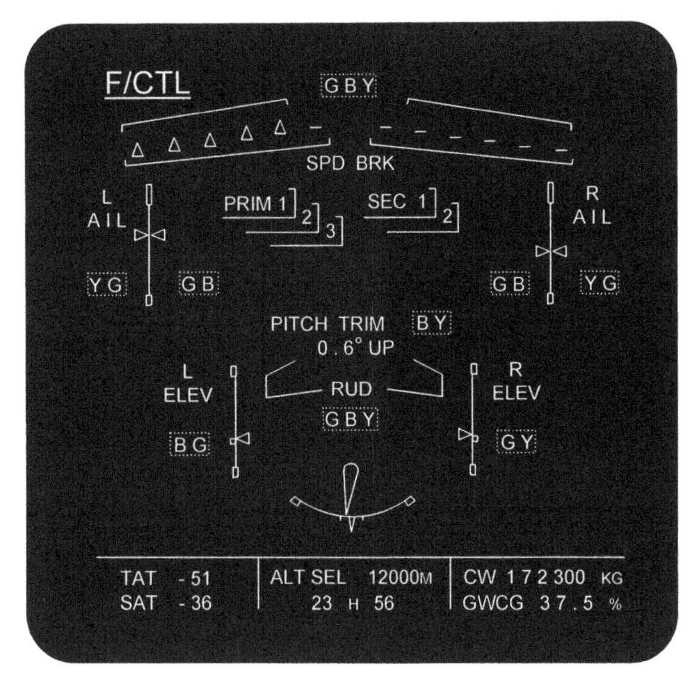

Figure 6.10 Typical arrangement of flight control system information in a modern multi-engine transport aircraft

The use of an HDD as a means of presenting shaded pictorial imagery began in the 1960s when weapons such as the Martell TV-guided missile were fitted to aircraft such as the Buccaneer. In this case the navigator used a hand-controller to slew the sensor in the missile head so that a central capture gate enclosed the intended target, and after the missile was released he continued to monitor the relayed TV image so that he could adjust the missile flight path. Subsequent developments have given a similar task to the single-seat Harrier pilot during ground-attack missions when using a laser-ranging system, and Eurofighter pilots will use the pictorial image from a narrow field-of-view thermal sensor as an aid to target identification. In all cases the display must present the monochromatic picture with greatest clarity, and overlaid symbols are kept to a minimum.

6.4.2 Display layout

Aircraft such as the Sepecat Jaguar, Panavia Tornado, McDonnell Douglas F-15 Eagle and the Hawker Harrier were conceived in the late 1960s and brought into service during the early 1970s. The cockpits of these types represent the pinnacle of electro-mechanical ingenuity. Although the Jaguar included a HUD and a projected map display, it was typical in having more than fifty display devices in the form of instruments and small dedicated CRTs. All aircraft conceived after this date, notably the F-18 Hornet and later versions of the F-15 and Harrier, were able to benefit from multi-function display systems.

The design teams were also able to draw on the outcome of studies conducted by aircraft companies and government research laboratories that had been set up to investigate the optimum exploitation of the inherent flexibility and re-configurability of the multi-function electronic displays. The aims were to devise configurations that provided the pilot with coherently assembled, relevant information and reduced significantly the number of dedicated indicators. In changing from an approach where all information was available, albeit amongst considerable clutter, to an approach that presented only an immediately useful sub-set, it was also essential to avoid penalising the pilot by giving him the extra task of switching frequently between display formats. As well as improving functional performance, the researchers also wanted to minimise the through-life costs by extracting as much as possible from modular digital electronics and modular programming techniques. They wanted to make efficient use of shared resources while allowing graceful degradation through redundancy. Novel self-test and fault isolation techniques were developed.

Figure 6.11 shows one arrangement of the instrument panel layout and the architecture of the supporting system. This was devised by the Hughes Aircraft Company as part of the Advanced Integrated Modular Instrumentation System (AIMIS) programme for the US Naval Air Development Centre[5]. It utilised three monochromatic HDDs, a HUD, a suite of electro-mechanical dial instruments and a colour moving-map projector that incorporated a CRT for overlaying monochromatic symbols. The design and characteristics of the last device are described in Chapter 7. The configuration was intended to ease the job of the pilot in a single seat combat aircraft and have some redundancy so that a mission could be continued should a display device or one of the programmable display processors fail in flight. It acted as a bread-board prototype for aircraft such as the F-18, and incidentally spawned much of the subsequent nomenclature. The master modes control panel, which used the valuable but shallow space left by the intruding HUD optics, became known as the up-front controller; this was the primary way for the pilot to input data into the system and select information for display.

Later aircraft, for instance Eurofighter, have extended the versatility of the layout by replacing the three monochromatic HDDs and the moving-map projector by three larger colour display devices, each of which has sufficient resolution to show a detailed picture such as a map. The electro-mechanical instruments have been replaced by a pull-out panel on the right of the console that can be employed in electrical failure to "get-you-home". This later generation of designs has also included arrays of switches with programmable captions that surround each of the HDDs. These enable the pilot to select alternative display formats, or for instance change the scale of the map, and they remind the pilot about the range of alternatives available. This can be problematical when the branching menu has more than five or six layers. To aid the pilot the formats shown on each display can be programmed to change automatically to show the most relevant information as the mission proceeds through its planned phases. Experimental studies suggest that automated mode changes can serve as useful reminders of the progress of the mission, but the design of the controlling software must enable the pilot to override the pre-determined set. In general, the management of displayed information is a current topic of research. Although it is mainly concerned with the rules for choosing what formats are presented on each display surface, the topic covers the removal of temporarily irrelevant information from formats to aid assimilation, the methods of controlling the interaction, such as by voice and eye rather than key-pressing, and the benefits of allowing individuals to tailor the rules to their preference.

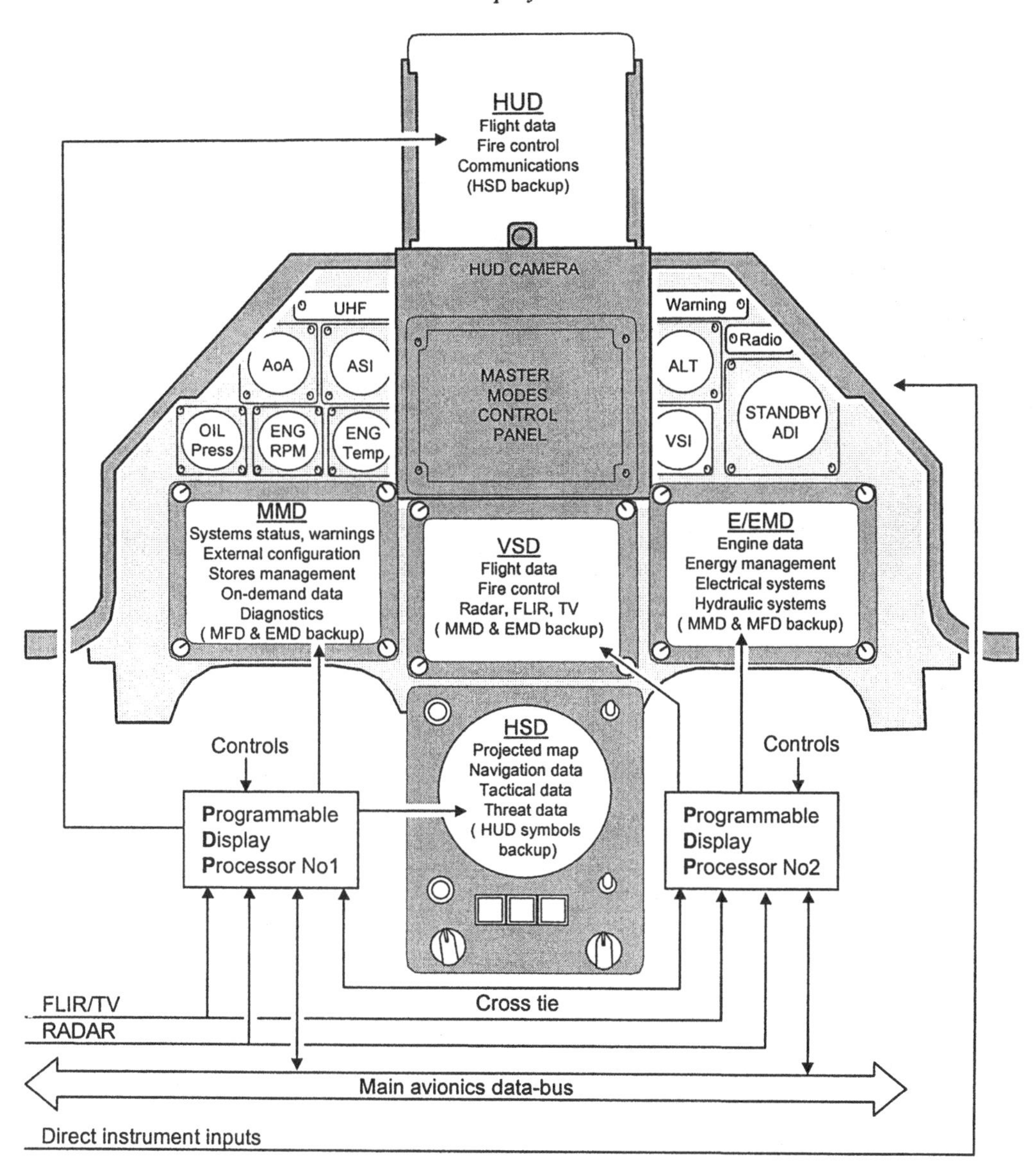

Figure 6.11 The baseline panel layout and system architecture devised by engineers in the Display Systems Laboratory of the Hughes Aircraft Company in 1975 as part of the Advanced Integrated Modular Instrumentation System (AIMIS) programme for the US Navy. The multi-format displays were called the Master Monitor Display (MMD), the Vertical Situation Display (VSD), the Horizontal Situation Display (HSD), the Engine and Energy Management Display (E/EMD) and the Head-Up Display (HUD). (Adapted from Weirauch M, *Integrated modular display system*, in Proc 2nd Advanced Aircrew Display Symposium, NADC Warminster, 1975)

6.4.3 Image generation

As shown in Figure 6.11, the HUD and HDDs are connected to a pair of Programmable Display Processors. These receive data from the on-board systems and generate the signals that the display devices transform into meaningful pictures.

The earliest forms of image generator were designed to drive the horizontal and vertical magnetic coils and the electron gun of a CRT, as illustrated schematically in Figure 7.4. The electronics took the form of a set of circuits, each of which produced a specific symbol (a complex Lissajous figure) by generating fine synchronised temporal variations of the deflection currents and the voltage that controlled the brightness of the moving spot. Other circuits selected the sequence of symbols to be displayed, activated the appropriate symbol circuits in turn and applied the relatively slow-changing currents that deflected the luminous character to the required position on the faceplate. The picture was maintained by repeating the sequence rapidly and by the persistence of the phosphor. The architecture consisted of an input interface, the timing and control circuitry, the symbol units and a set of output buffers, all of which were specific to a particular CRT device. In retrospect, these cursive waveform generators seem a marvel of technical ingenuity, especially in placing symbols with the accuracy demanded in a HUD for registration with features in the real world.

The next generation of symbol generator also wrote characters cursively on a specific CRT. The main advance was gained by using an embedded microprocessor to cycle through a programme that interpreted the incoming data and generated the required currents and voltages. The basic electronic architecture consisted of an input interface/data store, a 68000 microprocessor or similar, memory for the cyclic program, an output data store and a set of digital-to-analogue converters and output buffers. The microprocessor was programmed in machine code.

Modern HDD generators employ digital processing throughout, and output a raster scanned image in the form of a video signal as used in conventional television, as outlined in Chapter 7 and Appendix C. In order to achieve the necessary image brightness, several generators that feed a HUD or a HMD produce cursive signals for driving the CRT in daylight and raster video signals for relaying a pictorial sensor image at night. A hybrid display system combines a raster pictorial image with a cursive symbol overlay; the latter drawn during the flyback period between successive raster fields. The hardware and software architectures of the machine are arranged to suit the sort of imagery presented by the display device. For instance, the fast moving but sketchy monochromatic 2-D symbols of a HUD are updated more rapidly than a map image on a HDD, but the latter requires a large data-base and is composed of complex coloured features.

The next generation of aircraft will benefit from the increased luminance and resolution of flat panel display devices, and it is likely that all displays - whether head-down, head-up or helmet-mounted - will be driven using a digital raster video signal with a common line standard. This simplification will enable the cockpit design team to use a set of identical image generators carrying identical software to supply signals to the display devices. Failure of a generator or display can then be accommodated by re-configuring their interconnections.

The internal architecture will be much the same as employed in a general purpose graphics processor, such as a machine that provides visual imagery in a simulator, with all programming done using a compiled high-level language. These machines draw 2-D symbols and 2-D diagrams on a display by specifying the location, orientation and size of the feature on the screen, moving the features on successive frames to produce dynamic changes. The computations needed to construct dynamic 3-D pictures, such as a perspective view of the terrain, are significantly more complex and require a machine with an architecture that can perform a sequence of operations analogous to the roles taken by an artist and apprentice. An input interface receives data from the on-board systems that, among myriad details, define the viewpoint and the orientation of the viewer. The role of the artist is performed by a geometry processor that accesses a large 3-D data-base defining the objects in the scene and selects those that are included in the field of view. The objects themselves are sets of flat polygons defined by surface qualities, such as colour and transparency, together with the 3-D location of their vertex points. The shape of the terrain is usually represented as a large mesh formed by contiguous triangles, for example the US Digital Terrain Elevation Data (DTED) is available at three levels of detail with vertices at 1000m, 100m and 30m spacing. The essential role of the artist is to create a perspective sketch by calculating where lines projected from the 3-D vertices intersect the 2-D image plane. The resulting display list of 2-D polygons is handed onto the electronic renderer that, like the apprentice, applies colour, shading and texture. The final picture is deposited in a frame store as an array of pixels. Successive lines of pixels are extracted, passed through a digital-to-analogue converter and sent with timing and synchronisation pulses as the video signal to drive the display device. The output frame store is normally split into two parts (double-buffered), one of which is addressed by the renderer while the other is read by the video processor, the two parts being swapped at the end of the video frame if the renderer has constructed the image in time.

Many sophistications have been developed to create realistic pictures, circumvent artefacts and update the picture at the video frame rate. For instance, complex objects can be defined by a set of models with differing numbers of polygons so that the level of detail, and the number of computations, dwindles when an object recedes. Anti-aliasing techniques are used to reduce the staircase appearance of lines and edges that cut across the raster scan of a CRT or the rectangular pixel array of a device like an AMLCD. The most common technique involves constructing images at a significantly higher resolution than can be displayed, typically a factor of four, and assigning luminance and chrominance values to a display pixel by averaging the values of the four sub-pixels.

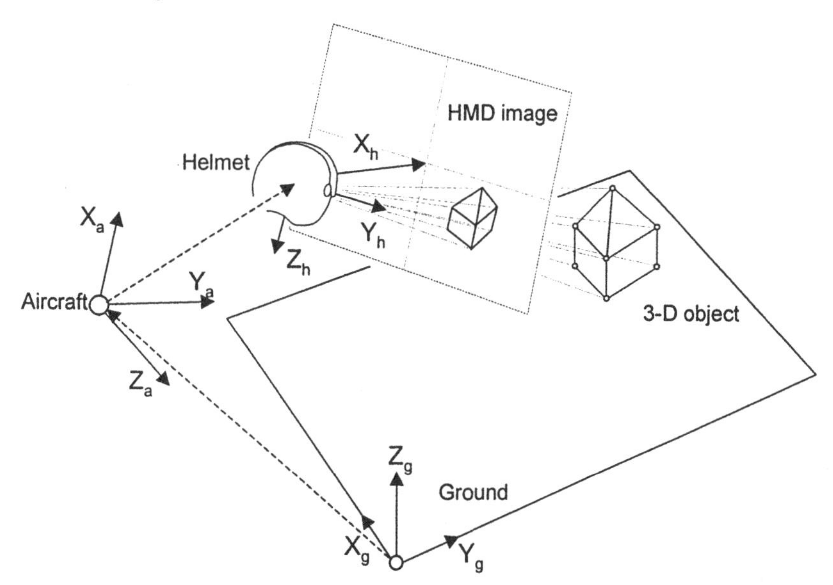

Figure 6.12 Computation of the 2-D image of a 3-D object viewed from a displaced and rotated position

In future, as described in Chapter 9, a wide field of view, full colour, partially-overlapped binocular HMD will place the greatest demands on the image generator. To provide a visually-coupled system the machine must import sensor imagery via a video analogue-to-digital converter and re-map the received pixels into the double-buffered output frame stores to match the logical display planes of the two channels of the HMD. In order to construct perspective images from a data-base where co-ordinates are defined relative to the ground, it is necessary to translate and rotate these into aircraft-referenced co-ordinates, and then translate and rotate the aircraft-referenced co-ordinates to get helmet-referenced co-ordinates. Figure 6.12 illustrates the case for a single HMD display plane. The ground-to-aircraft transforms need accurate, noiseless navigation and inertial data, and the aircraft-to-helmet transforms need accurate, noiseless helmet tracking data. In practice the series of transformations is done at the start of each picture construction cycle by concatenating the series of transformation matrices to produce a single 3 by 3 matrix that can be applied to all vertices[6]. Additional requirements of concern to the design of HMD systems, in particular the need to counter distortions introduced by the optical system and for fast updating of image content with minimal through-put delay, are dealt with in Chapter 9.

Recently, the availability of compact computing hardware with adequate processing power and memory capacity, and the concurrent availability of space-efficient AMLCD display devices, has enabled designers to integrate an image generator into the display unit itself. This has given new-found options to those devising the display suite. The most affordable are substitutes for dedicated mechanically indicating instruments, such as the ASI and the AI. These are packaged so that they can be retro-fitted to replace the original in the instrument panel. More comprehensive versions are equivalent to MFDs, and incorporate the peripheral keys for scrolling through the menu options and selecting alternative formats. The most complex incorporate inertial sensors, barometric transducers, interfaces to a magnetic sensor and GPS receiver, and have built-in back-up battery that can power a limited range of functions in the event of electrical supply failure. These can act as a complete avionics suite, and are likely to be found in most new light aircraft.

6.4.4 Control and display layout

The location of control and display devices within the cockpit reflects the pilot's priority for head-out awareness and his concomitant need for quick access to information and means of control. Although the priorities vary between aircraft types and roles, most western, single-seat combat aircraft that have been designed in the last two decades have tended to follow a fairly well optimised convention. In descending order:-

Above glare shield	HUD, AoA and G meters
Edge of glare shield	Attention-getter and warning lights
Face of HUD intrusion	Data entry panel (Up-front controller)
Upper front panel	Threat displays
Main front panel	HDDs
Lower front panel	Standby instruments, weapons and stores panel
Forward end of left console	Undercarriage control panel
Forward end of right console	Caution and advisory lights
Side consoles	Dedicated control panels

Individual types incorporate beneficial idiosyncrasies. For instance the designers of the Mirage 2000-5 and the later Rafale make better use of the HUD intrusion by integrating a set of optics to create a Head-Level Display (HLD). This provides a non-see-through collimated 20° by 20° colour image contiguous with the HUD field of view. The pilot can glance down to see a sensor image without the delay needed to refocus between distant and near accommodation states. The HDDs are also placed as high as possible, just beneath the glare shield, to minimise the diversion of attention, and these devices have integral touch-sensing face-plates for direct interactions. In addition to the main task-related displays and controls, a large number of utility systems require individual control and display panels. These are generally accessed before take-off and after landing, but they must also be useable in-flight, especially during emergencies. They are allocated a place on the side console that reflects the likelihood and the required speed of the emergency response.

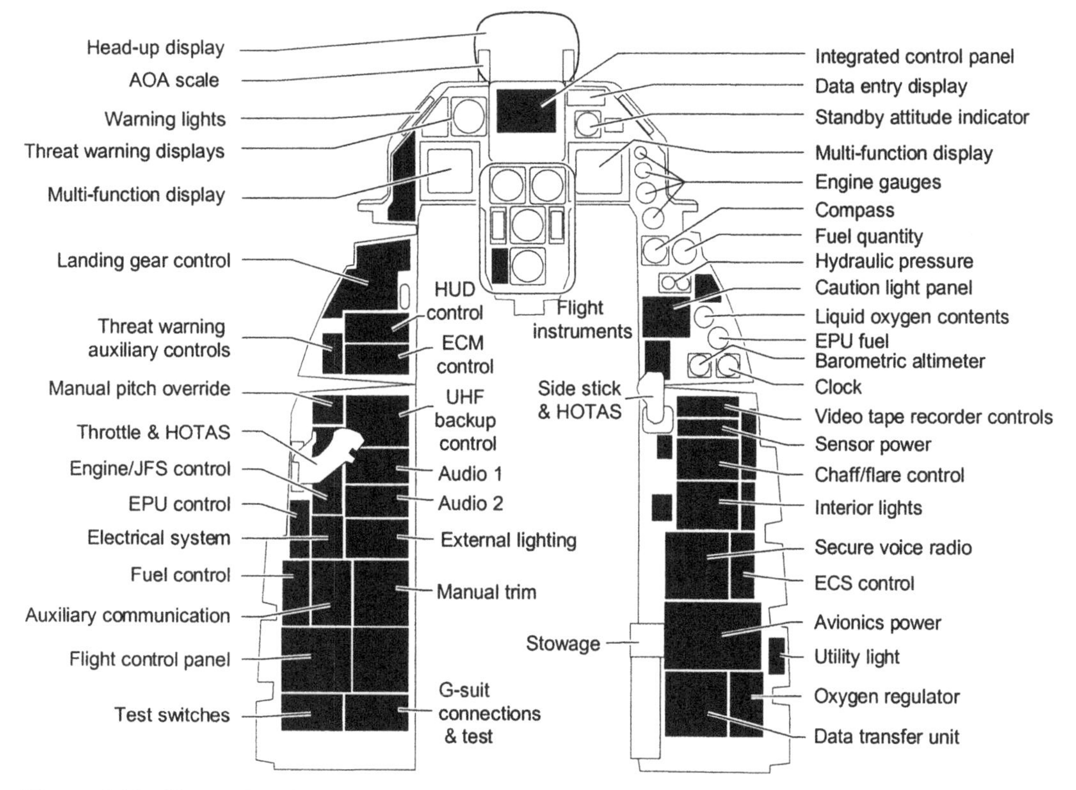

Figure 6.13 Example of the panel layout used in the F-16C (Reprinted with permission from Lockheed-Martin)

Figure 6.13, summarising the layout used in the F-16C, shows how the designers chose to allocate space on the left and right side consoles to the array of control panels in that aircraft type. There are, in addition to those identified on the figure, switches and levers to raise, lower and lock the canopy, raise and lower the seat, operate the de-icing and de-misting etc. In older aircraft types without a flight control system, the throttle would be operated intermittently and the stick almost continuously. The left hand therefore had greater freedom to release the grip on the throttle, and if it were necessary to use the right hand for operating a panel-mounted switch, the pilot would change hands on the stick grip. Keypads were therefore placed on the left side of the glare shield and the higher priority panels were placed on the left side console. However, since both the stick and throttle grips are now populated with HOTAS switches controlling a range of aircraft systems, both the pilot's hands are occupied. The arrangement adopted in the F-16 reflects an equal reluctance to reach for a panel-mounted switch with either hand.

In tandem-seat aircraft, the auxiliary panels are placed on the side consoles of the two cockpits according to their relevance to the defined crew role, and many controls and displays must be duplicated. With side-by-side seating, the central console can house panels that the crew can share. This is common in most civil fixed-wing and rotary-wing types, and extends to the use of overhead panels, typically for interacting with the fuel, electrical, hydraulic and anti-icing systems.

6.4.5 Warnings

Many of the aircraft-carried systems are devised or programmed to detect, as part of their control function, conditions that are of immediate concern to the pilot. These vary between aircraft types and range from general hazards, such as fire, to idiosyncrasies of specific aircraft types, such as the imminent exhaustion of the water injected into the Harrier engine to boost thrust. The cockpit design team must install a means of making the pilot take notice of warnings and give him a precise and concise indication of the cause of the alert so that he can quickly decide how to respond. The physical and procedural design of this audio-visual warning system is therefore a crucial part of the processes involved in devising an operating philosophy for the aircraft and configuring the display suite.

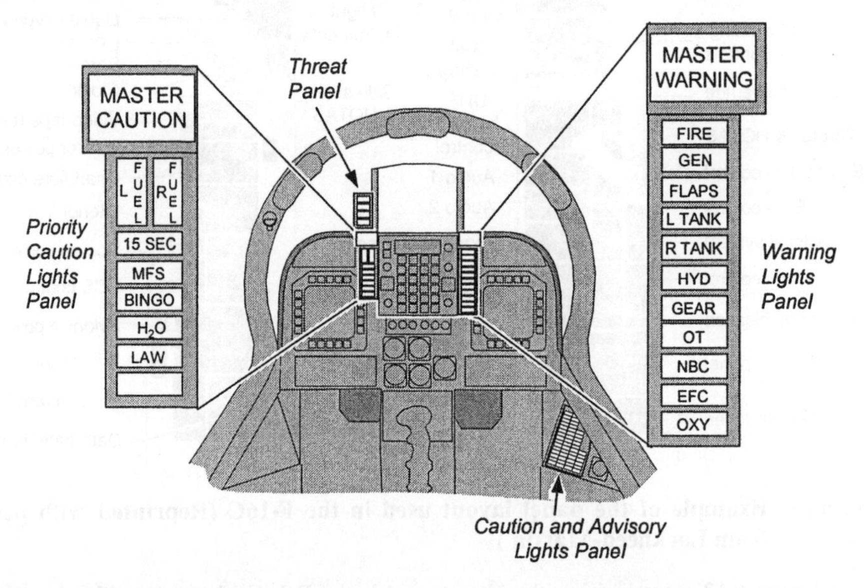

Figure 6.14 A typical warning and caution system in a modern fast jet

The design of the warning panels, the placement of the panels in the cockpit and the nature of the electronically-generated sounds injected into the pilot's earphones all differ between aircraft types. Aircraft of UK origin usually have a single, centralised panel in which the colour of the back-lit caption indicates the priority of the condition. Other installations, such as that shown in Figure 6.14, use a set of panels; a pair mounted close to the intruding HUD display unit alongside the Up-Front Controller, a separate threat warning panel fastened above the glare shield alongside the HUD combiner glass and a fourth placed at the forward end of the right side console. The last conveys cautionary and advisory information in the form of a panel containing about 50 illuminable captions. The back-illumination of the captions are intended to be bright enough for adequate contrast, and in this case the illuminant is blue-green for compatibility with NVGs.

The nomenclature and the manner in which an alert is announced indicate the rapidity with which the pilot must respond. For instance, illumination of the large, back-lit, attention-getting "master warning" caption signifies that an aircraft system has suffered a critical malfunction and the pilot should respond within seconds. The pilot can cancel the attention-getting light and the sound by pushing the master warning light, and then look down the set of smaller captions to see what has been lit to decide what to do. A similarly speedy response is needed if any of the "threat warning" lights is lit. These show the output from the aircraft's defensive aids suite and the direction of the threat. If the "priority caution" attention-getter is lit, the pilot has slightly more time to decide on a response. The "caution" captions signify that a system needs attention soon, and perhaps some follow-up action. The "advisory" captions are mainly provided as an accessible indication about the normal state of a system.

In general, it is assumed that the pilot is familiar with the working principles of all of the electrical, hydraulic, fuel, propulsion, flight control, navigation and communication systems, understands the repercussions of all of the warning conditions and remembers the appropriate corrective actions. The principal concern for the designers is to alert the pilot to the specific condition without inducing stress and confusion. To achieve this aim it is important that a particular class of warning captions and accompanying auditory tones are used consistently. The pilot can then know whether he must respond immediately or can delay and give a more considered response. Also, he must have a way to set the loudness of the auditory tones. If they are replayed into his earphones at a level that he cannot fail to hear, the effect may be more annoying and startling than informative. False alarms must be rare; they cause unnecessary diversions and can make the pilot distrust the system and delay his response.

In-service experience of such foibles can be used to refine the system, and as discussed in Chapter 10, much can be done using alerting sounds that penetrate the aircraft noise without excessive amplification. It is also possible to vary the pitch, tone and rhythm of the sound to provide an almost instantaneous indication of the urgency and nature of the condition, followed immediately by a voice message to give confirmation if the pilot is unsure.

6.5 Cockpit lighting

6.5.1 General cockpit illumination

Before the development of practical night vision systems, as described in Chapter 5, military aircrew conducted night missions during the fuller phases of the lunar cycle by relying on moonlight to illuminate in the outside scene. The crew prepared for the flight by adapting to the dark ambience, and remained adapted by avoiding exposure to a significantly more luminous field of view.

When bathed in full moonlight, a light coloured object would have a luminance of less than 0.1cd.m^{-2}, so the typical level of adaptation was an order of magnitude below this level. However, the pilot also needed to be able to read information from non-emissive displays, instruments, switches, control panels and printed information such as a map or a flight reference card. Ideally, for good clarity, this would be accomplished by bathing the cockpit with a luminous flux of about 300 lux to give a comfortable 100cd.m^{-2} surface luminance. In practice, the internal illumination had to be a compromise between the need to see the external scene, the need to see inside the cockpit and the need for rapid adjustment between the two conditions; a level of no more than 1cd.m^{-2}. The spectral composition of the illumination was also important. Under such mesopic adaptation, the pilot's vision relied on both the rods and the cones in the retina, and in comparison with normal daylight (photopic) vision, acuity and colour discrimination were poor and, as illustrated in Figure 2.12, fewer shades of grey could be distinguished. It was therefore more important for the illumination to maintain the perceived contrast of features in the cockpit than their colour.

In the 1940s, when this requirement became an operationally significant issue, two distinctly different approaches were possible; red lighting and ultra-violet lighting. The RAF and the US Navy adopted dim red lighting, with a wavelength above 600nm, in preference to white light of equivalent luminance because it had minimal effect on rod adaptation. Although the USAF converted to red in the 1950s, they and the Luftwaffe initially made use of ultra-violet. Although accompanied by a faint bluish glow, this was largely invisible but could stimulate emission from fluorescent paint on instruments and fluorescent ink or paper for maps. Both forms of illumination were incompatible with conventional maps, and special aviation series were printed[7]. The use of red illumination gave maps a monochromatic appearance in which red ink had little contrast against the white paper and blue appeared black. Features that were important for night navigation, such as lakes, rivers and woods, were rendered in shades of blue and green, while others that were less useful, such as roads, were printed in red. The coloured map then appeared serendipitously decluttered when used in the cockpit at night. Although almost all cockpit lighting is now brighter and whiter, to enable the pilot to discern colour differences, these conventions have been maintained. So too has the use of dark blue to overprint aeronautical information.

A wide variety of techniques have been used to illuminate the cockpit. Sources have been placed inside the instruments, in small floodlights, in even smaller stanchions that protrude from panels, and in a central housing that distributes the light via light-guides to instruments and panels. Transparent light "wedges" can be placed over the front of instruments. These are illuminated at one point on the edge and distribute the light uniformly over the rest of the diffusing edge by total internal reflection. The source itself has usually been some form of small incandescent filament bulb, but beta-lights, electro-luminescent plates and light-emitting diodes are also used. All can be set up to provide a controllable luminance. All but the bulb produce a limited spectral range and hence appear coloured, but because the spectral characteristics of the incandescent bulb depend on the filament temperature, an increase in luminance is invariably accompanied by a shift towards the blue end of the spectrum. Carefully chosen absorption filters have commonly been used to adjust emission characteristics.

The bezel areas of switch panels installed in the fascia and side consoles of modern combat aircraft are constructed from a flat plate of a transparent base material, such as acrylic or polycarbonate, that is laminated or coated so that the outer surface is matt grey or black. Characters and markings are produced by engraving the laminate or by selective screen printing of the coating. The difference between the low reflectivity of the coating and the high reflectivity of the cut away areas give the characters excellent contrast in daylight. At night this contrast is maintained by "edge-lighting"; light from incandescent bulbs or LEDs around the edge of the plate that is distributed by scattering and reflection from the internal plate surfaces and is emitted through the cut-away characters.

To enable the pilot to switch on and adjust the brightness of individual lights, or clusters of lights, a dedicated control panel is fitted to most aircraft types, usually towards the rear of a side console. The dimming usually follows a logarithmic law to make the brightness change appear linear, and ranges over two orders of magnitude from about 3cd.m^{-2} down to about 0.003 cd.m^{-2} before extinguishing. The main exceptions[8] are the Master Warning lights, which must be conspicuous and are illuminated to about 350cd.m^{-2} during the day and about 50cd.m^{-2} at night, and captions on the caution and advisory light panel, which must not be extinguishable but can be dimmed to about 0.15cd.m^{-2}. In general, the pilot switches the lighting on fully during dusk and applies dimming progressively as night falls. Forward pointing floodlights are installed close to the pilot's shoulders in most combat types to fill in shadows at dusk and for reading paper maps and reference cards. They are also available should the main system fail and when flying with nuclear flash protection goggles (described in Chapter 13) to compensate for the attenuated view into the cockpit.

In aircraft with emissive CRTs and back-illuminated LCD panels, the luminance of such devices is also settable, usually by rotating a knob on the device beneath the screen. The pilot can adjust the apparent brightness of all of the individual elements of the cockpit to suit his needs, for instance setting the side consoles - which are not covered by a glare shield and tend to reflect from the transparency - to be much dimmer than the front panel. However, a combat aircraft has about thirty elements, and the need to manage their relative conspicuity can become a significant part of night flying workload. In aircraft such as the Eurofighter and JSF, the lighting control panel includes an automatic mode that sets the luminance of all elements in response to the measured ambient light level. To minimise canopy-reflected glare, all of the side console panels are edge-lit, rather than flood-lit, and the levels are set to give the appearance of a black void from which adequately contrasting information is seen at a minimum, uniform luminance.

6.5.2 Lighting for night vision goggles

NVGs intensify the luminance of the external scene by at least three orders of magnitude. Any display device, instrument or panel bright enough to be seen by the naked eye would ordinarily overload the goggle if brought within the field of view. Early flight tests investigated the idea of dimming the cockpit to a luminance comparable with the external scene so that the pilot could view the instruments and displays through the goggles. This entailed fitting auxiliary lenses to shorten the focal length of the goggle objectives to focus on the panel rather than the distant scene, and the lenses were swung into position either manually or by an automatic mechanism whenever the pilot inclined his head downwards. However, the pilots disliked the complication and delay of the auxiliary lenses, the need to make exaggerated head movements to bring instruments and displays within the field of view, and the poorly resolved, monochromatic nature of this view.

An alternative approach was needed that exploited the pilot's adaptation to the goggle luminance of about $10 cd.m^{-2}$, which enabled him to discern detail and colour, and allowed him to see into the cockpit either through the transmissive eye-pieces of an indirect view goggle or by peering beneath the eyepieces of a direct view ANVIS-type goggle. This was made possible by a technique, devised at RAE Farnborough[9], whereby cockpit lights and displays were filtered to remove red and infra-red emissions, and complementary blocking filters that only passed deep red and infra-red light were fitted to the goggle objectives. The approach was soon standardised and adopted by most operators.

As shown in the lower graph of Figure 5.13, Gen-III night vision goggles are more sensitive to infra-red light than visible light, and such devices are more compatible with the complementary filter scheme than the older Gen-II devices. The resultant spectral sensitivities of the two classes of device are re-iterated as curves (b) and (a) in Figure 6.15. The effects of the sharp edged filters on Gen-III goggles specified for used in US fast jets and helicopters are shown by curves (c) and (d) respectively. The Gen-III NVG response is reduced to less than 1% of the peak by a "Class A minus-blue" filter at 595nm and a "Class B minus-blue" filter at 625nm, the latter allowing the retention of a broader range of colours in the cockpit. Curve (e) shows the characteristic of a filter tailored to pass a small amount of green light so that the pilot can see HUD symbols that have been dimmed appropriately before intensification. It is also possible drill a small hole, about 2mm diameter, in the centre of the 22mm diameter standard filter to create a "leaky" filter to pass a small amount of green light from the HUD.

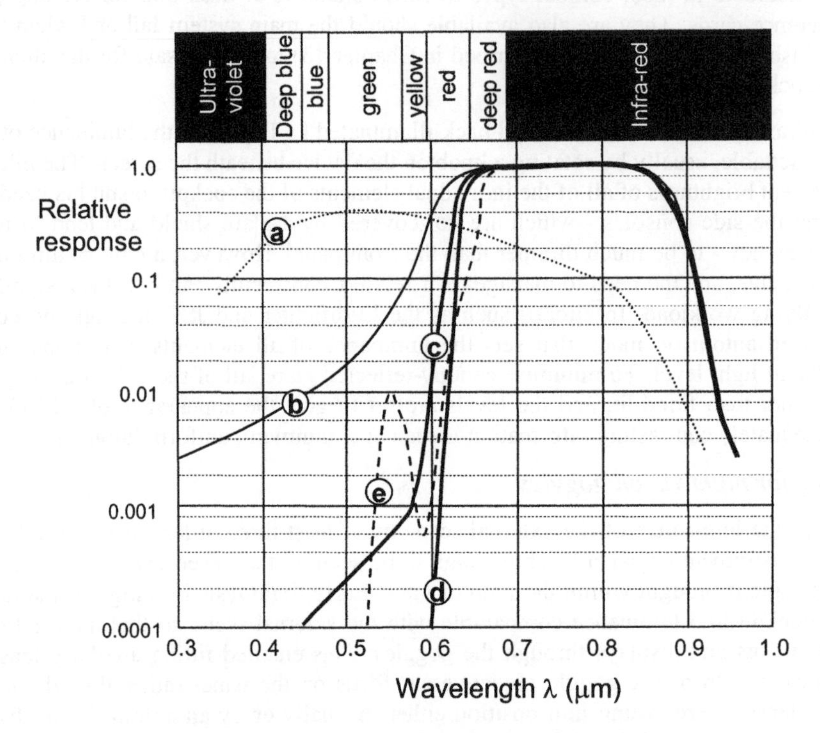

Figure 6.15 Spectral sensitivities of night vision goggles. (a) Gen-II multi-alkali photocathode, (b) Gen-III gallium arsenide photocathode, (c) Gen-III with MIL-L-85762A Class A (625nm) filter used mostly in US helicopters, (d) Gen-III with MIL-L-85762A Class B (675nm) filter used mostly in US fast jets and (e) Gen-III with "green leak" filter used in some fast jets equipped with an HUD (Reprinted with permission from Korry Electronics)

The interference caused by a light source in the cockpit is described in the UK by a non-dimensional figure of merit called the "Green/Red ratio", which is the ratio of the effect of the source on the eye in relation to the effect of the source on the intensifying goggle. The numerator is the integrated radiance of the source after multiplication by the normalised photopic response function of the eye, and the denominator is the integrated source radiance after multiplication by the normalised spectral sensitivity of the filtered goggle. The ratio can be measured using an appropriate green filter to convert the intrinsic spectral response of a photometer to match the eye and another IR filter to convert the intrinsic spectral response of a photometer to match the goggle.

The MIL-L-85762A procedure used in the US specifies the maximum visual luminance of a source and the maximum "NVIS Radiance"; the integrated radiance of the source multiplied by the normalised spectral sensitivity of the filtered goggle. The luminance of most displays should be less than $0.34 cd.m^{-2}$ and the NVIS Radiance should be less than $1.7x10^{-10} W.cm^{-2}.sr^{-1}$, but warning lights can be about two orders of magnitude brighter.

There are several ways to arrange that cockpit lights and displays contain very little of the red and infra-red wavelengths that are intensified by the filtered NVGs. The choice of the source is crucial, as shown in Figure 6.16. The most pragmatic form of cockpit installation, which has been used to convert older aircraft with electro-mechanical instruments, has involved turning most of the normal illumination off. Compatible lighting is then arranged by fitting filters to a few of the existing floodlights, installing additional filtered floodlights and fastening electro-luminescent strip-lights on the underside of the glareshield. Most of the filters are made by casting acrylic base material, loaded with proprietary dyes, between polished steel flats to form a layer about 3mm thick. Glass filters containing a mixture of metallic oxides are also used. Reflections from the canopy are reduced by controlling upward scatter from instrument faces, for instance by fitting micro-louvre filters as described in Chapter 7, and by wearing matt black clothing.

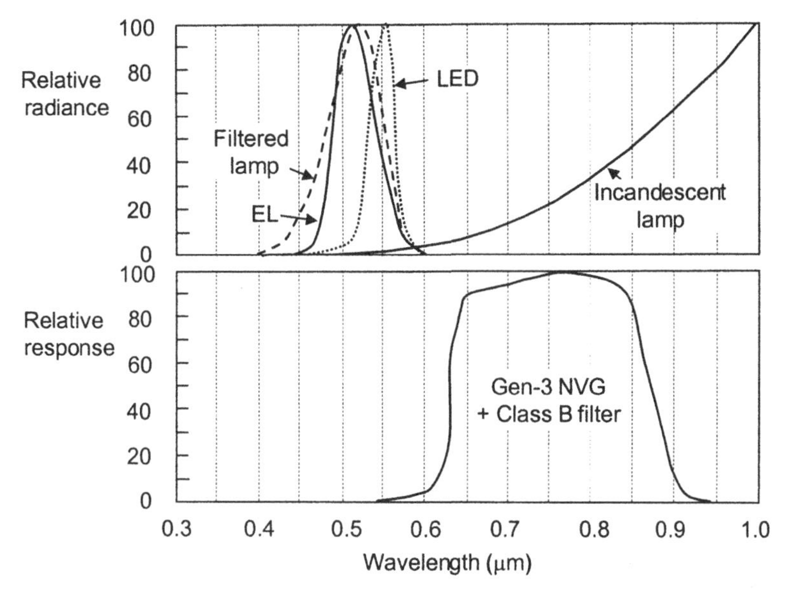

Figure 6.16 Typical spectral characteristics of light sources and the response of a filtered NVG. The phosphor of the electro-luminescent (EL) lamp and the doping of the light-emitting diode (LED) can be chosen to avoid deep red and infra-red emissions. Any incandescent lamp must be filtered strongly to remove the large proportion of incompatible emissions. (Reprinted with permission from QinetiQ)

Attaining a good figure of merit and virtually eliminating deep red and infra-red, is best done when the cockpit is at the design stage. In such cases, exemplified by Eurofighter and F-22, it is possible to specify that all instruments and switch panels are internally illuminated by compatible sources, as too are the floodlights. It is also necessary to ensure that all emissive HDDs either include NVG-compatibility filters or have night modes for conveying the requisite information without exciting deep red emissions. The approach depends on the nature of the HDD. For instance, with a CRT it is possible to restrict the colour gamut of a map picture. However, most backlights employed with LCDs emit a small amount of infra-red light. As this is transmitted whatever the intended colour of a pixel, without effective filtration the screen would appear to be a uniformly glowing rectangle.

To avoid overloading aircrew goggles, external lights must also emit very little red and infra-red wavelengths. Conventional aircraft navigation lights and anti-collision strobes can be fitted with filtered covers, and it is also possible to install covert IR lighting that can only be seen by other NVG users. It is however no small exercise to modify the landing lights, deck lights and transit markers fitted to aircraft carriers and helicopter-carrying vessels to achieve the required figure of merit.

6.6 Peripheral awareness displays

Spatial disorientation, a dangerous uncertainty or illusion concerning the orientation of the aircraft, is most likely to arise when the pilot has lost visual contact with a clear horizon. The condition mainly occurs at night and in poor visibility, or during combat manoeuvring, or when the pilot has been diverted to attend to a head-in task. To resolve the uncertainty, or dispel the illusion, the pilot must fixate to bring acute *foveal* vision to bear on a display such as the AI or the HUD. These present information within a small visual angle and cannot stimulate the complementary *ambient* visual pathway. As noted in Chapter 2, ambient vision has an extensive visual field, does not require eye movements, is intrinsically faster and is normally the main contributor to the perception of body orientation.

Several display devices have been devised that are intended to stimulate ambient vision with the specific aim of reducing the chance of spatial disorientation. They can also be regarded as belonging to a broader class of devices that have been investigated as ways to off-load central vision, particularly during instrument flying. The experiments of de Florez and Forbes, mentioned in Chapter 10, of using binaural tones to convey banking, pitching and airspeed control demands to enable the pilot to exercise control over the aircraft when deprived of visual cues, can be regarded as early examples of this broader group. Although in these experiments the auditory signals acted as director cues that the pilot followed to maintain a desired flight path, it would also be possible to have auditory tones represent the state of the aircraft. Hasbrook and Young[10] suggested that continuous bank attitude information could be fed to the pilot in a number of ways: by aural signals, by pressure against the side of the body, by rotary or lateral displacement of the control wheel or stick and by differential expansion of the hand grips on the wheel or stick. Peripheral and/or central visual signals could also be employed. They used a simulator to investigate the use of a pair of green and red lamps, mounted on the left and right of the handwheel, to give a peripheral bank angle cue. In their implementation, no light was lit if bank was less than 3°, the appropriate green light flashed slowly between 3° and 10°, more rapidly between 10° and 18°, and was on steadily between 18° and 22°. A flashing red light signified a bank between 22° and 90°. They concluded that the pilots adapted quickly to these cues, that they reduced significantly the fixation times on the instrument panel, and improved flight performance. The benefits were greatest when additional tasks were added to the pilot's flight duties.

More recent attempts to supply non-foveal attitude cues are illustrated in Figure 6.17. The Malcolm horizon, or Peripheral Vision Display[11] (PVD), projected a bright line of light onto the instrument panel from a 3mW helium-neon laser, and because the projector was mounted overhead, the device was most suited to rotorcraft and transport types with no overhead transparency. The bar of light was driven by bank and elevation signals to be centred in front of the pilot and remain parallel to the real horizon. Versions have been evaluated in simulators and flight trials aircraft, mainly rotorcraft. In the most recent implementation, made by Garrett Airesearch, the pilot could adjust the line brightness and the bank and elevation trim positions. The elevation scaling factor could be set to 2:1, 1:1 or 1:2. Reports by the trial pilots indicated a few problems, for instance that the line contrast was insufficient in a bright ambience, and that the obliquely incident line of light could appear jagged when intersecting instrument dials that protruded from the panel. Several hours of flying were needed before the pilot learned the relationship between the projected line and the real horizon, and given such familiarity it was possible to fly without fixating on the line and use it unconsciously and peripherally. Pilots reported reduced workload, better flightpath control and the ability to spend more time looking out in poor visibility conditions.

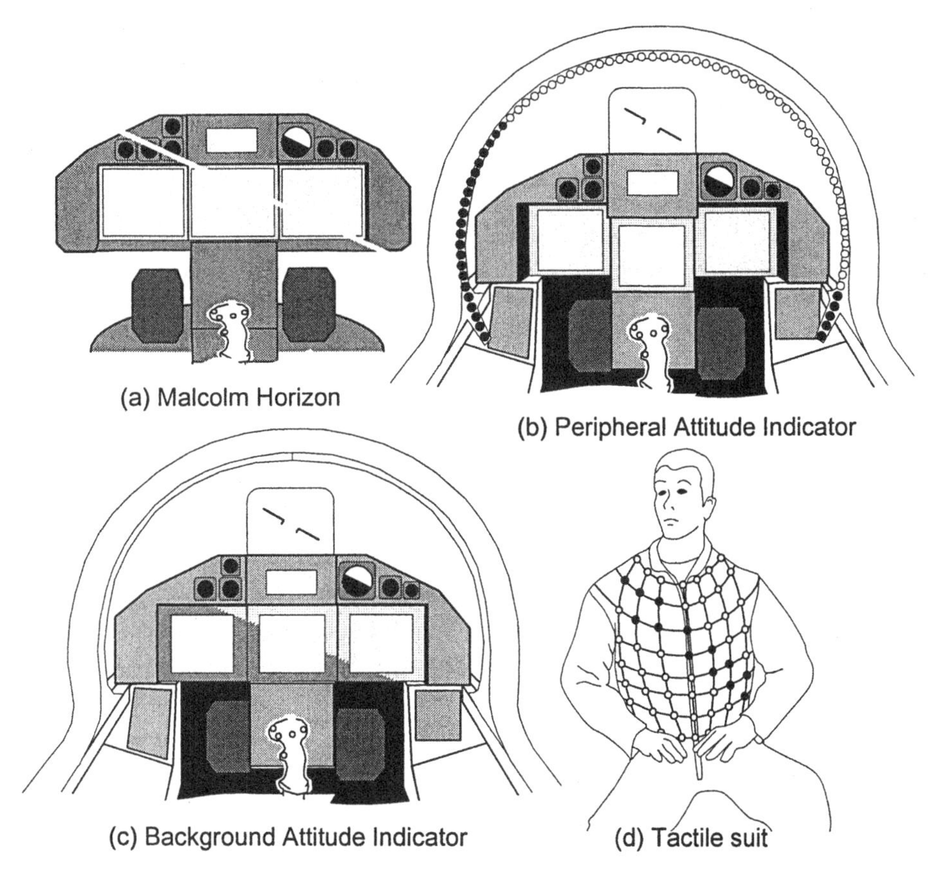

Figure 6.17 Suggested means of conveying aircraft attitude information. (The tactile suit illustration is adapted from Rupert A H, Guedry F E and Reschke M F, *Use of a tactile interface to convey position and motion perceptions*, in *Virtual interfaces: research and applications*, AGARD-CP-541, 1994)

The Peripheral Attitude Indicator[12] illustrated in Figure 6.17(b) consisted of an almost complete ring of light emitting diodes that were illuminated in sectors to delineate a bright sky portion and a contrasting dim ground portion, much like the "Sky-Arc" HUD attitude format described in Chapter 8. The ring was intended for mounting on the canopy arch of single seat or tandem seat combat aircraft. The Background Attitude Indicator[13], illustrated in Figure 6.17(c), was a similar idea but assumed that the HDDs mounted in the main instrument panel would be contiguous and of generous size. In this case the attitude information could be provided by dedicating the border of each HDD to supplying ground and sky areas separated by a clear horizon line. Both of these display ideas were assessed in simulation experiments, and although they had a limited angle range and an implied datum, both were considered capable of giving the pilot a non-attention-demanding awareness of the aircraft bank and elevation attitude.

The final system illustrated in Figure 6.17(d) exploited the under-utilised sensations of touch. The prototype device[14] consisted of a lycra vest containing a 8 by 24 matrix of vibrotactors; small moving coil loudspeakers like those used in earphones. Exciting an appropriate linear group with 150Hz pulsed waveforms could impart a sensation of orientation to an accuracy of about 5° over an elevation range of ±15° and bank range of ±45°. An alternative sensation of movement, similar to the perception of a directed flow of fluid over the torso, was conveyed by stimulating firstly the column pointing into the flow direction, then in sequence the three pairs of columns to each side of the first column. The successive column excitations felt like a wave running over the torso, and the interval between the column excitations could be shortened to suggest increasing flow velocity. This device seems to be a satisfactory way to give a pilot a non-attention-demanding awareness of the aircraft orientation. To avoid adding further complexity to the already cumbersome layers of aircrew clothing, it would be necessary to develop miniaturised vibrotactors that could be coupled to a body-mounted control unit using a thin, flat cable. However, unless aimed towards a wider application than military aviation, further development is unlikely. It must compete with other ways of conveying peripheral orientation cues, such as the use of appropriately stabilised symbols in a wide field of view HMD.

References

1. Jones R E, Milton J L and Fitts P M, *Frequency, duration and sequence of fixations during routine instrument flight*, AF-TR-5975, USAF AMC, Wright-Patterson AFB, Dayton, Oh, 1949
2. Collinson R P G, *Introduction to Avionics*, Chapman & Hall, 1996
3. Welch J F (ed), *Modern Airmanship*, Van Nostrand Reinhold Co, NY, 1981
4. Reising J M and Emersom T J, *Colour display formats: a revolution in cockpit design*, in *Advanced avionics and the military man/machine interface*, AGARD-CP-329, 1982
5. Weirauch M, *Integrated modular display system*, in *Proceedings of the 2nd Advanced Aircrew Display Symposium*, NADC Warminster, 1975
6. Foley J D and Van Dam A, *Fundamentals of interactive computer graphics*, Addison-Wesley, 1984
7. Taylor R M, *Colour coding in information displays : heuristics, experience and evidence from cartography*, in Gibson C P (ed) *Proceedings of a NATO workshop on colour vs monochrome electronic displays*, RAE Farnborough, 1984
8. Pincus R P, *Night lighting and night vision goggle compatibility*, in *Visual effects in the high performance aircraft cockpit*, AGARD-LS-156, 1988
9. Lloyd G, *Cockpit lighting standards and techniques for use with night vision goggles*, Unpublished MOD report
10. Hasbrook A H and Young P E, *Pilot response to peripheral vision cues during instrument flying tasks*, FAA Report AM-68-11, 1968
11. Malcolm R, Money K E and Anderson P, *Peripheral vision artificial horizon display*, in *Vibration and combined stresses in advanced systems*, AGARD-CP-145, 1975
12. Green R G and Farmer E W, *Attitude indicators and the ambient visual system*, in Sorsa M (ed), *Report of the XVI conference of the Western European Association for Aviation Psychology*, Helsinki, Finland, 1985
13. Liggett K K, Reising J M and Hartsock D C, *The use of a background attitude indicator to recover from unusual attitudes*, Proc Human Factors Soc. 36th annual meeting, 1992

14. Rupert A H, Guedry F E and Reschke M F, *Use of a tactile interface to convey position and motion perceptions*, in *Virtual interfaces: research and applications*, AGARD-CP-541, 1994

Chapter 7

Display Technology and Head-Down Displays

7.1 Introduction

Glass display screens are the most prominent and eponymous feature of the modern "glass cockpit". Originally introduced in the 1940s to show the output from a nose-mounted radar, these Head-Down Displays (HDD) have become the main way to show the output from all of the on-board systems, a predominance that has been driven mainly by the need for versatility. As shown in Figure 6.11, the screens not only present pictures from sensors, but they also show the graphical images generated by the specialised computers linked to the highly automated on-board systems.

None of this would have been practical had the teams designing the cockpits not been able to call upon display devices that would function satisfactorily in the cockpit environment. This chapter has three aims: firstly, to review briefly the required attributes of HDD systems; secondly, to give an overview of the physical phenomena exploited by the device developers; and thirdly to describe the engineered form and characteristics of their products. The overview of the alternative technological approaches also covers devices that are incorporated in the HUD and HMD systems described in Chapters 8 and 9.

7.2 Requirements

The general requirement for a versatile HDD is the ability to present a wide variety of pictures, including diagrams, maps and tonal sensor imagery, so that fine detail and colour are portrayed reliably in all viewing conditions. The fundamental quality is adequate resolution, as described in Appendix B, the ability to render fine features with discernible contrast. For an airborne display this contrast must be maintained in all operating conditions, of which bright daylight is most problematical.

The main factors that degrade the perception of the picture presented on a HDD during daylight are illustrated in Figure 7.1. Although the eye adapts to the luminance of the display screen when looking at the screen, a wider, brighter skylight can have a dominant effect. The discernible dynamic range of spatial luminance variations that the viewer can distinguish, the range between subjective black and subjective white shown in Figure 2.11, is dictated by a combination of the ambient field and the display. The range of luminance variations that carry the picture information in the displayed image must therefore fall within this instantaneous span. For a display device to remain legible in the worst case, where the sun is close to the line of sight, it is necessary to take account of the reduction in contrast produced by the internal scattering of light within the eye. Also, because most pilots would relieve the glare by wearing dark spectacles or by lowering the dark visor, the characteristics of these devices are important. The glare wash-out that occurs when ambient light falls onto an emissive device, such as a CRT, can be illustrated using the transfer characteristic of the device, described in Appendix C. In such cases the glare luminance of the screen, which is measured by switching the device off and exposing it to the ambient light, is added to whatever light is emitted from contrasting dark and bright areas.

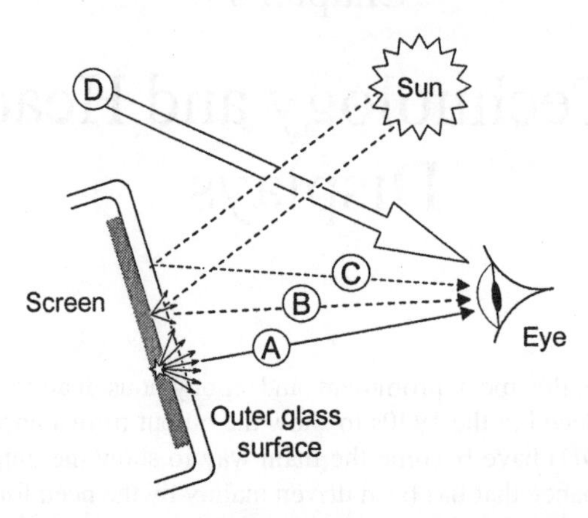

Figure 7.1 Factors affecting the visibility of a picture displayed on an HDD. (A) information-bearing light intentionally emitted or reflected from the screen, (B) sunlight and skylight scattered diffusely from the screen that combines with (A) and lowers feature contrast, (C) sunlight and skylight reflected specularly from outer glass surfaces that causes glints and further contrast reduction, and (D) ambient light that dictates the adaptation state of the eye

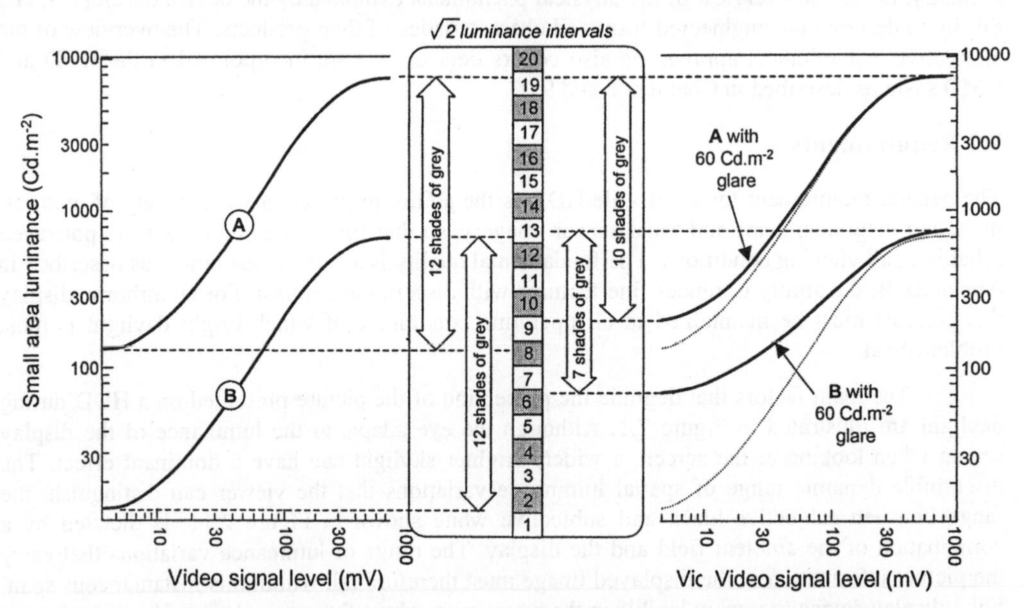

Figure 7.2 Illustrating the number of distinguishable shades of grey produced by two emissive displays. The left graph shows the logarithmic transfer characteristics of the devices in a dark ambience. The right graph shows the changed shape of the transfer characteristics produced by adding a constant value of ambient glare, and the marked reduction in dynamic range of the less bright device

Figure 7.2 shows how glare of about 60cd.m^{-2} would be added to the emission luminance and affect the dynamic range of two different displays. The definition of the number of shades of grey that can be distinguished within the dynamic range is described in Appendix C.3. For a bright display capable of a peak luminance of about 8000cd.m^{-2}, denoted by (A) in the figure, the glare would distort the perceived transfer characteristic, mainly by raising the luminance of the two darkest shades. These would not be distinguishable, and the total number of shades of grey would be reduced from twelve to ten. The same glare scattered from the face of (B), a device with only a tenth of the output of (A), would distort the characteristic markedly and reduce the brightness range from twelve to only seven shades of grey.

The primary defence against the reduction in the image contrast due to impinging light is obviously to prevent, or at least limit the amount of sunlight or scattered skylight reaching the screen. Specular reflections from the glass outer surface can be reduced dramatically using an anti-reflection coating. Glare shields are effective on the flight decks of transport aircraft with a relatively small windscreen, but these can be by-passed within the glasshouse of a combat aircraft. Here, unless the display is viewed through an optical system that limits the angular range of light that can reach the emissive screen, as in the Chin-Up Display of Rafale and the projected map display of the Tornado and Jaguar, the necessary attenuation is best done using an appropriate filter.

Such contrast-enhancement filters operate using a variety of phenomena. For instance, a thin micro-louvred transparent material embodying transverse light-absorbing planes, like a micro-miniaturised venetian blind, allows most of the emissions to pass straight through but blocks obliquely incident light. Another approach is to place a neutral absorbing filter in contact with the tube face, or use an absorbing glass for the tube face. These transmit a proportion T of the CRT emissions, but because the ambient light must pass through the filter twice, the glare luminance is reduced by a factor of T^2. If, for instance, T is 50%, the screen luminance is reduced by 50% but the glare is reduced by 50% of 50%, i.e. to 25%. Polarisation can also be used. As explained in Appendix A, a quarter wave plate on the inside of a linear polariser has no effect on the CRT emissions, so only a little more than half the light need be lost through the polariser. However, ambient light becomes circularly polarised on entry, and it is backscattered from the glass or phosphor with a reversed sense of rotation. On traversing the quarter wave plate for the second time it meets the linear polariser with a crossed plane of polarisation and is absorbed strongly. For monochromatic devices that emit a narrow wavelength band, a filter that absorbs all other incident wavelengths can achieve a remarkable improvement in contrast. Combinations of these techniques are possible, although the cost can be a concern, particularly the cost of applying anti-reflection coatings to all the surfaces of all of the layers that are placed between the eye and the image source.

Although monochromatic devices have been excellent for showing cursive line drawings and symbols, and for rendering pictures using mono-tonal shades, many forms of information[1] beg to be displayed in colour. It has proved advantageous, for instance, to show groups of entities such as threats, friendlies and unknowns on a tactical summary or a radar picture in red, green and orange respectively. Alternatively, changing symbols from green to red can draw attention to hazardous states, such as low fuel, low airspeed, excess rotor torque or low height above ground. Detailed mapping information places the most demanding requirement on a colour display.

As well as reducing the available contrast and the tonal range, ambient glare also reduces the range of distinguishable colours produced by an emissive display. The range can be described using the CIE 1976 definition of Uniform Colour Space, summarised in Chapter 2 in Figures 2.12 and 2.13. All that is needed is the measured chromaticity co-ordinates of the primary elements of the display device, for instance from the red, green and blue varieties of phosphor that form the dots on the screen of a shadow-mask CRT. Because the co-ordinates of any colour produced by a display can be calculated by summing the chromaticity values of the primary components in proportion to their luminance, the range is bounded by the triangle joining the primaries.

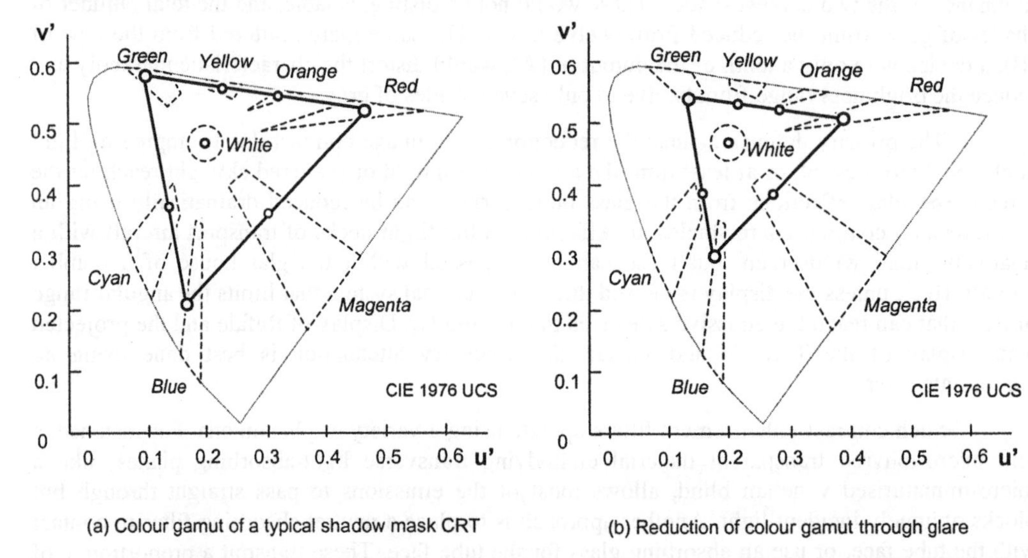

Figure 7.3 The range of colours produced by a display plotted on the CIE 1976 UCS diagram. The background shows the colours produced by narrow-band emitters (the spectrum locus) and the recommended maximum chromatic extent of seven colours plus white[2]. The bold triangle in (a) shows the typical range the colours that can be formed by the red, green and blue emitters of a display, and (b) shows how this gamut is shrunk by additional white glare light

The reduction of the range of colours that a display can produce can be assessed by adding the co-ordinates of the glare source to the primaries, again in proportion to their luminance. Broad band (white) sunlight de-saturates the whole gamut, and as illustrated in Figure 7.3, some of the colours - particularly colours formed along the sides of the triangle, such as cyan and yellow - can be whitened to the extent that they become easily confused with white.

The notion that both the colour difference and the luminance difference from the surroundings determine contrast and hence the visibility of a symbol presented on a display has been used as the basis for several evaluation procedures and Standards. One approach, devised by engineers at BAE Systems[3] and used during the development of the Eurofighter, computes the number of Perceivable Just Noticeable Differences (PJND) from the geometric sum of the number of Luminance Just Noticeable Differences (LJND) and Chrominance Just Noticeable Differences (CJND):-

$$\text{PJND} = (\text{LJND}^2 + \text{CJND}^2)^{1/2}$$

The number of LJNDs is related to the number of shades of grey between the symbol and the background. The number of CJNDs is related to the distance in CIE 1976 colour space between the symbol colour and the background colour. The computation takes into account the magnitude of the specular and diffuse reflections, the fall-off of human contrast sensitivity with reduced adaptation luminance, the effect of visor attenuation, the site of the display within the cockpit and the reflectivity of the clothing. It is done for three problematical conditions; flying toward the sun at dusk; flying with the sun over the shoulder illuminating forward cloud; and flying towards the sun and sunlit cloud. Experimental evaluations involving aircrew have established the number of PJNDs that are desirable. The scale starts at 40, the number needed for presenting low priority information, and it extends up to 120, the number needed for an attention-getting display.

In practice, these criteria are very exacting. As an example, at present it would be very difficult to make a device having a maximum luminance of 1400cd.m^{-2} and a white-to-black contrast of 140:1 that reflects 1% of light specularly and 0.15% diffusely. Such a display would provide between 50 and 73 PJNDs, and would be regarded as suitable for presenting complex static, but not dynamic, characters.

7.3 Display characteristics

HDDs are now largely based on commercial display systems, although much development work is generally needed to achieve the required performance and ensure that the package survives the cockpit environment. As well as the display device itself, which transforms electronic signals into a visible image, a system also needs a symbol generator, which specifies the symbols to be displayed, and a set of electronics to transform this picture definition into signals acceptable by the display device. This combination of symbol generator, electronic drive unit and display device must be designed as a whole, and the elements can either be distributed around the airframe, or as in more modern practice, integrated into the same box as the display screen. Another characteristic shared by all versatile HDD systems is a capacity to show dynamically changing pictures that are conveyed as video signals. As summarised in Appendix C, time is broken into a sequence of "frame periods", during which temporal variations of the signal are used to change the elements of the display and show a fresh picture. A picture element, or pixel, is the smallest area of the screen that can be changed, and display devices are characterised by the mechanism employed to effect the change. The alternative ways of selecting the area to change by steering a beam or by active and passive addressing are described later, within the sections dealing with the technologies. However, it is useful at this stage to delineate the alternative ways of controlling the luminance and colour of the instantaneously addressed pixel.

Consider firstly the method for controlling luminance. As shown in Figure 7.2, the shade of grey that a CRT spot can take depends upon a continuously variable grid voltage, which is governed directly by the signal voltage. Unlike this intrinsically analogue form of drive, some digital devices such as the ferro-electric LCD have cells that are in one of two states; *off* or *on*. In this case the luminance of a cell is controlled by splitting the frame period into intervals and controlling the proportion of intervals for which the cell is in the *on* state; a process referred to as "temporal dithering". The technique involves accurately timed switching of cells, sometimes in synchronism with pulses of illumination. "Spatial dithering", the alternative way of controlling the apparent brightness of such two-state pixels, divides each into structurally smaller sub-cells and varies the proportion of the sub-cells that are switched *on* during a frame period. A combination of spatial and temporal dithering is sometimes used.

A similarly wide range of techniques has been devised to solve the problem of changing the colour as well as the luminance of a display element. The "sub-pixel array" or colour triad, is probably the most common approach at present. It is akin to spatial dithering for brightness modulation in making use of sets of smaller elements, but in this case the sub-elements emit controllable amounts of red, green and blue light.

Table 7.1 Characteristics of airborne display devices

Type
- -Emissive
- -Reflective
- -Transmissive

Size
- -μ (~20mmϕ)
- -S (~150mmϕ)
- -L (>300mmϕ)

Grey shade technique
- -Analogue Modulation
- -Digital Spatial Dithering
- -Digital Temporal Dithering

		Device	Type	Size	Addressing	Colour mechanism	Grey shade technique
Intrinsically pixelated	Liquid crystal	Twisted nematic (TN)	T, R	μ, S	AM, PM	SI, SF, SS OC, SP	An
		Ferro-electric (FE)	R, T	μ	AM, PM	SI, SF, OC	TD (SD?)
	Light-emitting diode	Conventional semi-conductor (LED)	E	μ	AM, PM	SP	TD
		Organic semi-conductor (OLED)	E	μ	PM	SP	TD
	Other mechanisms	Vacuum fluorescent (VF)	E	L, S	PM	SP	TD
		Electro-luminescent (EL)	E	μ	AM	Ph	TD
		Plasma discharge (PD)	E	L	PM	Ph	SD, TD
		Digital micro-mirror (DM)	R	μ	AM	SI, OC	TD
Scanning spot		Scanned laser (SL)	E	μ	RS (CS?)	ML	An
	Cathode ray tube	Single phosphor (CRT)	E	μ	CS, RS, HS	SF	An
		Penetron (P-CRT)	E	S	CS, RS, HS	Ph	An
		Beam index (BI-CRT)	E	S	RS	Ph	An
		Shadow mask (SM-CRT)	E	S,L	RS (CS?)	Ph	An

Addressing
- -Active Matrix
- -Cursive Scanning
- -Hybrid Scanning
- -Passive Matrix
- -Raster Scanning

Colour mechanism
- - Sequential R,G,B Illumination
- - Sequential R,G,B Filter
- - Subtractive C,M,Y Stack
- - Optically Combine R,G & B
- - R,G & B Modulated Lasers
- - R,G & B Sub-Pixel Array
- - R,G & B Phosphors

The other popular class of techniques employs "sequential colour". This is similar in principle to the temporal dithering technique for brightness modulation, but the frame period is divided into three sub-intervals during which red, green and blue (RGB) component images are shown in sequence. In the case of an emissive display, the appropriately coloured emitters are excited during the sub-intervals. In the case of a transmissive or reflective device the picture information must be impressed on the appropriate elements in synchronism with bursts of red, green and blue illumination. Some devices combine brightness and colour adjustment by such rapid switching and synchronised illumination, and this allows the grey-shade adjustment intervals to be mixed with the colour adjustment intervals to ameliorate the problem of colour separation with moving imagery. Sequential colour can also use a RGB filter, constructed either as a rotating filter wheel or as an electronically-switched cell.

Another approach involves placing a sandwich of three very thin transmissive display devices together to form a "subtractive stack" illuminated by a bright white backlight. Elements in the first layer absorb a controllable proportion of red, those in the second a controllable proportion of green and the third a controllable proportion of blue. An area of the screen appears white if the aligned stack of elements do not absorb the white backlight, and black if all absorb fully. A range of colours and luminances is produced by suitable adjustment of the degree of absorption in each layer.

Any device can be described using a state-space-time diagram that summarises the possible states that the cell and the illumination can take during a frame period. To perceive a cell at the intended colour and shade of grey, the viewer should respond only to the average effect over the pixel area during the frame period and be unaware of internal state-space-time manipulations. On the other hand, to discern image detail and fine movement, the viewer must sense differences between neighbouring picture elements and between successive frame periods. The engineer must tread close to human perceptual limits to attain an affordable and practical balance between these criteria, and like a conjuror, know what is noticeable and what is not. The penalty is a visual artefact that is likely to affect the legibility of the displayed information. In practice, such artefacts also characterise the display device.

The CRT, in its several guises, has been the mainstay HDD and the source of the visible image in most HUD and HMD systems. It is largely the search for practical, economical substitutes for the CRT that stimulates investigations into a wide variety of exploitable electro-optical phenomena. Table 7.1 summarises the main characteristics of devices that have proven useful, or offer the prospect of becoming useful, in aircraft cockpits. The list includes devices that can be used in HMDs and HUDs as well as HDDs, so the size distinction between micro (μ), small (S) and large (L) is primarily intended to indicate suitability for these respective applications. The type indicates whether the device is intended to produce a visible image by emitting light, or by transmitting or reflecting light from a source. Reflective devices can operate using available ambient light or carefully directed pulses of illumination. The use of a scanned laser in HMD systems is discussed in Chapter 9. Some phenomena that seem unfeasible for video imagery, such as light scattering by liquid crystal materials, or electro-phoresis and electro-chromism, and others approaches that seem intrinsically unwieldy, such as oil-film light valves, are not included. However, the list is far from exhaustive. Given the current high level of commercial investment in research and development, technological approaches that are under investigation for other purposes may prove adaptable.

7.4 Cathode ray tubes

An understanding of the way the cathode ray tube (CRT) has been developed to produce a visible image is fundamental, if only because the characteristics of this ubiquitous form of display device have dictated the way picture information is encoded as a video signal. Although the "shadow mask" gained rapid acceptance in the 1960s as the main technological approach to the manufacture of colour CRTs for domestic and commercial use, early experiments with off-the-shelf devices showed them to be too frail for use in combat aircraft. The principal difficulty[4] was in attaining an adequate combination of luminance, resolution and reliability. Instead, manufacturers concentrated on two relatively obscure approaches; the penetron CRT and the beam-index CRT. The monochromatic device is introduced first, followed by short descriptions of four techniques that give colour images.

7.4.1 Monochromatic CRT

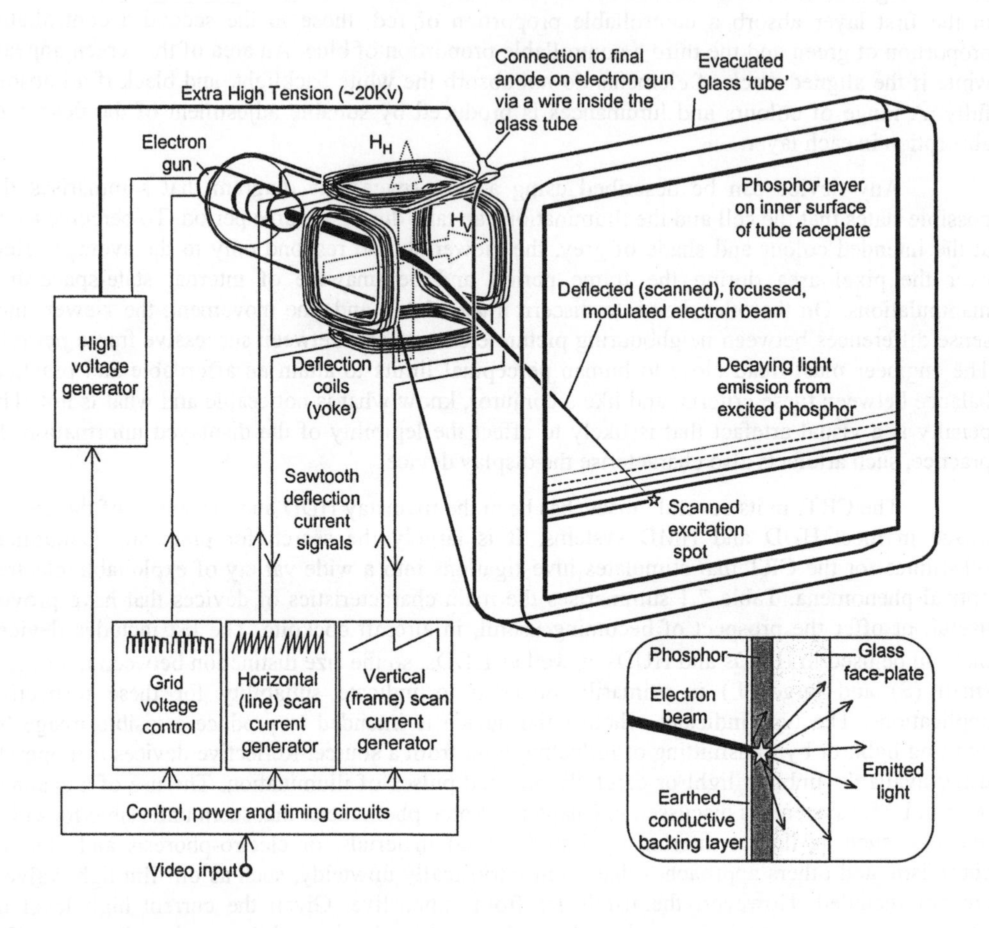

Figure 7.4 The essential elements of a monochromatic CRT display

As illustrated in Figure 7.4, the basic CRT display is a sealed evacuated glass envelope shaped like a narrow-necked flask with a flat base on which is deposited a thin phosphor coating. An electron gun in the tube neck contains a cathode of a low work function refactory material such as cerium oxide, which is heated to give off electrons (the cathode rays) that are accelerated to high energy by the attraction of the anode held at a potential of about 20 kilovolts. Most of the emitted electrons are channelled by a series of electro-static field lenses held at lower attractive potentials to form a narrow beam that passes through a hole in the final anode and converges to a fine focus on reaching the phosphor. The latter arrests the electrons and emits light in proportion to the electron energy and flux. In general, phosphors are complex compounds, such as terbium activated aluminium garnet (which is usually called YAG:Tb or its normal JEDEC[5] designation, Phosphor No53) that are chosen for their electron-to-photon conversion efficiency, emission spectra and emission persistence.

A picture is painted onto the phosphor screen by simultaneously deflecting and modulating the beam flux, and this is repeated with sufficient regularity to give the appearance of a constant picture. To avoid a flickering image the picture must be re-scanned at a rate which is above the critical flicker fusion frequency for the human eye, in general about 50 times per second, and the phosphor should persist in emitting light for a reasonable proportion of the refresh period, say 5msec of the 20msec period. If the persistence of the emission is longer than the refresh interval, displayed elements that are repositioned on successive scans will appear smeared. The beam is deflected by sets of magnetic coils, and it is modulated in brightness by controlling the emission from the cathode by applying a small negative potential to a control grid close to the cathode. The CRT is thus a "flying spot scanner" which, as shown in Figure 7.4, needs electronic units to generate the fixed high voltages, produce the cyclic currents that cause the repeated magnetic field scan excitation and transform the input picture information into the synchronised grid control signal. Circuitry is also needed to control the overall timing and, should the beam deflection fail, switch the beam off to prevent excessive excitation to a small area that would burn (change the chemical and physical structure of) the phosphor.

The path taken by the flying spot depends on the nature and source of the information to be displayed. In the original radarscope the spot was moved across the screen in a pattern that was analogous to, and synchronous with, the repeated scanning action of the radar dish in the aircraft nose, the instantaneous spot brightness showing the strength of the detected radar signal. However the two most common approaches in aviation are either to deflect the beam to write individual symbols on the phosphor or to scan the whole area as a pattern of parallel horizontal lines. The words "cursor" and "raster", Latin for runner and rake respectively, convey the ideas nicely.

The first of these has been essential for showing bright images in the head-up display. Only a small proportion of the refresh period is spent in moving the flying spot between the end of one symbol and the start of the next, and the intense electron flux can be moved relatively slowly when painting each symbol. The second, which is assumed in Figure 7.4, is however more usual because it has been adopted for broadcast television where the general requirement is to show tonal camera images. The scanning and modulation circuitry for the two are entirely different. The first requires the generation of vertical and horizontal deflection current waveforms that move the spot very precisely to trace out individual symbols. The second needs a very rapid sawtooth-shaped current waveform to produce a horizontal line, and a slower sawtooth-shaped current waveform so that successive lines are displaced vertically, together with rapid, synchronised control of the gun grid for varying the spot intensity to produce the tonal differences of the picture.

7.4.2 Penetron colour CRT

The principle of the penetron is illustrated in Figure 7.5. The tube was constructed like a monochromatic CRT except that the faceplate was coated internally with thin green-emitting and red-emitting phosphor layers. Alternating red and green images were produced by switching the electron beam from an energy level that excited the green phosphor to a slightly higher level that penetrated through to the red phosphor. Providing the images were refreshed rapidly, i.e. the pair were addressed as frequently as a monochromatic image, the viewer fused the stimulation temporally and perceived a single image in which the colour of each spot on the screen was the resultant of the red and green excitations. The colour gamut could in principle encompass any chromaticity along the red-orange-yellow-green locus joining the pair of primaries, as plotted in the CIE colour space, and the device was referred to as "polychromatic" or "bi-colour".

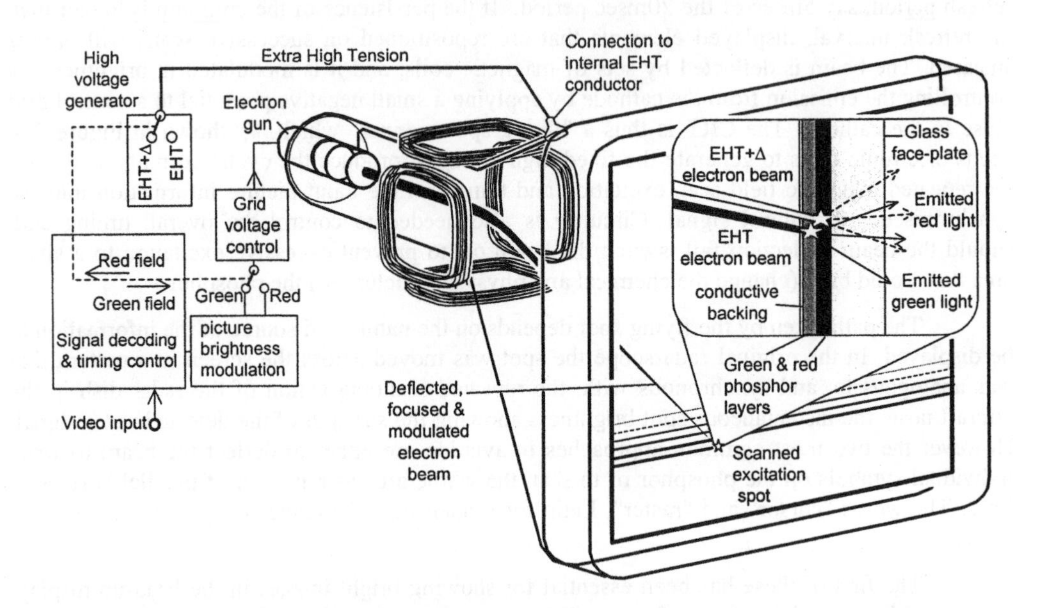

Figure 7.5 The idea of the penetron CRT display

Considerable care was needed in the design to ensure that the two images overlaid precisely despite the slight differences in focusing and deflection requirements. It was also necessary to arrange for sequential video modulation signals at twice the normal bandwidth, which invariably required storing each video frame, and for the horizontal and vertical scan to operate at twice the normal rate. The versatility of the basic CRT was not compromised; cursive addressing could be exploited to gain high brightness images, and raster scanning could give dimmer pictorial images. However, it was necessary to consider the colour attributes of the elements carefully, since for instance fine moving yellow symbols tended to separate into red and green components. In practice the technique lent itself best to a hybrid mode in which, say, a green pictorial raster image could be overlaid with red, yellow and orange cursive symbols, but only the red symbols could be moved with any rapidity.

7.4.3 Sequential shutter colour CRT

A more modern variant of the penetron display was the "sequentially-shuttered CRT", in which the phosphor emitted light over the visible spectrum, and blue, green and red component images were presented in succession by switching a filter covering the faceplate. Either a rotating three-colour wheel containing passive absorbing filter material, or a set of pleochroic liquid crystal shutters, have been used. The latter exploited polarisation phenomena, described briefly in Appendix A. The technique has been used in several prototype avionic helmet-mounted display systems, and it had the advantage over the penetron that the exact superposition of the component colour images was not upset by switching the high voltage to the gun anode. However, as with the penetron, the modulation and deflection systems were required to function at three times the normal rate, and small areas that moved across the screen would separate into their primary colours. Thus a vertical white line would be shown as a set of red, green and blue lines when moved horizontally, the displacement between the component lines equalling the product of the motion rate and the time interval between component refresh periods.

7.4.4 Beam index colour CRT

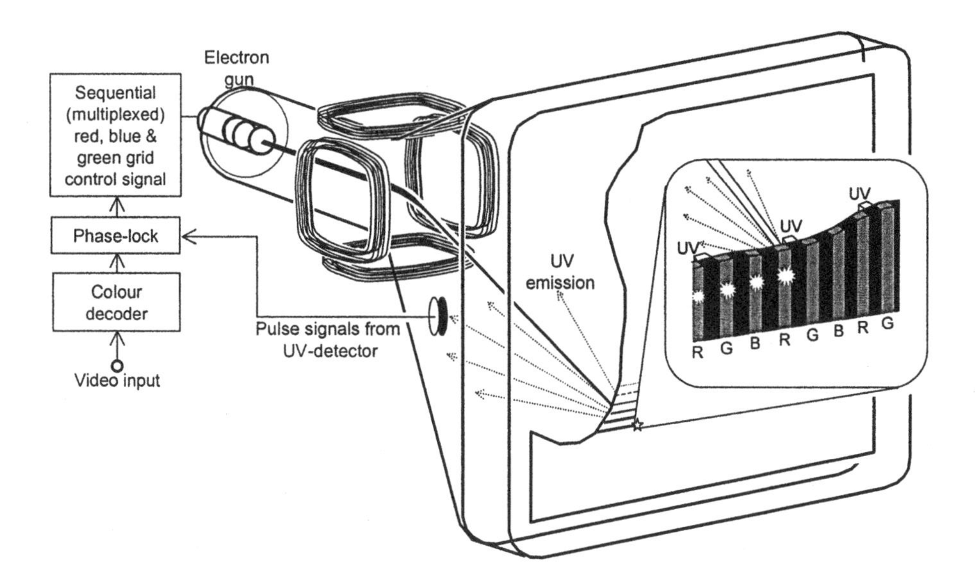

Figure 7.6 Principle of the beam index CRT display

The idea of the beam index colour CRT display, as shown schematically in Figure 7.6, was to avoid the sequential mis-registration problem and the HT switching inherent in the penetron. The manufacture of the tube itself was significantly more complex as it required the very accurate deposition of fine vertical stripes of red, green, and blue phosphors on the glass faceplate, and for these to be separated by black stripes of non-emissive material. In addition, vertical "index" stripes of a short-persistence, UV-emitting phosphor were deposited on the innermost surface. In operation, the focused electron beam activated visible light emission as it is was scanned across the R, G, and B phosphor stripes, and the necessarily precise synchronisation of the multiplexed signal for modulating the beam depended on detecting pulses of UV light emitted as the beam struck the indexing stripes. Although the idea achieved the principal requirements, the horizontal resolution depended fundamentally on the number of stripes across the tube face, the device could be addressed only in a raster mode, and the design of the circuitry to handle the high bandwidth of the modulating signal was problematical. Only a few manufacturers have produced suitable CRT tubes.

7.4.5 Shadow-mask colour CRT

The most common form of colour CRT display used in aviation uses a taut shadow-mask. It is a derivative of the conventional CRT that has been the primary form of display device used in many millions of TVs and computer systems. The scheme is outlined in Figure 7.7. Unlike the penetron and beam-index CRT, the shadow-mask tube has three separate, but physically identical, electron guns. All three beams are modulated and scanned in synchronism, and all three are brought to the same focal spot.

The means by which a mask is held in the beam path so that the red electron beam only excites red-emissive phosphor, the blue beam only excites blue-emissive phosphor and the green beam only excites green-emissive phosphor, is illustrated in the inset in the figure. The pattern of holes in the mask is identical to the pattern of the phosphor dots for one colour on the screen, and is held away from the screen by a distance that depends upon the inter-gun spacing and the dot separation. Essentially the idea exploits the geometry of similar triangles. Although Figure 7.7 shows the guns grouped as a triad, other configurations are used, for instance some manufacturers arrange the guns in line horizontally with phosphors printed as vertical stripes.

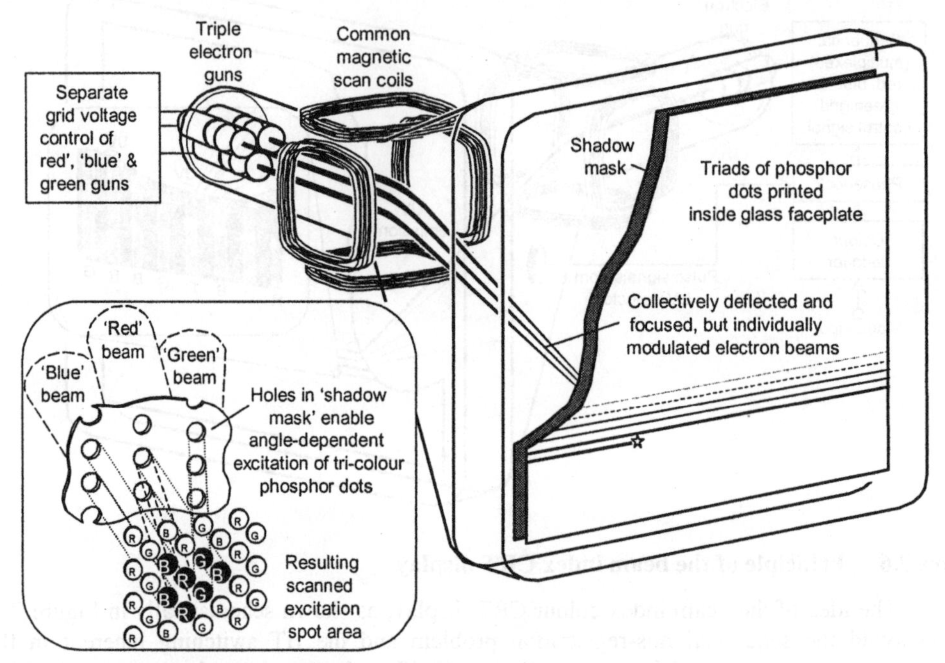

Figure 7.7 Principle of the shadow-mask CRT display

A considerable fraction of the electron beam flux is absorbed by the shadow mask. For domestic and commercial devices this inefficiency has been tolerable, but for avionic applications that require a screen luminance about fifty times greater, the heating causes the mask foil to expand and buckle. This upsets the fine alignment between the holes in the mask and the phosphor dot pattern, causing local colour anomalies. The production of high brightness devices was made practical by the adoption of a stiff mask frame, pre-stretching of the mask foil, and very careful thermal design. Displays have been produced in a variety of sizes, from about 100mm to 300mm diagonally, mainly with 1:1 and 4:3 aspect ratios. A typical taut shadow-mask tube uses an anode voltage of 25kV, produces a peak raster luminance of 800cd.m^{-2} and paints lines of about 0.18mm width on a screen made with 0.05mm diameter phosphor dots separated by a black matrix. The faceplate is made from flat glass with integral anti-glare filters and anti-reflection coatings.

In general, to meet the environmental and longevity requirements, the glassware would be packaged into an aluminium alloy mounting frame, the scan coil assembly would be potted and sealed onto the tube neck, and the whole of the device from the mounting frame to the gun would be enclosed within a tight-fitting mu-metal magnetic shield. Ruggedisation would also include anchoring and multiple insulation for the HT lead, and shielding the set of wires that carry the fixed voltages and signals to the electron guns. Unlike the production lines that supply commercial CRTs, the manufacture of avionic tubes has been a specialist activity. The devices developed for use in head-up displays and helmet-mounted displays must meet peculiar packaging constraints and give exceptional performance. The miniature HMD tube, described in Chapter 9, is made in small batches; each tube is almost hand-built by specialists.

7.5 Liquid crystal display (LCD)

The on-going search for devices to replace the ubiquitous CRT is motivated by the need to show coloured, detailed, dynamic pictorial images for a wide range of applications, from television and computers to airport terminals. The devices must also be lightweight, robust, flat, efficient, easily packaged, stable, cheap and avoid engineering complications such as high voltage excitation. The sheer strength of the commercial demand means that replacements for the CRT to satisfy the comparatively esoteric environmental demands of aviation can only be done by modifying devices intended for wider commercial use. At present the available "flat panel" HDDs are largely based on commercial active matrix liquid crystal displays (AMLCD) that have been adapted to operate satisfactorily in the cockpit environment.

The AMLCD has so little in common with the CRT that it represents a revolution in technological culture. The stream of hot, energetic electrons that excites a strange glow in an evacuated glass bottle is replaced by a matrix of tiny cells controlled by low voltage semiconductor micro-circuitry sandwiched between thin glass sheets. The devices exploit the polarisation state of light, which is explained briefly in Appendix A, and the phenomenon of liquid crystalinity, a somewhat paradoxical state exhibited by a number of organic materials.

7.5.1 Liquid crystal types

Liquid Crystal (LC) materials behave like viscous liquids in which the molecules can flow past each other, but they exhibit a crystal-like anisotropy because the molecules are polar and become aligned and layered. A wide variety of materials show different states of order over different temperature ranges to produce different electro-optical effects. The term "nematic" describes the phase in which there is only orientation order. "Smectic" and "cholesteric" phases are aligned and also layered, and in the latter there is a small angular twist between molecules that produces a spiral structure with a characteristic pitch. In all types, the "director" describes the local alignment of the molecules. This is important because the angle between the director and the plane of polarisation of the light (the direction of the E-field in the propagating electro-magnetic field) determines the effect of the material on the light. In general, the display devices are contrivances for applying an electric field to a small region within a liquid crystal layer in order to orient the local director and rotate the plane of polarisation of the emergent light so that more or less of it is passed by an analyser.

The most common form of AMLCD uses a nematic material in which the polar molecules spontaneously align with molecules on a contact surface or with an imposed electric field. The structure of the cells, illustrated in Figure 7.8, places a thin layer of the liquid crystal material in contact with thin polymer layers that have been rubbed in orthogonal directions so that the director of the nematic molecules twists by 90° across the cell; hence the descriptive term "twisted nematic". Immediately outside the polymer rubbing layers are thin layers of transparent conductive material, usually made by vaporising indium/tin oxide (ITO) in a vacuum and causing it to deposit on the glass substrate. These are used to apply an electric field across the thickness of the material. The device functions by rotating the plane of polarisation of light. The back-light is usually a compact fluorescent lamp packaged in a flattened form with a reflective backing and a diffusing front to give a uniformly bright surface. The light emitted by this surface reaches the AMLCD backing glass through a sheet of polarising material.

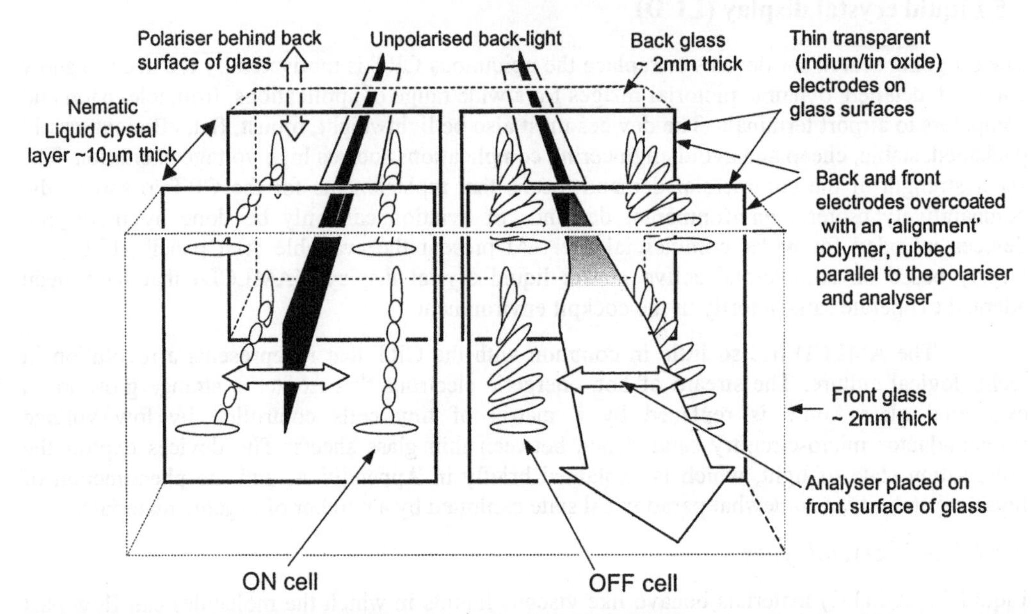

Figure 7.8 Operating principle of the active matrix liquid crystal display. The orientation of the ellipsoids represent the "director" of the nematic molecules. Two cells are shown. A small voltage is applied to the electrodes on the cell on the left

As shown schematically on the right side of Figure 7.8, without an applied electric field the nematic LC molecules align with a gradual 90° twist of the director between the crossed rubbing axes of the polymer surface layers. This acts as a stack of birefringent layers with incrementally reoriented optic axes that interact with the linearly polarised light via a complex mixture of helical optical activity and birefringence - with the latter effect predominating - to rotate the plane of polarisation of the transiting light through 90°. With this orientation, most light passes through a second orthogonal polarising layer, the "analyser", and in this state the cell appears bright.

As shown by the cell on the left of the figure, application of either a steady or an alternating voltage of about 2V to the ITO electrodes produces an electric field that aligns the director of most of the molecules in the cell to remove the twisted birefringence. In this state the linearly polarised light is not affected by the cell and it is propagated with little change to the plane of polarisation, so when it encounters the crossed analyser it is absorbed strongly and the cell appears dark. The process is completely reversible and removal of the field allows the cell to relax back to the transmissive state. Above a threshold voltage, there is a progressive increase in the degree of alignment of the director with the transverse field, and hence a progressive reduction of the rotation of the plane of polarisation. The amount of light emerging from the cell can therefore be adjusted by controlling the applied voltage; the cell behaves like a light valve.

7.5.2 Active matrix LCD

The means by which cells in a two dimensional array are selected depends upon the nature of the imagery displayed. In relatively simple devices the cells can be switched between two states using orthogonal stripes of ITO on the two glass plates, as shown in Figure 7.9.

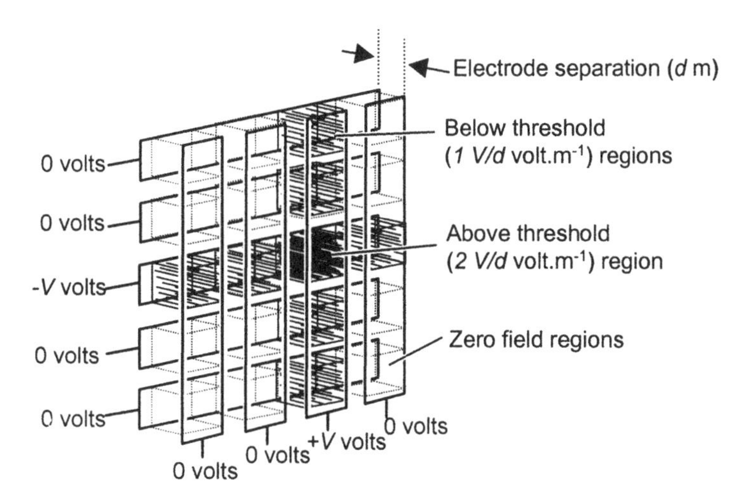

Figure 7.9 Simple crossed electrode addressing

Here an individual cell is switched to the *on* state when the electric field exceeds the molecular alignment threshold, and this requires the simultaneous application of signal voltages to the row and column electrodes that intersect at the cell. The pattern of cells that are switched *on* to show a symbol can be addressed by activating one row electrode and simultaneously activating the column electrodes for the *on* cells along that row. Then the next row is activated and the voltages applied to the column electrodes for the required *on* cells along that row, and so on until all the rows have been switched, whereupon the process starts again with the first row. This scheme is useful for activating emissive devices such as light-emitting diodes (LED) and for simple arrays in which the number of display elements is less than the number of connections. It is unsuitable for a twisted nematic LC device that requires proportional control over the cell transmission and which has a large number of connections. Also, as each cell can be *on* only for a line period during each frame period, most of the back-light would be wasted. Active addressing of the pixels is therefore used.

Figure 7.10 illustrates one method of using individual field-effect transistors (FET) to apply a relatively precise signal voltage to each pixel in a matrix. A pixel is electrically equivalent to a capacitor that is charged when its transistor is turned *on* by simultaneous pulses to the gate line and the drain line. The common ITO electrode on the upper glass is connected to an a.c. supply to prevent polarisation of the LC material. Gate and drain lines are addressed using shift registers formed as an integral linear array of FETs along two adjacent edges of the array. The whole panel usually has two broad edges to accommodate these elements and a line of relatively large gold-plated pads for connecting the device to the supporting circuitry.

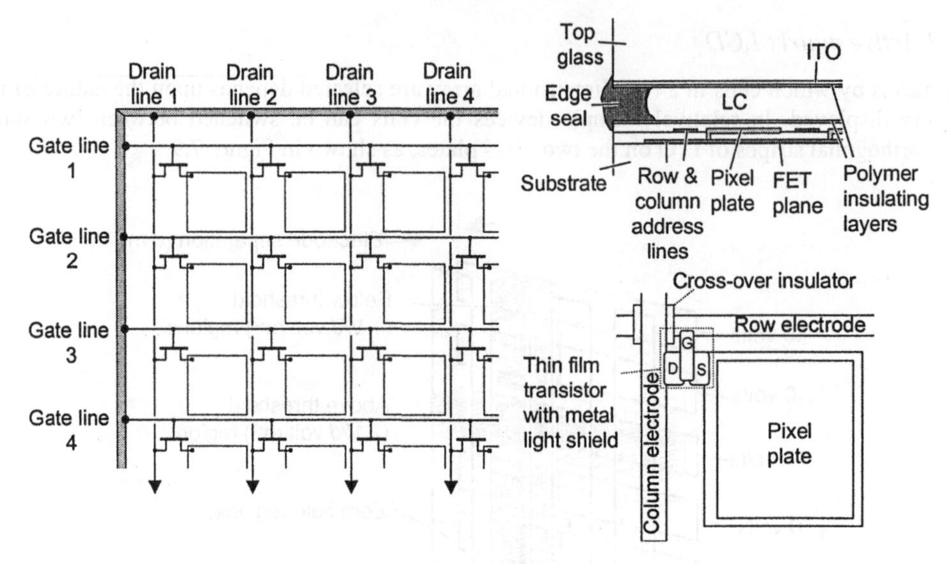

Figure 7.10 Simplified schematic for actively addressing pixels in a matrix display

Given the large number of pixels, the number of states each can take on and the rapidity with which the switching must occur, the manufacture of a large matrix has proved to be a severe technical challenge. For instance, to achieve the quality of a monochromatic CRT display for broadcast TV, the array has 585 by 780 pixels that can be set to transmit fractions of the back-light for at least six tonal shades, and each pixel must be addressed every 20msec. To give colour images, a triad of separately addressable red, green and blue sub-pixels replace each monochromatic pixel, so the array must have 585 by 2340 pixels, or an equivalent number divided more equally between the rows and columns. For a typical screen size of say 180mm by 240mm, each pixel is about 250μm square, and of this the transparent pixel plate would be about 150μm square. The margins of each pixel accommodate the FET elements and the gate and drain address lines. These are formed in a thin deposited film of amorphous or polycrystaline silicon by masking, etching, doping and deposition, the standard processes for micro-circuit production that have been extended to cover the large area of the glass plate rather than a few mm^2 of silicon wafer.

The whole back plane array is formed on the backing glass, as shown in Figure 7.10, by covering the etched active silicon layer with a transparent insulator, usually silicon dioxide, through which conductive pads make connections to the deposited ITO pixel plates. The thin polymer rubbing layer is finally deposited on the top. Manufacture of the top glass is comparatively simple; it is coated with a continuous ITO film and a polymer layer, rubbed in the perpendicular direction. The sandwich is put together by sprinkling glass micro-spheres onto the upturned backing glass, to hold the plates at the precise spacing, placing the top glass coated side down on top, and sealing the edge with UV-setting cement. A small gap is left in one edge seal, and the LC mixture is drawn into the pre-formed void by placing the sandwich in a vacuum chamber, to evacuate the void, and immersing the feed gap in a trough containing the LC material. Filling usually takes several hours. The final processes involve sealing the gaps and bonding polarising film, described briefly in Appendix A, to the outer glass surfaces.

The are many variations on the arrangement described. For instance the analyser can be rotated through 90° to reverse the opaque and transmitting states. The display can also be used in a reflective mode to modulate the brightness of incident light, in which case the back-light can be replaced by a polarisation-preserving, reflective diffuser. Alternatively, the diffuser can be partially transmissive and the back-light used only at night. Unlike an emissive display such as a CRT, the AMLCD has the distinct advantage of modulating the brightness of transmitted or reflected light, and it is therefore relatively immune to a loss of contrast from ambient light. It is mainly necessary to manage the unwanted light to avoid specular reflections that arise from the abrupt changes in refractive index at the layer interfaces, and to mask the inactive areas of the array with black material. Such masking is needed in any event to prevent light reaching the thin film transistors, which are intrinsically photo-sensitive.

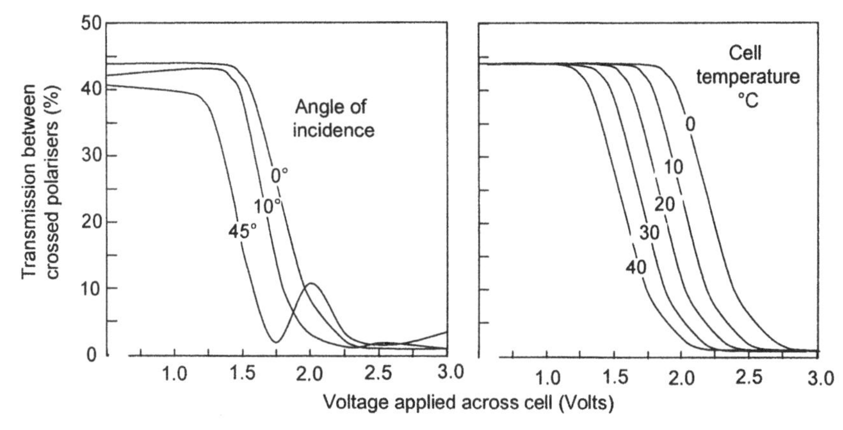

Figure 7.11 Typical transmission vs voltage characteristics for a twisted nematic liquid crystal display, showing the dependency on the angle of incidence of the light and the cell temperature (Reprinted by permission of QinetiQ)

There are several factors that degrade the useful qualities of this form of display, for which Figure 7.11 shows typical transmission-voltage characteristics. For normal incidence (0°) and room temperature (20°C) the transmission decreases from about 44% at a threshold of about 1.6 volts and reaches minimum of about 2% when the voltage is increased to about 2.4 volts. The ratio of the maximum to the minimum transmission is the dynamic range of the display under these conditions, and this governs the number of shades of grey that can be presented. The shape of the curve is important in converting the video drive signal into the cell activation voltage, but as shown in Figure 7.11, the shape of the curve changes with both temperature and angle of incidence, with a marked shift along the voltage axis. Thus the appearance of a cell, particularly one driven only slightly above threshold, is strongly dependent on the angle from which it is viewed. It is also necessary to build in an automatic adjustment to cater for the shift of the threshold voltage with temperature.

The cell temperature also affects the turn-on and turn-off transition dynamics of the cells. The turn-off is a comparatively slow relaxation of the molecular alignment, so it is two or three times slower than the field-induced turn-on. Both of the changes slow down markedly as the LC viscosity increases with a drop in temperature, and thinning agents are introduced into the chemical mix to extend the working temperature range. This is particularly necessary for avionic devices, which must function satisfactorily over a wider temperature range, typically –40°C to +60°C. Thus among the modifications that are needed to make commercial devices suitable for the cockpit environment, such as improved sealing and the addition of anti-glare filters, it is also usual to incorporate a thermostatically controlled heating film.

7.5.3 Colour techniques

"Sequential colour" techniques are rarely used with AMLCD devices because the cells cannot be switched sufficiently rapidly. The main technique for the manufacture of coloured displays involves sub-dividing the pixel structure into triads that are overprinted very accurately with red, green and blue "Bayer" dye filters. The pattern of coloured elements resembles the phosphor dots on a shadow-mask CRT, but with aligned rows and columns rather than a hexagonal grid. A quad arrangement with two green, one red and one blue pixel in alternating rows of -R-G-R-G-R- and -G-B-G-B-G- is sometimes used. This is mainly done to gain resolution, but because balanced white requires a greater proportion of green there is also a small gain in luminance, and some manufacturers use two different green filters to spread the colour gamut.

A variety of small AMLCD devices, with screen dimensions of about 25mm and pixels of about 25μm, have been developed for use in projectors and other commercial products. In these applications coloured images can be produced by techniques similar to those described in section 7.6.1 for digital micro-mirror devices. Several manufacturers have used the intrinsic transmissive qualities of AMLCD devices to develop subtractive colour displays, mainly in miniature form for use in helmet-mounted systems.

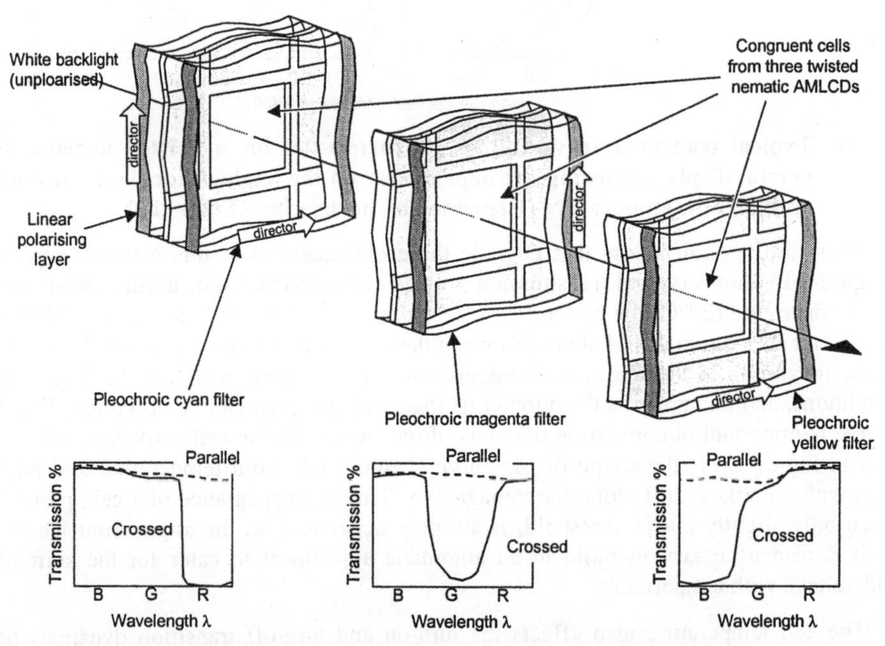

Figure 7.12 The idea of using a subtractive stack of AMLCD panels for modulating the colour and brightness of individual pixels

The operating principle of this device is shown in Figure 7.12. Intense white backlight is made parallel and directed through a linear polarising layer and then through a stack of AMLCD panels. These are aligned accurately so that light rays encounter a sequence of windows containing filters that control the transmission of red, green and blue light. The control depends on the pleochroic properties of aligned dye materials that, as described briefly in Appendix A, absorb light over a selective portion of the spectrum in proportion to the cosine of the angle between the plane of polarisation of the light and the material director. The first AMLCD panel is used to twist the plane of polarisation through a controlled angle so that the required proportion of the light is absorbed by a red-absorbing filter bonded to the outer face of the panel. The inset graph in the figure shows the high pleochroic transmission for light with parallel polarisation and the extinction of the red waveband for crossed polarisation. The same phenomena are used in the second AMLCD panel to absorb a controllable proportion of green, and in the third AMLCD panel to absorb a controllable proportion of blue. If the attenuation of each stage is set to the minimum most of the light emerges and the cell appears white, while maximum absorption at each stage produces black. Shades of grey result if equal amounts of light are absorbed in each stage, and colouration occurs when the proportions are unbalanced. For instance, setting maximum red absorption with the other stages clear makes the transmitted light cyan, while maximum blue and green absorption produce yellow and magenta respectively. The stages are also known by these complementary names.

The lightis largely emitted in a direction perpendicular to the screen, so the apparent luminance would fall off quickly if the device were viewed obliquely, This would be unimportant in a combat aircraft, but would prevent cross viewing in a side-by-side cockpit, and a diffuser or a lens array would be needed to scatter the light laterally. The main difficulty is one of ensuring that as little light as possible enters neighbouring cells *en route* through the stack. This is particularly difficult if the AMLCDs are small, as for instance in a head-mounted device, because the dimensions of the individual LC pixels (about 20μm) are very much less than the transit distance across the intervening layers of glass and plastic filter (about 2000μm). Even if the layers are aligned within several microns, the "aperture ratio" of a cell - the transparent area as a fraction of the total pixel area - is inefficiently small, typically less than 0.5. The diffraction of the light into neighbouring pixels causes further inefficiency and loss of contrast.

Pleochroic materials are also employed in simpler devices such as the sequential colour filter fitted onto a CRT that has a broadband phosphor. In this case the filter covers the whole CRT faceplate and there are few problems of managing light propagation *per se*. Also, the device is not required to generate tonal shades, so the stack can be switched between well-defined states using liquid crystal layers, each of which can either let the incident light pass with little change or rotate the plane of polarisation through 90°. This rotation is produced by a half wave plate (as described in Appendix A) in which the plane of polarisation of incident light makes an angle of 45° to the birefringence axis and the material introduce π radians phase shift between the ordinary and extrordinary wave components. Such active switching layers are therefore commonly called π-cells. In practice both twisted nematic and ferro-electric materials are used.

A very effective and economical device has been constructed that produces four states using only two LC layers and two polarisation-sensitive pleochroic filters. In this case one of the filters switches between transmitting blue (B) and its complement yellow (Y), which is the G+R portion of the spectrum, and the other switches between transmitting red (R) and its complement cyan (C), which is the B+G portion of the spectrum. No light is transmitted if the LC cells are set to produce polarisation states that induce the first cell to (B) and the second to (R). The other three combinations transmit the three primaries; blue results from (B) with (C), green from (Y) with (C), and red from (Y) with (R). Problems result mainly from the large LC cell area and consequent high capacitance, necessitating carefully shaped drive pulses to minimise the slow speed of response. For large screens, the cell area can be divided into smaller segments that can be switched from one state to the next during the field period as soon as the CRT line intensity has decayed to an unnoticeable level.

7.5.4 Ferro-electric LCD

The ferro-electric liquid crystal display is constructed in a similar way to the twisted nematic device. An active transistor backplane applies selective control voltages to individual cells that modulate the proportion of polarised light passed through an analyser by rotating the plane of polarisation to align with the director of the analyser. The main difference is that the molecules in the ferro-electric LC material have a permanent electrical dipole moment, and in a manner analogous to the collective realignment of magnetic dipoles in ferro-magnetism, the material adopts a collective dipole moment. Placed in a cell between parallel plates, application of a sufficiently strong field will align the material, and this alignment can be swung rapidly through 180° by applying a momentary field of sufficient strength in the opposite direction. Thus the material can be switched between two stable states by applying brief pulses of alternating polarity, typically of about 3μsec duration and about 2V in strength. As the director of the material is not parallel to the dipole axes, the birefringent anisotropy can be engineered to induce a rotation of the plane of polarisation equivalent to the *off* and *on* states in Figure 7.8. Although devices that exploit this mechanism cannot operate by fine adjustment of the cell transmission in response to a continuously variable control voltage, they can achieve the equivalent by temporal dithering.

AMFLCD devices that operate by modulating reflected bursts of synchronised tri-colour illumination from pulsed LED emitters have been constructed to give an exceptionally wide range of colours and shades, and the aperture ratio is maximised by placing the pixel plates above the control transistors. Miniature devices having this configuration, called Liquid Crystal on Silicon (LCOS) are mainly used in image projectors.

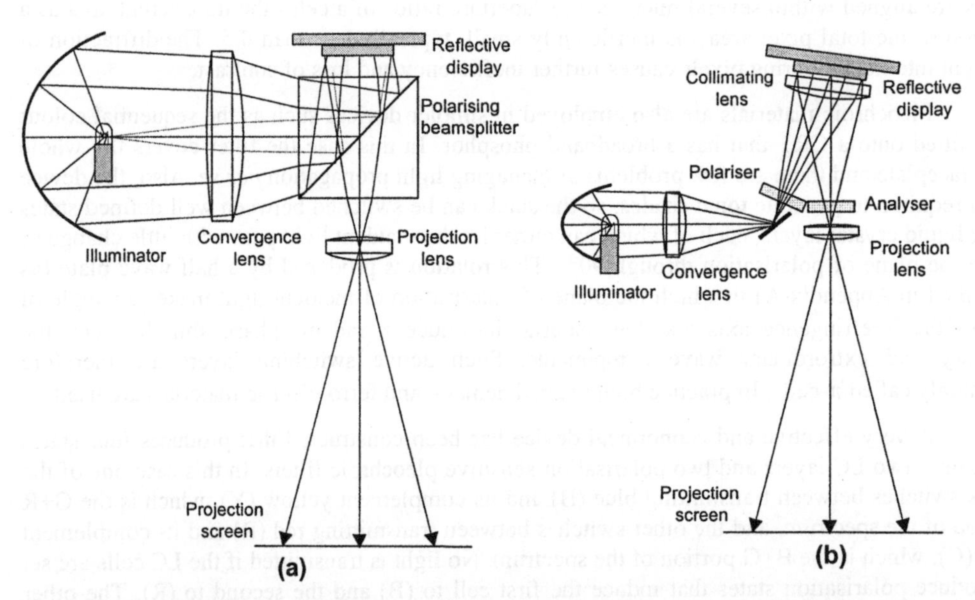

Figure 7.13 Typical arrangements of the illumination and projection optics for a miniature reflective AMFLCD display device: (a) axially, using a polarising beamsplitter, (b) obliquely, using a collimating lens

Figure 7.13 shows two optical arrangements that illuminate the mirror-like pixel plates and capture the reflected light efficiently. The convergent illumination illustrated in (a) is reflected from the display device, collected by a projection lens and focused to form a real image on a back-projection screen. The polarising beamsplitter may be produced as a carefully deposited multi-layer reflectance filter embedded in a glass cube prism, but the more modern technique uses a metal layer deposited as a set of fine parallel micro-wires. Both of these reflect the s-polarised incident light (see Appendix A) towards the display device, and transmit only the light that has been reflected from the display with a crossed plane of polarisation, i.e. light from the *on* cells in the matrix. The illumination optics, which must be larger than the panel, must illuminate the whole panel evenly to give uniform brightness across the picture, but the projection lens can be relatively small as most of the light is directed to converge at its centre. The scheme illustrated in figure (b) uses divergent illumination that is made parallel by a collimating lens placed almost in contact with the panel. This collimating lens also converges the reflected light through the projection lens. Although the illumination optics are more compact, the interposition of the collimating lens and the small tilt of the panel introduce some geometric distortion.

The use of a ferro-electric material carries a complicating penalty. Although the material is switched to one state by applying a short positive pulse for about 3μsec, and to the other by applying a short negative pulse, repeated application of pulses of the same polarity would cause a build up of charge in the pixel. The pixel would remain locked in that state. It is therefore vital to ensure that pulses alternate in polarity. At present the most common technique is to follow every switching event by one of opposite polarity by creating a complementary phase during which every cell that was *on* is turned *off* and vice versa. At first sight this seems an unlikely cure as each cell would therefore be on and off for equal time periods and the whole picture would be uniformly grey. However, this problem can be overcome when using pulsed illumination by suppressing the illumination during the complementary period, accommodating the loss of half the light and the consequent drop in image brightness. If, however, the device is configured to function with continuous illumination it is necessary to introduce a switchable retarder into the illumination-reflection path so that the modulation mode is inverted during the complementary phase.

7.6 Other technologies

7.6.1 Digital micro-mirror display (DMD)

The axial illumination technique illustrated in Figure 7.13 (a) has also been used with other miniature reflective display devices such as the Digital Micro-Mirror Display (DMD), a unique development by Texas Instruments. This device has a matrix of individual aluminium mirrors deposited over, but separated from, the active switching elements on a silicon wafer, and it is made by adapting techniques normal to integrated circuit manufacture. Each square micro-mirror, about 15μm by 15μm in area, is separated from its four neighbours by a gap of less than 1μm and stands on a central yoke on a torsion hinge that is anchored to the substrate beneath the diagonal mirror corners. The other corners are positioned above pads that are switched to apply sufficient electrostatic force to strain the torsion hinge and tilt the micro-mirror through an angle of about 12°. Although only the relaxed state is stable in the absence of power, the mirrors can be switched in about 15μsec between this and the tilted state. The individual cells can be addressed as an active matrix to apply temporal dithering for grey scale modulation, much like a ferro-electric AMLCD. The addressing is organised as static random access memory (SRAM), simplified by taking advantage of a bias voltage that can latch mirrors in the tilted state when the stronger switching signal is removed.

To attain good contrast, the divergence of the light reflected from a micro-mirror in the off-state must be less than twice the tilt angle, otherwise the unwanted light would not be tipped away from the optics that collect the untilted reflection. With the parallel or convergent illumination illustrated in Figure 7.13 the tilted reflections are prevented from entering the projection lens by a small aperture placed in front of this lens. The incident light that enters the gaps between the micro-mirrors and penetrates to the underlying silicon can induce disruptive photo-electronic noise and heating as well as contribute to unwanted reflections, so this is minimised by creating minimal gaps and masking the under-layer with non-reflective material.

Also the mirror edges must be sharp, because any rounding would scatter light into the collection aperture. Devices that achieve these high manufacturing qualities produce particularly high contrast images, because the residual brightness of a black pixel is governed solely by the small amount of diffuse reflection from the mirror surface and by edge diffraction. The device is also remarkably efficient because little light is wasted and none is discarded by polarisation. Thus the chief difference between the use of an AMLCD and a DMD in the schemes illustrated in Figure 7.13 is that the latter device obviates the need for a polarising beamsplitter or polarising filters.

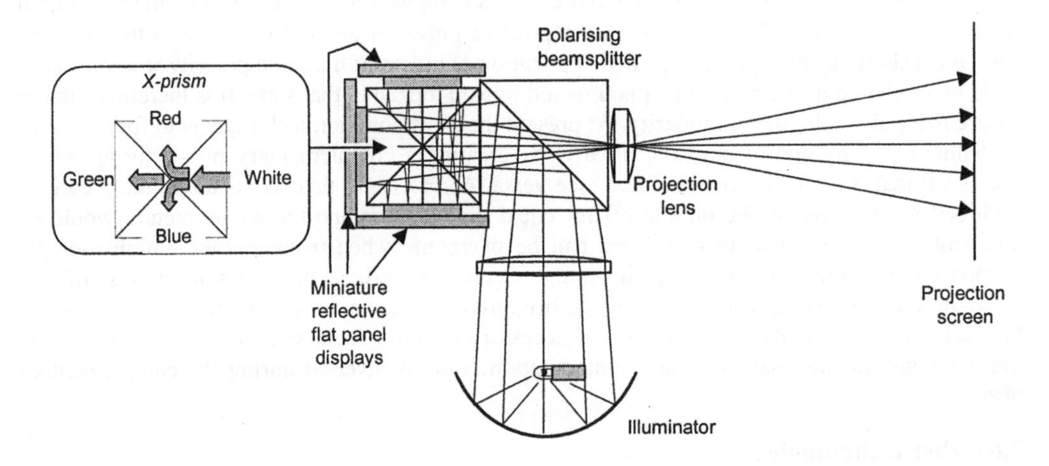

Figure 7.14 Production of full colour images using three miniature reflective flat panel display devices and a cubic X-prism

Figure 7.14 illustrates how triplets of reflective devices such as the twisted nematic and ferro-electric AMLCDs, and the DMD, can be arranged to give bright full colour images. The three separate miniature flat panel devices that are used to present the red, green and blue pictures are illuminated with light of the appropriate colour using a pair of red and blue reflecting filters formed as intersecting planes within a cubic glass prism. The same pulse width modulation is therefore applied to the three colours simultaneously, and so avoids the colour fragmentation problem of the sequential colour sub-field technique.

7.6.2 Electro-luminescent display

Table 7.1 includes several types of flat panel display that have either been used, or may in future be used, in the cockpit. Most are emissive and the main concern is to provide sufficient brightness, and little reflected glare, so that the image can be viewed in daylight. One approach uses electro-luminescence; the excitation of emission from a phosphor by electrons flowing within the phosphor, rather than the impingement of high energy electrons used in a CRT.

The general construction of an electro-luminescent device, with row and column addressing electrodes, is illustrated in Figure 7.15. The image is produced by exciting emission within the phosphor layer by causing a current to flow within the phosphor when a voltage is applied to the electrodes on either side of a cell. The presence of layers of insulator on both surfaces of the phosphor seems paradoxical, as these should prevent current flow. However, without the insulation, the self-exaggerating effects of current flow and temperature rise in the semi-conducting phosphor can cause the phosphor to melt, so the device would have a very short life. The insulating layers transform each cell into a capacitor with a semi-conducting core. The application of a voltage charges the capacitor, which produces a displacement current within the phosphor as the charge carriers migrate rapidly to the opposite surfaces of the material, and the cell produces a pulse of light during the displacement current flow. It is therefore necessary to alternate the polarity of the voltage to swing the charge carriers from one surface to the other, otherwise repeated applications of the same polarity would merely produce a small current to replenish any leaked charge. The brightness of a cell depends upon the pulse rate, and therefore the alternating excitation frequency, as well as the magnitude of the voltage swing. Typically, a voltage in the order of 50V and a frequency of 60Hz gives a luminance of about 50cd.m^{-2}.

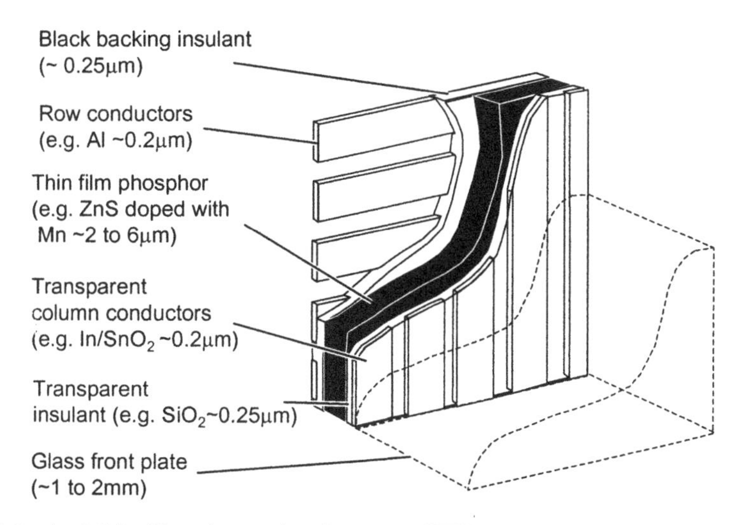

Figure 7.15 A typical thin film electro-luminescent (TFEL) display

Current TFEL devices tend to be dim, and several are manufactured as miniature panels with active matrix addressing. They produce monochromatic images using temporal dithering for grey shade modulation, and are used as image sources in night vision systems. The high theoretical electron-to-photon conversion efficiency of phosphors makes development worthwhile, so daylight-viewable brightness levels and full colour are considered possible.

7.6.3 Plasma display

As illustrated in Figure 7.16, the alternating current (AC) plasma panel is similar in principle to the TFEL except that the emission is produced by exciting a discharge within a low pressure gas, such as a mixture of neon and argon, sandwiched between glass plates. The reactions that occur in response to the sudden application of a voltage across a cell are complex. Emission is the result of the avalanche of collisions between gas ions and electrons attracted in opposite directions by the electric force field. Both the electron-atom interactions and the recombination of ions and electrons result in photon emission. The spectrum produced consists largely of sharp lines corresponding to transitions between the characteristic electron energy states in the constituent atoms.

Like the TFEL, emission is related directly to the excitation current, and a decaying current pulse results from the application of a field rather than a steadily maintained field. It is therefore necessary to alternate the field direction, effectively interchanging the cathode and anode, to create a sequence of emissive discharges. Unlike a TFEL, the gas must be maintained in a partially ionised state so that the threshold of the avalanche is reduced to a controllable voltage. This is done by building a set of sustainer electrodes into the device. Early commercial panels were constructed to produce 600 by 400 pixels, each about 300μm square, having the characteristic argon-neon orange colour, a luminance of about 200cd.m^{-2} and a contrast of about 20:1. These were useful only for presenting symbolic information as each pixel could either be "off" or "on", and the response was too slow to allow temporal dithering.

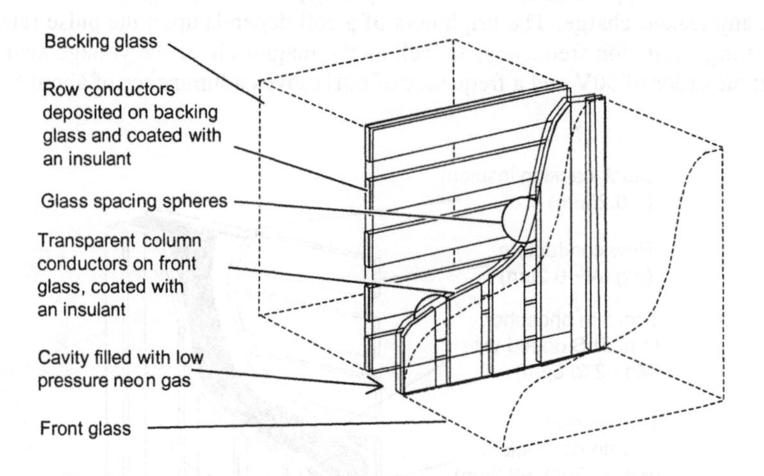

Figure 7.16 Section through an AC plasma discharge panel

Development of AC plasma panels has sought to improve all of the parameters, principally by using integrated driver circuitry, effectively transforming the device into an active matrix, and by treating the gas discharge as a source of excitation for a phosphor layer. The latter is applied as a thin coating to the inside of the outer glass, and capitalises on the previously unwanted UV emission as well as the visible light and the electron-phosphor interaction to improve efficiency. Large, high resolution, colour devices are now manufactured in commercial quantities by a number of companies, but the limited brightness and susceptibility to glare make them unlikely candidates for use in combat aircraft.

7.6.4 Light-emitting diode display (LEDD)

Semi-conductors are solid elements, such as germanium and silicon, or compounds such as cuprous oxide and gallium arsenide, that are insulators at absolute zero but conduct electricity at room temperature. Unlike metals, the highest energy electrons are bound to their atoms and need to be agitated thermally to gain the energy that frees them to drift through the material under the force of an electric field. The "band gap", the difference in energy between the bound (valence) state and the free (conduction) state of these electrons is a characteristic of the material, ranging from 0.27eV in lead selenide through 0.67eV in germanium, 1.14eV in silicon, 2.42eV in cadmium sulphide to 5.33eV in diamond. As noted in Appendix A, light can be regarded as a flux of photons with energy inversely proportional to wavelength, and an impacting photon can transfer sufficient energy to an electron to raise it to a state in which it can drift under an applied electric force field, i.e. contribute to an electrical current. This property is exploited in a wide range of photo-conductive cells. The reciprocal phenomenon is equally useful because a photon of wavelength λ_c (= $h.c/E_{BG}$ where h is Planck's Constant, c is the velocity of light and E_{BG} is the Band Gap energy) is emitted whenever a conduction electron drops into a bound state.

Natural semi-conductors can be alloyed with elements of a similar atomic weight, with the "dopant" atoms substituting for host atoms in the crystal lattice. If the dopant has a greater valency than the host, for instance phosphorus in silicon, electrons that are not needed to make the inter-atomic crystal bonds become free to act as "minority" electrons. These contribute to conduction in addition to the majority carriers generated either thermally or by impinging light, and because these extra carriers are negatively charged the material is called an N-type semi-conductor. The addition of dopant atoms that lack the number of electrons for complete inter-atomic bonding creates "holes" that, somewhat surprisingly, also move between atoms under an electric force. Holes are also minority carriers, and they move as though carrying a positive charge, so an indium or aluminium alloy is called a P-type material. At a junction between an N-type and a P-type material the minority charges attract each other to produce a standing electric field. This is used in diodes (devices with two electrodes) to create asymmetric barriers to conduction and in transistors (three-terminal devices) to create conduction barriers that can be controlled.

The Light-Emitting Diode (LED) makes use of the photon emission that results from the re-combination of electrons and holes in a P-N junction when driven by a small applied voltage. Normally a junction is made on the surface of a pure single-crystal chip of semi-conductor using photo-lithographic processes, vapour deposition and chemical diffusion under vacuum, to produce a structure illustrated schematically on the left of Figure 7.17. Interconnected arrays of LEDs, with on-chip addressing transistors, have been made into a wide variety of display devices, the simplest presenting fixed sets of alpha-numeric characters, the most complex presenting full grey-shaded pictorial images. The principal drawback, particularly for the latter devices, are the limited efficiency that arises partly because an element only emits when energised and partly because of an intrinsically low efficiency, typically less than 10 lumens/watt, which is about the same as a conventional incandescent lamp. It has also proved difficult to attain short wavelength emissions at sufficient luminance to produce a full colour gamut.

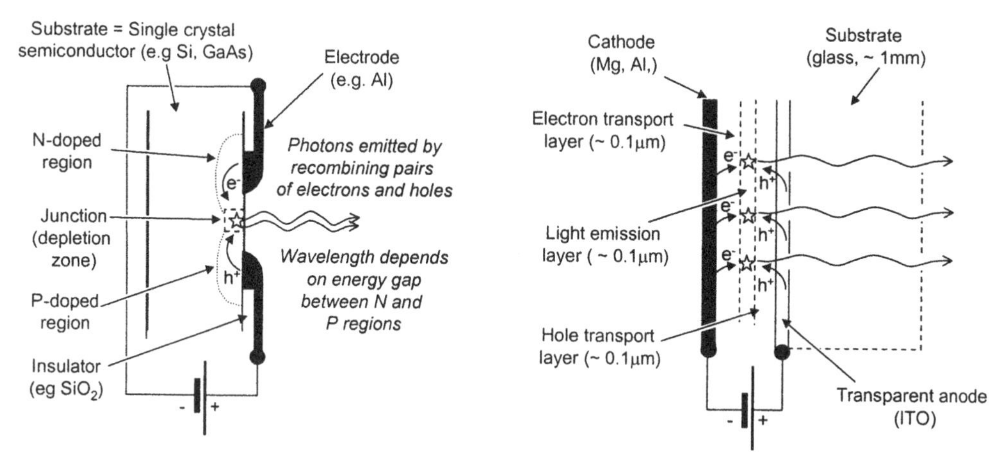

Figure 7.17 Contrasting the general form (left diagram) of a conventional Light Emitting Diode (LED) element formed on the surface of a pure crystaline inorganic semi-conductor with (right diagram) a typical Organic Light Emitting Diode (OLED) element employing chemically deposited semi-conductors on a coated glass substrate

The discovery of several classes of organic light-emitting semi-conductor materials in the late 1980s has, however, brought a revolution. As shown on the right of Figure 7.18, in Organic Light-Emitting Diode (OLED) devices, the inorganic crystaline P-N junction is replaced by a set of layers; a hole transport layer (HTL), electron transport layer (ETL) and light emission layer (EML). Although chemically complex, these thin uniform layers can be formed on a transparent substrate, at present glass but in the future a flexible plastic, by spin coating in a clean environment at room temperature. The relatively expensive vacuum chamber is needed only for pre-coating the substrate with a transparent electrode, such as indium-tin-oxide (ITO), and for depositing the outer electrode of magnesium/aluminium alloy.

The three classes of OLED exploit different photon emissive phenomena:-

a) Organic molecular semi-conductors, such as 8-quinolinol-aluminium complex (Alq^3) and 9,10 bis-styrylanthacene (BIS), behave most like inorganic materials and produce photon emission by the direct recombination of electron-hole pairs.
b) Organic polymeric semi-conductors, such as poly(p-phenylenevinylene) (PPV), produce photons by the radiative recombination of excited states produced by electron and hole collisions. The wavelength of the emission is governed by the transition between exciton states, and this can be adjusted chemically to produce effective red, green and blue spectra.
c) Organic rare earth metal complexes, or "chelates", which are organic ligands with multiple bonds to atoms of Lanthanide metals, each of which produces emission of a different characteristic wavelength. For instance, red light is emitted from Europium, green from Terbium, blue from Cerium and UV from Gadolinium.

Table 7.2 Properties of some experimental OLED devices

Structure					**Efficiency (lu/W)**
Anode	**HTL**	**EML**	**ETL**	**Cathode**	
Organic molecular semi-conductor					
ITO	HTL	Alq^3	ETL	Mg/Al	0.1 - 10
Organic polymeric semi-conductor					
ITO	-	PPV	-	Mg/Ag	22 (green)
ITO	-	PPV	-	Mg/Ag	3 (blue)
Organic rare earth metal chelates					
ITO	TPD	Tb-chelate	-	Al	60 (green)
ITO	TPD	Eu-chelate	Alq^3	Al	15 (red)

Table 7.2, which summarises the constitution and properties of some experimental devices, shows that OLED-based devices are very efficient in comparison with other emissive displays, although they cannot at present be operated at high luminance for more than several hours. Commercially available micro-devices, using organic molecular semi-conductors with active matrix addressing, provide 852 x 600 full colour pixels, 15μm by 15μm square and give a luminance of about $100cd.m^{-2}$ and a contrast ratio of 50:1. The current aim is to understand how manufacturing impurities and in-service degradation govern the trade-off between luminance and lifetime.

Given the level of commercial investment and the wide range of applications, such as a flexible display[6] that can be rolled-up when not needed, manufacturing difficulties will undoubtedly be solved and OLED-based displays will become commonplace. Commercial devices will be adapted for use in the cockpit, in particular for helmet-mounted systems.

7.7 Projected head-down displays

The unusual demands placed on the designers of head-down display systems have stimulated a number of developments that exploit projection optics. Two examples illustrate the ingenuity of the design teams. The first is the Combined Optical Map and Electronic Display (COMED^{TM}) devised by Ferranti engineers in Edinburgh (now part of BAE Systems) to meet the need for a comprehensive "horizontal situation indicator". The second is the on-going work at Kaiser Electronics (now part of Rockwell Collins) to develop the large area display for the JSF.

7.7.1 COMED^{TM}

In earlier aircraft the pilot maintained his knowledge of the aircraft position and the course to steer by following a route marked on a hand-held map. The development of doppler and inertial navigation systems, which could keep track of the aircraft position with less than half a mile error, opened up the possibility of operating without a co-operative ground station in unfamiliar terrain and in relatively poor visibility conditions. The COMED was developed in the late 1970s to make use of the navigation data and act as an electronic equivalent of the finger-on-the-map. The device was part of the AIMIS suite of displays, shown in Figure 6.7, and has been employed in several combat aircraft including the F-18 and the Tornado. In the rear cockpit of the Tornado the device was given pride of place above the glare shield directly in front of the navigator, and the CRT showed a number of formats including an overlaid radar picture. In the front cockpit, it was placed directly beneath the pilot's HUD.

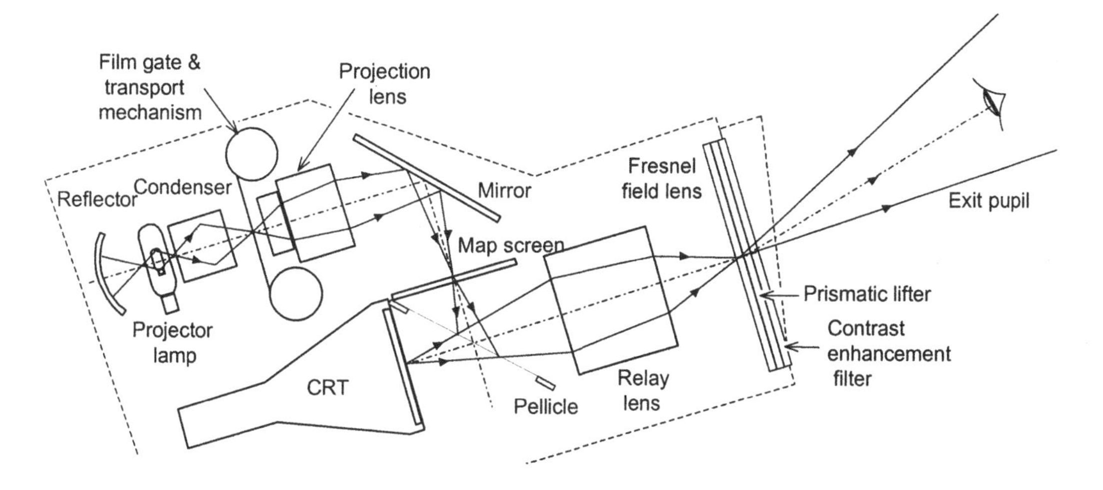

Figure 7.18 A simplified schematic of the Ferranti COMED^{TM} (Adapted from Boot A, *The F-18 Horizontal indicator optical system*, in Advanced avionics and the military aircraft man/machine interface, AGARD-CP-329 1982. Reprinted with permission from BAE Systems and Thales Optics)

The scheme for superimposing bright green CRT symbols that marked the aircraft position and the planned route, plus timing, speed and heading information, onto a moveable colour map is shown in Figure 7.18. The maps were areas of Air Navigation charts at 500,000:1 and 2,000,000:1 scale that had been photographically shrunk onto 35mm film. The film was held in a transport mechanism that could move rapidly between frames to select the appropriate area, and which could be shifted vertically and horizontally to place the mapped ground location beneath the aircraft position symbol shown on the CRT. The whole transport mechanism could also be rotated about the optical axis to enable selection of either a "north-up" or a "track-up" map orientation.

As shown in Figure 7.18, the optical arrangement, designed by specialists at Pilkington Perkin-Elmer[7] (now Thales Optics), was uniquely complex. A concave mirror and lens group condensed the flux from the 50W lamp onto the film plane, and a very bright, well focused map image was relayed onto a glass screen by a projection lens. The rest of the optics enabled the pilot to see a magnified combination of the CRT symbols and this map. The superposition was done by a pellicle beamsplitter, a semi-reflecting mirror formed on a plastic film that reflected the light from the screen along the same path as the light emitted from the CRT faceplate.

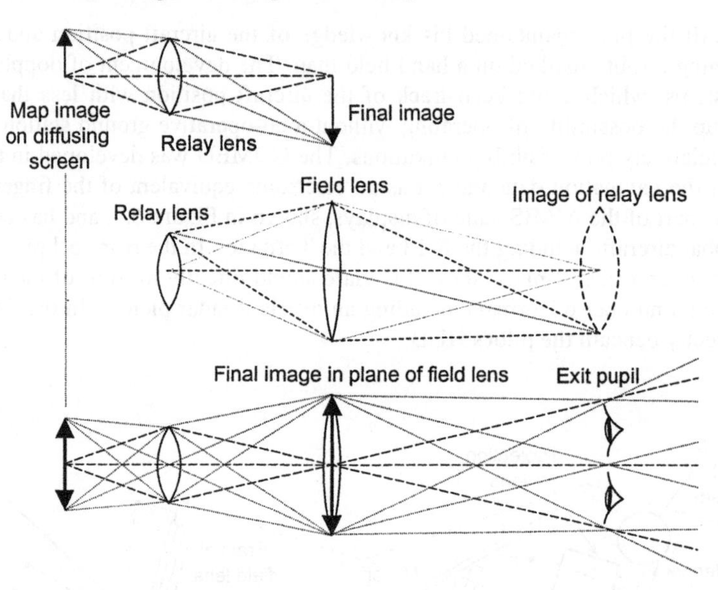

Figure 7.19 COMED viewing optics. A relay lens formed the final image in the plane of a field lens that directed the image-forming light towards the viewer (Adapted from Boot A, *The F-18 Horizontal indicator optical system*, in Advanced avionics and the military aircraft man/machine interface, AGARD-CP-329 1982. Reprinted with permission from BAE Systems and Thales Optics)

The downstream viewing optics consisted of a relay lens and a field lens that, as explained in Figure 7.19, produced the final image and limited the range of viewing positions to a volume called the "exit pupil". The final image filled the field lens, about 175mm in diameter, and the exit pupil - the image of the relay lens at a viewing distance of about 750mm - was about 200mm in diameter, which was sufficient in a single-seat combat aircraft. The high brightness of the magnified image, about 3500cd.m^{-2}, resulted because the field lens directed the image-forming light toward the viewer, much like a diffusing screen with a very high gain. The field lens also reduced ambient glare markedly because the element scattered much less than a diffusing screen and the viewer's head blocked most of the cone of ambient light that could reach the CRT and intermediate map screen. The directed light also produced minimal canopy-reflected glare at night.

The design of the necessarily large diameter relay lens was simplified because it was mainly necessary to control field curvature and distortion for the small bundles of light collected by a pair of eyes within the exit pupil rather than the whole image-forming bundle. The designers had experience of dealing with such requirements in devising collimating lenses for HUDs, as discussed in Chapter 8. However, the configuration presented the designers with interesting manufacturing problems. The pellicle was formed on a nitro-cellulose film, which was highly flammable and easily de-tensioned by absorbed solvent vapour. It was necessary to install this at the last stages of assembly. To direct the light into the viewing optics, the underside of the intermediate map screen was impressed with a fresnel lens, and it was necessary to vibrate the screen rapidly in a circular motion in its own plane to remove scintillation and the obvious concentric fresnel rings. The large field lens was also made as a fresnel structure, in this case by diamond machining a polycarbonate substrate. Despite the efficiency of the anti-reflection coatings, the reflection of the pilot's sun-lit face reduced the image contrast noticeably. This was solved by tipping the outer facet down by 12°, and as shown in Figure 7.18 a prismatic plate was installed to lift the optical axis by an equal angle to restore the position of the exit pupil. The horizontal facets on this plate were also formed by diamond machining.

7.7.2 Panoramic HDD

The Lockheed-Martin team developing the JSF cockpit have made a number of brave decisions regarding the arrangement of the display and control suite. In essence, the pilot will perform head-out tasks using HOTAS, a speech recogniser and a HMD, and he will perform head-in tasks using a single, large HDD fitted with a touch-sensitive screen. Kaiser Electronics have been appointed to develop this panoramic HDD. Having engineered a range of HDD systems using ruggedised versions of commercially-available AMLCD glassware, many as form-fit replacements for obsolete and unobtainable CRT systems, then finding that the suppliers of the adopted glassware had ceased production, the development team were acutely aware of the problems of obsolescence. Knowing that the aircraft will remain in service for considerably longer than the supported life of any commercially developed components, this concern is shared by the aircraft design team.

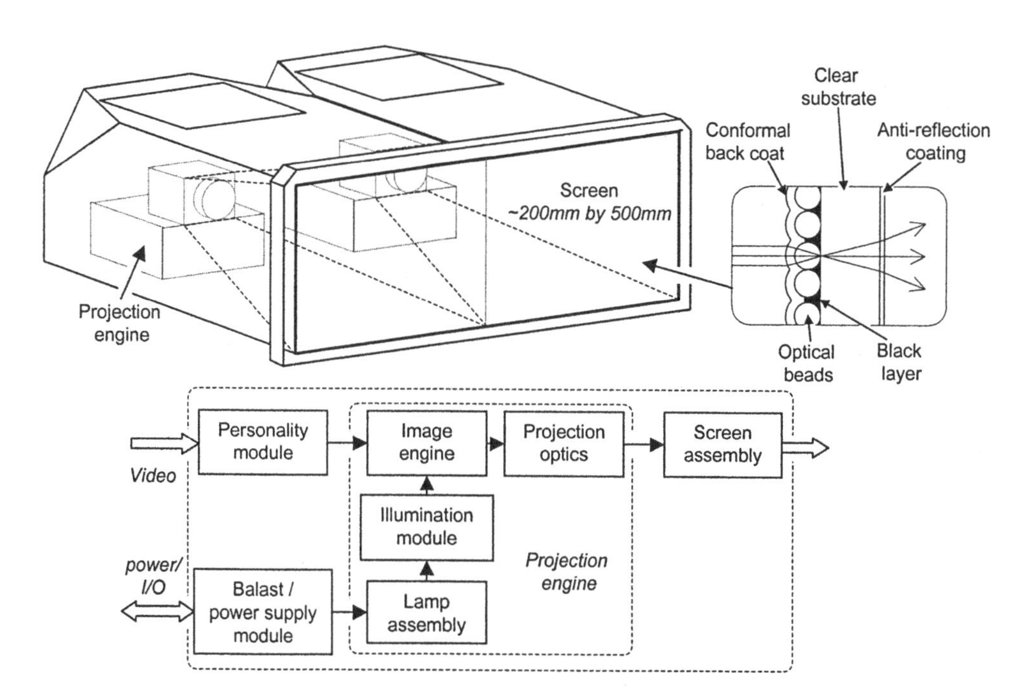

Figure 7.20 Modular construction of the panoramic HDD in the Joint Strike Fighter (Adapted from Kalmanash[8] and Wolfe *et al*[9])

The configuration of the proposed HDD is determined as much by commercial reality as the usual triad of technological practicalities, human limitations and operational needs. As shown in Figure 7.20, the pictures on the 200mm by 500mm screen area of the device are produced as a seamless combination of fields from a pair of projection engines, each of which is composed of modules. The importance of the modular construction is that it allows the designers to use an available set of components for the construction of a prototype, and select alternatives when these are superceded. The use of two independent projection units, combined with the duplication of less reliable modules, also allows the system to continue to provide information in the event of component failure.

The heart of the device is a set of reflective microdisplays, such as described in 7.6.1 and illustrated in Figure 7.14, together with lamps, driver electronics, power control electronics, illumination optics and projection optics. The active picture-producing modulators, whether based on DMD or LCOS mechanisms, are very suitable. They are reliable and produce adequate contrast and resolution. 1024 by 1280 devices are commonplace, and 1560 by 2048 devices are under development. Unlike large area AMLCD glassware, it is not necessary to apply heating layers or make other expensive conversions. The main difficulty is in attaining the necessary luminance and dimming range, and controlling the spectral composition of the output for compatibility with night vision goggles. The developers have therefore put effort into packaging ultra-high pressure (UHP) mercury lamps with a reflector to create a lamp module, and converting this output in an illumination module using lenses, quarter-wave plates, polarisers and filters into a reasonably uniform, linearly polarised, spectrally constrained flux that can be dimmed by a factor of 10^4.

The screen, known as a Blackscreen™, is also used in a number of ground-based back-projection systems and achieves an immunity to glare in high ambient light levels because 95% of the screen area is black. The image-forming, back-projected light is made perpendicular to the screen using a fresnel lens, and then channelled by a micro-lens array to pass through pin-hole apertures forming the remaining 5% of the black layer. The lens array and aligned apertures are made, very cleverly, by a dense layer of small, spherical glass beads in contact the glass screen, covered by a conformal transparent resin layer. Work is underway to eliminate the anomalous "grain" and "speckle" modulation of the imagery, which is produced by interference from internal reflections within the beads and by interactions between the partially coherent light emitted from adjacent beads.

References

1. Krebs M J and Wolf J D, *Design principles for the use of colour in displays*, Proc. Society for Information Display Vol 20 (1) pp3-9, 1979

2. Laycock J, *Recommended colours for use in airborne displays, in Proceedings of the workshop on colour vs monochromatic electronic displays* at RAE Farnborough, NATO Defence Research Group DS/A/DR(84)431, 1984

3. Martin K W, *Display visual performance prediction in a military cockpit environment*, in *Proceedings of the workshop on colour vs monochromatic electronic displays* at RAE Farnborough, NATO Defence Research Group DS/A/DR(84)431, 1984

4. Banbury J R, *Electronic displays for use in military cockpits*, in *Proceedings of the workshop on colour vs monochromatic electronic displays at RAE Farnborough*, NATO Defence Research Group DS/A/DR(84)431, 1984

5. *Optical characteristics of cathode ray tube screens*, JEDEC Electron Tube Council, Publication No 16-C, 1975
6. Hack M, Chwang A,Lu M H, Kwong R, Weaver M S, Tung Y-J and Brown J J, *Flexible low power consumption OLED displays for a universal communication device*, in Hopper D (ed) *Cockpit Displays X*, Proc. SPIE 5080, 2003
7. Rogers P J, *Transfer lens and head-down display using the same*, Patent No GB2049217, 1982
8. Kalmanash M H, *Panoramic projection avionic displays*, in Hopper D (ed) *Cockpit Displays X*, Proc. SPIE 5080, 2003
9. Wolfe C R, Kinosian K, Lewis K, Vance D and Vu J, *Reducing speckle and grain in Blackscreen™*, in Hopper D (ed) *Cockpit Displays X*, Proc. SPIE 5080, 2003

Further reading

Castellano J A, *Handbook of display technology* 3rd edition, Academic Press Inc. 2001

Chapter 8

The Head-Up Display

8.1 Introduction

This chapter introduces the idea of the head-up display (HUD) as a development of the gunsight. The requirements and optical principles employed by engineers to present information in this way are stated briefly. The limitations of early electro-optical projection hardware and the problems of installing units in the confined fast jet cockpit lead to a description of more modern techniques and the use of holographic optical elements. Alternative configurations are mentioned, such as the devices fitted to civil aircraft.

Because the benefit of superimposing symbols on the view of the world can be outweighed by masking the view of the world, symbols must convey their intended information with exceptional efficiency. The design of a display format is therefore a compromise between attaining a parsimonious way of presenting information and the possibility that the symbols can be misinterpreted, sometimes dangerously, by the pilot. The chapter gives the flavour of the history of format development by describing briefly the main varieties of symbols and their dynamic behaviour for the differing requirements of low level flying and aerial combat. It ends with an overview of operational issues.

8.1.1 The gunsight

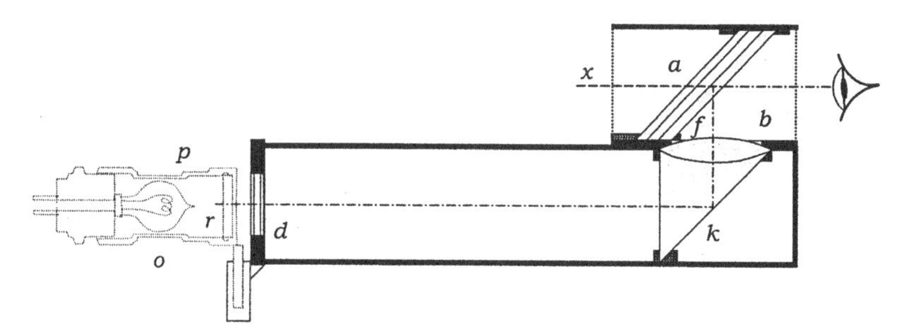

Figure 8.1 Sir Howard Grubb's collimated gunsight. The viewer saw in direction *x* a distant image of a back-illuminated graticule *d* placed at the focal point of a lens *f* by reflection in a prism *k* and a protected half-silvered mirror *a* that also allowed the viewer to see the target

Grubb devised the collimated gunsight[1], as illustrated in Figure 8.1, to project a distant virtual image of a back-illuminated aiming graticule that could be harmonised to the bore of the gun. The gun could then be aimed, with due offset for the gravity drop and windage of the projectile, by superimposing this graticule over the distant target. This was an advance on the usual need to align the target with a back sight and a foresight, both of which would be at different distances from the eye, and out of focus. The principle was used widely, and was introduced in compact form into fighters in the early 1920s where it was known as the "reflector gunsight". The dimensions of the aiming graticule, a small dot and a concentric circle, mimicked the use of a ring foresight in enabling the pilot to gauge the range to the target and the angle of offset.

The gyro gunsight was devised by Maurice Hancock and developed at RAE Farnborough[2] in the early 1940s. It was brought into service in 1943 in RAF Spitfire and Hurricane fighters, and as illustrated in Figure 8.2, was really a pair of aligned, but independent sights. One was a folded version of the Grubb invention that projected a fixed aiming cross. This was used to check the second sight and provide a standby. The second had an adjustable aiming symbol that shifted across the line of sight by the angle needed to compensate for the distance the target would move during the flight of the bullets. This lead angle actually depended on the aircraft speed, altitude, attitude and turn rate, and on the bullet velocity and the target range. However, the embodied approximation, which sensed the turn rate and estimated the target range, was very effective.

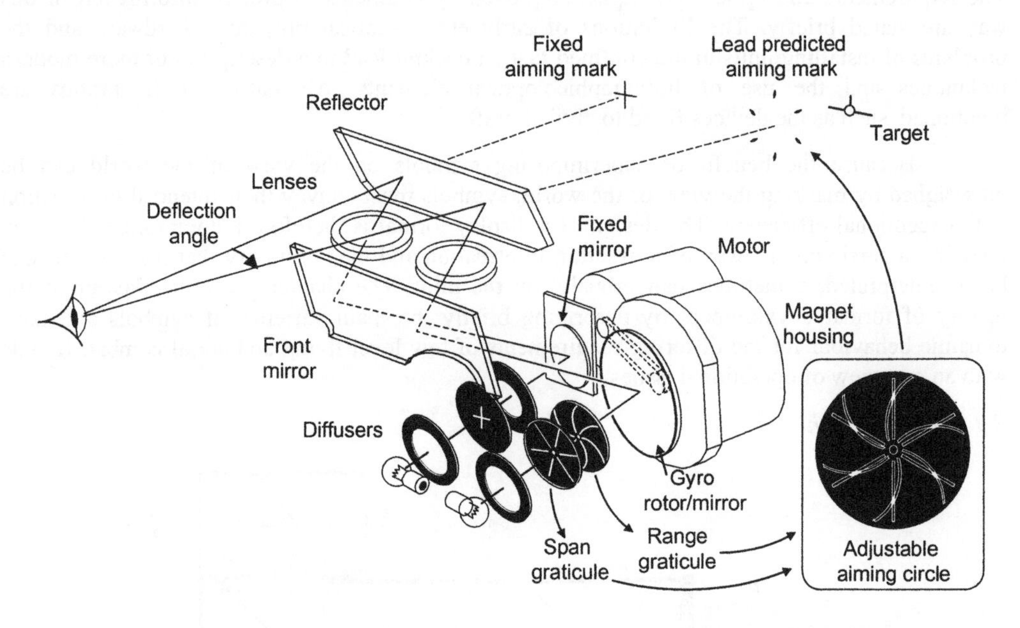

Figure 8.2 Schematic of the essential elements of the Mk2 RAE gyroscopic lead-predicting gunsight

As shown, light from the graticule was reflected from the mirror surface of a freely mounted gyro rotor that was spun around the aircraft X-axis by an electric motor. As described in Chapter 6, this rotor would tend to maintain a fixed orientation in space, and any aircraft manoeuvre would rotate the airframe and gunsight relative to this orientation. The optical path was arranged so that the resultant change in the angle of reflection from the mirror surface of the spinning gyro caused the projected aiming symbol to lag behind the changing aircraft orientation. Ordinarily the rotor would soon hit the gymbal limit with a force that would immediately induce a violent precessive topple, but the spinning rotor was backed by two sets of electro-magnets. These exerted a force that, in a steady manoeuvre, limited the excursion to a small angle proportional to the rate of turn.

The complementary range information was provided by a "stadiametric" technique. The aiming symbol was a set of six diamonds formed as a circle where light was transmitted through the intersecting segments of a pair of graticules. A "span" knob on the face of the device and a twistgrip on the throttle, linked via a sprung Bowden cable, rotated the graticles and changed the diameter of the diamond circle. As soon as the pilot engaged the enemy he would use the knob to set the target's wingspan, and during the attack he would use the twistgrip to make the circle just enclose the target. The position of the twistgrip was linked to a set of rheostats that adjusted the currents through the electro-magnets so that the deflection was limited progressively with reducing range.

Great skill was needed to keep sight of the enemy while co-ordinating the smooth adjustment of the diamond circle with the control of the aircraft manoeuvre, especially in an aircraft like the Spitfire where the nose commonly obscured the forward view. The device brought a significant improvement to the effectiveness of the aircraft, and it remained in RAF service for more than twenty years. Later versions were coupled to the aircraft radar, which supplied better range information. Modern variants, fitted mainly to trainer aircraft, receive control signals from a separate gyro sensor, and either the mirror deflection is controlled by an actuator or the back-illuminated graticule is replaced by a small, bright CRT.

8.1.2 Rationale of the head-up display

The head-up display (HUD) was devised in the late 1950s as a derivative of the gyro gunsight with the main objective of expanding the field of view and supplying basic flight information in addition to a weapon aiming symbol. This was intended to obviate the need for the pilot to glance down, accommodate to about a half a metre to read the panel-mounted instruments, and then have to re-accommodate on returning to attend to the distant external world. In the same way that the aiming graticule showed the possible bullet trajectory, the facility also made it possible to project dynamically scaled, shifted and rotated symbols such as a line parallel with the real horizon that could supply a reference in the absence of a visible horizon. The two main virtues of the HUD were therefore to supply information conveniently and conformally.

8.1.3 Requirements

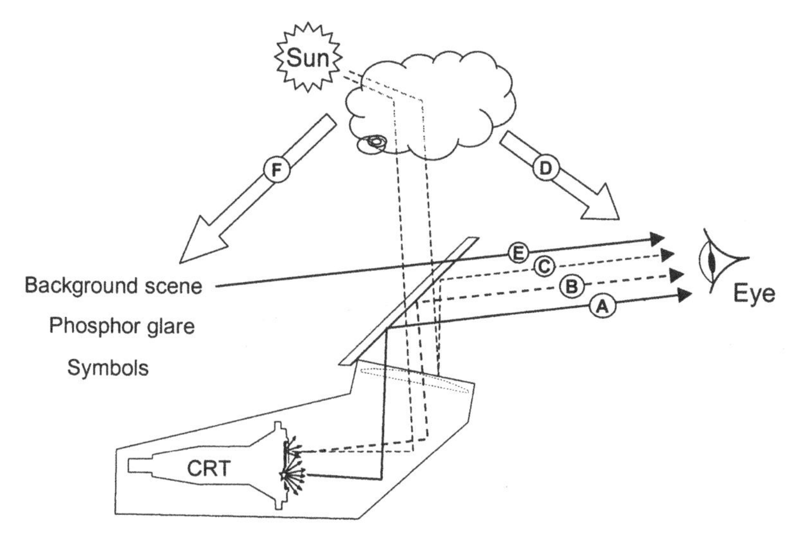

Figure 8.3 **Factors affecting the contrast of HUD imagery: (A) the luminance of the phosphor emission, (B) glare caused by back-scatter of ambient illumination from the phosphor, (C) glare caused by specular reflection from optical glass surfaces, (D) the adaptation state of the eye in response to ambient illumination, (E) the luminance of the scene on which the HUD symbols are superimposed and (F) the illumination of the scene**

The clarity of the symbols presented by the HUD depends on the contrast between the symbols and their background, as discussed in Chapter 7. That the symbols are distant, untouchable, virtual images makes them no less susceptible to glare, and unlike HDD images the background on which they are superimposed is commonly very bright and complex.

The factors that contribute to the loss of image contrast in strong ambient light are illustrated in Figure 8.3. Some glare (B) is produced if ambient light is gathered by the collimation lens and reflected down the reversed optical path so that it illuminates the phosphor screen. The condition is equivalent to the ambient light falling on a HDD, but in a HUD the collimation optics transform this, as well as the symbols, into a distant virtual image. The tube face is therefore seen as a ghostly disc that reduces the contrast of both the symbols and the forward scene. In early devices it was particularly noticeable if the sun or a bright cloud was above the aircraft. The problem has been alleviated in more modern designs by ensuring that the combiner reflects almost all of the narrow waveband emitted by the phosphor, and a matching narrow-band pass filter is placed in the optical path to absorb all other wavelengths before arrival at the tube face.

The optics themselves can produce glare by reflection (C) from the air-to-glass boundaries, and all surfaces must be given an anti-reflection coating. However, the main factor that affects image legibility is the symbol luminance (A) needed for adequate contrast against the background (E), which is the view of the world attenuated by the windscreen and the combiner. To produce an image with at least 20% contrast (see Appendix B for a definition of line contrast) so that it can be discerned easily against the tops of sunlit clouds at about 100,000cd.m^{-2} – the worst practical case – it is necessary to achieve a symbol luminance of about 20,000cd.m^{-2}. To date all systems have employed a CRT with a highly efficient narrow band phosphor and cursive beam deflection, coupled with efficient optics, to attain this luminance and simultaneously maintain image quality. It should be noted that when the background is very bright, which occurs in clear air when flying towards the sun or when the sun is within about 30°of the aircraft nose, the intra-ocular glare can reduce the perceived contrast of HUD imagery appreciably. Pilots in HUD-equipped aircraft have an extra problem when landing in a westerly direction just before dusk. At night the luminance of the symbols must be reduced dramatically to allow the pilot to maintain a minimum photopic adaptation and avoid overloading night vision goggles. A 10,000:1 dimming range is needed.

Conformal symbols must be positioned within the field of view so that they are superimposed accurately on the appropriate features in the world. Greatest accuracy is demanded by weapons such as unguided bombs that must be aimed within a few meters of the target. A maximum of 2mrad static symbol misplacement (about 0.12° or 7 minutes) is accepted. It is also necessary for the symbols to shift smoothly in the field of view and maintain registration during rapid manoeuvres. For this the image must be updated at least 25 times per second to avoid jerkiness and there must be minimal lag between the onset of the manoeuvre and the change in the position of the displayed symbol. Such dynamic accuracy places a speed demand on the whole system, from the inertial and air-mass sensors, the data transfer bus, the HUD symbol generator through to the process of updating the display itself. For instance, when the aircraft is rolled at 50°/sec, a throughput delay of 1/50sec (20msec) produces a 1° error in the position of the horizon line. As the computational cycle and display refresh take at least two frame periods, i.e. 40msec for a 50Hz display, and the other processes are carried out at the maximum rate that the component systems allow, the throughput latency can be about 100msec. To minimise dynamic errors, it is necessary to use a higher frame rate. Some installations present conformal symbols at the maximum rate, transferring the crucial data directly from the sensors as analogue signals. Fixed-position symbols are updated less frequently via the data bus.

8.2 Optics

8.2.1 Optical principles

As illustrated on the left of Figure 8.4, a lens held at a distance less than the focal length from a real object produces a virtual image of the object. When the separation is equal to the focal length the light entering the eye is parallel, or collimated, and the virtual image is infinitely distant.

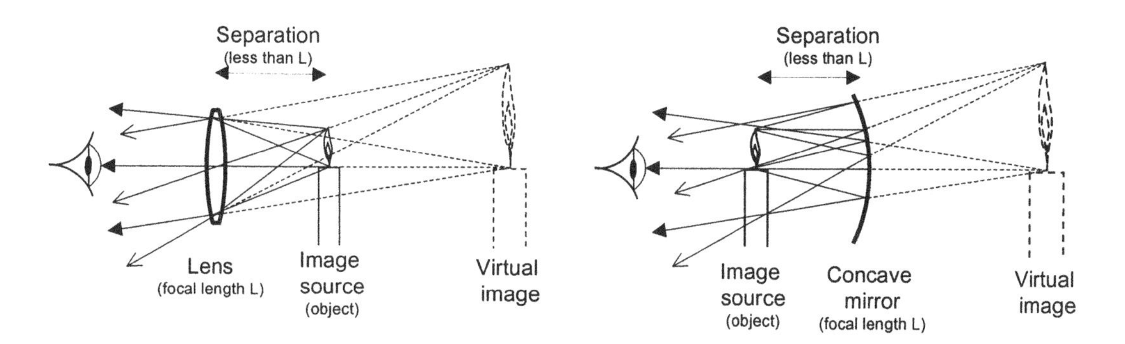

Figure 8.4 Use of a convex lens or a concave mirror to form a virtual image

The arrangement in this figure, with an additional partially reflecting combiner mirror that folds the light path to remove the lens and image source from the forward view, is used in the gyro gunsight and the conventional refractive forms of HUD. As shown on the right of Figure 8.4, a concave mirror can also be used to form a distant virtual image, and it is this principle that has been exploited in more recent reflective devices.

8.2.2 Refractive collimator designs

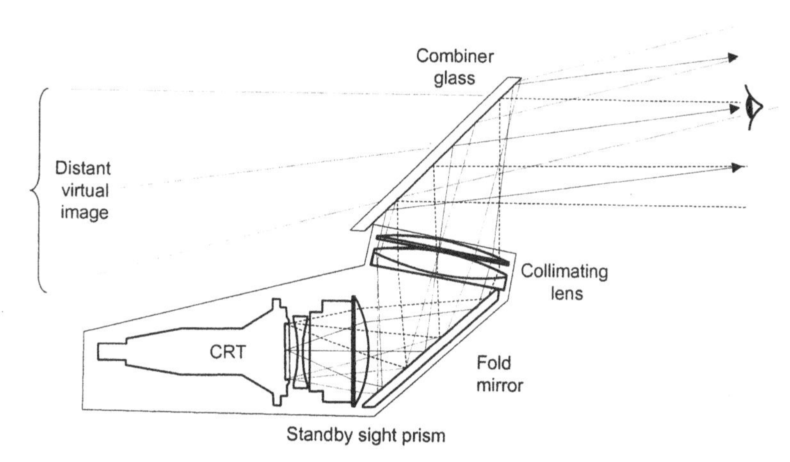

Figure 8.5 Optical arrangement of an early HUD display unit (Reprinted with permission from QinetiQ)

The optics of an early HUD are sketched in Figure 8.5. The symbol format was drawn on the circular faceplate of a high brightness CRT. The emitted light was collected by a set of collimating lenses and a fold mirror, and about 75% was reflected into the pilot's forward view from a glass combiner plate. The CRT and optics were invariably housed in a stiff, sealed cast aluminium housing, and the reflective coating on the combiner was sealed to form the filling in a glass sandwich that was supported by a pair of stout alloy braces. The display unit included a prism that reflected light from a circular reticule back-illuminated by a small lamp, and this could be switched on to provide a fixed aiming circle if, as was not uncommon, the CRT or the drive electronics failed.

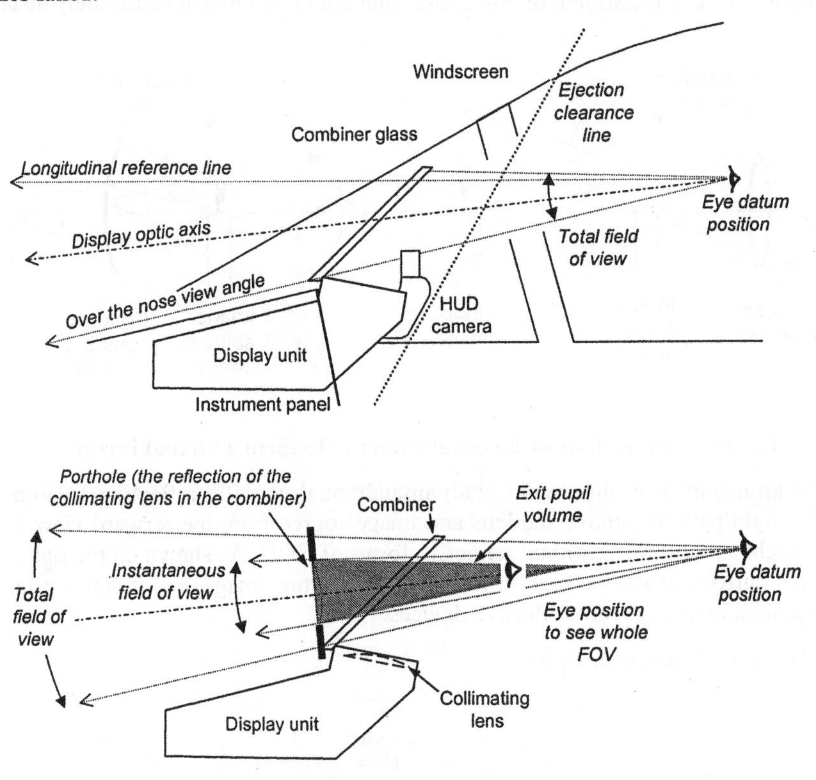

Figure 8.6 Installation of an HUD display unit in a cockpit

Although the collimating lens was essentially a simple magnifier that made the light appear to have come from a great distance, considerable refinement was needed to show well-focused symbols in the required direction to the required accuracy no matter where the eyes were placed within the light path. The allowable error of 2 milliradians (about 0.1°) was roughly the width of the line used to form the symbols.

The form and shape of the projection unit[3] was constrained by the cockpit geometry, as shown in the upper sketch in Figure 8.6. The space to accommodate the combiner between the front glare shield and the windscreen was limited, and the space in the instrument panel beneath the glare shield was valuable for the head-down displays. Also, the whole device could not intrude much into the cockpit lest it hinder the clear passage for the pilot's feet in an emergency ejection. The optics imposed other limitations. In particular, the optic axis had to intersect the datum position for the pilot's eyes, and the displayed image had to cover the lower part of the pilot's view over the aircraft nose, generally about 15° down from the longitudinal datum. It was therefore necessary to install a display with a typical vertical field of view of about 20° so that the optical axis was depressed 5° from the aircraft longitudinal datum. This gave an unequal split of 15° below the datum and only 5° above.

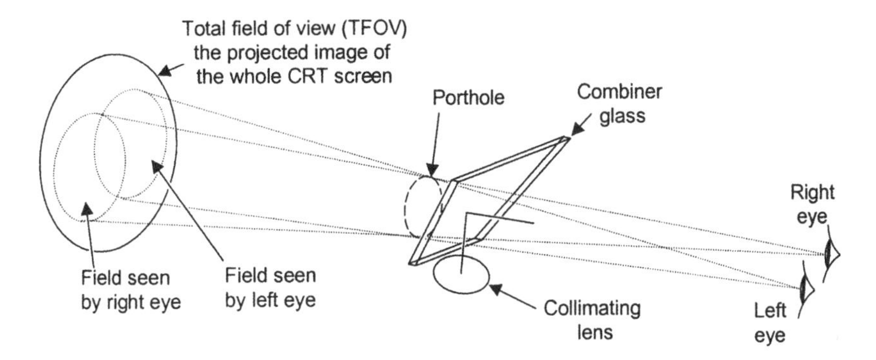

Figure 8.7 Overlapping vignetted fields of view seen by the two eyes

The most evident limitation was that only a portion of the field of view could be seen from the design eye position. This, as indicated in the lower sketch of Figure 8.6, was a direct result of making the display fit in the space available, which limited the diameter of the collimating lens to about 110mm. The reflection of the collimating lens, as seen in the combiner, defined the "field stop" or "porthole" for the optics, and with the eye at the datum position the field of view of the porthole was much smaller than the angular size of the projected image of the CRT screen. In the parlance of the optical design engineer, the total field of view was vignetted by the collimating lens. Not only did the pilot have a limited "head box", but he had to move his eyes within this box to see all of the symbols.

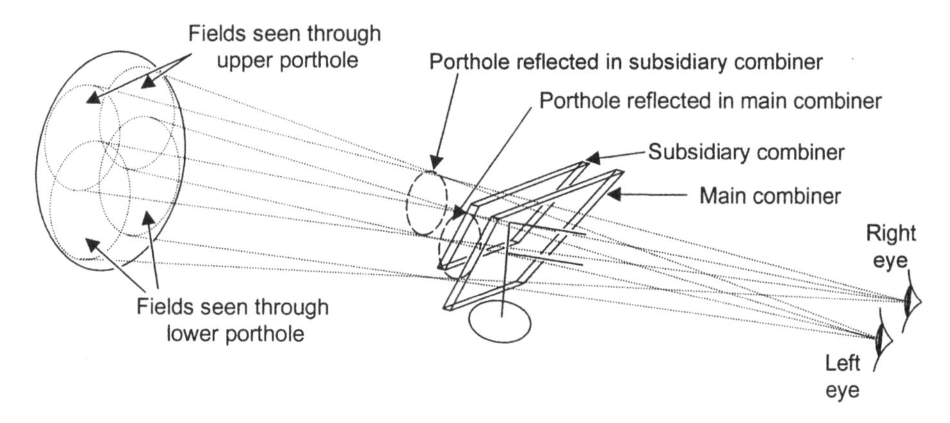

Figure 8.8 Extending the vertical field of view by adding a second combiner

As shown in Figure 8.7, the left and right eyes had slightly different views through the porthole, so the lateral truncation was less evident than the vertical. The simplest method of extending the vertical field was to install a second combiner, parallel to the first but offset towards the windscreen, and as shown in Figure 8.8, this reflected a portion of the image-forming light and formed a second porthole above the primary porthole. Although the proportion of the image forming light transmitted and reflected by the main combiner was graded over the vertical field of view to get uniform image brightness, the pilot could place his eyes where the overlap produced a marked lateral band. This could occasionally be mistaken for the horizon.

In most systems a small forward-facing sensor measured the background light level so that the display brightness could be adjusted automatically. The symbol contrast was then maintained at a constant level without involving the pilot. Pilots generally accommodated other problems. For instance, during a bombing run the pilot would slacken the harness straps and lean forward so that both eyes were closer to the porthole. In this position, all the displayed symbols could be seen, and the struts supporting the combiner glass also intruded less into the view of the world.

For systems in aircraft such as the Tornado GR4 and the Harrier GR7 in RAF service, the symbol generator was arranged to supply three forms of output. In the daylight mode, to provide the necessary symbol luminance, the symbols were displayed cursively and the writing period occupied most of the 20msec refresh period. In the night-time mode, when a pictorial image from a thermal sensor was imported by the symbol generator for display on the HUD as a raster image, the symbols were also written cursively, but because less luminance was required, this was done rapidly during the 1.6msec field fly-back period. The third form of output produced symbols as a raster image that were keyed onto the raster scene image for recording on the aircraft video cassette recorder during night flying. In daylight the same recorder was used to capture the forward view with the overlaid symbols as sensed by the HUD video camera. The overall system was not only complicated and expensive, but it reduced in-service reliability and required the HUD to be treated as an exception to the HDD devices. As noted in section 8.2.4, this is a further incentive to develop raster-based display devices that can provide the combination of high luminance and excellent clarity that are currently only available as stroke-written symbols on a CRT.

8.2.3 Reflective collimator designs

Figure 8.4 showed the principle of projecting a virtual image by placing an object in the focal plane of a concave mirror. In this position the object interrupts the view of the virtual image.

Figure 8.9 shows a variety of ways in which this interruption can be circumvented. A semi-reflecting mirror is used in (a), and although no HUD has employed this arrangement because the image source would be exposed to ambient glare, an inverted version has been used extensively in simulation systems to produce a distant outside scene. Embedded in a suitably shaped glass or plastic prism, the design is also used in several compact head-mounted displays. Scheme (b), sometimes called the "birdbath", is a slight rearrangement of (a) that has the obvious advantage of placing a single partially-reflecting plane mirror in the direct view. The image-forming light must however pass through the plane mirror, reflect from the concave mirror and then be reflected from the plane mirror, which is a problem because a mirror cannot normally transmit and reflect efficiently. A flightworthy prototype device employing this configuration was developed at RAE Farnborough[3] and called the "Z-HUD". In this embodiment the light from a low-mounted CRT was relayed to the focal plane of the concave mirror, as in (c) and (d), using a relay lens and a horizontal, downward-facing plane mirror. The problem of attaining efficient transmission and reflection was circumvented by filtering the CRT emission to select a narrow waveband, introducing a slight tilt to the optic axis and by holographic coatings on the plane and concave mirrors. (The combination of angle and wavelength selectivity afforded by holographic layers is described below.) It is of note that the upper mirror in combination with the selectivity of the coating on the concave mirror blocked almost all glare. The only practical detraction was the need for an automatic mechanism to swing the concave mirror upwards to clear the ejection line.

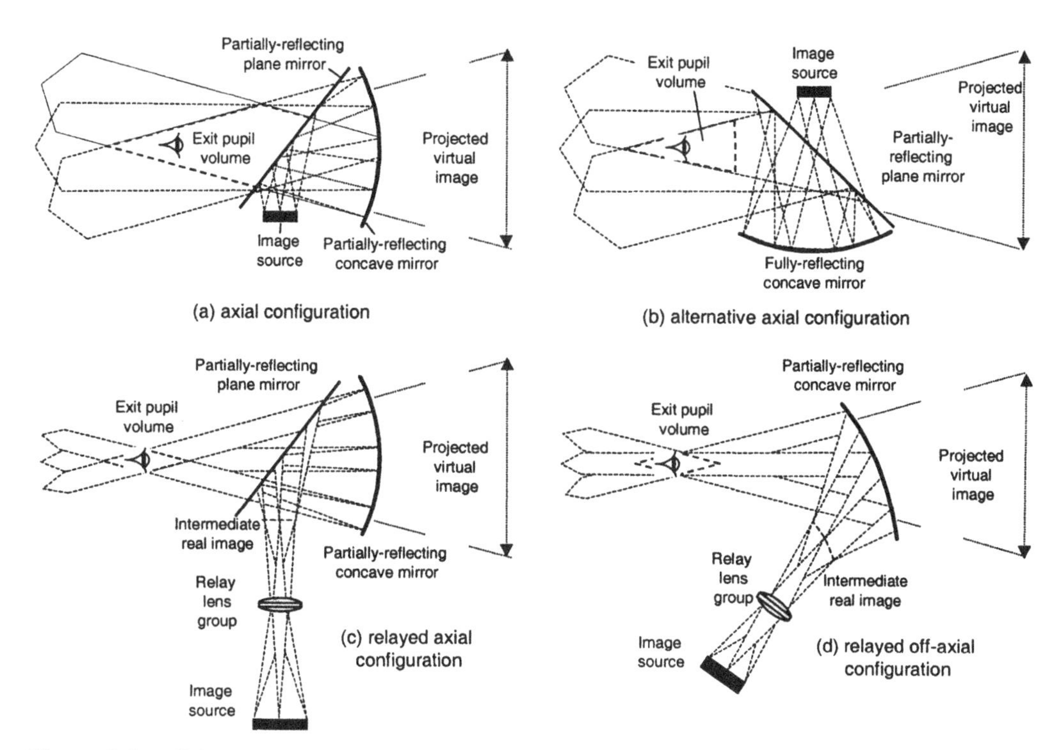

Figure 8.9 Schemes for avoiding obstruction by the image source when using a mirror collimator

The third case (c) is the same as (a), but it includes an additional relay lens that transfers the image-forming light from the luminous source to the focal plane of the concave mirror. This scheme has not been used for a HUD but, scaled down, it is the basis for numerous HMD configurations. The fourth case (d) also uses a relay lens and has only one element in the light-paths for the imagery and the direct view. The principal disadvantage is that the optical axis is reflected through a large angle at the concave mirror, so the relay lens must counter the strong spherical aberrations that arise. In practice the corrections for asymmetric astigmatism, coma and distortion entail a larger number of optical elements in the relay, most of which are eccentric and tilted to the optical axis.

Recent advances in computer-aided ray-tracing have transformed the process of optimising such complex optical systems. For instance, it is now routine to athermalise the design by choosing combinations of optical materials that also compensate for likely temperature variations in service. It is also usual to predict how image qualities are affected by manufacturing errors. The knowledge is essential for solving the problems of making and cutting-out asymmetric lumps of polished glass, and assembling them into a barrel to an accuracy of about 20μm. The last is the art of fine threads, ring spacers, anti-rotation grooves and laser alignment aids. With such techniques versions of (c) and (d) have been used extensively in head-mounted displays, and (d) is the basis of most modern HUDs.

It can be seen from the simplified ray traces in Figure 8.9 (c) and (d) that the relay lens limits the size of the ray bundles emitted from the source that intersect to produce a volume some way down the optical axis in which the viewer can place his eyes to see the whole virtual image.

It is the widest part of this exit pupil volume that is termed the "projected porthole"; effectively the image of the relay lens aperture produced by the collimating mirror.

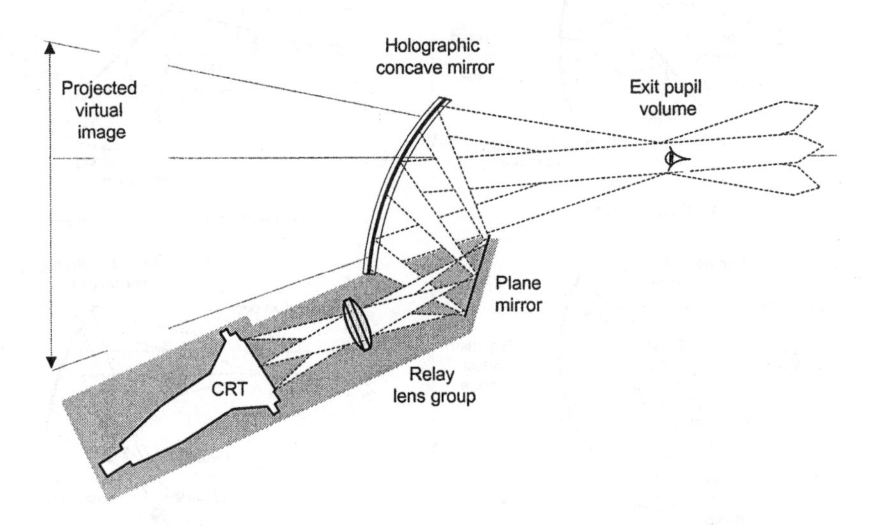

Figure 8.10 Simplified optical arrangement of a wide field-of-view holographic HUD

The most successful of the later designs, used for instance in the F-16, SAAB Viggen and the Eurofighter, have employed the configuration shown in Figure 8.9(d) with an additional mirror to fold the optical path so that the unit can be installed under the glareshield, as shown in Figure 8.10. This approach enables the total field of view to be stretched from the conventional 20° limit to about 30° vertically by 40° horizontally. The designers have ensured that the projected porthole was more than wide enough to accommodate both eyes, the rhomboidal exit pupil volume allowing about 40mm of eye motion before either eye lost sight of a portion of the image. Pilots are however surprised to find that unlike a conventional refractive HUD they see a progressively smaller portion of the virtual image if they move towards the mirror. The arrangement is not as resistant to ambient glare as the Z-HUD, but the reflective layer in the combiner is formed holographically to transfer image-forming light efficiently and minimise the attenuation of the transmitted light for the direct view.

The use of a hologram needs some explanation. The term is normally associated with a layer that can diffract light of a precise wavelength and incident angle to make it appear that a three-dimensional object is embedded in the layer. Such holograms illustrate wonderfully the wave nature of light. They are made by illuminating a real object with a beam of parallel coherent light, invariably produced by a laser, so that the reflections from the object create a complex pattern of interference fringes when they interact with a parallel reference beam obtained from the same laser. The hologram is formed by exposing a photographic plate to this interference field for perhaps 20 minutes, during which all of the elements are held on a massive vibration-isolated optical bench so that the sub-microscopic interference pattern is stable. When the developed photographic plate is subsequently illuminated with light that matches the reference beam in wavelength and angle, it redirects the waves as though they have been reflected from the original object, giving the appearance that the object is embedded in the plate. The most efficient holograms cause the light to be diffracted by introducing microscopic variations of refractive index that introduce phase shifts to the propagating waves. Most are formed in dichromated gelatin[4] (DCG), which is a complex mixture of amino acids. The exposed

DCG is however developed by a wet chemical process similar to that used in conventional gelatin film photography.

The HUD combiner is a rather special form of hologram in which the fringes are formed by the interference between a parallel beam and a coherent spherical beam; the fringe pattern of a point object. The result is usually termed a holographic optical element (HOE) or diffractive optical element (DOE). It is employed because the HUD combiner must transmit as much of the external light as possible while simultaneously reflecting as much of the image-forming light as possible. A hologram is exploited to achieve this precise filtration because a DCG layer can have a diffraction efficiency of more than 90% for the narrow band of wavelengths emitted by a CRT phosphor, such as P-53, over the small range of angles needed by the optical design. In this case the fringe structure is a comparatively simple set of parallel planes of undulating refractive index, spaced at about half a wavelength, making a hologram of about 20 μm thickness.

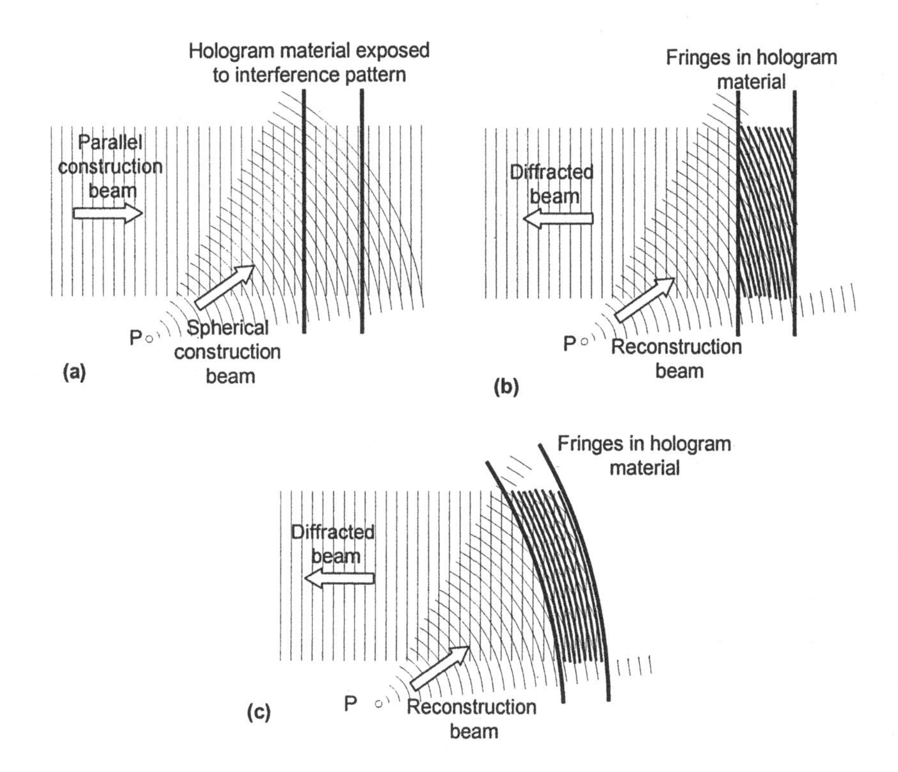

Figure 8.11 The idea of a holographic optical element

Figure 8.11 (a) and (b) illustrate a planar HOE that can be formed on a flat glass substrate and diffract light as though it has been reflected from a concave mirror. This seems a nice way of breaking conventional geometric constraints, but as shown in (b), in such a "non-conformal" HOE the fringes are not parallel to the flat substrate, and they intersect the layer surface to create a periodic structure rather like a two-dimensional diffraction grating. Thus light having the combination of incidence angle and wavelength that meets the diffraction condition for this surface grating structure will be strongly diffracted. A non-conformal HOE may therefore cause strong diffraction spectra, particularly from direct sunlight, to be directed towards the viewer when illuminated from certain directions. If however, as shown in (c) the substrate is parallel or "conformal" to the hologram fringes, there are no surface irregularities and this side-defect is removed. Accordingly, the substrate must be constrained to provide the optical properties of a concave spherical surface, and the HOE is best considered as a highly efficient filter.

The disadvantages of forming a filter holographically are largely associated with the expense of the optical production facilities and the need for very careful processing. This is complicated by the hygroscopic character of DCG, and the consequent need to dry the HOE carefully so that it has the correct fringe structure before it is sealed between the substrate and an accurately formed glass cover plate. Although alternative methods of manufacture using photopolymer materials on plastic substrates are possible, none has yet provided the diffraction efficiency of DCG without a slight accompanying milkiness.

8.2.4 Other configurations

Engineers have sought alternative electro-optical HUD configurations that are resistant to glare from sunlight, have a large exit pupil and adequate field of view, and show images that are bright, geometrically accurate and well resolved. They have produced a variety of designs that are, for instance, suitable for combining in a package with a collimated HDD and for use in civil aircraft. The idea of the dual combiner, illustrated in Figure 8.8, has been extended to make the combiner become a glass block incorporating a large number of suitably-coated, parallel, inclined reflective interfaces that produced a considerably enlarged porthole, albeit at some expense and increased weight. One particularly ingenious design[5], the "Perihud" was devised by engineers at Marconi-Elliot Avionics Ltd as a compact rearrangement that removed some of the installation constraints of the conventional HUD by placing the CRT and the collimating lens directly on top of the glare shield. A series of mirrors, offset to the sides, acted as a pair of periscopes that enabled the pilot to look around this obstruction and see the HUD imagery laid over an undistorted forward view. In this case the porthole produced by the collimating lens was much closer to the eye, so the instantaneous binocular view of the symbols was almost unvignetted. The only obvious detraction was that some parts of the forward scene were seen by one eye.

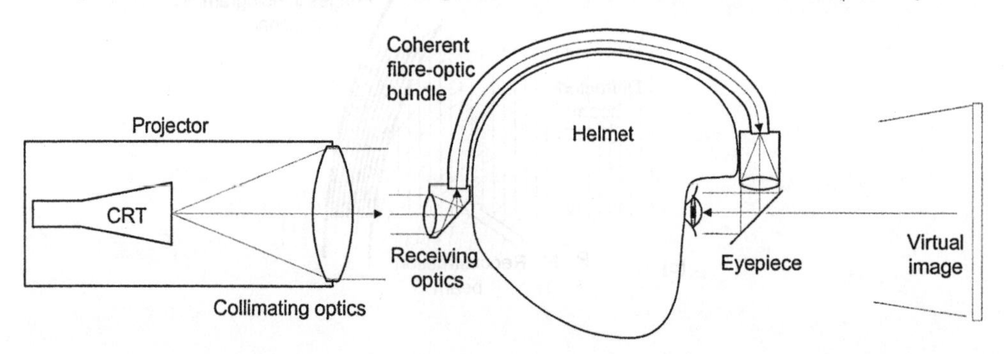

Figure 8.12 The idea of the "see-back-a-scope"

Another ingenious device that projected a virtual image onto the forward view was called a "see-back-a-scope" by the inventor, Fred Baggs of RAE Farnborough. As shown in Figure 8.12, the source of the image was a projector unit, sited behind the user, that contained much the same ingredients as a conventional HUD, with the exception of a combiner glass. If the user were to turn around and look into the large collimating lens he would see the intended symbols as a bright, collimated, inverted image at 180° from the aircraft boresight.

The helmet was fitted with optics at the nape that received a sample of the projected light and formed a real image on the input face of a coherent fibre optic bundle. This relayed the image to the focal plane of an eyepiece, arranged like a scaled down HUD, which re-projected the light to form an erect virtual image in a direction precisely opposite the received direction. Two sets of helmet-mounted optics were used to give biocular images. With suitably corrected optics, and providing the user kept the receiving optics within the exit pupil volume of the projector, the user could rotate his head and see clear, aircraft-stabilised, geometrically accurate virtual images. The approach obviated some of the disadvantages of a HUD. For instance, the front instrument panel and windshield remained clear of intruding optics, the helmet blocked most of the glare light from entering the projector, and a dark visor could be deployed to attenuate the brightness of the scene. It also obviated some disadvantages of a HMD system, as discussed in Chapter 9. In particular, the passive helmet-mounted optics required no electrical connection or orientation measurement, and there was no consequent symbol misplacement due to the system latency. The main disadvantages arose from the need to trade image quality and field of view for head mobility, combined with the limited field of regard produced by the projector unit. No cockpit design team has considered that the unique merits of the system outweigh these detractions and addressed the problem of accommodating the projector unit behind the seat.

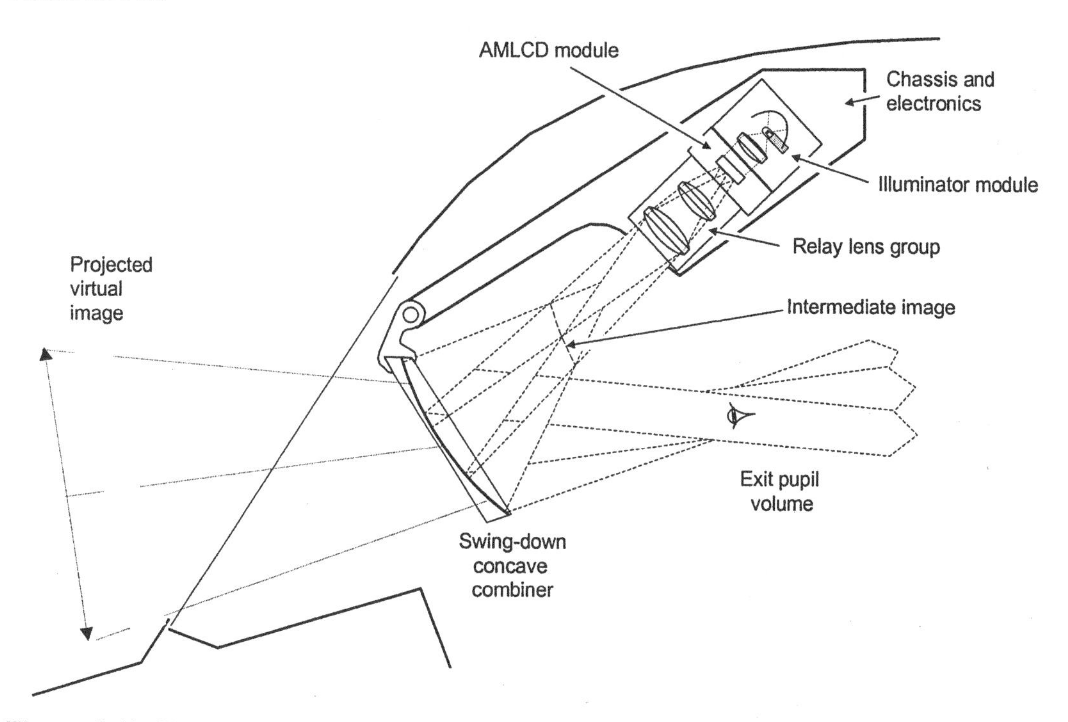

Figure 8.13 Simplified schematic for an overhead HUD installed in a helicopter or transport aircraft (Adapted from Brown R D, Modro D H, Quast G and Wood R B, *High resolution LCD projectors for extra-wide field of view head-up display*, in Hopper D (ed) Cockpit Displays X, Proc SPIE Vol 5080, 2003)

Experimental systems that provide coloured symbols have been produced using penetron CRTs as the image source. However there is no pressing need for coloured symbols, and the colour of a symbol may not be perceived as intended when superimposed on a naturally occurring bright background. The overriding requirements for high image luminance, fine symbol width and narrow spectral band are met using the highly-developed monochromatic CRT, but as discussed in Chapter 7, there are considerable incentives to replace such devices by cheaper, commercially available flat panel devices. Several manufacturers have developed substitute units employing miniature, back-illuminated AMLCDs. Figure 8.13 illustrates this idea in a package suitable for installation in aircraft without an overhead transparency. In this case the large, swing-down concave mirror is embedded in a parallel-sided glass block to obviate see-through distortion, and the pivot mechanism contains a detent that allows the block to detach or move forward if struck by the pilot's head. Because the provision of virtual imagery in military fast jet aircraft is increasingly dependent upon helmet-mounted displays, future HUD development is most likely to be driven by applications in commercial aircraft that favour this configuration. The most obvious problems arise when the pilot is also using direct-view NVGs, in which case the pilot must pull his head backwards to keep the goggle objective lenses within the HUD exit pupil volume.

8.3 Symbol design

8.3.1 Early symbol set

The earliest form of HUD put together by Naish and his group at RAE in England in the late 1950s was mainly intended to help the pilot maintain a continuous forward look out on the approach to a landing[6]. Altitude and airspeed were given by digits, and the aircraft attitude was shown by the sketchiest possible representation of an AI instrument; a fixed reference symbol and a shifted and rotated line. A small circle was moved left or right, and up or down, relative to the fixed symbol to guide the pilot in the same way as the localiser and glide-slope needles were moved within the head-down instrument. The symbols were not intended to overlay corresponding entities in the world, and the device was regarded mainly as a means for projecting a convenient group of virtual instruments.

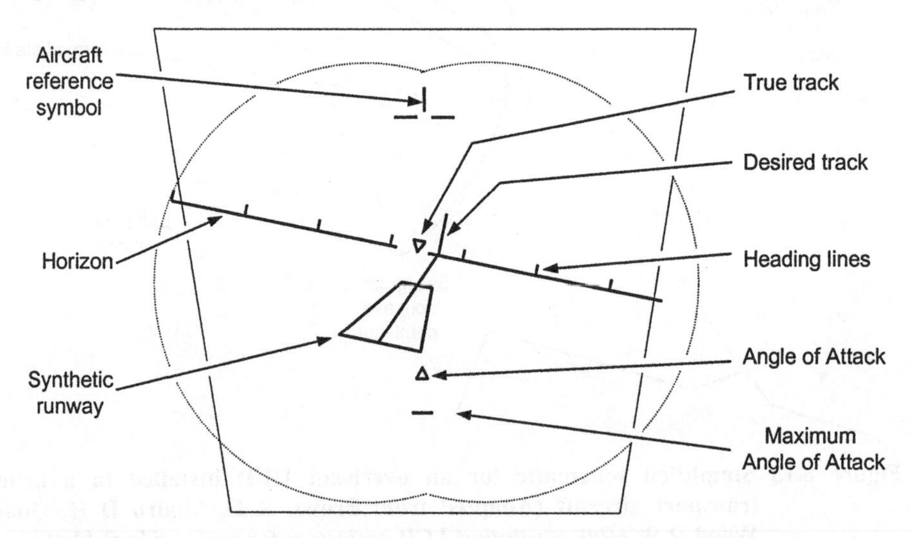

Figure 8.14 The Contact Analogue horizon and runway, and crucial aircraft performance symbols (From Klopstein G, *Rational study of aircraft piloting*, Thomson-CSF report ca 1966. Reprinted with permission from Thales Avionics]

Other originators, for instance the group led by Klopstein[7] at Thomson-CSF in France, were keen to exploit the unique ability of the device to project "contact analogue" symbols that overlaid corresponding entities in the world. As illustrated in Figure 8.14, their designs included a horizon line marked with short vertical heading markers and a perspective trapezium representing the runway. A fixed reference symbol was drawn near the top of the field to show the aircraft pointing direction, and another triangular symbol was displaced downwards towards the bottom of the field of view to show the angle of attack.

The advantages to be gained by presenting conformal symbols can be appreciated by considering the variables that describe the dynamic manoeuvring state of the aircraft, as shown in Figure 8.15. The longitudinal fuselage datum (LFD) is the fore-aft axis used by the designers of the aircraft. The location of any part of the airframe is defined with reference to this X-axis and the corresponding lateral (Y) and vertical (Z) axes. In flight, the Heading, Elevation and Bank angles form an Euler set that describe the airframe orientation and the LFD pointing direction relative to the local geographical [North, East, Down] frame of reference. Another set is the Track, Climb/Dive Angle (CDA) and Bank angles that describe the airframe orientation and the Velocity Vector (VV) relative to the same local geographical axes. For a fixed-wing aircraft in still air, the difference between the directions in which the airframe is pointing (the LFD) and actually travelling (the VV) depends on the angle of attack (α) and the angle of sideslip (β).

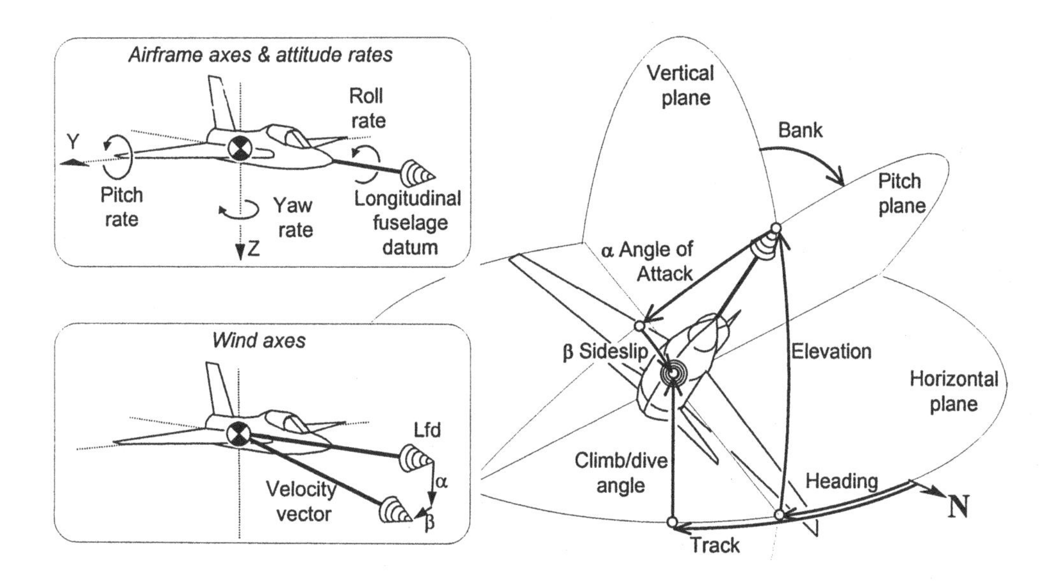

Figure 8.15 Terms describing the relationship between the aircraft orientation and the aircraft flight path in relation to the local earth surface. The velocity vector – the instantaneous direction of travel – points out of the paper in the right diagram

The earliest military use of the HUD was in ground attack aircraft operating at low level, such as the Buccaneer in the RAF and the A7E in the USAF, where even momentary inattention to flight path control could put the aircraft at risk, especially when flown over hilly terrain with vertical obstacles. The form of the information presented within the field of view, about 15°, of an early RAF HUD is shown in Figure 8.16.

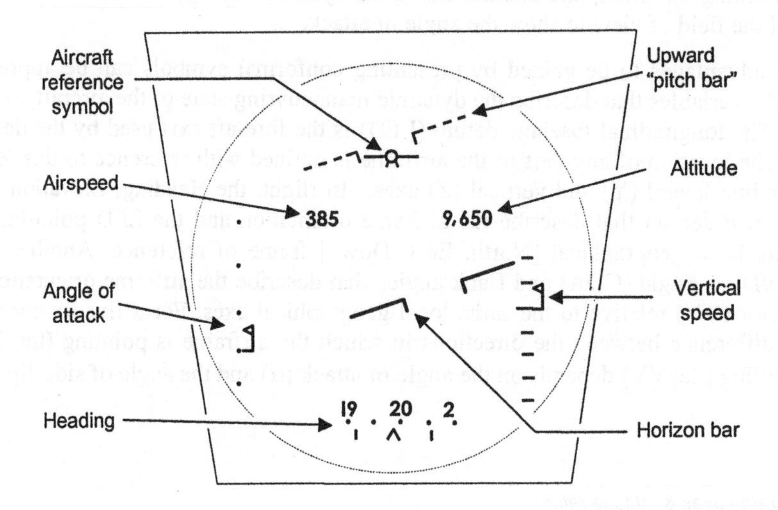

Figure 8.16 Example of the symbols presented on an early HUD (Reprinted with permission from QinetiQ)

Additional symbols representing the computed ballistic path of bullets or a bomb would be added when these weapons were selected. The digits gave airspeed as IAS to the nearest 5kn, and barometric altitude in feet, usually to the nearest 10 feet. The heading scale was arranged as though the pilot was looking at the edge of a strip compass where the digits, marking tens of degree intervals, together with the intermediate 5° markers would move left and right across the field of view. The value was read against a small fixed caret shaped like an inverted "v". The angle of attack and vertical speed scales were erected vertically, with the caret showing the instantaneous value moving vertically up and down the fixed scale. In Figure 8.16 the central symbol, a circle with a pair of short protruding lines, represented the forward-pointing direction of the airframe. This was not the actual LFD, but it acted as the reference point for judging the aircraft attitude relative to the horizon line and other lines parallel to the horizon that represented up and down pitch angles. The pair of shorter lines drawn dangling from the aircraft symbol were used to show when the undercarriage was down and locked. None of the attitude information or scales were conformal.

Several features of the design can be noted. Firstly, the geometrical arrangement of the information remained consistent with the standard T of the blind flying panel; airspeed was top left, attitude was central, altitude was top right and heading bottom centre. Secondly, the symbols were pared to the minimum. There were no annotations, explanations or units, and the pilot was expected to have no doubt about the meaning of the symbols. It should be noted that in diagrams such as Figure 8.16 (and in figures depicting more modern HUD symbol designs) although the area occupied by a scale was not marked, the portions of any moveable lines and symbols that encroached within the scale area were not drawn. Also, in normal flight, the pointers, digits and attitude bars constantly moved or changed, and this gave each parcel a dynamic coherence that helped the pilot to attend to each item of information selectively.

8.3.2 Use during low-level flying

The pilot of an aircraft manoeuvring close to the ground is mainly concerned to adjust the flight path accurately in relation to obstacles and ridges. He can gain most benefit if the aircraft symbol is driven down the HUD field of view by the instantaneous angle of attack, sideways across the field of view by the sideslip angle and horizontally by the crosswind drift angle. The aircraft symbol then shows the velocity vector, the direction of travel of the aircraft.

The position of the aircraft symbol in relation to the horizon line shows the bank angle and the CDA, and this, for instance, enables the pilot to execute a level turn by keeping the aircraft symbol in alignment with the horizon line. Such accurate manoeuvring, to about half a degree, is particularly helpful for steadying the aircraft during weapon aiming and the approach to landing. So, for low-level flight and air-to-ground operation, such as undertaken by the A7E, the "VV mode" is the preferred mechanisation of the attitude symbols. The consistency also extends to the symbol that shows the ballistic trajectory of a bomb that, when released, starts out with the same velocity vector as the aircraft, and in the absence of a crosswind drops vertically down to finish at ground level. From the pilot's perspective, as displayed on the HUD, the bomb falls along a vertical line drawn downwards from the VV symbol. It is also usual at low level for the direction scale to be positioned as high as possible, to avoid cluttering the view of the terrain, and to show Track rather than Heading. The altitude read-out shows the instantaneous height above ground measured by the radar altimeter, the readout having an "R" symbol ahead of the digits as a rare reminder of this changed significance.

8.3.3 The pitch ladder

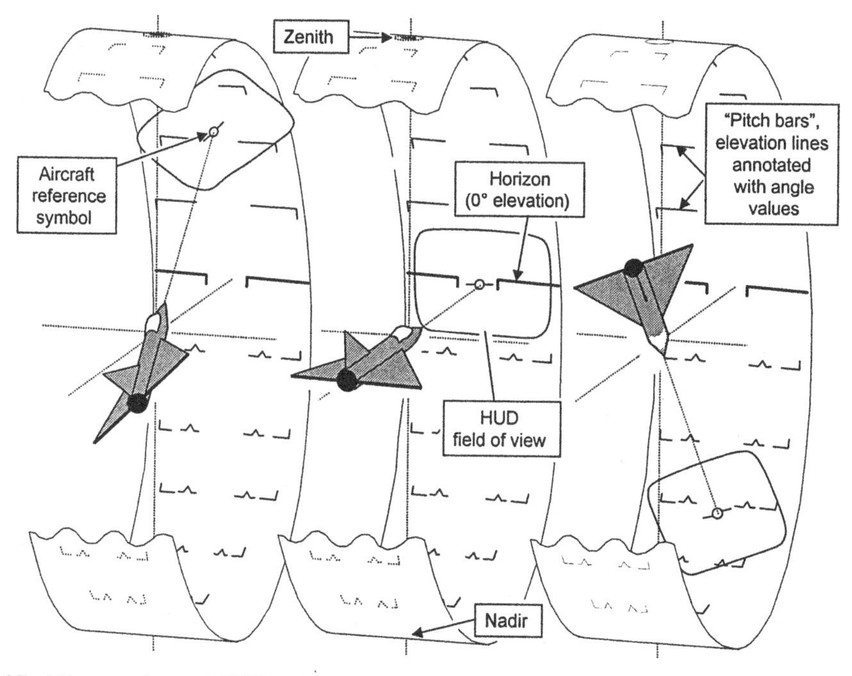

Figure 8.17 The conformal HUD pitch ladder. Note that in practice bars would be at 5° intervals and each would be marked with the elevation angle

The extension of the horizon bar to include parallel lines above and below the horizon to mark elevation angles, from the Zenith at 90° up right through to the Nadir at 90° down, gave rise to the notion of a "pitch ladder". Although it would have been more consistent to call it an "elevation ladder", the logic of this way of providing an accurate, conformal, spatial reference is illustrated in Figure 8.17. This diagram takes three examples of an aircraft at different orientations and shows how the HUD conveyed the aircraft attitude to the pilot. The mechanisation could be understood best as an artificial horizon instrument that had been expanded hugely to form a virtual ring that was perpetually vertical and centred on the aircraft. The HUD provided a small window that gave the pilot sight of a portion of this ring. Many alternative designs for the individual elements have been devised. The essential requirements were to ensure that at least two pitch bars from the ring would be in the field of view at any time, that each could be identified – particularly the horizon – and most importantly, that it was absolutely clear which way was up and which was down.

8.3.4 Use during air combat

The requirement changed significantly for air combat. Here the pilot would want to spend most of the time looking out and around, and the aircraft would be manoeuvred, occasionally at high-G and usually at higher altitude, at extreme positive and negative elevation and bank angles. In these circumstances, the pilot would want to glance at the HUD to check height, airspeed and attitude, and assess their rate of change, mainly to ensure that he was exploiting the best part of aircraft's performance envelope and that there was little danger of running out of fuel or hitting the ground. As well as needing additional information about G-level and fuel contents, the combat pilot found that the representation of aircraft attitude given by a conformal pitch ladder was not usable. With the more extreme manoeuvres the horizon bar would rarely be within the HUD field of view, and the pitch bars would also move across the field too rapidly to be legible. He wanted something that was more like the head-down attitude ball, and certainly something less frenetic. This was arranged by gearing the pitch ladder.

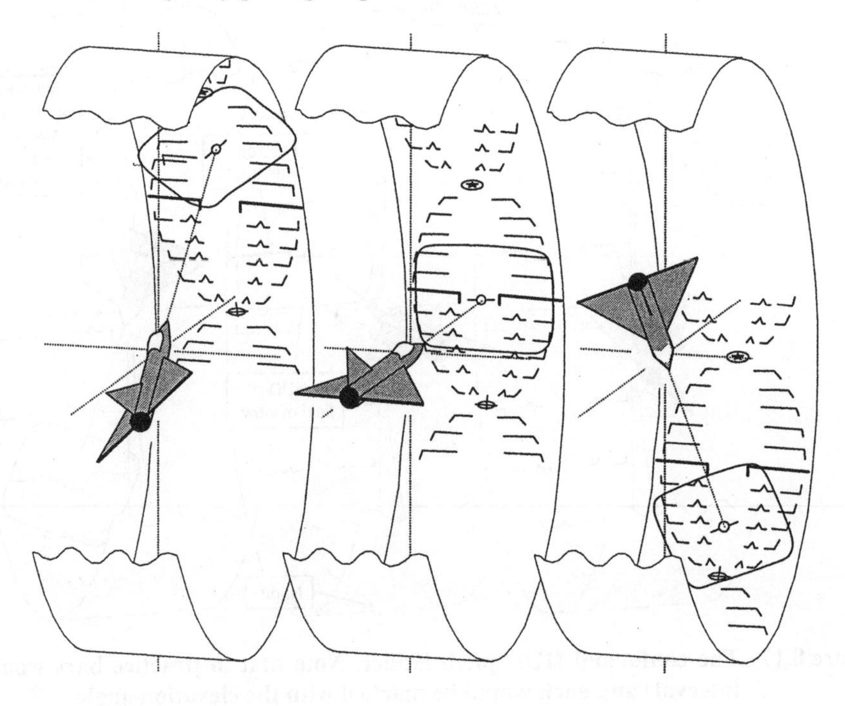

Figure 8.18 The geared HUD pitch ladder. Note that in practice each bar is marked with the elevation angle

Figure 8.18 extends Figure 8.17 to illustrate the logic behind gearing the pitch ladder. The ladder was positioned in the forward field-of-view to show the correct elevation and bank angle relative to the aircraft symbol. However, the ladder was shrunk by a factor of, say, three to one so that bars that represented true elevation intervals of 15° would be shown at only 5°. This had the advantages of including more bars within the HUD field-of-view, and during vertical manoeuvres the bars would move across the field-of-view at a third of the fully conformal rate. The other notable difference was that the combat pilot wanted most to be able to gain a rapid appreciation of attitude, and was less interested in the precise flight path. He therefore preferred to have the aircraft symbol mark the fixed LFD in the field-of-view rather than the mobile velocity vector. Again, for an aircraft that used a gun for air-to-air combat, the LFD provided a logically consistent reference. Although, like the bomb fall line, the bullet path would be a vertically descending parabola, the initial direction would be aligned with the airframe.

8.3.5 Alternative symbol sets

Recent designs of HUD attitude information that are based upon accurate and timely inertial, airmass and navigational data have attempted to provide a compromise between the two ways of presenting the pitch ladder and the alternative forms of aircraft reference. One scheme[8] that has been accepted in the UK uses conformal bars up to 30° of either side of the horizon, and a variable gearing that changes linearly from 1:1 to 4.1:1 between 30° and the 90° extremes. As illustrated in Figure 8.19, the LFD can be marked with a small fixed cross, the bar-and-circle symbol – the pivot for the pitch ladder – can move along the aircraft pitch plane by a "quickened" angle of attack to show a stable VV. A small diamond can show the actual velocity vector that takes account of crosswind and the instantaneous angles of attack and sideslip.

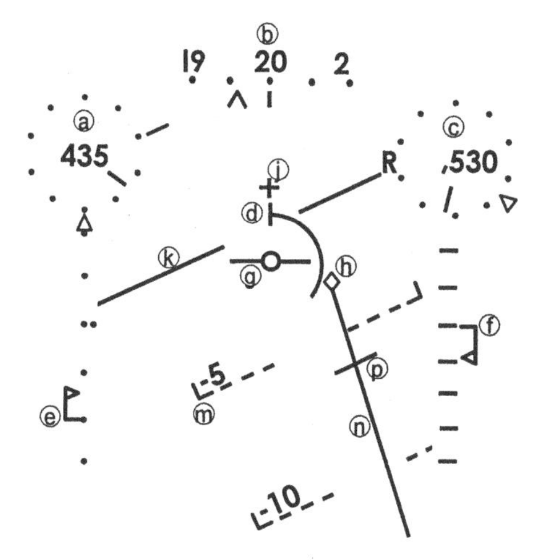

a. Airspeed counter-pointer *(bug set to 500kn)*
b. Heading scale
c. Altitude (radalt) counter-pointer *(bug set to 350ft)*
d. Unwinding time-to-target circle *(about 20 seconds remaining)*
e. Angle of attack scale
f. Vertical speed scale
g. Aircraft symbol *(quickened pitch plane velocity vector)*
h. Velocity Vector
j. Longitudinal fuselage datum
k. Horizon line
m. An elevation line *(5° below the horizon)*
n. Bomb fall line
p. Computed impact point

Figure 8.19 Example of a modern ground-attack symbol set (From Hall J R, *The design and development of the new RAF standard HUD format*, in Combat automation for airborne weapon systems: man-machine interface trends and techniques, AGARD-CP-520, 1993. Reprinted with permission from QinetiQ)

The studies undertaken to refine this conventional approach to the design of the set of HUD symbols, and to the modes that show information relevant to other phases of the mission, have addressed a host of detailed issues. These have included the character size and font, how the symbols behave on reaching the edge of the field, whether it is necessary to blank an interfering line, whether numbers should rotate or remain upright, and the length, solidity and inclination of the pitch bars. Many of the difficulties have been alleviated by the recent optical designs that remove the porthole restriction and increase the field, and by the general improvement in reliability of the CRT and the data chain. A wide range of symbols has been devised to direct the pilot to use a preferred flight path and airspeed, for instance the idea of the contact analogue has been taken into three dimensions as a "pathway-in-the-sky" that can lead the pilot to touchdown[9]. Symbols can also indicate tactically valuable information, for instance the range and direction of targets and threats, and the manoeuvring limits of a missile. Although some symbols convey information of relevance to specific aircraft and roles, for instance in the VSTOL Harrier and a variety of rotary-wing types, much of the symbol content is common. To avoid a proliferation of aircraft-specific designs, shorten the necessary training period and help pilots who must operate a variety of types, customers can insist that the symbols conform to a standard specification. Newman[10] gives a definitive review of HUD usage, symbol design issues and relevant standards.

The subject is far from cut and dried. In newer types the HUD has become the primary indicator of flight information, and electro-mechanical flight instruments are retained only to supply reinforcing indications and a back-up should the HUD fail. Both the design of the HUD hardware and the arrangement of the symbols are therefore safety-critical. As described in Chapter 3, the main requirements are to minimise the possibility of Type 2 (unaware) spatial disorientation, and should the pilot suffer Type 1 disorientation, the HUD should give a clear indication of an unusual aircraft attitude so that the pilot can quickly apply a corrective input to the flying controls. Several research groups have sought to achieve this aim by devising novel ways of conveying the primary flight information so that it can be assimilated more rapidly and with greater certainty.

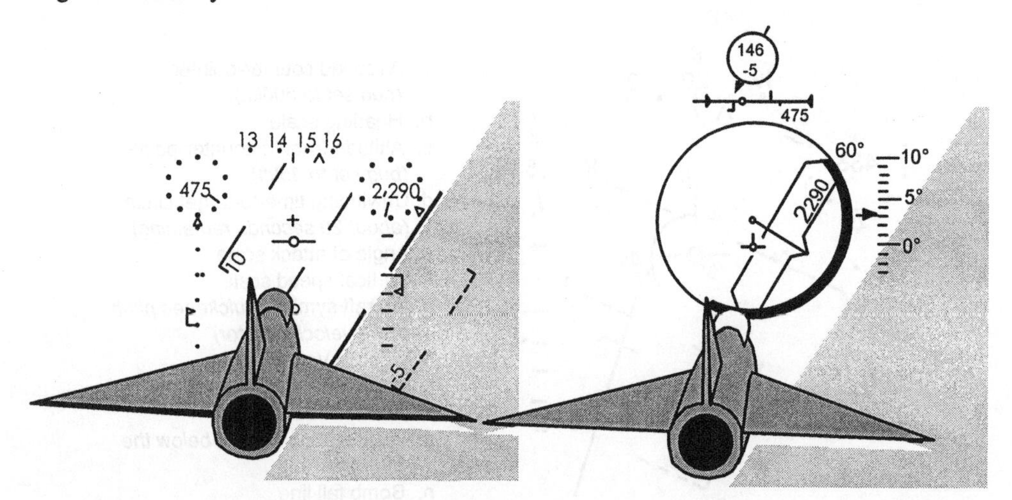

Figure 8.20 Comparison between a conventional HUD format and the Sky Arc format

Figure 8.20 illustrates one novel approach[11], called the "Sky Arc", showing how it would convey the same information as the more conventional design. In this case aircraft bank and elevation are shown by segmenting a bold circle, centred on a fixed aircraft centreline, into a "sky" portion and an "earth" portion.

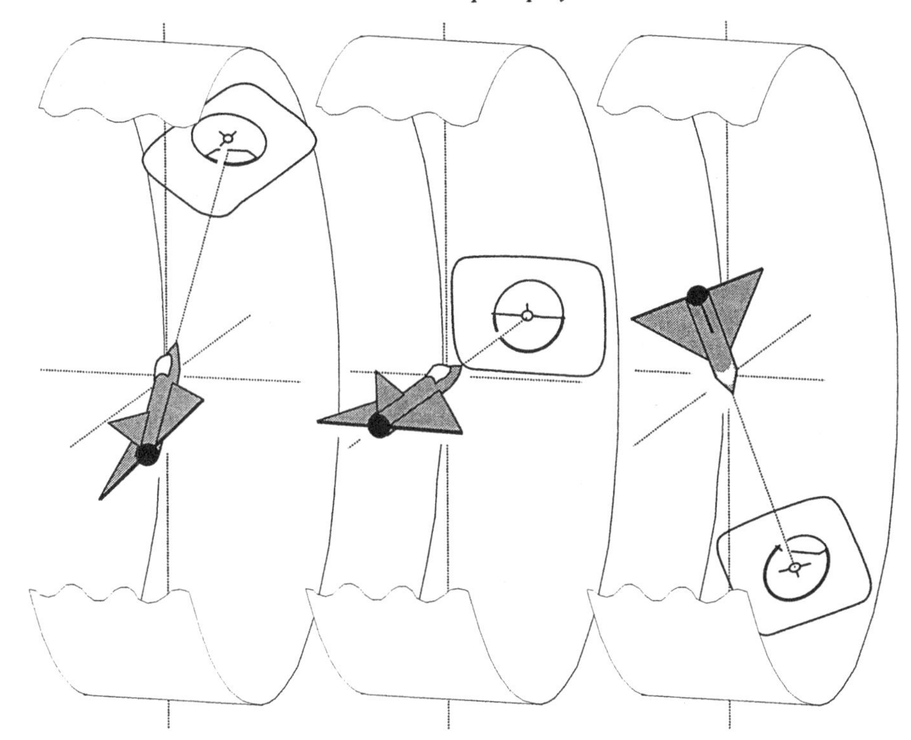

Figure 8.21 Logic of the Sky Arc attitude symbol

As illustrated in Figure 8.21, the axis of symmetry is kept vertical and the lengths of the sky and ground arc segments are proportional to the positive and negative elevation angles. High rates of roll and pitch do not cause dramatic gyrations of the arc, and the pilot can recover to a safe wings-level state by rolling to bring the sky segment uppermost in the HUD and then pitch up or down to make the portions equal. Zenith and Nadir symbols are formed as light and dark circles when either the sky or ground arc is complete.

The height information is also intended to be intuitive in the sense that increasing height extends the length of the vertical line from the circle centre to the line drawn parallel to the horizon; the "earth reference line". The arrowhead at the base protrudes up or down to show the rate of climb or descent. Airspeed is conveyed on a scale marked with symbols that move to show the best turn and maximum rate of energy gain speeds for the current altitude and load, between the minimum (stall) speed and the maximum speed, as described in Chapter 3. The instantaneous airspeed is given as a digital readout alongside the scale, and digital values are also given for altitude, bank, elevation and heading to enable precise manoeuvring. The heading value is contained in a circle that shows a North-pointing arrow as though presented on a compass rose, while a fine elevation scale is shown on the right as a moving tape.

8.4 Operational issues

The HUD has received much attention from researchers in the human factors community over the thirty or so years that versions of the device have been in squadron service. The main work has been directed towards the optimum form and mechanisation of the displayed symbols, but several groups have investigated visual anomalies[12] experienced by pilots. One oddity is that the collimated symbols - which are optically very close to infinity - commonly appear to be much closer to the viewer than the scene. This occurs whether the scene is viewed directly or is relayed as a thermal image on the HUD, and in the latter case the symbol and scene image are definitely identical in colour and focal distance. The consensus in this case is that the misleading perception is a consequence of the brightness and contrast of the symbols and the way they move over the scene, which are powerful cues that indicate the relative nearness of the symbols. Also, perhaps reading material is expected to be less distant than scenery. Another common observation by pilots is that the two forms of visual information cannot be assimilated simultaneously and instead it is necessary to switch attention from one to the other. This attentional focusing would be expected when, for instance reading the speed digits, but it can also occur when using conformal symbols, such as when judging the position of the predicted bomb impact point in relation to the target.

These oddities are readily apparent to the pilot, but other visual anomalies are more subtle. It had been expected that the tendency for the eyes to relax to focus at about a metre if there is little to stimulate a more distant focus, would be alleviated by giving the pilot the HUD symbols to look at. This "empty field myopia" was considered to depress the already reduced probability of seeing a hazard in poor visibility conditions or at night. However, experiments conducted in daylight by Iavecchia and associates have shown that the collimated HUD imagery does not induce the eyes to adopt a comparably distant focus[13]. This group has also raised a concern that the very presence of the virtual imagery, and perhaps the HUD combiner and supporting framework, may induce the eyes to focus inappropriately somewhere between the near and far states. They suggest that such an insidious condition induces the pilot to underestimate the size of entities in the real world and overestimate their distance, and this accounts for some otherwise inexplicable instances of "controlled flight into terrain" (CFIT)[14]. They have also given a warning that the misperceptions would occur with other devices that form virtual images, such as helmet-mounted displays and night vision goggles. Other investigators have however taken a different stance. For instance, Weintraub has suggested that the increased risk reflected in the accident statistics is intrinsic to high-speed low-level flying, and perhaps the provision of the HUD could have altered the pilots' perception of this hazard[15]. He drew an analogy with the added risk of driving a car at night; if you chose to drive at night, none would judge headlights to be dangerous.

The temptation to rely on the HUD indications when the visibility of the terrain and the horizon is really too poor for contact flying under visual flight rules must be resisted. It is also possible to become preoccupied with the imagery displayed by the HUD at the expense of attending to the real world. The piloting community recognise that these problems are intrinsic to flying and human nature, and that no technological prescription can alleviate overconfidence, attention capture or "perceptual tunnelling", except a wariness derived from training and experience. In practice pilots cope with the issue by adjusting the brightness of the HUD very carefully so that they can exercise selective attention to see the symbols, or see through the symbols, as they choose. The debate has largely abated in the context of the HUD, having been overtaken by a wider concern over the perceptual oddities experienced by pilots using helmet-mounted display systems. These are discussed in Chapter 9.

References

1. Grubb H, UK Patent No 12,108 1900
2. Hancock M, *The aiming problem in aerial gunnery*, The Aeronautical Quarterly Vol VIII, pp. 31-48 1957
3. Banbury J R, *Wide field of view head-up display*, in *Advanced avionics and the military man-machine interface*, AGARD-CP-329 1982
4. Swift D W, *Diffractive optics for avionic displays*, in *Advanced avionics and the military man-machine interface*, AGARD-CP-329 1982
5. Lewis C J G, *The Peri-HUD and a comparison with a conventional HUD*, Proc 2nd Aircrew Advanced Displays Symposium, NATC Warminster, 1975
6. Bently L C and Naish J M, *Means for displaying navigational information to the pilot of an airplane*, UK Patent No 891255 1959
7. Klopstein G, *Rational study of aircraft piloting*, Thomson-CSF report ca 1966
8. Hall J R, *The design and development of the new RAF standard HUD format*, in *Combat automation for airborne weapon systems: man-machine interface trends and techniques*, AGARD-CP-520, 1993
9. Reising J, Barthelemy K and Hartsock D, *Pathway-in-the-sky evaluation*, Proc 5th Symposium on aviation psychology, Columbus, Oh 1989
10. Newman R L, *Head-up displays: designing the way ahead*, Avebury Press, Aldershot UK 1995
11. Voulgaris, E Metalis S and Mobley S, *An integrated primary flight display: the sky arc*, in *Helmet- and head-mounted displays and symbology design requirements II*, SPIE, Vol 2465 1995
12. Biberman L M and Allusi E A, *Pilot errors involving head-up displays (HUDs), helmet-mounted displays (HMDs) and night vision goggles (NVGs)*, Institute of Defence Analysis, Virginia IDA Paper P-2638 1992
13. Iavecchia, J H, Iavecchia H P and Roscoe S N, *Eye accommodation to head-up displays*, Human Factors 30(6) pp. 689-702 1988
14. Roscoe S N, *Judgements of size and distance with imaging displays*, Human Factors 26 (6), pp. 617-629 1984
15. Weintraub D J, *HUDs, HMDs and common sense; polishing virtual images*, Human Factors Soc. Bulletin pp. 1-3 Oct 1987

Further reading

Weintraub D J and Ensing M, *Human factors issues in head-up display design : the book of the HUD*, State of the Art Report, Crew System Ergonomics Information Analysis Centre, Wright-Patterson AFB, Dayton, Ohio, 1992

Chapter 9

Helmet-Mounted Display Systems

9.1 Introduction

The pilot's helmet has been used as an anchorage for useful aids since the early days of aerial combat. A German airman in WW1 was reputed to have attached a dental mirror close to his brow to give him the crucial edge of all-round awareness; eyes in the back of his head. Although the veracity of the story and the success of the device are both unknown, it provides a nicely allegorical introduction to the display devices that have been devised subsequently. Like the HUD, described in the previous chapter, helmet-mounted displays (HMD) project virtual images, but the images are aligned with the orientation of the head rather than the orientation of the aircraft. A HUD mounted on the helmet not only frees the pilot from the need to glance down and re-focus to see information presented within the cockpit, but he no longer has to glance forward. The intention has been to extend the convenience and conformality of a HUD to provide the imagery wherever the head is pointing. Although the size of the visible image is limited to the display field-of-view, the total visual envelope in which imagery can be presented (the field of regard, FoR), is limited only by the mobility of the head.

This chapter does not deal directly with the problems of mounting a display on the pilot's headgear. These are considered in Chapter 13 in the broader context of the need to integrate other devices such as NVGs and sensing transducers as well as displays. Instead this chapter can be regarded as a four-part discussion of underlying factors. The first part reviews the uses of image projection devices that are mounted on the pilot's helmet. The second discusses the engineering issues that must be addressed to avoid artefacts and anomalies that can affect this somewhat peculiar way of presenting information. The third describes some of the electro-optical configurations that have been devised, and the final part emphasises the need to take account of the whole system, not just the part mounted on the helmet.

9.2 Applications

9.2.1 The helmet-mounted sight

The idea of the helmet-mounted sight (HMS) was devised in the early 1960s by US scientists at Wright-Patterson Air Force Base, Ohio, as a means of aiming short range missiles[1]. The idea is shown schematically in Figure 9.1. The pilot of a combat aircraft was supplied with a helmet orientation sensing system (described in Chapter 11) together with a simple monocular "reticule" display that indicated the helmet pointing direction. These components were integrated into the weapon control system so that the helmet orientation signals could be sent directly to the seeker of a lock-before-launch missile, such as the infra-red sensitive AIM-9 Sidewinder. The pilot could designate an external target by moving his head so that the aiming reticule was superimposed over the target, listen for the change in the tone that told him when the missile had locked to the target, and then pull the trigger to release the missile. This was in contrast to the normal aiming technique that required the pilot to use more extreme manoeuvres to point the aircraft so that the target was brought within the small field-of-view of the HUD. Essentially the missile launch success zone (LSZ) expanded from the HUD field-of-view, a cone of about 10° half-angle, to one of about 30° half-angle. This enabled the pilot to exploit the inherent missile agility and attain earlier weapon release to win the combat.

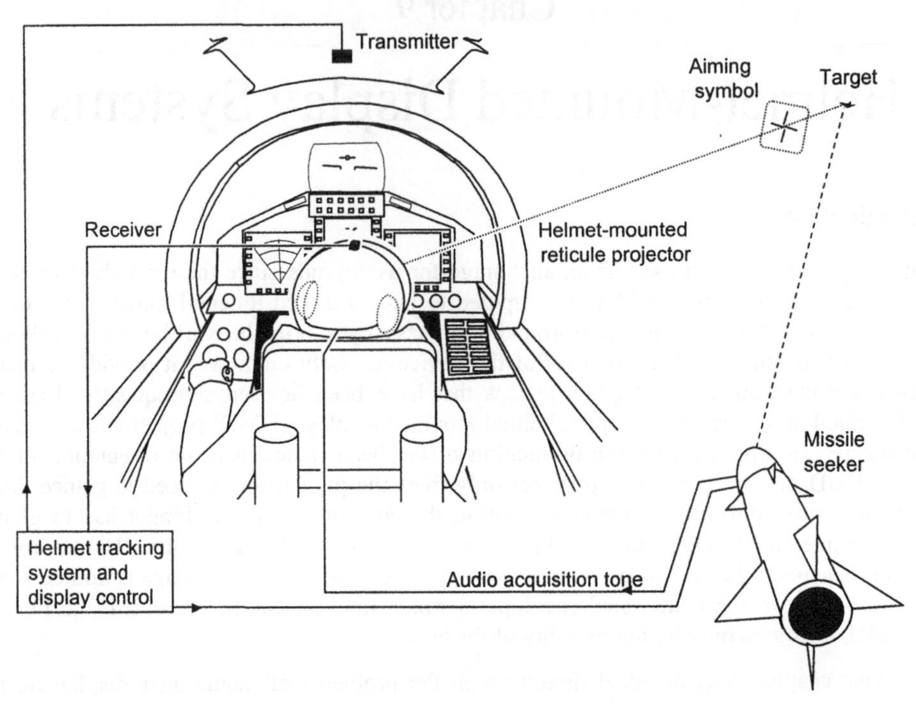

Figure 9.1 The idea of the helmet-mounted sighting system (Reprinted with permission from QinetiQ)

Slight sophistications brought further benefits. The signals from the helmet sensing system could also be used to point the aircraft radar so that the target range and range rate could be measured and the target G-level computed. Additional symbols in the reticule projector could then be used to tell the pilot whether the dynamically fluid relationship between the two aircraft represented a robust firing opportunity or merely a transitory chance shot. If the pilot looked away the radar would remain locked to the target, and arrow-shaped symbols alongside the projected aiming symbol could be illuminated to cue the direction in which he should move his head to re-acquire visual contact. A similar cueing arrangement could also help one crewmember point out the target to another crewmember.

Many factors affect the pilot's ability to use an HMS; vibration-induced head motion, the difficulty of accurate head pointing at high-G, optical distortions produced by the windscreen and canopy, and the accuracy, update rate and head box size of the helmet tracking system. The practical pointing accuracy available with early HMS systems, such as the Honeywell Visual Target Acquisition System (VTAS) deployed in a USAF squadron of F-4 aircraft in the early 1970s, was comparable to the capture field of an infra-red missile, a cone of about 3°.

9.2.2 Dynamic monocular helmet-mounted display

Since the 1970s a steady technological improvement by a number of avionic systems manufacturers has established the helmet-mounted sight as an essential facility in combat aircraft operated by many NATO Air Forces and the Air Forces of Israel and the former Soviet Republics. The sight will be a standard fitment in future combat aircraft such as Eurofighter and Rafale. However, in most of these installations the aiming symbols are one element of a more complex set. Additional symbols give flight data to enable the pilot to maintain an awareness of the state of the aircraft while looking around. During air combat, symbols also provide target cueing, target range and missile state information to enable the pilot to assess the robustness of the firing opportunity while he fixates on the target during the few seconds needed to make the crucial decision to release the missile. A similar set of targeting cues is provided when engaging ground targets.

Some projection requirements are the same as those for a HUD. Symbols must be designed and positioned to avoid masking the view of the external world and be legible against a variety of backgrounds including bright, sunlit cloud. Because the contrast and colour appearance of overlaid symbols depend on the background, the use of monochromatic symbols not only simplifies the system design but obviates misinterpretation of cues conveyed by colour differences. Other requirements are subtly different from the HUD. The contrast of the projected symbols can be enhanced by deploying an attenuating visor and the imagery can be supplied to one eye only, although luminance must be increased by about 40% to have the same contrast as a binocular device. It is necessary to decide how individual symbols should be positioned within the field-of-view. Those that must register with corresponding entities in the world place great demands on the accuracy of the optics and they must be re-positioned within the field-of-view speedily to cater for rapid head movement. Fixed symbols can be composed into a format that occupies a field-of-view of about 15°, but spreading them any further stretches the pilot's perceptual span, as discussed in Chapter 2, and unlike the HUD, eye excursions needed to scan the symbols cannot be shortened by small head movements.

9.2.3 Visually-coupled system

Another established use of a helmet-mounted display is to relay the image from a head-slaved sensor in a visually-coupled system (VCS), as described in Chapter 5. This image is overlaid with symbols similar to those supplied in a HUD, and to date the main operationally fielded example is the AH-64 Apache attack helicopter where the VCS gives the pilot a view of the world at night and in some weather conditions that obscure visible light, but pass the medium infra-red wavelengths received by the thermal sensor.

The requirements for helicopter night flying are very different from those of the combat jet. The pilot relies on the imagery to be aware of the surroundings and any hazards, and the safety of the aircraft depends on the qualities of the system. Good resolution, a wide dynamic range and a wide field-of-view are paramount, but a bright image is unnecessary and the chance of glare from ambient light is minimal.

9.2.4 The "virtual HUD"

The most recently devised application for a HMD is to supply the imagery that would ordinarily be projected onto the forward view through the windscreen by a HUD.

This usage pre-supposes that the HMD included in the display suite would be able to project bright symbols in daylight and a have a much wider field-of-view than the HUD, sufficient for instance for use in a VCS at night. It is proposed by the designers of the JSF as a way to avoid the expense, weight, complication and geometrical intrusiveness of a HUD. As discussed later, to provide a properly stabilised substitute for HUD imagery, the HMD system must display sensed data with negligible delay. Also, as the virtual HUD symbols are only presented on the HMD when helmet is pointed in the vicinity of the boresight direction, the pilot cannot glance forward to control the aircraft flight path with the aid of the symbols if his head is oriented elsewhere.

9.3 Design considerations

The two established uses, the HMS and the VCS with a symbol overlay, illustrate nicely the rationale for helmet-mounted display systems. However, as well as manifesting the intrinsic problems of the HUD, the transfer of the image projector from a mounting under the windshield to a mounting on the helmet has brought into play many new human, technological and operational issues. The acceptability of the augmented headgear, the visibility of the image and the usefulness of the information all depend upon the design of the system and its intended use. Figure 9.2 gives an overview of the pitfalls. These are amplified in the following sections.

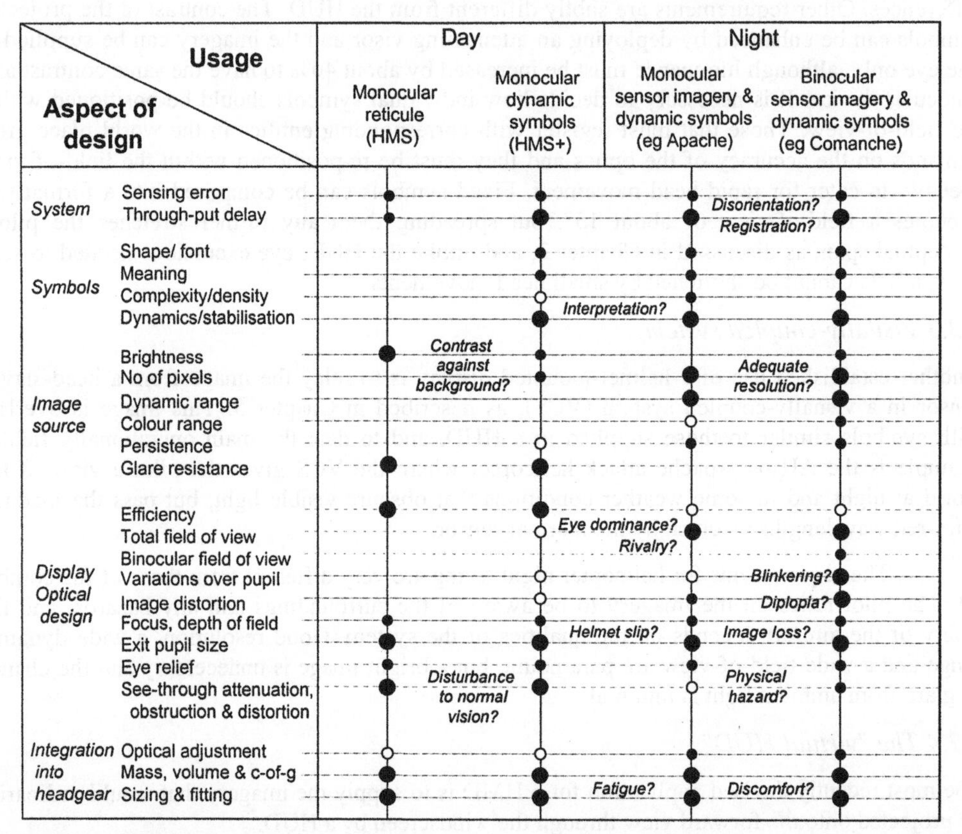

Figure 9.2 Summary of human factors issues associated with helmet-mounted display systems

9.3.1 Interpreting HMD symbols

The problems of assimilating information supplied by a HMD can be caused by difficulties of understanding and interpretation that are similar to the symbol ambiguities encountered with a HUD, as outlined in Chapter 8. However, the projection of the symbols into a head-slaved region of visual space can bring extra complications. For instance, the set of symbols implemented in the AH-64 Apache IHADSS resembled an early HUD format in that the symbols were intended to be interpreted as convenient virtual indicators rather than spatial references. Many symbols, such as the numerals giving height, speed and heading, could be understood no matter where the head was pointed, but the horizon line was superimposed on the real horizon only when the head was erect and pointing along the longitudinal fuselage datum. With his head oriented elsewhere the pilot had either to ignore this prominent line or, in order to judge the aircraft roll and pitch attitude, he had to imagine the line viewed with his head erect and pointing along the longitudinal fuselage datum. This trick of mental spatial rotation was complicated by the need for the pilot to remember that the thermal scenic image, which was the primary view of the outside world at night, was also an unreliable spatial reference. The sensor platform only followed the helmet in azimuth and elevation and could not be driven to roll about the line of sight in concert with the helmet. Thus if he rolled his head when looking forward, the pilot saw the scenic image and the horizon line symbol roll inappropriately against the real horizon. Confusion and spatial disorientation were obvious hazards.

Rather than have the pilot suffer from the head-pointing freedom of the HMD, it would be better to move the symbols within the HMD field-of-view so that they are located appropriately in visual space. To achieve this conformality, the appearance and position of symbols should be transformed spatially so that they are locked to appropriate space, ground, airframe, cockpit or helmet frames of reference, as illustrated in Figure 6.9. In the case of the IHADSS format, the attitude information would then be displayed correctly relative to the airframe rather than helmet frame of reference, and the pilot would be spared the effort of performing this transformation imaginatively.

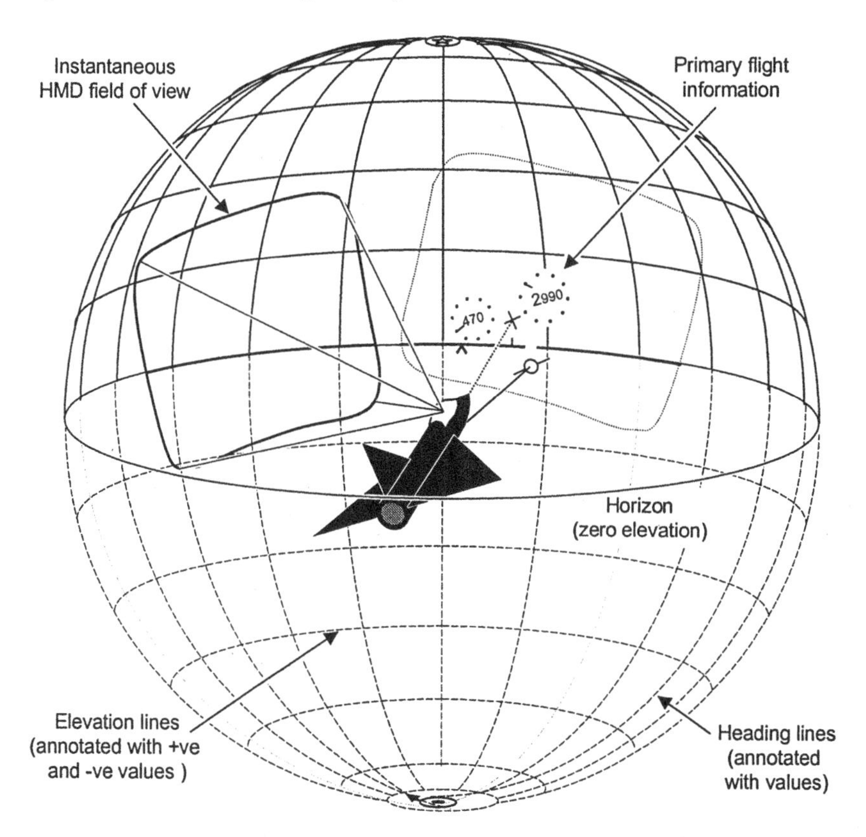

Figure 9.3 Possible method of conveying unambiguous flight information using an HMD

Rather than treat the HMD as just a convenient floating display surface, the display designer should consider the system as providing a limited window into an enveloping world of virtual symbolic entities. Figure 9.3 illustrates one such design approach that surrounds the aircraft with a space-stabilised virtual sphere marked with heading and elevation lines and a prominent horizon line. This set of symbols can be regarded as an enlarged version of the conformal horizon devised by Klopstein, shown in Figure 8.13. The aircraft LFD and VV are stabilised to the airframe, and so also are the symbols giving airspeed and altitude. The latter could however be locked to the helmet frame of reference and, as in the Apache format, remain perpetually in the field-of-view. These egocentric design principles form the basis of the "virtual cockpit", discussed in Chapter 14.

9.3.2 Seeing HMD imagery

A second cognitive problem has also been illustrated by the Apache IHADSS. In this installation the pilot saw the scene and the overlaid symbols on the monocular HMD with his right eye, and although this eye had a vestigial sense of the cockpit through the transparent combiner, the predominant view of the HDDs and panels was gained through the left eye. So a glance into the cockpit to check the fuel or change radio channel, or any other transfer between a head-out and a head-in task, needed a switch from the perception of the surroundings supplied by one eye to that supplied by the other. Most pilots selected to fly the aircraft have been able to adapt to these peculiar visual circumstances, but they can suffer headache, double vision, blurred vision and disorientation[2]. They may also have difficulty controlling what they see, and towards the end of a flight they commonly report that when looking out at the external scene the unwanted view of the cockpit intrudes, and vice versa. It is as though the unattended eye over-rides the mechanism of voluntary attention and the eyes seem to be in competition; the phenomenon is called "inter-ocular rivalry". Although the most practical solution is to close the offending eye, such forced winking is wearisome.

The condition experienced by Apache pilots is similar to many ordinary experiences, such as when looking out through a window at dusk. If the lit interior of the room reflected in the window is comparable in brightness with the outside scene it is possible to look around and examine details in the scene and notice the room interior only as annoying visual clutter. The room itself is not seen as a coherent structure. It is also possible to chose to look around the reflection of the room, whereupon the outside scene becomes visual noise. In this case both the scene and the room are imaged on the retinae of both eyes, so it is not possible to exclude the intrusive image by closing one eye. The phenomenon, known as "retinal rivalry", is similar to the experience of Apache pilots using a monocular display in that acutely dissimilar sources of stimulation, the scene and the room, seem to compete for attention. The general failure to project collimated symbols using a HUD, a display NVG or a HMD so that they seem to be embedded in the background scene may be another manifestation of the same condition[3]. Symbols and scene are perceived as separate entities; one fades if attention is paid to the other.

The similarity between inter-ocular rivalry, retinal rivalry and symbol-scene rivalry invites a common explanation, but none is generally accepted. (Appendix D gives the bones of an explanation in which perception is treated as the erection of a transiently useful representation, and attention is the selection of the elements relevant to instantaneous intentions.) Designers of HMD systems have proceeded pragmatically, avoiding inter-ocular rivalry by supplying the head-out images to both eyes, and minimising retinal rivalry by blanking the head-out imagery where it overlays the view into the cockpit. In recent programmes, such as the Comanche and Tiger helicopters and the Eurofighter fast jet, the designers have adopted a biocular see-through HMD. The satisfactory in-service experience of binocular night vision goggles has contributed greatly to this choice. However, the competition between symbols and background seems to be an irremovable aspect of superimposition, and voluntary attention transfer could perhaps be eased if symbols and the scene differed in colour.

9.3.3 Attaching a display to a helmet

The design and integration of the helmet-mounted part of the HMD requires attaching a scaled-down HUD electro-optical unit in front of one eye for a monocular device, or attaching two separate electro-optical units for a binocular device. To be merely acceptable for wearing in a combat aircraft the design must satisfy many mechanical criteria, for instance the components must not add significant mass, upset balance or make the headgear uncomfortable. The sheer proximity of the components to the eyes brings the concern that the view of the surroundings can easily be restricted, distorted, coloured and attenuated. There is also the possibility that the eyes could be damaged by optical elements, especially if these are delicate and easily detached in an accident. These issues are considered in Chapter 13.

Some of the factors that affect the visibility of the displayed images are similar to the HUD. For instance, the image must have contrast against a bright background, and ambient glare must be minimised. However, the mounting brings new problems. The most directly upsetting phenomenon is that any movement between the helmet and the head tends to shift the eye pupil outside the display optical exit pupil, commonly referred to as the "eye box". Small movements can cause the image to dim, and with grosser movement the image is lost altogether. Thus the stability of the helmet on the head and the provision of adjustments for accurate optical alignment to suit the individual wearer are as important as the size of the exit pupil[4].

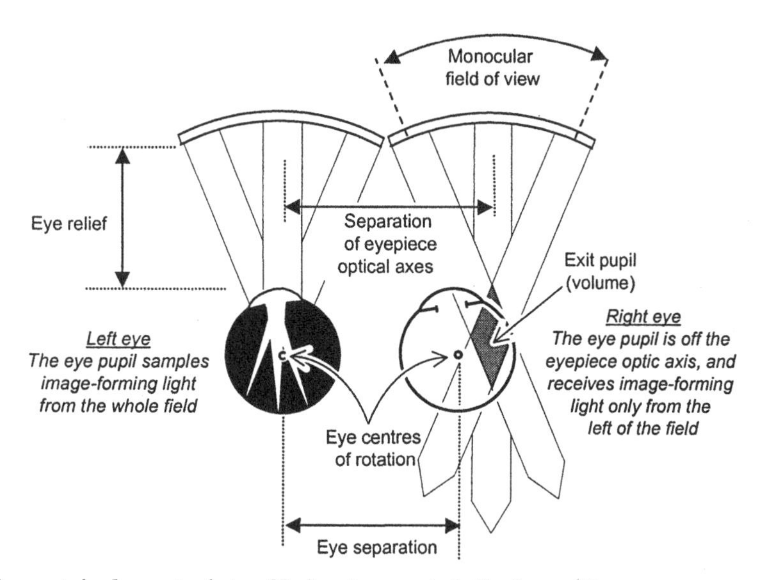

Figure 9.4 Geometrical constraints of helmet-mounted display optics

As shown in Figure 9.4, if the optical axes of the display eyepieces are not aligned with the pupils of the user's eyes, light from some directions may not enter the eye and these parts of the image field are not seen. The viewer can also have the paradoxical sensation of seeing the whole field when looking straight ahead, but a feature near the edge of the field disappears when he takes an interest and fixates on it. This occurs because the eye pupil, which is eccentric from the centre of rotation, is swung out of the bundle of light rays that come from the edge of the field when the eye is rotated towards the edge of the field. It is evident from the figure that the size of the display exit pupil should be sufficient to provide light to fill the eye pupil, say 6mm, plus that needed for helmet slip, say ±5mm, plus that needed to accommodate eye rotation and allow for any lack of positional adjustment. Typically, 15mm is just adequate. Figure 9.4 also shows that the size of the eyepieces is governed by the exit pupil, the eye relief and the field-of-view. Any increase in these parameters is therefore likely to increase the mass of the optics.

9.3.4 Inter-ocular differences

Individuals vary in their sensitivity to differences between the images available to the two eyes. Some inequalities seem to be unimportant. For instance, a noticeable colour mismatch, particularly of a monochromatic display, can generally be accommodated. A 30% mismatch in luminance may just induce a noticeable Pulfrich effect - an apparent error in the distance of a crossing object - but neither this anomaly nor the brightness difference itself are likely to cause concern. However, designers of helmet or head-mounted display systems that supply binocular images need to proceed carefully in controlling most of the other factors that influence the attributes of the images.

Relatively small geometric differences between the images supplied to the two eyes can give rise to a variety of perceptual oddities. The most apparent condition, double-imaging (diplopia) is most prevalent when the user can see the outside scene clearly and therefore has a strong stimulus for eye co-ordination and vergence. In this circumstance the outside scene is fused to form a coherent structure and the user will be particularly sensitive to disparities in the direction of corresponding features in the two virtual images. A vertical offset greater than about 1mrad (3.4 minutes or 0.1 prism dioptre) is likely to be uncomfortable, and if offset vertically by more than about 2mrad the feature or symbol will not be fused and will instead be seen as a diplopic pair[5]. Divergent horizontal disparities cause much the same effects, but a convergent horizontal disparity is more likely to give rise to an apparent displacement in depth, and diplopia ensues only if the error is greater than about 4mrad.

With a see-through HMD at night, or in any non-see-through system such as a NVG in current UK service, these strictures can usually be relaxed slightly because the outside scene is absent and binocular co-ordination is stimulated only by the projected virtual imagery. The possible consequences range from a delay or an error in reading information, perhaps accompanied by a mild visual discomfort, and perhaps headache and nausea. Much depends upon the individual and the nature and duration of the task being performed. In this regard, aircrew have a right to demand high specifications because their survival can depend on what they can see. They cannot afford the short periods of respite available to microscopists and other users of binocular equipment.

9.3.5 Partial overlap

As noted in the description of night vision goggles in Chapter 5, systems that project imagery into visual space so that both eyes see identical images are known as 'bi-ocular'. Those that can project images into different regions of visual space and have independent means for controlling the content of the imagery are 'binocular'. The former can be regarded as a special case of the latter. The two eyepieces need not be aligned, and it has been particularly advantageous for some applications to install optics with each eyepiece rotated outwards about the helmet vertical axis, as in Figure 9.5. This outward splay gives a wider binocular field-of-view, albeit at the expense of the overlapping portion, and the inside edges of the optics are less likely to clash when adjusted for narrow set eyes.

As shown in Figure 9.6, when the imagery is distant, which is the usual case for head-out symbols and pictorial imagery, the separation between the two eyes can be ignored and the horizontal visual geometry is simple. The overlapping binocular region at the centre of the visual field is flanked by two regions that are seen by the left and right eyes monocularly. If the splay is less than an eyepiece field-of-view (EFOV) and the two channels are matched, the total field-of-view (TFOV) is equal to one eyepiece field-of-view EFOV plus the splay (S). The TFOV is also the sum of the binocular overlap (BOFOV), the left monocular field-of-view (LMFOV) and the right monocular field-of-view (RMFOV). Thus these monocular fields are as wide as the splay angle.

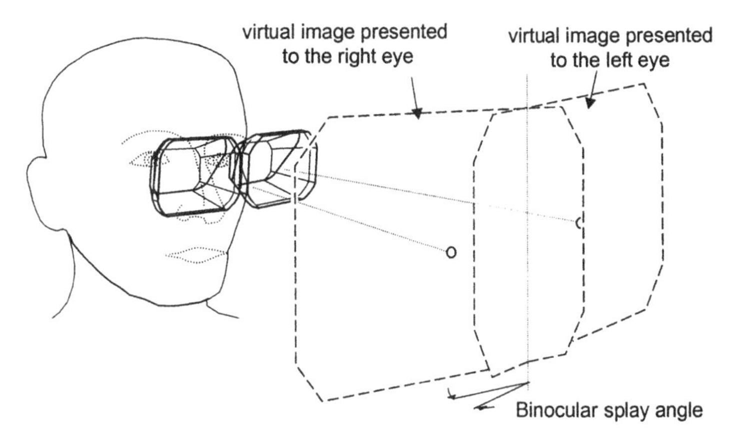

Figure 9.5 Binocular display with outward-turned eyepieces

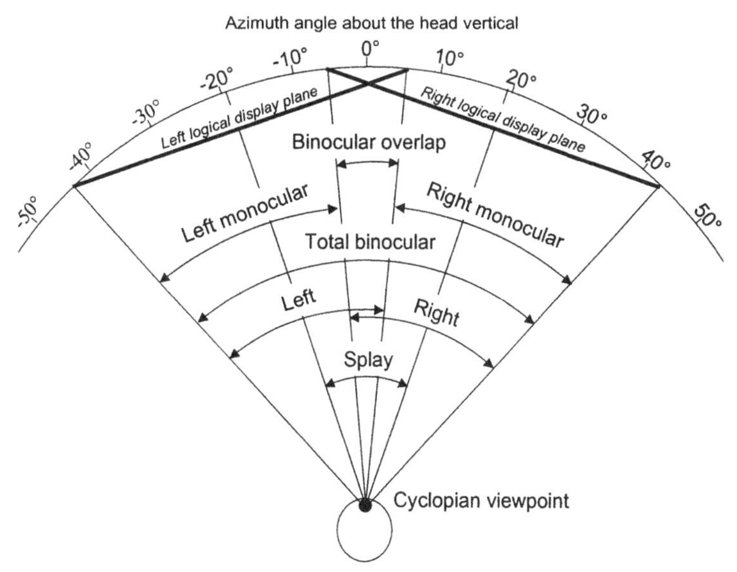

Figure 9.6 Terms used to describe the regions within the horizontal visual field for a binocular display where the separation between the two eyes is very small in relation to the distance of the imagery

Although it is unadvisable, the greatest lateral field is attained with no overlap when the splay is equal to the EFOV, but several factors suggest that this amount of splay should not be used. In this condition no stimulus is supplied that can enable co-ordination of the two eye's pointing directions – particularly for convergence – and the images can seem to drift laterally. There is also the increased likelihood of binocular rivalry if it is necessary to use one monocular field for an enduring period; the other eye's view can intrude even if there is nothing to see. As noted above, these effects can be disturbing and quite tiring. On the other hand, partial overlap generates other peculiarities. For instance, "binocular summation" of the stimuli from the overlap generates other peculiarities. For instance, "binocular summation" of the stimuli from the overlap area causes this to appear about 40% brighter than the monocular peripheries. The abrupt change in brightness at the boundaries is sufficiently intrusive that it has been called "luning" in recognition of the crescent-shaped monocular areas that result from partial overlap of circular EFOVs.

9.3.6 Logical display planes

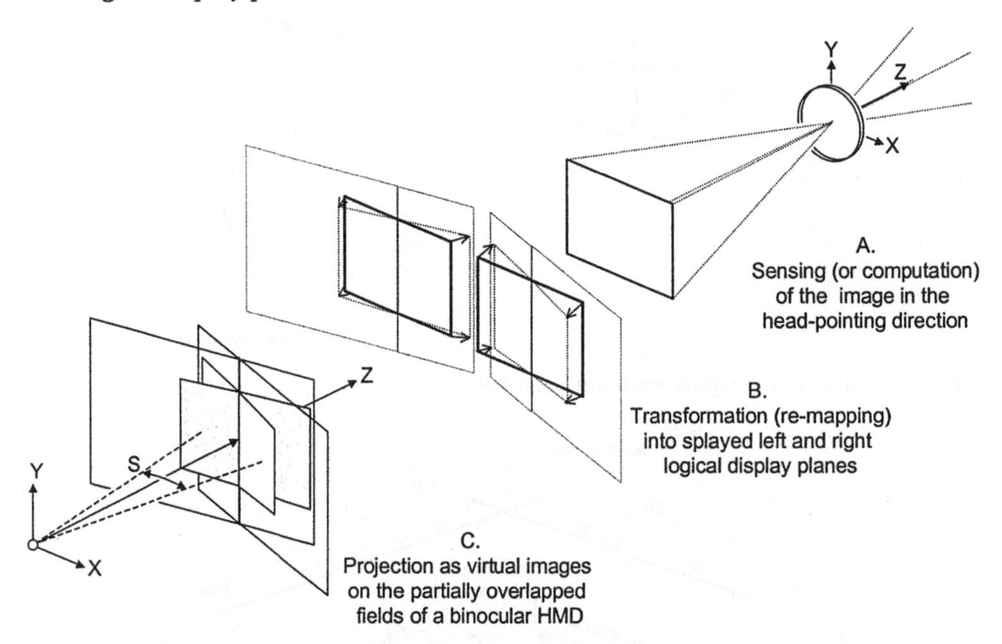

Figure 9.7 Re-mapping an image from a head-pointed sensor for correct presentation on a binocular HMD with partially overlapping fields

Figures 9.5 and 9.6 also show that it is necessary to treat the virtual images produced by splayed optics as though they define two "logical display planes" (LDP) that intersect at the splay angle in the depth plane. It is imperative that the form and content of the left image is constructed to be appropriate to the left LDP and that for the right image is appropriate to the right LDP. It is not possible for instance to generate, or use a camera to sense, a larger picture and merely extract rectangular portions separated by the splay angle that can be fed to the rectangular array of pixels that make up the displayed picture. It is also necessary to introduce a distortion that takes account of the different direction in visual space of corresponding pixels in the sensor focal plane and pixels in the display focal plane. Figure 9.7 illustrates the case of the transformations that are needed to re-map a sensor image having a LDP perpendicular to the head-pointing direction so that the image is geometrically correct when presented on a partially overlapped HMD. In this case a simple linear scaling, proportional to the depth difference between intersecting planes, changes a rectangle in the plane of the sensor into a trapezium on a display so that it appears in visual space as the original rectangle.

9.3.7 Image distortion

The overall requirement to present an image in its correct region of visual space invariably requires some consideration of the distortion introduced by the collimating optics. The optical designer aims to attain sharp, accurately focused images over the whole field-of-view and exit pupil, and in practice some geometric distortion inevitably arises. It is defined as the error between the actual and the geometrically correct visual angle of a pixel that cannot be corrected by simple position and magnification adjustments of the image content. All HMD systems do not transform a rectangular source image into a rectangular virtual image, but the degree and complexity of the distortion varies between optical configurations. For narrow field on-axis systems the errors are small in magnitude, for instance about 2%, and they are symmetrical about the optic axis. The image usually appears "barrelled" because the magnification decreases slightly with increasing eccentricity. However for large field-of-view off-axis visor-projected systems the error can be as much as 10% of the field angle, and it can vary over the field-of-view so that a rectangle is transformed into a complex shape resembling an axe-head.

The approach taken to correct distortion depends upon the characteristics of the device that forms the source of the image. For instance a reticule made as a back-illuminated transparent mask, or a simple fixed pattern of LED elements, can be designed with a counter-distorting shape. A CRT image source invariably includes the electronic circuitry to enable fine adjustment of the current waveforms that drive the magnetic scan coils to set the shape of a raster image. The correction of optically induced defects involves only a small extension to this facility; a simplification that has contributed greatly to the employment of CRTs in more versatile HMD systems. On the other hand, as discussed in section 9.6, a miniature flat panel, such as an AMLCD, has an immutably perfect geometric pixel array. Here the only viable solution is to include additional electronics that can counter-distort, i.e. warp, the image content.

9.3.8 Dynamic defects

Most electronic display systems that are fixed to the cockpit, the HDDs and the HUD, have the same requirements for speed and timing as in other installations. The designers are mainly concerned to avoid flicker and to update the picture at a rate that makes dynamic elements move smoothly; concerns that have largely been met by in formulating the picture standards for broadcast television and computer graphics. However, the usefulness and legibility of information presented on HMD systems is also affected strongly by other temporal factors. The main effects are caused by the sheer rapidity of head motion and the consequent shift of the imagery across visual space during the time taken to paint a frame or update the content of the image.

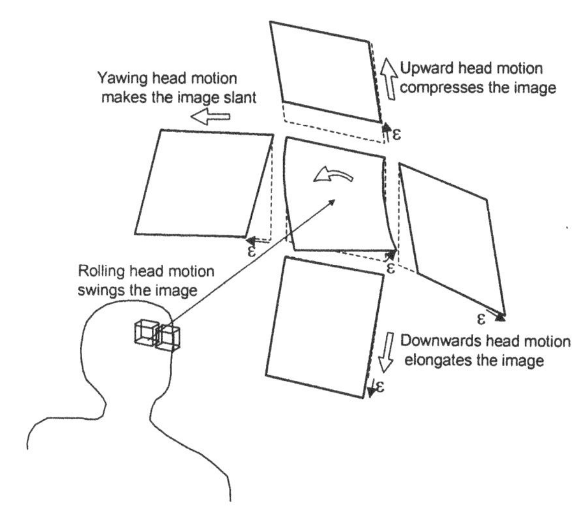

Figure 9.8 Illustrating the manner in which an HMD image can become distorted in visual space when the user rotates his head rapidly

Figure 9.8 illustrates the way in which a rectangular video image, which is addressed as a series of raster lines starting at the top and finishing a field period later at the bottom, can seem to be distorted in visual space. These forms of distortion are most noticeable if the image source is a CRT with a short persistence phosphor and conformal imagery is viewed against a real world background. The magnitude of the distortion ε is the angular excursion of the head during the frame or field period, for instance ε = 1° for 50Hz CCIR video and a head rotating at 50°/sec. With interlaced video, the two fields of each frame that are intended to have a vertical shift of single line width are seen instead as separated images. For instance if the viewer shakes his head, he will see a vertical line as a pair of parallel lines that move apart and swing, the effect being most evident when the head is moving most rapidly. Although the oddity is quite subtle, it is a contributing factor to mis-registration between symbols and the corresponding features in the world, and it can affect symbol legibility, particularly when the head is induced to move involuntarily by aircraft vibration. It should however be noted that with long-persistence image sources, such as an AMLCD back-lit by a fluorescent panel, the break-up and distortion is much less evident than a general smearing of the pixels across the visual field.

The more obvious temporal defect is usually caused by the delay between the instant at which the helmet orientation is measured and the instant at which the image content is changed in response to this helmet orientation.

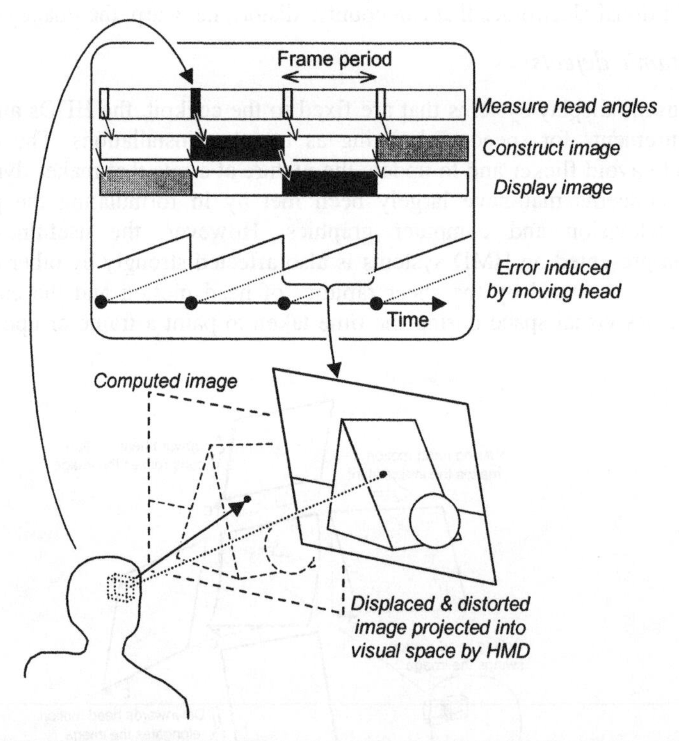

Figure 9.9 Simplified timing diagram showing the processes that contribute to "head lag". The main effects are a gross positional shift of the image due to the sequence of delays and a distortion because the bottom of the image is delayed more than the top

Ideally, the HMD should show the image from an accurate, fast-response head-slaved sensor, driven by a correspondingly fast and accurate helmet tracking system, with the output signal from the sensor coupled directly to the display device. In this case there would be little discernable mis-registration between features in the virtual imagery and their real counterparts. However, this ideal is rarely attainable as it is usually necessary to manipulate the image, for instance to counter the mapping or optical distortions, and some image content - such as overlayed symbols - must be generated in a computer. The system latency, called the "head lag", is the resultant of the sequence of separate sensing, data transfer, image construction and image read-out delays.

Figure 9.9 shows the case where the image is constructed in a computer that employs a double buffer frame store to avoid a conflict between the construction and read-out processes. Here the two parts of the buffer are interchanged at the end of each frame period and the analogue video signal is generated by converting the pixel data held in the area of the frame store updated during the prior frame. The minimum computation delay is therefore one frame period; 40msec for CCIR interlaced video and 16.6msec for most non-interlaced computer-graphics standards. In practice this is stretched to multiple frame periods by the need to construct complex imagery, and head measurement and data handling are not instantaneous. The overall delay can extend to 100msec or longer. As the angular error between the positions of real and virtual entities is the product of the delay and the rate of head rotation, mis-registration is significantly more noticeable. The viewer's experience is that the virtual entities seem to be dragged across the real world and fall into place only when head movement ceases. At best this is a subtle motivation to minimise head movement. At worst it can affect the legibility of symbols and induce nausea or disorientation.

9.3.9 Stereoscopic imagery

A binocular HMD can project separate images to the left and right eyes, and if the content of the two images satisfies the basic geometric rules of stereoscopy, the viewer can be given the sensation of viewing solid 3-D entities at a precise distance rather than 2-D pictures that float somewhere in 3-D space. The powerful sense of depth results from the small differences in direction of corresponding entities in the pair of images, as outlined in Chapter 2.

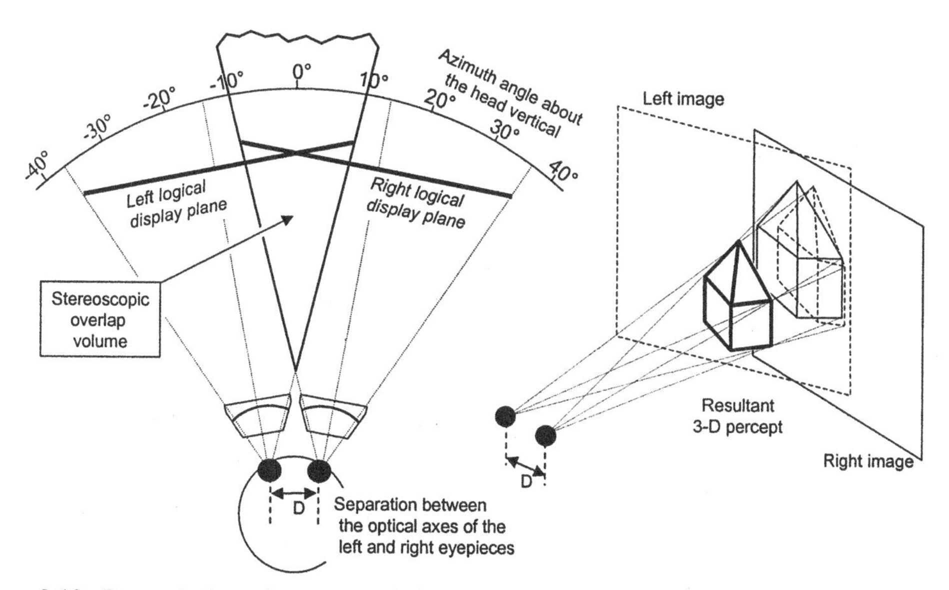

Figure 9.10 Presentation of stereoscopic imagery using a binocular HMD

Figure 9.10 illustrates the geometric requirements of a HMD when it is used to project stereoscopic pairs of images. As shown in the plan view on the left, it is only possible to create disparate images in the portions of the overlapping visual fields, and the resulting volume of 3-D space is also dependent upon the spacing between the optical axes at the two eyes. As illustrated in the sketch on the right, the stereoscopic image pairs that are needed to create the sensation of a 3-D object are those necessary to provide the requisite retinal disparity and convergence cues. The geometrical rules for constructing each image are therefore the normal perspective rules for projecting a 3-D object onto an 2-D image plane, but the points of projection must coincide with the eye's viewpoints and the 2-dimensional image planes must be the LDPs for the two eyepieces.

An HMD can still give the viewer the sensation of seeing 3-D objects if the projected images are not ideal stereoscopic pairs if the errors are sufficiently regular for the disparities to be preserved locally. The human visual system is surprisingly tolerant to lateral shifts of several degrees and will still fuse images that differ in magnification by about 2%, or when one image is rolled by about 4°. However objects invariably appear at inappropriate distances, and are therefore seen as scaled in size in proportion to the distance error. The perspective composition of the images must be appropriate to the viewpoint and the inter-ocular separation. Errors of 1mm in the latter are readily discerned. The spatial resolution of the projected pixel arrays is also important. If too coarse, a virtual object appears to consist of a stack of slices rather than a continuous entity.

Other human factors need to be taken into account. As noted in Chapter 2, the reflexive linkage between the eye muscles that control eye pointing for vergence and the eye muscles that adjust the power of the lens tends to make the eyes focus more closely when converging on a close object. It would therefore be expected that objects that are displaced in depth from the focus distance of the HMD optics would be blurred. Fortunately the depth of focus of the eyes, the limited resolution of the displayed images and the comparative looseness of the accommodation-convergence linkage largely nullify the anticipated difficulty. Problems are only likely when viewing close images, at less than 0.5m distance, and most people will tolerate a disparity of ±1dioptres between the two parameters[6]. It must however be noted that the perceptual conditions change when the HMD imagery is superimposed on a directly viewed scene, which could act as another stimulus for eye focus and vergence control. Military aviators do not have "stereo-blindness", the inability to extract depth cues from binocular disparities, which affects about 10% of the broad population.

9.3.10 Coloured imagery

Although systems fielded in current aircraft are monochromatic, it is possible that the next generation of airborne systems will supply coloured images. As with the HUD, during daylight there is little benefit from using colours as a redundant cue to emphasise the importance of a virtual symbol because the colour appearance would depend on the instantaneous brightness and colour of the background. The main incentive occurs during operations at dusk, dawn and night when there is only a weak competing background and colour can help differentiate between the significance of symbols and distinguish overlaid symbols from a tonal sensor image.

The requirement for colour is met by introducing a coloured image source and by designing the relay and eyepiece optics to handle the wider range of wavelengths. Although purely reflective elements are intrinsically insensitive to the wavelength of the light, the properties of most refractive materials and optical coatings, and certainly all diffractive optical elements, depend upon the wavelength. Thus an achromatic optical arrangement that handles wavelengths over the full spectral range from about 400nm to 700nm invariably includes a fair proportion of elements whose only function is to compensate for the dispersion inherent in the main image-forming elements. Colour-corrected optics tend to be more complex, bulkier, heavier and more expensive than a monochromatic equivalent.

The designers of colour HMD devices must also be careful to chose image sources in which the spatial registration of the pixels is preserved over the frame period during rapid head rotation. For instance, if colour is produced by painting red, green and blue pictures in rapid succession, these component images will be seen to be misplaced by whatever angular excursion is induced in the short interval between the separate colour presentations. Consider a system that employs an image source of 780 by 640 pixels, spread over a field-of-view of 40° by 30°, where the individual colours are painted at 180Hz to preserve the overall frame rate of 60Hz. In this case, rotating the head at a rate of only 10°/sec causes the colours to separate by a single pixel, and although most symbols and features in a scene are wider than a single pixel and the bulk of their colour is unaffected, they soon acquire coloured edges. Other temporal dithering techniques, discussed in Chapter 7, can also induce similar disruptions to the projected image. As the illegibility is exacerbated by rapid head motion, the user tends to hold his head still when it is necessary to assimilate fine detail.

9.3.11 The number of pixels: A trade-off between field-of-view and resolution

The density of individually alterable elements within a picture governs most of the perceived qualities. If the reproduction of fine detail is important, the HMD designer will compress the limited number of pixels in the image source to occupy a narrow field-of-view so that the elements cannot quite be discerned by the user. On the other hand, if it is necessary to provide a sensation of immersion in a synthetic visual environment, the scope of the picture is important and there is considerable incentive to spread the pixels widely. Most HMD systems are a compromise between these driving requirements.

The intrinsic trade-off between the provision of image size and image detail is illustrated in Figure 9.11 for several HMD systems. The axes of the graph are the two limiting parameters; the number of pixels along the vertical and horizontal edges of the picture and the corresponding extent of the vertical and horizontal fields of view. Systems are represented by points marking the (V)ertical and (H)orizontal subtence of a pixel, and those that fall on the same gradient line have the same resolution. The resolution of a system can be expressed as a fraction of the limiting acuity of the eye. If this is taken as one cycle per minute of arc, the individual pixels will not be resolved in a device that presents them at a density greater than 0.5 pixels per minute of arc. In practice, although the individual picture forming elements are just discernable, a device that presents pixels at about 1.0 pixel per minute of arc can render detail that enables the user to read characters of the size on the 20/20 line of the Snellen acuity test chart. Point (A) therefore represents the maximum FoV of a 625 line TV picture where eye-limiting details can be shown. (B) represents the resulting 20/60 equivalent resolution chosen for the binocular 30° by 40°Eurofighter HMD. (C) represents the Comanche HMD; resolution is improved by the 1024 by 1280 pixels of the miniature flat panel image source, and the horizontal FoV is extended by splaying the eyepiece optical axes. (D) represents a non-see-through, partially overlapped, colour HMD devised for simulator use. (E) represents a typical NVG. (F) shows the considerably extended horizontal FoV of a Panoramic NVG.

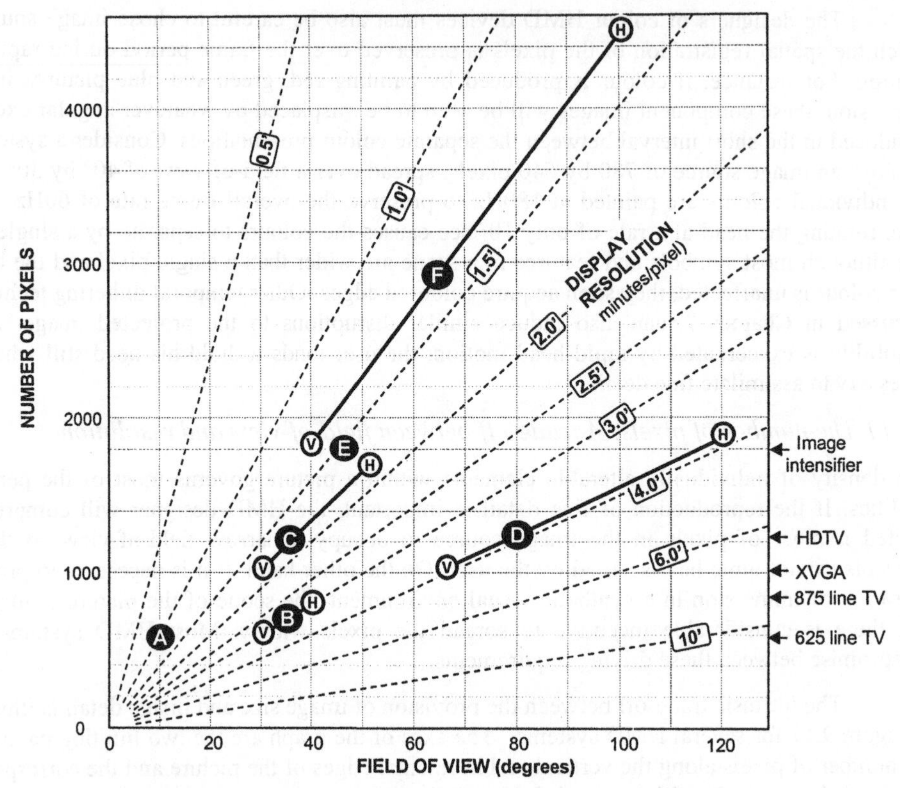

Figure 9.11 Trade-off between the field-of-view (FOV) and resolution for several HMD systems (Reprinted with permission from QinetiQ)

Figure 9.11 makes several points. Firstly, as marked by point (E), the number of pixels provided by the channel plate of an image intensifier tube in a night vision goggle is far in excess of the number of pixels available from any current video-addressable image source. Although the inter-channel contrast is commonly less than the inter-pixel contrast, a goggle having the same field-of-view as a visually-coupled system can supply images of significantly better resolution; about 70% of human acuity rather than about 50%. The figure also shows just how much development is needed for imagery projected by a HMD to approach the limits of human vision. An image that would fill the almost hemispherical human forward field-of-view, illustrated in Chapter 2, Figure 2.17, would need an image source of about 20,000 by 20,000 pixels, (i.e. $4x10^8$) to meet the human acuity limit over the full field. This is more than the number of rods and cones in the eye, but because the eye pointing direction rarely extends beyond about 25° from the nominal forward axis, the density of pixels could be reduced considerably outside this region.

9.3.12 Safety issues

Recognising that exposure to bright light, and in particular the emissions from lasers, can be damaging to the eye, many national and international standards organisations have established guidelines for assessing the risks, and have given recommendations for the maximum permissible levels of irradiance. In general, all wavelengths across the ultra-violet, visible and near-infra-red bands can be damaging, and the external surfaces – particularly the cornea – are mainly at risk from ultra-violet radiation. If such invisible radiation were emitted by a display source, a blocking filter would be vital. For the display designer the main concern is the visible and IR wavelengths that pass though the ocular media and are focused to some degree at the retina and could cause local heating and photo-chemical changes. Both damage mechanisms depend greatly on the wavelengths and intensity of the impinging radiation, but other parameters, such as the angular extent of the source and the size of the eye pupil, affect the retinal intensity[7]. Different standards organisations give different formulae for calculating the limits. For example, BS-EN 60825-1:1994, the British and European standard for the safety of laser products, recommends that the irradiance at cornea should be less than:-

$$\text{Maximum Permissable Exposure (MPE)} = 10^{-2}\ C_3\ C_6\ \text{Wm}^{-2}$$

where C_6 and C_3 are parameters that allow greater irradiance if the light source is spread over a reasonable angular size and the emissions have a wavelength greater than 550nm. Specifically, $C_6 = 91\alpha$ (for $\alpha < 0.1$ steradians) or $= 9.1$ (for $\alpha > 0.1$ steradians), and $C_3 = 1$ (for $400\text{nm} < \lambda < 550\text{nm}$) or $= 10^{0.015(\lambda-550)}$ (for $550\text{nm} < \lambda < 700\text{nm}$).

This MPE formula can be used to calculate the maximum safe luminance of a display by considering the contributions from small bands across the emission spectrum and transforming these contributions from radiometric into photometric units using the CIE luminous efficiency curve (explained in Chapter 2). The calculation is much easier for a monochromatic display and it serves just as well to illustrate the point. If V_λ is the normalised luminous efficiency that has a maximum value of unity at 550nm, the maximum safe luminance of a small area display is $618.8\ C_3\ V_\lambda / \alpha$ cd m^{-2}. This curve is plotted in Figure 9.12.

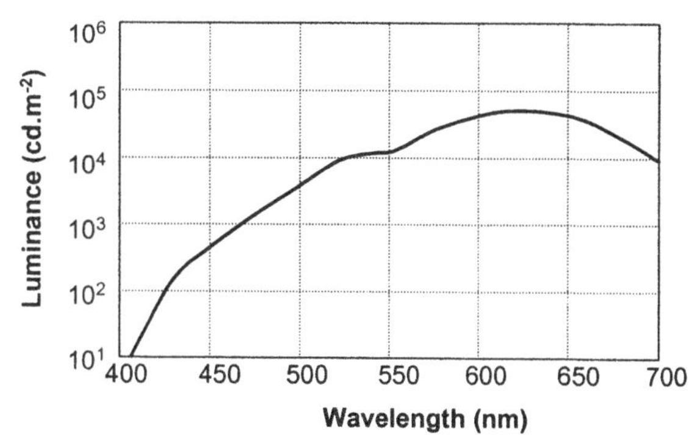

Figure 9.12 Maximum luminance of a 30° FoV monochromatic display complying with BS-EN 60825-1:1994

Several factors are apparent in Figure 9.12. Firstly, the risk of damage is greatest when the eye is exposed to short wavelengths. This "blue light hazard" reflects the vulnerability of retinal structures to photo-chemical degradation from photons of high energy. Secondly, the safest monochromatic display emits light between about 600nm and 650nm where the eye is on balance most sensitive and least vulnerable. Thirdly, the maximum imposed by the standard would not preclude the design of a very high luminance display, particularly one producing a yellow-orange image.

The stringent limit recommended by BS-EN 60825 arises because the standard is intended to cover the safety of laser products that may inadvertently irradiate the eye rather than display devices that deliberately illuminate the eye. The formulae quoted above take account of diffusely reflected laser beams that behave like relatively extended, non-coherent sources, a condition that corresponds best with a display. However, the standard states most explicitly that under the conditions in which a laser product is used the pupil could be as large as 7mm diameter, and the retinal irradiance is based on this assumption. In the case of a scanned laser display, such as the virtual retinal display (VRD), described in section 9.5.5, the greatest risk would seem to arise if the beam scanning mechanism were to fail. In this eventuality the energy that would normally be spread over the retinal area for the whole field-of-view would suddenly fall on the area occupied by a single pixel. Somewhat surprisingly, if scan failure were detected and the beam quenched within 250msec, or the user were to shut his eyes equally rapidly, the exposure would probably be less than the dose from viewing the display at peak brightness for eight hours.

Other standards bodies, such as the International Commission of Non-Ionizing Radiation Protection (ICNIRP) and the Gezondheidsraad, the Health Council of the Netherlands, would allow higher luminance than BS-60825. In this instance the recommendations of the American Conference of Governmental Industrial Hygenists (ACGIH), which is the principal, relevant body in the USA, are the same as the ICNIRP recommendations. However, no standard covers exposures of more than 24 hours, and designers of high brightness displays that are intended to be used for weeks and years on end recognise that devices should be kept well within the recommended limits.

9.4 Summary of HMD requirements

As no practical HMD can project perfect images, or avoid affecting normal external vision, the process of designing a system invariably begins with the user and supplier agreeing upon a set of requirements. The usual range of attributes, and the methods used to check that a system meets the requirements, are summarised in Table 9.1. The list is fairly extensive, and this is just to ensure that the visual qualities of the displayed images and the view through the display optics are satisfactory. These are only part of the problem facing the design team, and they combine with other ergonomic, safety, protection, life support and control requirements in any practical configuration, as considered in more detail in Chapter 13. It should be noted that all the requirements depend greatly on the equipment usage. The "typical" quantitative values for each attribute represent targets that could be set for the next generation of systems, and they are rather more challenging than those set for programmes such as Eurofighter and JSF. For instance these programmes require fields of view of 40° by 30° and 53° by 30° respectively. Developers use a mixture of general purpose and bespoke test equipment to assess characteristics[8]. When delivered, systems are usually checked and maintained by operational units using robust, automated test equipment.

Table 9.1 Summary of the required attributes of HMD systems

	Attribute	Typical requirement	Typical test method
A1	Image field-of-view : per eye	60° horizontal by 45° vertical	Assessed comprehensively using a viewing telescope that can be rotated in elevation and azimuth about the centre of a 5mm sampling pupil. The angular image boundaries are plotted with this sampling pupil set in a range of locations within the exit pupil.
A2	Image field-of-view : total	90° horizontal by 45° vertical	Obtained by combining measurements obtained in A1
A3	Optical exit pupil : size	>15mm (All image qualities must be maintained for a 5mm eye placed within a circle of this diameter)	Assessed comprehensively by measuring A1, A2, A6....A15 using a 5mm sampling pupil displaced selectively from the optical axes.
A4	Optical exit pupil : position adjustment range	27-38mm laterally and ±3mm vertically for each eye independently relative to the helmet datum	Usually assessed in the dark by displaying a uniform field, placing small diffusing screens at the design eye locations and measuring the range of movement of the luminous patches.
A5	Eye relief	>22mm	System is fitting to testers and adjusted to satisfy A1 and A2. The distance from the cornea to the nearest

	Attribute	Typical requirement	Typical test method
			optical surface is measured photogrametrically.
A6	Image resolution (See Appendix B)	(a) target features of 1.0 cycles/mrad must be visible, (b) equivalent to Snellen acuity of 20/60 (c) MTF > 10% at 0.2cycles/min (d) resolve all 1240 by 1028 pixels.	Tests (a) and (b) require the display of a test pattern (e.g. the USAF horizontal & vertical diminishing 3-bar, or Snellen diminishing characters) which allow a set of testers to judge the minimum size of a resolvable feature. Test (c) requires an objective instrument to measure the contrast of a sinusoidal grating pattern of adjustable spatial frequency. Tests usually check for variations over the field-of-view and the exit pupil, at different brightness and contrast settings. Tests can also be done with dynamically moving imagery.
A7	Image contrast ratio, and grey shade rendition (see Appendix B)	Contrast >100:1 Dynamic range >14 grey shades. Both against a black background	A spot photometer is used to measure the luminance of black areas relative to white areas in a simple test pattern. A grey scale test pattern is displayed and the luminance of the stepped increments measured, usually across the field and with different brightness and contrast settings.
A8	Image colour gamut (see chapter 7)	Encompass the triangle formed by:- green [0.12, 0.65] blue [0.17, 0.18] red [0.46, 0.53] [u',v'] co-ordinates in CIE (1976) colour space	A spot spectro-photometer is used to measure the chromaticity of small areas of saturated colour at different grey shade levels in a colour test pattern. The variation over the field-of-view, and with different luminance settings, is assessed.
A9	Image colour registration	To maintain A7 & A8	This is usually assessed by testers observing a colour test pattern containing vertical and horizontal line features of saturated and mixed colours. The colour appearance should be as intended, and should not be affected by moving the features across the field, or by head motion.
A10	Image luminance	<20% variation over the field and continuously adjustable between 0 and 2500 cd/m^2, maintaining colour balance	A spot spectro-photometer is used to measure the luminance and chromaticity of sample areas over the field-of-view.
A11	Image geometric distortion	(a) pixel misplacement < 1mrad vertically and 2mrad horizontally (b) <2% combination of barrel, pincushion, magnification, skew etc	Assessed objectively by measuring the elevation and azimuth angles of a grid test pattern with a sighting telescope as in A1. Can be assessed by testers viewing the virtual grid test pattern superimposed on a real test grid at the same apparent distance.
A12	Image focus distance	Optical infinity on axis +0, -0.01D, at edge of field +0, -0.06D	Using a sighting telescope or a dioptrescope (a telescope with minimal depth of focus, and therefore needs precise focussing, and a calibrated focus adjustment).
A13	Alignment of binocular images	The optical axes must align within 1.0 mrad vertically and +0.5, -1.5 mrad horizontally. Edge error from image rotation <1 mrad V or H	Measured using a sighting telescope, as in A1, which is moved accurately between the two eyepieces. Can be assessed by testers viewing the left and right virtual grid test patterns superimposed on a real test grid at the same apparent distance.
A14	Image artefacts	Speckle, flicker, jitter, smear, stray light, noise, pixelation, raster structure etc must not be noticeable. Ghost images must be <5% main image luminance	These are usually assessed by a panel of testers observing test imagery in a variety of relevant conditions. The intensity of any ghost image is measured using a spot photometer.
A15	Differences between left and right images	A10 & A13 difference <10% A12 difference <25% A11: differences in pixel	These are assessed from the measured attributes of the left and right images.

	Attribute	Typical requirement	Typical test method
		misplacement < 1mrad vertically and 2mrad horizontally	
A16	Image gamma	Switchable between 1 & 2.4	The facility should be included in the display electronic unit. Performance is checked as in A7.
A17	Image stability	Maintain image qualities indefinitely after <1min warm-up period	Checked by repeating measurements over a long term.
A18	Addressing modes	Raster only	Defined by electronic design.
A19	Refresh rate	50Hz	Usually defined by input video standard e.g. 1250line HDTV
A20	Tolerable attenuation and colouration of external vision	Transmission > 75% between 400 and 700nm Discolouration < 0.03 chromaticity units in CIE 1976 colour space.	Assessed using a spectroradiometer viewing an external source from the design eye location over a range of visual angles. Changes to colour discrimination can be assessed by testers examining standard (e.g.Munsell) hue swatches.
A21	Tolerable obstruction of external vision	No obstructions within frontal region (e.g. within 40° of axis). Within remaining visual envelope, monocular obstructions are acceptable if < 4 in number and <0.05sr in size.	Checked by testers wearing the device while looking at a 'perimeter' (a hemispherical screen marked with visual angles) set to align with each eye in turn. The boundaries of obstructions are mapped onto the perimeter and quantified.
A22	Tolerable prismatic distortion of external vision (see Chapter 5)	< 0.1D in frontal region (e.g. 40° of axis), and < 0.2D vertical/ 0.3D horizontal elsewhere. Inter-ocular differences <0.18D.	Usually assessed over a wide range of angles by shining a laser outwards onto a measuring screen from the design eye position, noting the shift introduced by the optic.
A23	Tolerable refractive distortion of external vision	Sphere & cylinder (astigmatism) < 0.1D in frontal region (e.g. 40° of axis) and < 0.2D elsewhere	Usually assessed over a wide range of angles using a telescope or dioptrescope with a ~5mm sampling pupil at the design eye position observing a distant 'spoke wheel' target containing a spread of radial lines. The eyepiece is adjusted to focus on the spokes in turn. The necessary range of dioptre values gives the cylindrical power and alignment. The mean value gives the spherical power.
A24	Tolerable polarisation & optical activity	The legibility of the HUD and HDDs should not be affected by viewing through the HMD optics	Usually checked by completing the HUD/HDD legibility tests with the testers wearing the (inactive) HMD, using a variety of head orientations.
A25	Tolerable glare and haze (see Chapter 5)	Haze < 2% Glare < 0.02 $cd.m^{-2}$/lux	Haze, the loss of contrast of the scene viewed through the display optics, is measured using a spot photometer and a high contrast (black/white) target. Glare is assessed using a photometer at the eye position measuring the apparent increase in the forward external luminance with a beam of known luminous intensity pointed at the display optics from a variety of directions.

9.5 Example electro-optical configurations

The optical systems that have found favour among designers of HMD systems are similar in principle to those employed in HUD systems. A few early experimental HMD systems, and several intended for use by ground forces rather than aircrew, have been based on scaled-down versions of conventional HUD optics mounted on the helmet brow. In these, the light from the downward-facing CRT or miniature electro-luminescent matrix has been collimated by a lens, with a focal length of about 40mm and a diameter of about 30mm, and reflected into the eye from a flat semi-reflecting combiner plate. However, almost all the practical avionic designs use a spherical, or nearly spherical mirror to form the distant virtual image, and are versions of the configurations investigated for HUDs illustrated in Figure 8.9. The off-axis design shown in Figure 8.9(d) is a particularly attractive way of making use of the helmet visor.

9.5.1 A simple helmet-mounted sight

The principal requirements for this application have been to provide a simple, monocular, bright, sharp reticule image over an adequately large exit pupil, and allow adjustment for the position of this pupil to suit the helmet wearer.

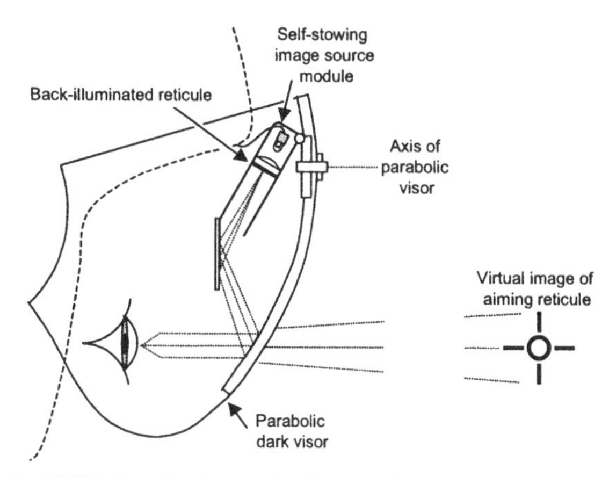

Figure 9.13 A sketch of the VTAS reticule projection optics

Fig 9.13 shows how these requirements were met in the Honeywell VTAS reticule projector by using a dark visor as an off-axis collimating combiner with a back-illuminated aiming reticule housed in a small tube attached to a pivot at the parabola axis. The parabolic shape of the visor removed most of the spherical aberration. The tube was hinged so that it folded into the gap between the helmet brow and the visor when the latter was raised, and the pivot could be rotated about the parabola axis to direct image-forming light towards either eye. In this case the narrow field-of-view of the image obviated the need for additional optical correction, and no coating was needed on the inner visor surface because the image was viewed against an outside scene attenuated to about 15% by the dye in the visor. The ghost image of the reticule reflected from the outer surface was attenuated to 15% of 15% by the double passage through the visor; and had about 2% of the brightness of the intended reticule. It was too dim to be noticed.

Many of the designs for a narrow field monocular helmet-mounted display system for use in combat jets have used the visor as a collimating combiner, but unlike the VTAS these have been required for flying in conditions that include poor light. This has mandated the adoption of a clear polycarbonate visor with a narrow band multi-layer reflective coating on the inner surface, and spherical or toroidal shape to enable co-axial mounting of an outer dark visor of slightly greater vertical radius. Typical systems incorporate a single miniature CRT addressed cursively for adequate brightness. Such a configuration has been chosen by Vision Systems International for the Joint Helmet-mounted Cueing System (JHMCS) that will be fitted to F-15, F-16 and F/A18 aircraft in US services. In this case the display, including the visor, has been packaged as a module that can be interchanged with a "night" module incorporating image intensifiers and an image source.

9.5.2 IHADDS

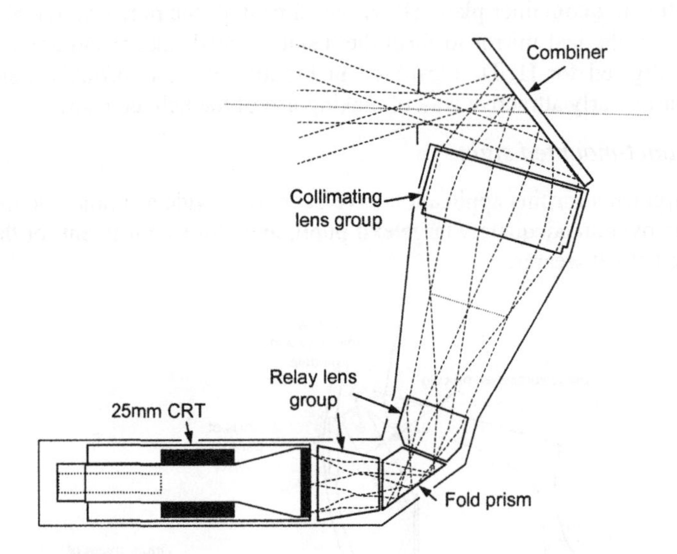

Figure 9.14 Sketch of the optical arrangement of the Integrated Helmet And Display Sight System (IHADSS)

As shown in Figure 9.14, Honeywell engineers created a design for the HMD to be used in the AH-64 Apache that was similar in concept to a conventional refractive HUD, but with a relay to stretch the optical path and project the porthole[9]. It employed a 25mm diameter CRT and rotationally symmetrical lens groups with a flat combiner to give a 40° by 30° field-of-view. The long cylinder containing the CRT was attached to the lower right of the helmet shell by a quickly-detachable coupling that also enabled adjustment for the small exit pupil position to coincide accurately with the wearer's eye entrance pupil. This robust design has been in service with little modification for more than 25 years.

9.5.3 Binocular HMD

Figure 9.15 illustrates several ways that optical designers have produced clear, sharp binocular images using miniature monochromatic CRT image sources, as well as meet the other requirements listed at length in Table 9.1.

The first example, also from Honeywell, is mounted low down on the side of the helmet shell using adjustable clamps in much the same way as the IHADSS device[10]. The "tilted catadioptric" eyepiece is a variant of the axial spherical mirror configuration shown in Figure 8.9 (c), and has semi-reflecting multi-layer coatings tuned to the narrow CRT emission, set slightly away from the 45° and 90°angles. This small tilt brings four small but worthwhile benefits; an increased proportion of image-forming light reflected initially by the flat mirror, an increase to the amount transmitted after reflection from the spherical mirror, an overall increase in the amount of light reaching the eye from the external scene, and slightly greater eye relief. The main detraction is the complicated relay lens group that uses several canted, eccentric elements to correct the off-axis aberrations introduced by the tilt.

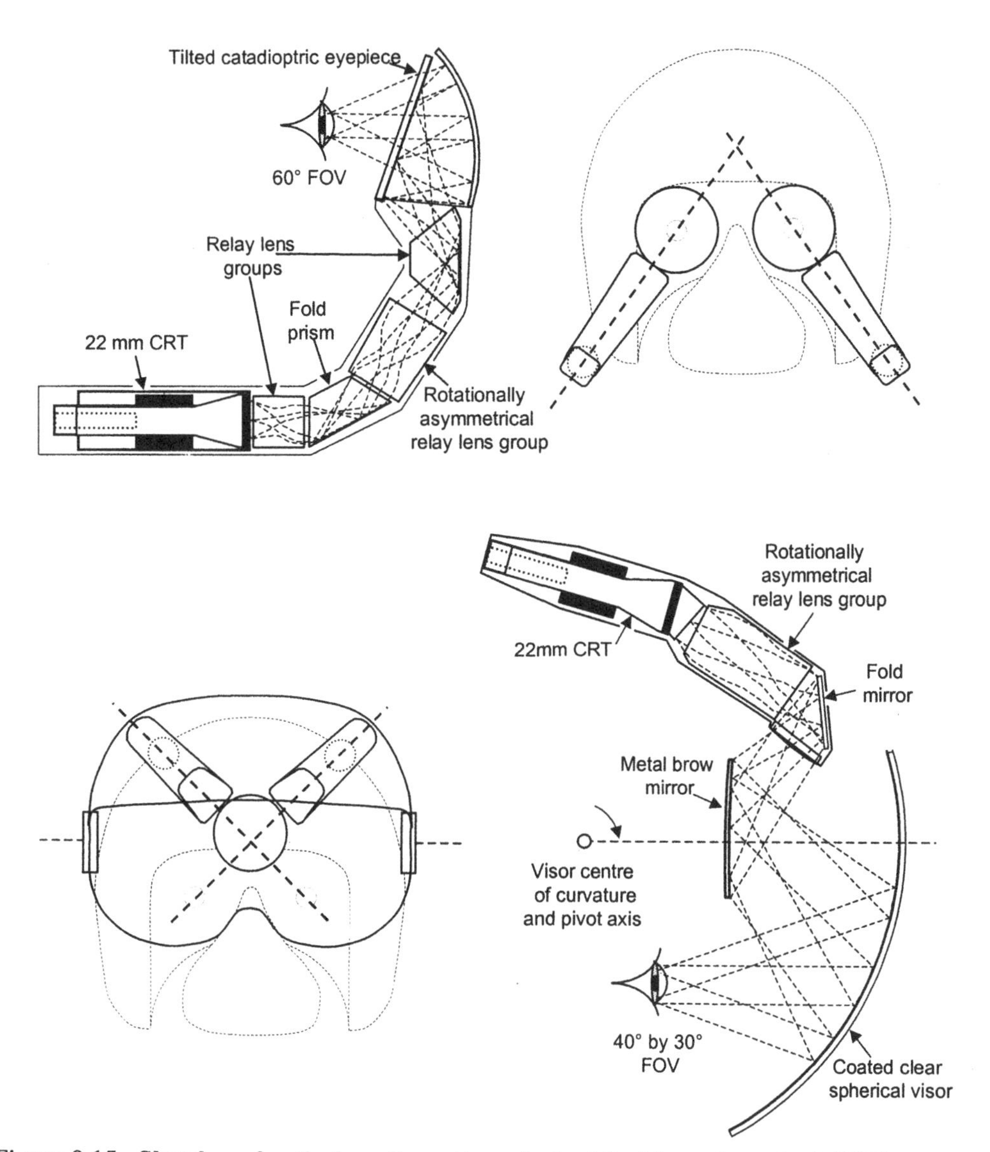

Figure 9.15 Sketches of optical configurations devised for binocular avionic HMD systems. The upper diagram is an arrangement designed by Honeywell that employed a tilted catadioptric eye-piece and a complex rotationally asymmetrical group of relay lenses. The lower diagram shows the configuration devised by GEC (now BAE Systems) for the Viper series of HMDs

A second example, shown in the lower half of Figure 9.15, uses a clear visor as an off-axis spherical mirror, with the inner visor surface coated with a broadband metallic film to reflect about 30% of the image-forming light[11]. This introduces little discolouration and attenuation, so the configuration allows almost unrestricted external vision, which is particularly important for daylight flying. The symmetry of the arrangement is exploited by packaging the two channels so that only one brow mirror is needed. This is done by directing the light from the CRT and relay lens assembly on the left temple so that it is reflected from the brow mirror, then onto the visor and into the right eye. The right electro-optical unit supplies light to the left eye by a matching arrangement. The visor is pivoted around a lateral axis that intersects the visor centre of curvature, so the quality of the images is unaffected by the rotational position of the visor around the pivot axis. As with the Honeywell design, the penalty is borne by the manufacturing complexity of the relay lens assembly in which the majority of the elements are de-centred and tilted. The most evident detractions are the proximity the brow mirror to the forehead, the need for a strong but precise scan manipulation to counter the off-axis geometrical image distortion and the complexity of the means of adjusting the position of the exit pupils to cater for a range of inter-ocular distances. As the adjustment is done by rotating the CRT-relay lens assemblies around the visor forward axis, the images undergo a small rotation and this requires further adjustment to the form of the distortion correction.

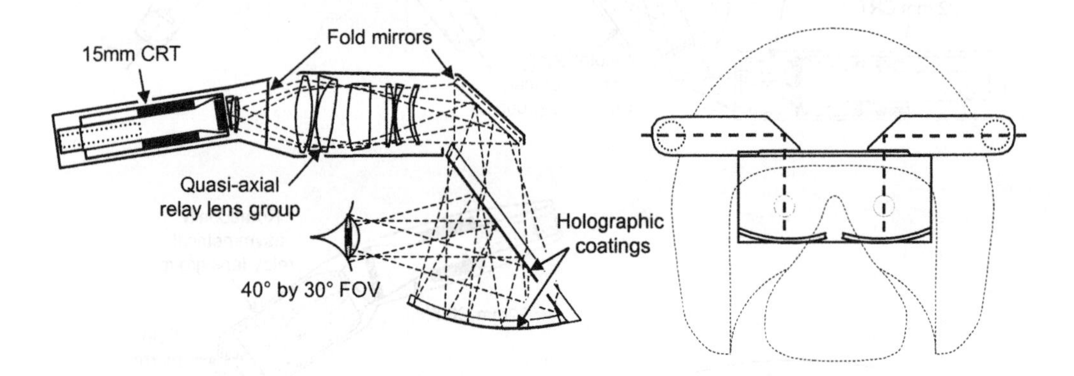

Figure 9.16 Experimental HMD developed at RAE Farnborough (reprinted by permission from QinetiQ)

A third system, illustrated in Figure 9.16, was devised by John Banbury[12] at RAE Farnborough and built by Pilkington Optronics in the mid 1990s. The configuration was similar to the "Z-HUD", described in Chapter 8, and was slightly off-axis for the same reasons as the Honeywell tilted catadioptric system shown in Figure 9.15. The arrangement employed small (15mm diameter) CRTs and a compact relay assembly for lightness; the mass of the whole image-forming assembly was less than 280g. The projected images were well resolved and the flat combiner gave a clear external view. In this case the flat combiner and the concave mirror were made in polycarbonate to obviate the weight of an additional blast visor, and the tuned reflective coatings for the image-forming light were formed within spin-coated photo-polymers as conformal holographic optical elements, as explained in Chapter 8. The ratio of the luminance of the projected image to the luminance of the image formed on the CRT was about 30%. This unusually high optical efficiency produced images with more than 6000cd.m^{-2} luminance without excessive tube excitation.

9.5.4 Scanned laser HMD

An alternative way to solve to the problem of providing a bright virtual image using lightweight, head-mounted, electro-optics was called the Virtual Retinal Display (VRD) by its inventors[13].

Figure 9.16 shows a simplified scheme for combining the beams from three independent red, green and blue solid state lasers to form a single beam that is scanned and reflected into the eye. The crucial element is the scanner that deflects the beam to draw horizontal lines. This must operate at the line scan frequency of the video signal, which would typically be about 30kHz for the asymmetric saw-tooth waveform used in a CRT, but is halved to 15kHz if the scan is symmetrical and the beam intensity is modulated during both directions of deflection. The line scanner developed for the VRD employs this principle and is in effect a miniature galvanometer in which a mirror, about 2mm square, is formed on a short torsional spring tuned to resonate at half the line scan frequency. Stable resonance is maintained by a feedback oscillator driving a miniature magnetic coil. The vertical scan, at 60Hz, uses a similar galvanometer, driven with a saw-tooth waveform to produce slower, linear deflection down the frame, and the beam is reflected back onto the line scanner to double the horizontal deflection angle.

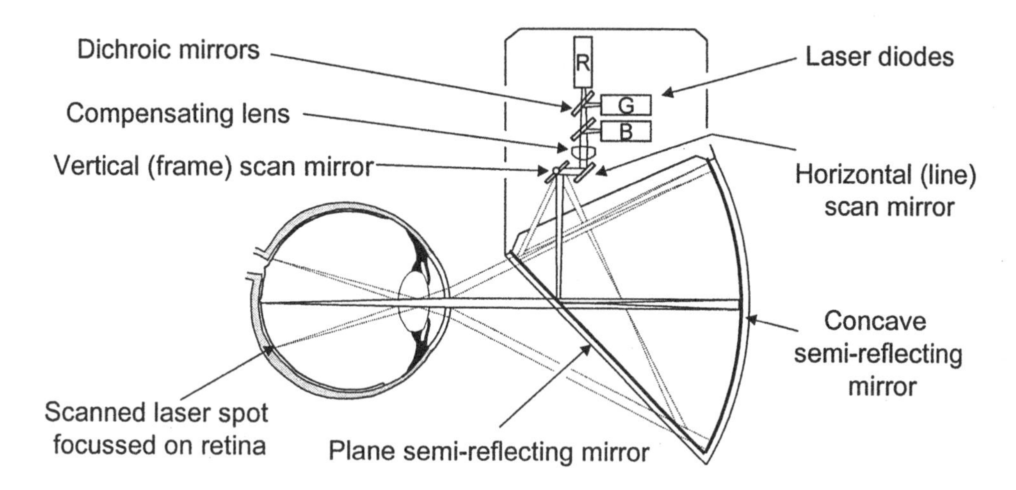

Figure 9.17 A simplified scheme for scanning a set of co-axial red, green and blue laser beams to create a bright colour image

In the arrangement illustrated, the concave mirror is concentric with the middle of the eye pupil, and the plane semi-reflecting mirror folds the optical path to put the pair of scanning mirrors optically at the eye pupil. The scanned beam hits the concave mirror normally, is reflected back through the plane mirror and enters the eye pupil to form an image on the retina. The compensating lens is needed to diverge the beam incident on the concave mirror so that the reflection is parallel and the image collimated. The design is effectively the reverse of an opthalmoscope, which is used to observe the retina through the eye pupil, and it can also be regarded as a version of that shown in Figure 8.9(c), where an intermediate real image is formed at the focus of the concave mirror.

The scanned laser display of this form was expected to have three exceptionally qualities; an excellent colour gamut, because the primaries were saturated spectral lines produced by lasers; excellent grey scale rendition because the lasers could be modulated between a genuine zero and a high peak output; and excellent brightness because almost all the light impinging on the cornea would enter the eye. It was also hoped that the device could be simple and compact. However, several of these advantages are lost in a practical implementation. With a beam as narrow as that shown in Figure 9.17, rotation of the eye to look towards the edge of the field-of-view would swing the eye pupil away from the beam convergence point, so that no light would enter the eye and no image would be seen. Although the originators proposed to measure the position of the pupil and shift the convergence point dynamically, it has been easier to create a sizable exit pupil to accommodate the shift between the optical axis and the eye pupil, as illustrated in Figure 9.4. Efficiency is then no better than a conventional eyepiece.

A practical scanned laser display based upon a resonant line scanner must incorporate a system that corrects the image distortion and brightness variations due to the sinusoidal deflection. Without correction, the beam would be stretched and dimmed when moving most rapidly at the centre of the line, and compressed and brightened at the end where the swing is reversed. It is also necessary to incorporate a reliable and fast means of quenching the beam excitation should either scanner fail. For instance, if one scanner were to fail the eye would be presented with a line image about 600 times more intense than the normal image, and should both scanners stop the image would be a spot about 250,000 times more intense. As noted in section 9.3.12, the effect need not be dangerous, but it would certainly be dazzling.

Creating an image by scanning a spot of light across the retina relies on the sensation of brightness remaining proportional to the time average irradiance. Fortunately, even with a display that contains about a million resolvable pixels and which illuminates each resolving retinal area for one millionth of a frame period during each frame period, this proportionality holds. The excitation follows Bloch's Law of reciprocity between pulse duration and intensity. However, the long interval between successive bursts of stimulation can create an anomaly that passes unnoticed with other forms of head-mounted projector in which the images, although periodically refreshed, have some persistence. With a scanned laser display, rapid movements of the head can cause large areas of the image to lose their solid coherence, and small features such as text characters can become illegible. It is suggested that such disruptions arise because the retinal receptive areas shift between bursts of excitation and this fragmentary stimulation is not accommodated by the normal perceptual processing that corrects for eye drift and other micro-movements during fixation. The most practical way to avoid this defect is to decrease the interval between stimulation instances by increasing the frame rate significantly.

9.5.5 The pancake window

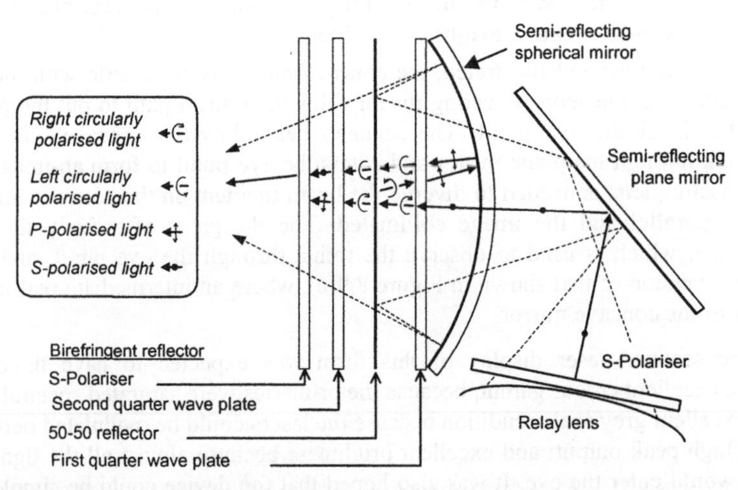

Figure 9.18 Principle of the pancake window

One of the most ingenious helmet-mounted display optics was devised by the Farrand Optical Company[14]. As illustrated in Figure 9.18, this arrangement used an on-axis, partially transmissive, concave, collimation mirror in conjunction with a thin multilayered birefringent reflector rather than the usual inclined plane beamsplitter. Image-forming light from the display screen passed through the relay lens group and a linear polariser to form an aerial image in the plane of focus of the concave mirror as reflected in the birefringent mirror. The portion transmitted by the concave mirror then encountered a quarter-wave plate, which (as described in Appendix A) transformed it into right hand circularly polarised light, some of which was reflected and some transmitted by the partial reflector. The transmitted part entered the second quarter wave plate - having a perpendicular director axis - where it was transformed into linearly polarised light of the opposite alignment. This light was therefore absorbed by the second polariser. The reflected part underwent a 180° phase shift, became circularly polarised with the opposing sense of rotation and on returning through the first quarter wave plate became p-polarised. Some of this was reflected from the concave mirror and was thereafter transmitted by the birefringent multi-layers. This apparently tortuous scheme ensured that only the collimated light entered the viewer's eye.

Although very inefficient, the concave mirror could be placed significantly closer to the eye than was possible using an interposed inclined beamsplitter, and this allowed the designer to widen the field-of-view without compromising the exit pupil and eye relief. Several binocular devices have employed the principle, most notably the 80° by 120° fibre-optic head-mounted electronic display (FOHMED) by CAE Simulation, in which the output from three cockpit-mounted, high luminance red, green and blue CRTs are combined and coupled to the head-mounted eyepiece using a coherent fibre-optical bundle. The arrangement has not as yet been employed in flightworthy equipment.

9.6 Implications for HMD system design

The HMD is a system that includes a symbol generator, electronics to drive the image source, the image source itself, optics to project the virtual image and the mechanical means of mounting the components on the helmet.

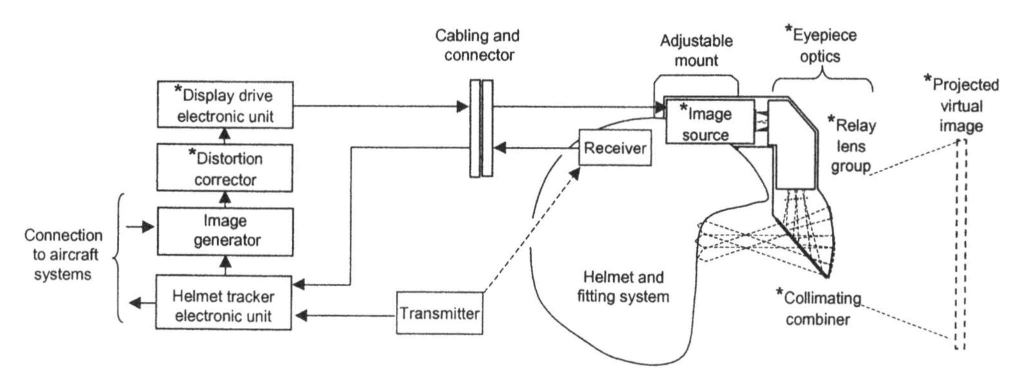

Figure 9.19 Elements of a HMD system

Figure 9.19 illustrates the linkages between the elements. The helmet tracker should be regarded as an integral component, and the elements marked with an asterisk would be duplicated for a binocular system. As well as optimising these components, the designers need to take account of the signalling delays that can contribute to the system latency, the in-service maintenance of the system and both the number and nature of the wires joining the helmet to the aircraft. The latter is not a minor problem because a crewmember cannot make a rapid exit from the aircraft on the ground without an easy, rapid means for disconnecting or removing the helmet. In a combat jet fitted with an ejection seat the disconnection must be part of the automatic sequence should the seat be ejected in flight. The high anode voltage required for CRT devices give particular problems. The intricacies of the helmet mounting and automatic cable disconnection are discussed in Chapter 13, and alternative helmet tracker techniques are discussed in Chapter 11.

Existing systems have employed a relatively narrow variety of miniature image sources. Back-illuminated reticules and miniature fixed-format LED arrays are most often used in simple monocular HMS systems, while the miniature CRT has been the mainstay of the HMD. This device has three remarkably useful qualities. Firstly it has the versatility to provide bright stroke-written images in daylight and dimmer raster-structured images at night. Secondly, the face-plate emits light in directions that match the acceptance cone of the optics, and thirdly the shape of the image can be adjusted relatively easily to counter the inevitable geometric distortions produced by the optics.

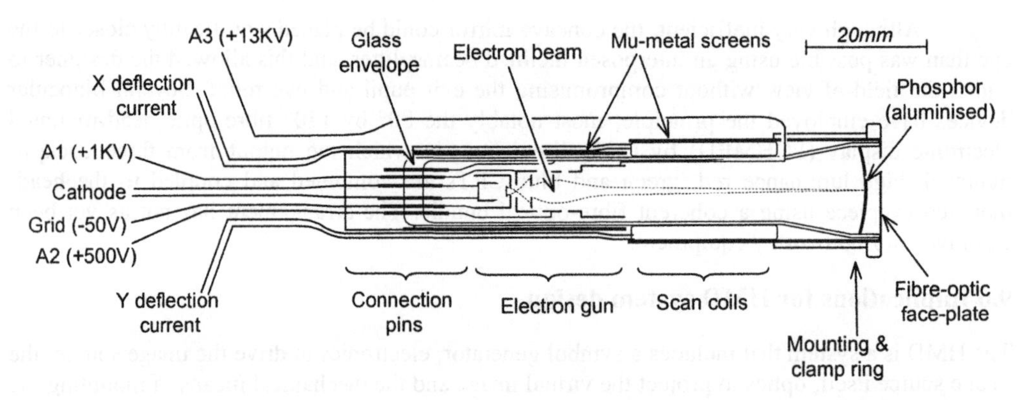

Figure 9.20 Packaging of a miniature CRT (Reprinted with permission from QinetiQ)

The miniature CRT shown sectioned in Figure 9.20 has a screen of 18mm diameter and is largely symmetrical about the long axis. The electron gun has a dispenser cathode and is similar in size and geometry to the gun employed in a larger avionic tube, but optimised to produce a bright spot of about 15μm diameter on the phosphor screen at a closer focus of about 40mm. The whole glass envelope and scan coils have a circular cross-section and are potted into outer cylindrical sleeves made from mu-metal for robustness and electro-magnetic shielding. The mass of the resulting package is between 50g and 100g. In some devices the face-plates are made from optical-quality glass, but a fibre-optic face-plate with a concave inner phosphor-coated surface is commonly used for accurate focusing of the electron beam over the whole area. It should be noted that CRT devices of this nature are made in comparatively small batches and can be tailored precisely to the application, for instance the exterior fibre-optic screen surface can be suitably concave or angled to match the requirements of the downstream optics. However, with the inherent variability of such small scale production, the manufacturer must carry out extensive tests on each device, and the scan sensitivity and voltage-luminance characteristics are given to the manufacturer of the HMD for setting the electronic drive circuitry to produce consistent image quality. Some CRTs are therefore supplied with the data in a characterisation PROM (programmable read-only memory), and the HMD drive electronics includes a facility to read and act on the parameters contained in this memory chip. Whenever a CRT is changed the maintainer must also ensure that the appropriate PROM is changed.

The high anode voltages, the requirement for safe disconnection and the variability between devices are certainly detractions. The device also has a limited service life, not usually resulting in catastrophic failure but more from a gradual deterioration of image brightness as a result of micro-structural changes in the phosphor. For instance, local overload can produce burns that cease entirely to emit image-forming light. These factors, combined with the drive to minimise headgear mass and supply coloured images, have prompted designers to consider alternative image sources. In principle any of the flat panel display devices described in Chapter 7 that can be manufactured with pixels of about 20μm and a screen of about 20mm may be suitable.

A wide variety of experimental flat panel HMD systems have been built, but to date only a few have been put into production. Rather than aviation, the main driver has been commercial and personal uses, such as "virtual reality" (VR), computer interaction, computer games and as a compact portable way of enabling a viewer to see a TV programme or replay a video recording. Most of these applications need a system that is lightweight, compact, affordable and binocular, and supplies coloured imagery at moderate brightness. The resolution is usually governed by the video standard, say 600 by 800 pixels for computer interaction and high quality conventional TV, spread over a field-of-view which allows easy viewing, typically 20° or 30°. Systems intended for high quality VR, such as architectural visualisation and mathematical modelling, require a larger visual field, in the region of 60°, and the higher pixel count available from computer-graphic workstations, such as the 1024 by 1028 SXGA standard. The images must also have sufficient geometric accuracy and focus constancy for stereoscopy. However, in all of these applications the user wants the competing view of the world to be masked. The designer can therefore choose from a wider range of optical configurations because there is no need to provide uncoloured, unobstructed, undistorted see-through, and it is possible to use relatively dim, glare-sensitive image sources because the optics act as a shield against ambient light.

The simplest approach involves the use of a miniature transmissive AMLCDs with an intense backlight, as the image formed on the outer glass surface can be substituted directly for the phosphor screen of a CRT. An intense backlight is needed because a fair proportion of the light is absorbed in the polariser and in the non-transmissive mask that covers the conductors and control transistor alongside each pixel. To maintain contrast it is also necessary for the projection optics to accept light that has passed perpendicularly through the polariser, cell and analyser, and reject obliquely transmitted light. The other current technological approach is to use a reflective ferro-electric LCD array with a front illuminator. In this case the panel itself is more efficient because a larger proportion of the cell area, typically 80% rather than about 40%, contributes to image formation. Several systems have been constructed using the reflective illumination schemes illustrated in Figure 7.14 packaged into a module similar in size to a miniature CRT, enabling the designers to retain the downstream optical configuration intended for a CRT.

Any of the techniques for producing colour images on head-down displays, as summarised in Chapter 7, can be used with helmet-mounted image sources, providing the device can be made small and the pixels shrunk proportionally. Miniature versions of a Penetron CRT have produced red/green imagery, and a sequential RGB colour shutters has been used with a white phosphor CRT. However, with both of these approaches the image is seen as a displaced pair of component colour images when the viewer moves his head rapidly. This anomaly, combined with the inherent complexity of the system and the other detractions listed above for monochromatic CRTs, has only added impetus to the search for flat panel substitutes.

The most ubiquitous commercial head-mounted displays have used miniature AMLCD devices with triads of printed RGB filters, and several avionic HMD systems have been designed using the same commercially available matrices. The problem of providing sufficient back-illumination has been tackled using airframe-mounted light sources, for instance a set of red, green and blue lasers each producing several watts of power, the output from which can be coupled to the helmet using a flexible fibre-optic bundle. High power (~500W) discharge lamps have also been used with appropriate optics to focus the flux into the fibre bundle. In principle, almost any passive flat panel described in Chapter 7 can be used to provide an intense image using such illumination, particularly the reflective FE-LCD device with triad cover-filters, the subtractive stack and the set of separate RGB digital micro-mirror devices. However it must be admitted that the bulk and complexity of the resulting systems, and the difficulty of disconnection, somewhat defeats the incentive for employing a flat panel.

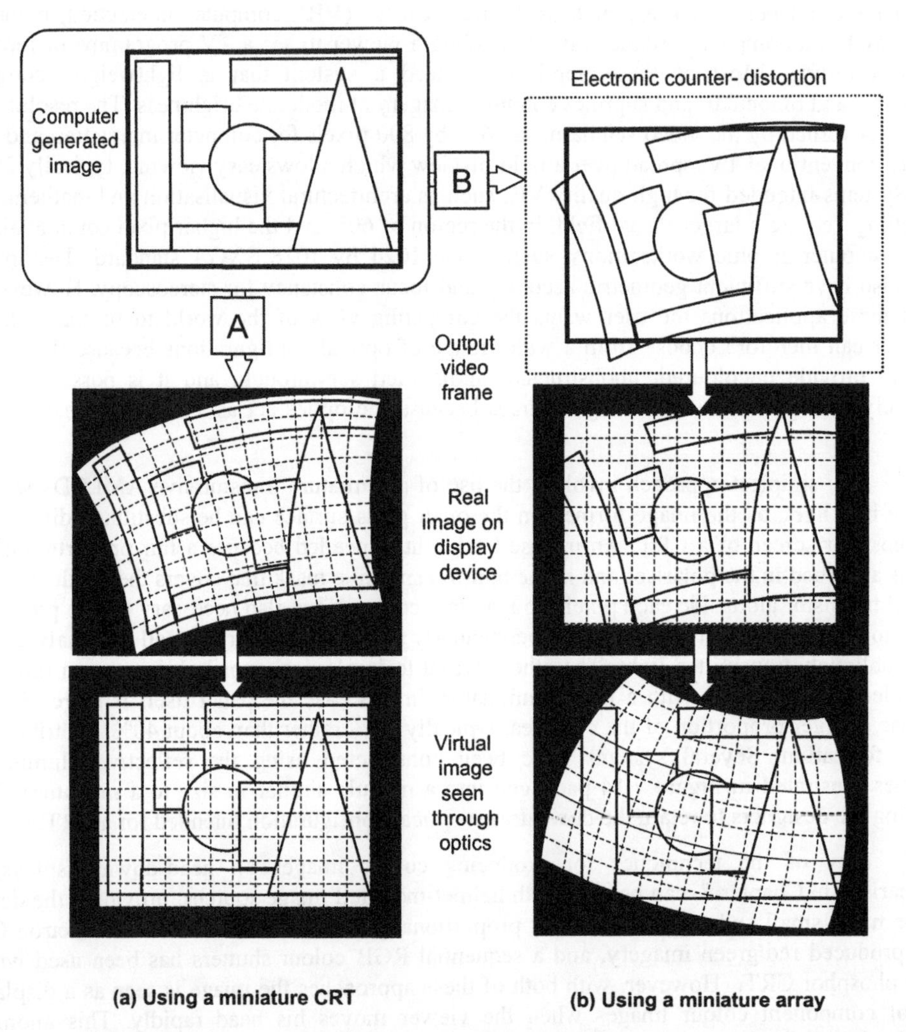

Figure 9.21 Illustrating the "warping" manipulations required to counter optical image distortion

As noted in Chapter 6, section 6.4.3, the computer that generates images for a binocular HMD needs a particularly powerful architecture. Not only must the machine import and re-map sensor images, construct overlaid symbols stabilised to several frames of reference and compute the output for projection onto left and right logical display planes, but it may also be required to counter optically-induced distortions. The last requirement depends on the type of image source employed in the HMD system. Figure 9.21 illustrates the difference in the approach for (a) a miniature CRT and (b) a miniature flat panel device such as an AMLCD. With the former, the computer-constructed image can be sent straight to the CRT drive electronics as a video signal, and the synchronised scan current waveforms for the horizontal and vertical beam deflection are manipulated to change the shape of the raster pattern. Although needing great precision and stability, the necessary pre-distortion of the image on the tube face can be treated as an extension of the need to counter the inherent pincushion produced by magnetic deflection.

However, it is not feasible to pre-distort a rectangular AMLCD pixel array, and as indicated in Figure 9.21(b), it is instead necessary to counter-distort the content of the image displayed on the pixel array. Several methods have been devised, and all involve digital manipulation of the pixels. One employs an extra electronic unit that takes in the computer-generated image, if necessary converting the picture information from an analogue to a digital representation, placing this in an input frame store and either manipulating the picture content prior to read-out or creating a counter-distorted video signal by addressing the appropriate sequence of pixels. The extra electronics introduce multiple conversions and increased throughput delay. It would be better for the originating computer to apply an appropriate counter-distortion when the images are constructed, or have the read-out from frame store of this machine address the pixels in an appropriate counter-distorting sequence. No solution has quite the elegance of CRT raster manipulation; some picture information is lost at the frame edges, and both resolution and brightness are degraded where the image is stretched.

The system latency problem is not easily solved. The most easily implemented counter to is to estimate the magnitude of the delay and construct imagery appropriate to the predicted orientation of the helmet. One scheme estimates the angular velocities to calculate the predicted orientation from differences between successive measurements of helmet orientation. Another measures the angular velocities by attaching miniature inertial sensors to the helmet. The latter is more accurate and less sensitive to measurement noise, but more expensive and complex. Neither approach accommodates variations in the delay, and even perfect prediction would not counter the within-frame distortion. Canceling "head lag" defects completely requires a treatment that repositions individual pixels rather than the whole frame. Two approaches can be envisaged; "dynamic pixel shifting" and "dynamic frame store addressing". The first would shift each pixel as it is displayed, for instance by rapid, fine, adjustment of the beam deflection in a CRT-based system. The second would leave the instantaneously addressed pixel in visual space where displayed, but extract the data for this location in visual space from the correct location in the image frame store. Both approaches require an extremely high rate of helmet orientation measurement to calculate the instantaneous error between the actual and the intended pixel locations.

Dynamic pixel shifting has been investigated as part of the Virtual Cockpit research programme at the Defence Evaluation and Research Agency (DERA) in the UK, using the CRT-based simulation equipment described in Chapter 14. Although it proved possible to eliminate pixel misplacement when the head was turned rapidly, the computation introduced a random jitter that reduced the legibility of the images. In practice it would be necessary to filter the computed error signals, and this would incur a small delay. As the approach is also incompatible with systems that employ miniature flat panel displays, it must be concluded that dynamic frame store addressing offers a better long-term solution.

References

1. Jacobs R S, Triggs T J and Aldrich J W, *Helmet-mounted display/sight system study*, USAF Tech Rep AFFDL-TR-70-83 Vol 1, 1970

2. Behar I, Wiley R W, Levine R R, Rash R R, Walsh D J and Comum R L S, *Visual survey of Apache aviators*, USAARL Report 90-15, 1990

3. Biberman L M and Allusi E A, *Pilot errors involving head-up displays (HUDs), helmet-mounted displays (HMDs) and night vision goggles (NVGs)*, Institute of Defence Analysis, Virginia IDA Paper P-2638 1992
4. Melzer J E and Moffitt K, *Head-mounted displays: designing for the user*, McGraw Hill, NY 1997
5. Kalich M E, Rash C E, van de Pol C, Rowe T L, Lont L M and Peterson R, *Biocular image misalignment tolerance*, in Rash C E and Reese C E (eds) *Helmet- and head-mounted displays VIII: Technologies and applications*, SPIE Vol 5079, 2003
6. Boff K R and Lincoln J E (eds) *Relationship between accommodation and vergence*, Section 1.231 in *Engineering data compendium: human perception and performance*, AAMRL, WPAFB, Oh 1988
7. Sliney D H and Wolbarsht M, *Safety with lasers and other optical sources: a comprehensive handbook*, Plenum Press, NY 1980
8. Speck R P and Herz N E, *The impact of automatic calibration techniques on HMD life cycle costs and sustainable performance*, in Lewandowski R, Haworth L A and Girolamo H J (eds), *Helmet- and head-mounted displays V*, SPIE Vol 4021, 2000
9. Ferrin F J, *An update on optical systems for military head-mounted displays*, in Lewandowski R, Haworth L A and Girolamo H J (eds), *Helmet- and Head-Mounted Displays IV*, SPIE Vol 3689, 1999
10. op. cit.
11. Cameron A A and Steward D G, *The Viper HMD - from design to flight test*, in Lewandowski R, Stephens W and Haworth L A (eds), *Helmet- and Head-Mounted Displays and Symbology Design Requirements*, SPIE Vol 2218, 1994
12. Banbury J R, *Helmet-mounted optical systems*, US Patent No 5,646,783, 1997
13. Furness T A and Kollin J S, *Virtual retinal display*, US Patent No 6,008,781 1999
14. La Russa J, *Infinite optical image-forming apparatus*, US Patent No 27,356 1972

Chapter 10

Auditory Displays

10.1 Rationale

Although the word "display" may seem strange when used to describe systems that make sounds, it is difficult to think of a better alternative that conveys the idea of engineering something to present clearly understandable information.

Early experimenters investigated the use of auditory signals for instrument flying. With surprising boldness, de Florez[1] showed that he could control a Fairchild 24 when blindfolded. He did this by turning left or right in response to the difference in the intensity of the signal fed to his ears, and increasing or decreasing the airspeed in response to a change in pitch. Forbes[2] followed up this effort using a Link training simulator. He devised a three-in-one auditory signal that sounded like the behaviour of the airplane; sweeping a tone repeatedly from left to right to convey a turning cue, using pitch variations as a tilt cue and a variable "-put-put-put-" modulation associated with the motor as an airspeed cue. He also pioneered the idea of a voice annunciator by commissioning the development of a fast-access multi-channel magnetic sound reproducer that carried the spoken messages on the edges of rotating aluminium discs. He concluded that properly designed tones could convey as many as four control indications without interference from, or interference to, occasional spoken communications and automated announcements.

Auditory cues have generally been regarded as a means of reinforcing or complementing visual indications. According to the multiple resources theory[3] outlined in Chapter 2, the areas in the human brain that mediate hearing, logical reasoning and speech form a pathway that is almost independent of those for vision, spatial reasoning and the other forms of muscular activation. An auditory display should therefore provide information without interfering with a concurrent visual task, and can gain access to the mysterious mental mechanism that selects instantaneous intentions. As hearing is spatially limitless the ears can also tell the eyes where to point. These complementary perceptual qualities are the rationale for the siren on an ambulance and the bell on an alarm clock. They have also caused airborne speech communication systems to evolve into a more comprehensive auditory system that warns the pilot and, in newer aircraft types, feeds back speech from the voice command system.

This chapter outlines the design considerations, the arrangement and the limitations of audio systems in combat aircraft. Although there are several helmet-mounted devices for protecting the pilot from excessive levels of cabin noise, as discussed in Chapter 13, in most combat aircraft the resulting level still constitutes a major disturbance. This chapter therefore starts with a short explanation of the deleterious effects of noise on the sounds that the listener wants to hear. It ends by describing some recent work to investigate the value and practicality of introducing cues that make a sound appear to originate from a controllable direction.

10.2 The intelligibility of speech

In almost all aircraft the annotations, abbreviations and acronyms applied to cockpit instruments, switch labels and displays are English in derivation, and so also is most on-board textual material such as flight reference cards and operation manuals. Voice communication between civil aircrew and ground-based controllers employed by national bodies presumes a limited vocabulary of English words. This may seem fortunate for English-only speakers, but spoken English is not ideal for conveying precise information concisely, especially between strangers, not least because words can be spoken with a variety of accents and pronunciations. The language encourages speakers to exploit a wide vocabulary (confirm, verify, repeat, say again.....GMT, UTC, Zulu), many thousands of words are ambiguous (taxi, slot, right....) and there are many homophones that signify unrelated constructs (for eight, four ate).

Aircrew and flight controllers therefore talk to each other using standardised "voice procedures", a sub-language with a minimal vocabulary of short words and phrases that can be articulated clearly. Numbers, which are commonly misheard, are pronounced in an exaggerated way ("wun", "tree", "fife", "sev-en", "niner"), others are carefully distinguished ("wun tree" and "tree zero") and decimals are explicit ("eight decimal fife six"). Unusual words are spelled out using the phonetic alphabet. The conversational interchange is regulated by key words ("roger", "over", "say again") and the listener is encouraged to repeat back what he has heard. All of this is done because meaning is easily lost when the context is unfamiliar, or the message is phrased in jargon or when the speaker or listener is tired and under stress. Physical factors are also important because the received signal can be corrupted by the communication channel and the listener is commonly immersed in a noisy environment.

Loud, unexpected noise is invariably startling, and a constantly noisy environment can be physiologically arousing in the short term but fatiguing in the longer term. The effects on voice communication are complex but the consequences are mainly understood. If both the speaker and the listener are exposed, both risk hearing damage, at least temporarily and perhaps permanently. Continuous noise induces the speaker to shout in an attempt to raise the vocal signal above the noise level, which may require unsustainable effort, while the listener tends to mishear and commonly needs the message to be repeated. Even when the listener receives clearly articulated and amplified speech at a level well above the background noise he is still likely to misinterpret subtle differences, for instance between "six" and "fix".

The reason is best explained with reference to Figure 2.7, which summarises the characteristics of normal speech. This shows that most vocal energy is in the main vowel formants that occupy a band between 300Hz and 1000Hz, while the transient consonant sounds that contribute most to the intelligibility of speech lie between 1000Hz and 3000Hz. As most speakers use a phrasing that produces about 40dB difference between the peak level and pauses, a speech transmission system should have a dynamic range of 40dB and a pass-band that extends from 300Hz to at least 3000Hz. This falls nicely inside the "auditory communication window" of the listener, shown in the right of Figure 2.7. There is little value in transducing signals below about 150Hz or above 6000Hz.

Noise reduces hearing sensitivity to a degree that depends upon the character of the noise. Broadband noise raises the listener's auditory threshold by about 10 to 15dB above the noise level, and the level for reliable identification of words is raised by at least a further 15dB. A similar effect is produced by a narrow band of noise, for instance the whine of a gearbox, but this "masking" has an effect only in the vicinity of the noise frequency. In general, a noise field consisting of a broad background with superimposed peaks raises the floor of the auditory communication window by about 25dB above the complex spectrum, compressing the window in much the same way as shown by the "effective service limit" on the right diagram of Figure 2.7.

If the speech sound level at the ear is amplified to maintain the speech-to-noise ratio, the intense parts become distorted, and perhaps painful. Aircrew are also aware, as noted in Chapter 13, that the amplified sounds contribute to the overdose inflicted on the delicate hearing organ; the tympanic membrane, ossicles and cochlea, so the avoidance of long-term hearing loss is a further disincentive against over-amplification. Aircrew generally set the amplification to a level that gives a balance between these dangers and the loss of the subtle speech information. In consequence the speech-to-noise ratio falls with increasing levels of noise, and the intelligibility of the signal is degraded. The chance of errors is, however, difficult to predict. In addition to the physical conditions, the probability of mishearing an utterance depends on many factors[4] including the size of the vocabulary, the acoustic similarity of the words and the familiarity of the talker and listener with the vocabulary. In tests using monosyllabic words, a listener who mistakes about 50% of what he hears when the speech and noise levels are equal, will recognise about 70% of the words correctly when the speech is raised by 3dB above the noise level, and about 30% when the speech falls 3dB below the noise. A speech-to-noise ratio of more than 12dB is needed for fewer than 10% errors.

Speech intelligibility can be assessed by recreating the cockpit noise environment in a sound chamber and asking a panel of test subjects to write down what they hear when a set of words are relayed at a prescribed level relative to the noise level at the ear. A variety of different vocabularies are used that range from lists of nonsense words to words that differ in only one speech attribute. One common set, composed of confusable, phonetically-balanced English words, makes up the Modified Rhyme Test (MRT).

An alternative approach, which can be used to characterise a communication channel, dispenses with human listeners and instead feeds the communication system with a test signal. This is composed of octave bands of noise, each of which is modulated with a sine-wave envelope. Any noise, dispersion or reverberation introduced by the channel reduces the depth of modulation, and a number between 0 and 1.0, called the Speech Transmission Index (STI) can be calculated by weighting and combining the measured modulation transfer ratios. An STI value less than 0.3 is bad, between 0.3 and 0.45 is poor, 0.45 to 0.6 is fair and above 0.6 is excellent. It is also possible to predict the overall probability of errors incurred in a speech communication system using a mathematical model. An accepted way of performing the calculation is given in American Standard ANSI S3.5-1997. This results in a figure of merit called the Speech Intelligibility Index[5] (SII) or Articulation Index (AI) that also ranges from 0 to 1.0. Values less than 0.3 are unsatisfactory, between 0.3 and 0.5 are satisfactory, between 0.5 and 0.7 good and a value above 0.7 is excellent.

10.3 Voice communications

Figure 10.1 shows a typical audio system installation in an in-service aircraft. This presents auditory signals from the central warning system and from the on-board defensive suite and enables the crew to speak to each other, as well as relaying speech from the crew to the transmitting radios and speech received by the radios to the crew.

Racal Acoustics Type 13100 moving iron microphones, which are used as the standard transducer in RAF oxygen-masks, have 210 Ω impedance, 20mV sensitivity and 300-6000Hz bandwidth. The earphones mounted in each of the earcups inside the helmet are balanced armature transducers with 150 Ω impedance and the same bandwidth as the microphone. Mask-mounted microphones have an external switch, and in most multi-seat aircraft this is usually turned off unless the user wants to speak so that the crew do not have to listen for radio communications against the amplified distractions of their fellow crewmembers inhaling and exhaling.

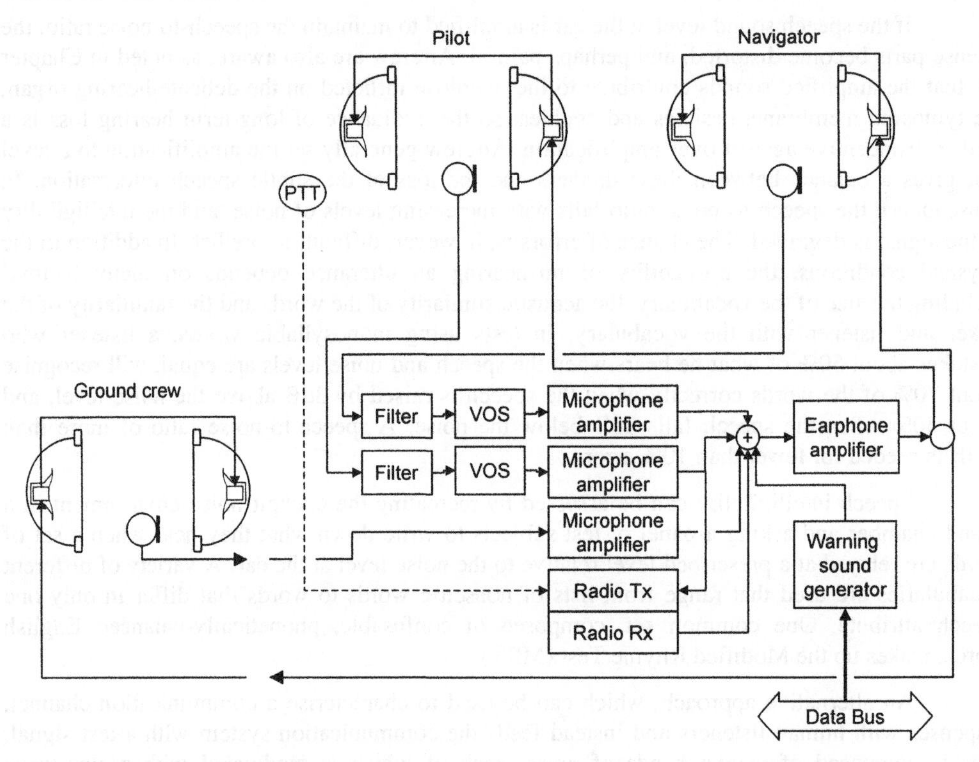

Figure 10.1 A typical audio communication system in a modern combat aircraft

In aircraft where the crew do not use a breathing mask, the microphone is almost invariably "noise cancelling" and mounted on a curved boom so that it can be brought close to the lips. Such devices have openings at the front and the back, allowing the sound wave to impinge on both faces of the transducer, inducing an output that depends on the instantaneous pressure gradient rather than the instantaneous pressure. As sound pressure drops with the inverse square of the distance from the emitter, and the noise field is emitted from sources within the cockpit that are more significantly more distant than the lips, the speech sounds have a higher pressure gradient than the noise field. Although the effectiveness of this form of microphone deteriorates at frequencies above about 1000Hz, the signal-to-noise ratio is improved by about 5dB at 500Hz and 10dB at 250Hz, with a corresponding improvement in dynamic range because the transducer is less likely to saturate.

It is also beneficial for the microphone amplifier to include a voice-operated switching circuit (VOS). This acts like the manual switch in a mask-mounted microphone and the benefit is particularly useful in aircraft with more than two crew; for instance four microphones pick up a background level of noise 6dB above one microphone. It can be noted that a microphones strapped close to the larynx, a throat microphone, can avoid picking up ambient noise, but it does not receive the high frequency signals generated by the mouth, lips and teeth. As these sounds convey so much of the information in the speech signal, a throat microphone is rarely used in an aircraft.

The speech signals from the microphone must be filtered, ideally in the microphone pre-amplifier, to remove components below 300Hz to avoid overloading downstream circuitry. It is also advisable to remove frequencies above the useful speech content, for instance above about 3000Hz for communications, and above 6000Hz if the signal is also fed to a speech recogniser. Automatic control of the microphone amplifier gain is rarely employed in combat aircraft, but it may be used in other aircraft where the signal is relatively unaffected by ambient noise. It is essential however that a "side tone" is fed to the earphones of the crew so that they can hear their own speech. This is done in recognition of the "Lombard effect", whereby the perceived level of the spoken sound influences strongly the vocal effort and the resultant characteristics of the speech signal. Too much side tone volume induces a low S/N ratio, while a low side tone can induce shouting or give the impression of equipment malfunction.

10.4 Auditory warnings

As noted in Chapter 6, the central warning system alerts the pilot by illuminating, and sometimes flashing one or more lights placed prominently at the front of the glare shield. These are intended to cue the pilot to look at other information presented on dedicated displays further down in the cockpit. Warnings of this nature vary in importance and priority, ranging from those that signify immediate emergencies, such as the detection of a missile attack, to those that indicate a hazardous trend such as low fuel. An audio signal generator has been introduced into more modern aircraft, as shown in Figure 10.1, to generate sounds that reinforce the visual attention-getters. Different sounds are used to indicate different states, a typical sound being a sequence of bursts at a single frequency or a series of alternating tones. These are played at a level high enough to guarantee that the pilot takes notice, and they persist until the pilot operates the switch to cancel the warning condition. The system described in section 6.12 is typical.

The implementation of audio warnings has not been altogether satisfactory. The significance of individual sounds is easily forgotten, their pre-set loudness is sometimes startlingly excessive and false warnings are common. In many aircraft additional sounds have been introduced to ameliorate in-service problems, and this has resulted in a large number that are arbitrarily assigned and easily confused. Because a sound can signify a different condition in different aircraft types, aircrew who fly different types, or who convert from one type to another, are particularly likely to misinterpret the significance of what they hear.

What are needed are auditory signals that do not add stress or interfere with communications but attract the pilot's attention and simultaneously give him easily interpreted cues about the nature of the concern. One line of research[6] instigated in the RAE in the UK, mainly for helicopter applications, has been aimed at exploiting digital sound generation techniques to devise sound-plus-voice warnings. These had the form of an attention-getting sound followed by a voice message that specifed the problem, and it was originally intended that they would be repeated at short intervals. The work started by grouping the conditions that could warrant engaging the pilot's attention into four categories:-

Priority 1: Immediate actions, where a response is required within 2 seconds
Priority 2: Immediate awareness, where the pilot would generally confirm the condition before responding
Priority 3: Awareness, where the pilot could need to modify longer term intentions
Priority 4: Information/status, to keep the pilot abreast of system states during the mission

For example, a low height condition would be Priority 1, an engine fire Priority 2, a low fuel state Priority 3 and confirmation of the selection of a particular flap position Priority 4. These corresponded to the distinction between warnings, priority cautions, cautions and advisories in the example described in Chapter 6.

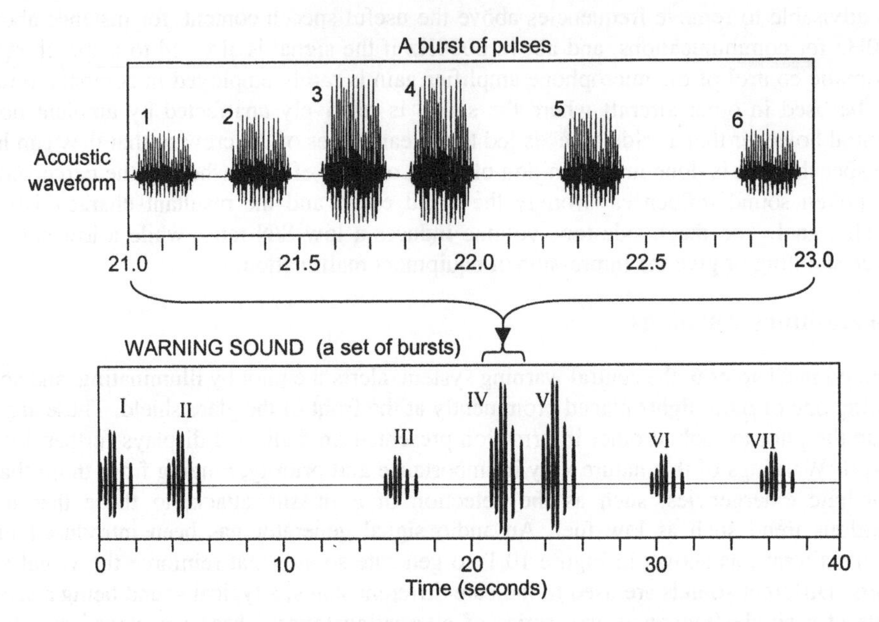

Figure 10.2 The building blocks for an audio warning. In this example a characteristic attenson comprises six bursts of pulses. This is followed by the voice message. The overall warning, shown in the lower diagram, starts with a pair of attenson+ voice announcements at moderate loudness (I and II), followed by a ten second silence. If during this interval the pilot does not respond and cancel the warning he is given a relatively quiet reminder (III). If he remains unresponsive, this is followed by two loud attensons+voice (IV and V). Relatively quiet reminders are given regularly thereafter (VI,VII...). (Adapted from James S H, *Audio warnings for military aircraft,* in "Audio Effectiveness in Aviation", AGARD-CP-596, 1997. Reprinted with permission from QinetiQ)

The approach taken by the RAE team was to create a repertoire of attention-getting sounds (attensons) composed of bursts of sound pulses, one of which is illustrated in Figure 10.2. Other attensons in the set had different combinations of pulse pitch, tempo and rhythm. The spectral characteristics of the pulses were chosen to avoid masking by the noise received at the ears, and this depended on aircraft type and the attenuation provided by the headgear. A set of ten audio warnings was considered the maximum number that could be remembered reliably, and their characteristics were adjusted in a series of simulator experiments to ensure that the sounds were distinctive and conveyed the requisite differences in urgency. The set included six Priority 1 conditions, all with individual attensons so that the pilot could respond before receiving the voice message as confirmation. All Priority 2 warnings had the same attenson with a female voice specifying the problem, such as "engine fire". Priority 3 warnings also shared the same attenson, and a general "caution" voice message was used to cue the pilot to look at the master warning panel to identify the problem. The lowest Priority 4 warnings had a single attenson and, if warranted, an explanatory voice message. However, "Low Height" fell outside the other priority categories and was assigned a dedicated attenson and voice message, largely because the pilot could set the trigger height and therefore the significance of the warning.

Further refinement and flight testing suggested that much of the verbal information was unnecessary, and a minimally intrusive warning set was devised in which - except for the low height case - all of the attensons and voice message are presented just once. This outcome reinforced the earlier conclusion of Wheale[7], who suggested that the way a pilot responds to a warning and the disruptive effects of the warning depend on several factors, such as the number of alternatives, the quality of the synthesised speech and the visual task loading. If the warning came from a set of more than four or five, and especially if the speech quality was poor, the pilot would generally check the legend on the central warning panel before responding. This confirmatory strategy was most likely to be adopted by civil pilots, partly because of their training and partly because they could afford to take their eyes away from the forward view. However, a military pilot operating in close proximity to the ground would be less inclined to look away from the external scene and would instead respond quickly and directly to the warning. The set of refined attensons devised by the RAE team has been adopted by the Naval version of the Merlin helicopter. If standardised, the set will be introduced into all UK service rotorcraft and to some fixed-wing types.

Experiments have proceeded in the UK to develop sounds that convey direct cues about trends and transient aircraft states, such as rotor overspeed and excessive torque. These sounds have been termed "trendsons" because they provide immediate feedback about the direction of change of a particular aircraft parameter and are intended to obviate the need for the pilot to look at the relevant visual display. An attenson and visual warning would still be used to indicate a dangerous condition when the parameter exceeds a safe limit. The problem has lain in finding sounds that were easily associated with the parameter, and with acoustic features that could be changed in an equally obvious and unambiguous way to convey changes to the parameter. Based on a series of iterative experiments, a set of five trendsons have been devised to indicate rotor overspeed, rotor underspeed, excessive engine power, excessive $+G_z$ and excessive $-G_z$. So far these have not been adopted for use in any helicopter or other aircraft.

10.5 Location cueing

At present the most active topic of research to improve the presentation of auditory information in aircraft concerns the technique recently developed in the US for modulating the signals fed to the binaural earphones so that sounds are perceived as originating from a definite direction. It was considered that the introduction of such 3-D sound could augment the pilot's general spatial awareness in a way that required no special training, and if the cues were reasonably precise they could reinforce visual cues to help him find and identify other aircraft and threats.

The work is underpinned by a psycho-acoustic explanation, originally put forward by Lord Rayleigh, for the human ability to perceive the location of a source of sound. Figure 10.3, illustrates the case where the two ears receive sound from an emitter at an azimuthal angle left of the forward head-pointing direction. The wave, travelling at about $340 m.sec^{-1}$, is received by the left ear about 5msec earlier than the right ear. This readily distinguished inter-aural time difference (ITD) is considered the main locational stimulus for low frequency sounds, below about 1000Hz, where the wavelength is much larger than the dimensions of the head and the intensities at the two ears are comparable. At high frequencies, where the time difference could be several pressure cycle periods, the ambiguity is resolved because the wave front is distorted and attenuated by head diffraction. In this case the right ear sits in a sound shadow and the inter-aural intensity difference (IID) provides the main cue stimulus.

In principle, the ITD and IID do not distinguish between sounds located at equal angles to an axis through the two ears, and this would create "cones of confusion" on either side of the head. However, it is considered that additional azimuth and elevation cues are introduced by directionally-sensitive resonance and interference effects in the outer ears (pinna) providing the sound contains wavelength components comparable with the dimensions of the pinna, typically sounds above 2500Hz. The relative strength of the effects can be gauged from the accuracy with which experimental subjects can assess the direction of a sound source. In optimum conditions – a quiet ambience and a steadily pulsed broad-band emitter – the source can be localised to an accuracy of about 5° in azimuth and about 15° in elevation over the forward head-pointing hemisphere. The mean location error is about two times greater in the rear hemisphere.

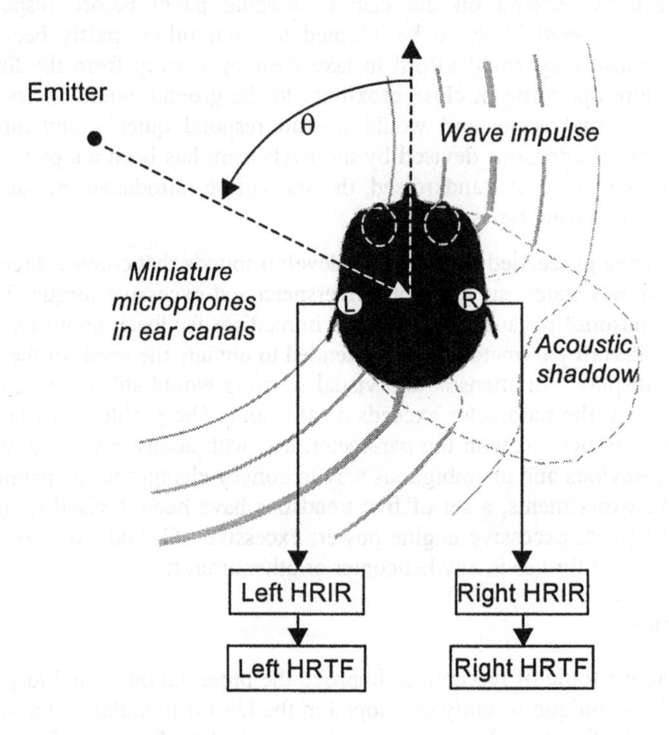

Figure 10.3 System for measuring the HRIR and HRTF for an emitter at a particular location in relation to the head of the listener (see text for an explanation of the abbreviations)

The technique for manipulating a sound so that it can be played through binaural earphones and made to appear as though emitted from a controllable direction depends on the precise measurement of the characteristics of the sound received by the two ears over the required field of hearing. As indicated in Figure 10.3, this is done by attaching miniature microphones near the entrances to the ear canals of the individual. A small linear loudspeaker, set at a required location at least two meters from the individual's head, is excited by a voltage impulse to emit a fairly loud pressure wave that impacts the pair of left and right microphones. The signals that are produced, the temporal Head-Related Impulse Responses (HRIR), are transformed by Fourier analysis to become a pair of left and right Head-Related Transfer Functions (HRTF). This procedure is repeated with the emitter placed at a large number of locations. To represent human location discrimination[8], about 250 equally-spaced emitter locations would be needed to cover the surrounding sphere. The subject's head must be held steady throughout, and the procedure is normally carried out in an anechoic chamber to avoid interference from ambient noise and reflected waves. The resulting set of measurements consist of a pair of left and right HRTFs for each of the emitter locations, with each HRTF describing the amplitude and phase of the received signal relative to the transmitted signal at intervals of at least 1/3 of an octave in frequency.

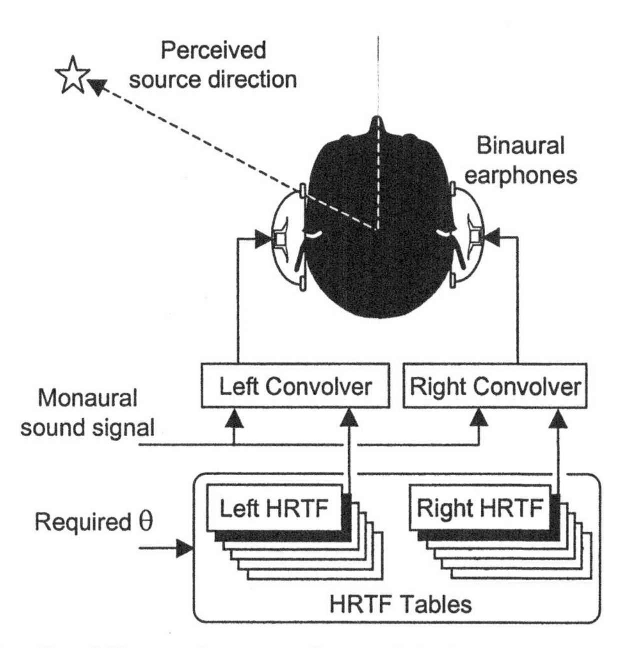

Figure 10.4 Basis of a 3-D sound system for modulating a monaural sound signal to impart a controlled directional sensation

Figure 10.4 outlines the way the HRTF tables are used to modulate a purely monaural signal so that a sound image is created at a particular azimuth and elevation location relative to the head-pointing direction when the modified signals are fed to the left and right earphones of a headset. The use of binaural disparities to impart a spatial cue is analogous to the use of appropriately disparate binocular views of a scene in a stereoscopic visual display. In this case the manipulation is done in real-time by convolving the audio stream with a finite segment of the HRIR obtained from the required HRTF. The technique was originally developed by Crystal River Engineering for NASA using a DSP-based electronic engine marketed as the "Convolvotron". Several manufacturers now offer a similar capability, and current systems can apply different spatial localisation modulations to about 32 sound signals simultaneously. Although the technique is commonly called 3-D sound, it should really be 2.5-D sound. The third dimension – the absolute range to the emitting object – could be conveyed by intensity, but this would require the user to learn an arbitrary relationship between sound level and range, which is impractical given S/N constraints. However, the relative distance of emitters can be made apparent by level differences, and dynamic level changes can indicate that an emitter is approaching or receding.

Several factors affect the accuracy and usefulness of practical systems. It is necessary to interpolate between HRTFs to minimise the size of the table and avoid abrupt discontinuities. Also, to fix the location of the sound in relation to the aircraft, the angular position of the emitter must take account of head orientation, and this is measured by a fast sensing system, as described in Chapter 11. Experiments conducted by Gilkey[9] *et al* have confirmed that, as expected from the psycho-acoustic theory, judgements about the location of a "virtual emitter" are most accurate in the left/right azimuthal plane, that there is significantly greater confusion between front-back and up-down locations, and that these differences are maintained over a ±12dB S/N range. Others by McKinley[10] *et al* have supported the expectation that the localised auditory cue would help the pilot find a target. In this case, a laboratory-based task involving a search for space-stabilised symbols presented using a HMD, the mean search time decreased by a factor of eight when a directional auditory cue was added.

For accurate reproduction, the sound reaching an ear canal should not undergo significant modulation by the earphone transducer, the earphone cavity or the pinna, and in principle this can best be done using a miniature wide-band earphone inserted directly into the ear canal. This is one topic of current research, which is complicated by the use of the ear insert for passive and active noise reduction, as outlined in Chapter 13. A second avenue of research is directed towards optimising the spectral characteristics of directionally-modulated sounds so that they avoid bands masked by noise but contain the frequency components that convey directional cues as well as the vowel formants and the transient consonant sounds of speech. A third topic has arisen from the general impracticality of creating a set of HRTFs for each user using accurately located sound sources in an anechoic chamber. As these recording facilities are rarely available it is of great practical value to know whether non-individualised data can be used to give adequate spatial resolution to the sound images.

This raises the core question - what spatial resolution is adequate? - a topic that is being addressed, mainly using simulators that enable pilots to fly complex missions in authentic noise environments. Pilots are asked to assess the practicality of dealing with multiple simultaneous sound images, such as threat warnings, speech communicated from other aircraft, high priority attensons and feedback from the voice command system. The location of the warnings and speech varies dynamically, while the attensons and feedback are emitted from fixed, but separated locations in the aircraft. The experiments will also explore practical ways of reducing front-back and up-down confusions by introducing subtle changes to the pitch and tone of the sounds that the pilot can easily learn. Given that future manned aircraft will be fitted with auditory display systems that are fully digital and capable of generating an arbitrary number of directionally manipulable sounds, it can be anticipated that this work will result in the erection of international standards that define the preferred usage and characteristics of the displayed information.

References

1. de Florez L, *True blind flight*, Journal of Aeronautical Sciences Vol 3, pp. 168-170 1936
2. Forbes T W, *Auditory signals for Instrument flying*, Journal of Aeronautical Sciences, pp. 255-258 May 1946
3. Wickens C, *Processing resources in attention,* in Parasuraman R and Davies R (eds), *Varieties of attention,* Academic Press, NY, 1984
4. Miller G A, Heise G A and Lichten W, *The intelligibility of speech as a function of the test material*, Journal of Exploratory Psychology 41, p 329, 1951
5. *Method for the calculation of the speech intelligibility index*, American National Standards Institute, ANSI S3.5-1997, 1997
6. James S H, *Audio warnings for military aircraft,* in *Audio Effectiveness in Aviation*, AGARD-CP-596, 1997
7. Wheale J, *Performance decrements associated with reaction to voice warning messages*, in *Advanced avionics and the military aircraft man/machine interface*, AGARD-CP-329, 1982
8. Begault D R, *3-D sound for virtual reality and multimedia*, Academic Press Inc. 1994

9. Gilkey R H, Simpson B D, Isabelle S K, Anderson T A and Good M D, *Design considerations for 3-D auditory displays in cockpits*, in *Audio Effectiveness in Aviation*, AGARD-CP-596, 1997

10. McKinley R L, D'Angelo W R and Ericson M A, *Flight demonstration of an integrated 3-D auditory display for communications, threat warning and targeting*, in *Audio Effectiveness in Aviation*, AGARD-CP-596, 1997

Further reading

Stanton N A and Edworthy J, *Human factors in auditory warnings*, Avebury, Aldershot UK, 1999

Chapter 11

Controls

11.1 Rationale

The team designing a cockpit regards the suite of controls as the counterpart to the suite of displays. The latter provide information from on-board systems to assist the crew to make decisions, while the former provide the physical agency for executing the decision-making and altering the state of the aircraft and on-board systems. Control devices should primarily enable tasks to be executed speedily and reliably.

Until the recent adoption of voice command and helmet tracking systems in some combat aircraft, and except for a few actions performed by the feet and legs, all control has been exercised using the arms, hands, fingers and thumbs. Aircraft were operated manually, partly because that was all that was needed, partly because that was all that could be arranged and partly because of precedence. Nowhere is convention more evident than in the set of levers that are pushed and pulled to control the flight path and airspeed. In fixed-wing aircraft these are the throttle operated by the left hand, the stick by the right and the pedals by the feet, the first pair being replaced by the collective and cyclic in rotary-wing aircraft. Other crucial controls have also kept to a convention, for instance in aircraft so equipped, the use of a short lever that is moved analogously down for lowering and up for retracting the undercarriage. The sheer familiarity of these manual devices eases the transfer of training between aircraft types and helps prevent errors when the crewman is under stress. However, the control stick and the undercarriage lever are distinctly different. The stick is in almost constant use, and as the very act of gripping supports the arm against high-G and transient disturbances, it has evolved to incorporate switches and other inceptors (another name for control input devices) that are embedded in the hand-grip. In contrast, the undercarriage lever, although mounted on a convenient front panel, on the few occasions that it is operated requires the crewman to find and reach for it in order to make finger contact. Not only must the crewman let go of the throttle, but reaching is invariably guided by looking, so there is a concomitant diversion from head-out attention that is exacerbated by any disturbing accelerations of the aircraft.

It is partly to avoid these detractions, and to decrease the reliance on manual devices for interacting with increasingly complex on-board systems, that has motivated the development of alternative devices. Such voice-, head-, eye- and brain-actuated modes of control cannot substitute for manual devices entirely, and their usefulness depends dramatically on the nature of the control action required of the crewman. Consider that actions performed by aircrew can be classified as:-

a) Vehicle steering (e.g. for flight-path control)
b) Sensor steering (e.g. for pointing a radar or an infra-red camera)
c) Rare emergency action (e.g. for jettisoning external stores or initiating engine fire suppression)
d) Frequent time-critical selection, (e.g for transmitting a radio message)
e) Routine selection (e.g. for changing the scale of a map)
f) Routine data-entry (e.g. for inserting the co-ordinates of a waypoint)
g) Cursor slewing, (e.g. to designate a position on a HDD)
h) External feature designation (e.g. to designate a visible target)

It is here that two other distinctions arise. The first is that military crew tend to perform all of these actions, while civil aircrew do not. Civil aircraft carry no systems to deal with targets and threats, and their crew rarely interact with external entities. The second distinction is that most civil crew enjoy a relatively quiet environment that allows some control facilities to be shared and rarely exposes them to high-G or vibration. Neither do civil crew suffer the burden of a tight seat harness, multi-layer clothing, gloves and heavy headgear that collectively hamper reaching movement, visibility and dexterity. Supplying satisfactory control mechanisms for combat aircrew is thus substantially more necessary and more difficult. As well as convenience, reliability and the least diversion of attention from looking out, they need controls that can be operated despite these disturbances and impediments. The operating environment, the relatively cramped workspace and the need to ensure compatibility with the ejection seat, clothing and headgear also generates engineering complexity.

Any control mechanism must give the crewman an unambiguous confirmation that he is using the intended control, and when the control is activated he must get feedback to tell him so. The alternative mechanisms differ greatly in the manner in which such confirmatory feedback is supplied. In this regard manual mechanisms can be engineered relatively easily to give the tactile (sometimes called haptic) sensation of movement accompanied by a resistive force gradient and a sudden yielding as the switch closes. However for other control devices, such as speech recognition systems, confirmation must be contrived using auditory or visual displays.

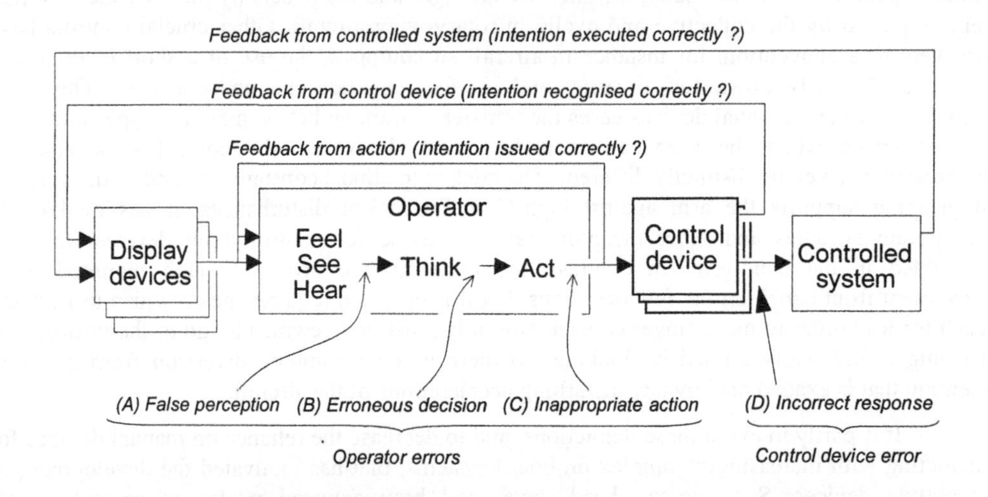

Figure 11.1 Simplified representation of the information processing stages, the varieties of feedback and the varieties of error in a man-machine interface (Reprinted with permission from QinetiQ)

Figure 11.1 indicates the general flow paths that give the operator information about his control actions, and summarises the loci and variety of errors that can arise. A control can only be regarded as error resistant if the net probability of all sources of error is very small. This must be distinguished from the reliability of the control system itself which depends upon it giving a consistently correct output signal in response to the action of the operator, irrespective of the correctness of the action of the operator. For this, error D should be highly improbable. The other forms of error, called false perception, erroneous decision and inappropriate action in Figure 11.1 although apparently due entirely to the operator, may in reality result from the way the control system and the feedback have been arranged. For instance the small movement and change of reactive force that are intended to give the tactile and kinaesthetic sensations of a key being triggered, could be too delicate and induce the crewman to make multiple presses. Or if a speech recognition system mis-classifies an utterance and jumps to an inappropriate branch of the syntax, the crewman could become confused and take the wrong corrective action.

The challenge for the developers of control devices, particularly those of the non-manual sort, is not only to make their device reliable in the physical cockpit environment but to understand how the device should be integrated into the overall vehicle operating system.

11.2 Manual controls

11.2.1 Throttle, stick and pedals

In most agile fixed-wing aircraft the stick, throttle and pedals are the accepted means of exercising control over the aircraft flight path and airspeed. In some heavy aircraft the stick takes the form of a double-handed rolling wheel on the top of a push-pull column. In both cases hand movement initiates a causal chain. The stick or column moves aerodynamic surfaces which interact with the airflow to produce rolling and pitching moments that alter the aircraft attitude and hence the direction of the air flow over the wings, tailplane and fin. This in turn changes the direction and magnitude of the resulting lift, thrust and drag forces on the airframe, and these in combination with gravitational and inertial forces determine the resultant aircraft direction and speed of travel. Although there are numerous complicating cross-couplings, which differ considerably between aircraft types and flying conditions, the relationship between hand motion and aircraft response requires the pilot simply to point the stick-top in the direction he wishes to travel. This action, which becomes as natural and automatic to the pilot as leaning in the direction he wants to walk, is largely maintained in the effects produced by displacements of the pedals and throttle. The pedals are linked to the moving surface of the vertical fin so that the aircraft nose swings in the direction of the extended leg, and the throttle is coupled to the engine so that forward movement increases the forward thrust. In the same way, levers used for deploying secondary aerodynamic control surfaces, such as flaps and slats, are arranged to conform with the pilot's mental model of the aircraft; down for deployment and up for retraction.

The cyclic of a conventional rotary-wing aircraft is similar in principle to the stick of a fixed-wing aircraft in that it primarily adjusts roll and pitch attitude. The pedals also control the yaw attitude. However, the left hand raises or lowers the collective lever to alter the pitch of the rotor, and hence the rotor lifting force, and the grip must be twisted simultaneously around the lever axis to open or close the engine throttle to maintain the rotor speed. Although an auto-throttle is common, co-ordination is more demanding, mainly because there are strong cross-couplings between the effects of control actions. For instance, upward movement of the collective and an accompanying throttle opening induce a strong torque reaction that yaws the aircraft fuselage in the direction opposite the rotor spin. The pilot must counter this rotation by pushing the appropriate pedal to increase the moment produced by the tail rotor.

Basic conventions were established early in the history of aviation when controls were coupled directly to the engine carburettor and the aerodynamic surfaces by rods, pulleys, cables and bell-cranks. These are still used in most modern light aircraft and gliders. Considerable engineering refinement was needed to lighten the muscular effort to move the aerodynamic surfaces whilst allowing sufficient reaction to be fed back to the stick through the rods and cables so that the pilot could feel the force of the airflow over the moveable surface. This called for judicious siting of articulation axes and the use of counterweights. In most aircraft it also helped the pilot to attend to other tasks if the aircraft would fly a reasonably steady course when he let go of the stick. This was done using springs that returned the stick to a central position. The designer had to use the lightest possible springs to avoid swamping the subtle reaction forces.

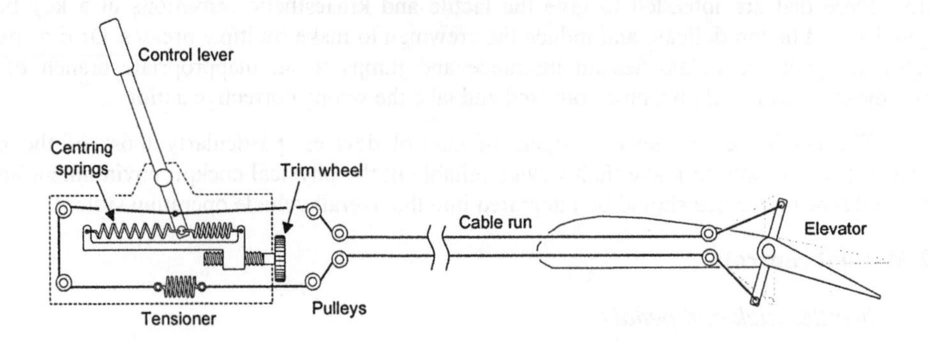

Figure 11.2 Schematic showing the main elements of a cable control mechanism

If the return springs were attached to a moveable anchorage, as illustrated schematically in Figure 11.2, the central position could be trimmed to remove an unwanted command when the stick was released. Accurate trimming reduced the effort the pilot would otherwise have to exert to hold the stick in a displaced position, which was essential in heavy aircraft and a considerable aid in all aircraft for accurate, relaxed handling. The hands-off position of the stick depended on the position of the resultant lift vector in relation to the aircraft c-of-g, which was affected mainly by the load distribution. As this would shift during flight, for instance as fuel was consumed, a trim wheel was supplied for in-flight adjustment. In some aircraft types it was only necessary to provide adjustment in pitch, while others could be trimmed in pitch, roll and yaw. In most current jets the pitch and roll trim mechanisms are driven by electric motors activated by a convenient five-position switch set into the top of the hand-grip. The interconnections are arranged so that pilot can use his thumb to annul a steady control force by rocking this trim switch in the direction of the force.

Throughout their development in both large and small aircraft, flight control systems have preserved the earlier co-ordination between hand movement and aircraft response. The early use of trim-tabs as aerodynamic servo-mechanisms and the later adoption of hydraulic rams and servomechanisms were essentially means of enabling tolerably light forces to be input through the pedals, stick, or column, particularly in large aircraft. However, the relationships between hand, arm and foot movement and the control forces has always been crucial because both the force and the movement conveys proprioceptive feedback about the aerodynamic state of the aircraft. For instance at very high angle of attack, just before the aircraft stalls, the pitch up reaction force on the stick usually falls, and at high airspeed it would be quite easy for the pilot to make unintentionally large control demands without the feedback of a strong resistive force. These could over-stress the airframe. Hydraulically powered flight control systems are therefore designed to have optimised non-linear control loading characteristics; the spring-like relationship between stick force and stick displacement steepens with increasing displacement, the stick is shaken as an artificial cue to the imminence of a stall and the gradient of the stick force-to-displacement is stiffened by decreasing the servo gain with increasing airspeed. The last factor is known as q-feel because the stiffening is made proportional to q, the dynamic pressure sensed by the pitot probe.

The introduction of automation into the control loop went a stage further. As discussed in Chapter 4, the earlier forms enabled the pilot to engage analogue electronics that received signals from the air data computer and the inertial platform and applied deflections to aerodynamic surfaces via electro-hydraulic valves so that the aircraft maintained a state that could be set by the pilot. Such auto-pilots maintained attitude, airspeed, heading or height, and had a control panel with switches for the pilot to set the required speed, height and heading, and engage the appropriate mode. The pilot could remain aware of the working of the system by monitoring visually the aircraft flight path and proprioceptively from the movements of the stick, control column or throttle. In some implementations the system would disengage if the pilot intervened and operated the control manually, and in others it would disengage if the pilot operated a cut-out switch, usually on the stick top. Similar forms of sensing-plus-actuating systems were also introduced to provide stability augmentation; essentially ways of altering the handling characteristics of the aircraft, largely to remove instabilities and cross-coupling. Electronically they were akin to an auto-pilot, with the pilot remaining in the loop and his control input acting as a dynamic adjustment to the set-point.

The modern amalgamation of hydraulic actuator, auto-pilot, automatic engine controller and stability augmentation system is the full authority fly-by-wire flight control system (FCS), which for safety and integrity is built around a triplex or quadriplex combination of sensors, digital computers and actuators. It brings all the benefits of the original component systems, together with optimised deployment of secondary flight control surfaces, care-free manoeuvring and, in combat aircraft, the viability of efficient but aerodynamically unstable configurations, such as the Eurofighter canard-delta.

In such aircraft the primary hand-grasped and foot-pressed controls provide input signals conveying the pilot's flight-path demands to the FCS, but the relationship between the input and the aircraft response, and between the latter and the force fed back to the control inceptor, can be whatever the designers can implement. In a fixed-wing combat aircraft the fore-aft and side-to-side forces and movements of the bi-directional lever operated by the right hand normally mimic the action of a conventional control stick, and the left-hand-operated lever acts like a throttle to control the engine thrust. However, instead of inducing aircraft pitch, roll and speed changes, a wide range of control modes is possible. For instance, the fore-aft and side-to-side displacement of the right hand inceptor could normally demand negative and positive G-levels and right-left roll rate respectively, but in an aircraft such as the Harrier when thrust-borne, the same movements could command forward-backward and right-left rate of change of position.

In an aircraft fitted with a digital flight control system the aircraft is to all intents flown by the embedded computer, which stabilises the aircraft, adjusts the trim, counteracts numerous aerodynamic cross-couplings, optimises the engine state, and deploys slats, flaps and airbrakes as needed. The manual levers and pedals have become devices for the pilot to signal his manoeuvre instructions to the computer. Their location and force-displacement characteristics can be tailored to enable accurate manoeuvring. Engineers aim to attain good handling quality ratings, as explained in Chapter 4, from the assessing pilots.

The F-16 fighter was perhaps the first aircraft to benefit from this liberation, and many lessons were learned. In this aircraft, the designers reclined the seat to about 30° and raised the pedals, mainly for the comfort of the pilot when manoeuvring at high-G. They mounted the stick on the right side console where it did not interrupt the view of the comparatively shallow front panel and could be adjusted in position to be grasped easily by aircrew of all sizes. In the prototype aircraft, the output signals from the stick were proportional to the side-to-side and fore-aft forces exerted by the pilot and were interpreted as roll-demand and G-demand by the FCS, scaled at about 20°/sec per pound and 1G per pound respectively. However, the stick itself was rigid and did not move. During early flying tests the pilots had difficulty controlling the aircraft. They could not assess what force they applied, know what proportion of the range they were demanding or distinguish between grip force and control force, particularly when the aircraft itself was manoeuvring rapidly. They had to rely solely on the observable response of the aircraft. The solution to the problem was simple and immediately effective; the stick was mounted on a compliant base to give a small movement, about 50mm, from the central position. The resultant articulation of the wrist gave sufficient proprioceptive feedback for them to exert accurate control forces.

It is interesting that some aircraft design teams, such as the Rafale and F-22 combat aircraft, the Comanche attack helicopter and the Airbus 330 transport aircraft have mounted the stick on the right side console. An equally eclectic sample, such as the Eurofighter and Harrier combat aircraft, Tiger attack helicopter, MV-22 Osprey tilt-rotor and Boeing 777 transport aircraft, have retained a central between-the-thighs location. All have a FCS in which the primary flight controls are purely electronic inceptors, so there can be no engineering preference for the location of the control. The choices adopted by the design teams reflect their assessment of the visual intrusiveness of a central stick versus the ability to reach the stick and the grip-top switches with the left hand. These, in turn, depend upon the layout of the HDDs, the proportion of switches on the right console that can only be reached by the right hand and the simultaneous need to operate the stick and the embedded switches.

Future developments are likely to improve the quality of feedback cues. Primary flight inceptors are under evaluation that have a travel of about ±30° and incorporate electric motors powerful enough to produce a force of about 450N. These can be programmed to provide a wide variety of haptic cues, for instance the breakout from a central detent that can go from zero to the full force, soft stops, hard stops, variable damping, variable friction, variable stiction, nonlinear and non-symmetrical force-deflection characteristics, q-feel and stick shaking. These can alert the pilot to flight envelope restrictions, for instance by placing a step in the force gradient when he demands a G-level that cannot be sustained, or when approaching stall, store carriage or structural limits. Backdriving in multi-seat aircraft can give the non-pilot the same haptic and visual cues as if the sticks were mechanically coupled.

11.2.2 Other hand-operated control devices

A wide variety of hand-and-finger operated electrical switches exist, for instance in toggle, rotary, slider, rocker, edge and push-button forms, all having different size, form, axis of movement, range of movement, operating force, surface friction and number of switching states. Switches also differ in the form of the feedback they provide (tactile, audible or an integral light) and in their action (for instance push-on/release-off, push-on-release/push-off-release). The demands of the military aircrew job, combined with the need for compatibility with aircrew clothing – in particular the glove assemblies worn by aircrew – and the cockpit environment, have tended to reduce the options. Most designers of combat aircraft cockpits have installed guarded toggle switches or push-buttons on dedicated panels for rare emergency actions, push-buttons in the stick and throttle grips for frequent time-critical selections, soft (variable function) keys surrounding the HDDs for routine selections, and compact keypad arrays for routine data-entry. In civil transport aircraft the less taxing environment, relative expansiveness of the flight deck and need to enable easy crew interaction have induced designers to use different options. Only a few switches are embedded in the control column handles, large keypad arrays are used for routine data entry and rotary switches are commonly used for routine selections. The latter are mounted in logical groups on front and overhead panels where they can be reached by either crewman.

All aircraft switches are manufactured to high engineering standards. For reliability they must be designed and tested to perform their function in the worst operating environment at least 10^7 times, considerably more often than the likely usage over the aircraft life. They are therefore made from durable materials, incorporate multiple non-oxidising contacts, usually gold, sealed within an inert gas envelope, and most have metal cases to reduce electro-magnetic emissions.

Some of the recommended ergonomic characteristics of switches are summarised in Table 11.1. Other characteristics are also recommended by these standards, for instance that for non-keyboard applications the switch surface should be concave or have high friction to prevent fingertip slip, that switches should snap or click when activated, that cover guards should be used to prevent accidental activation and that mechanical interlocks or barriers should be used if the recommended separation can not be arranged. STANAG 3869 A1 also recommended arraying the keys of a numerical keypad so that the digits from 1 to 9 form a matrix with 0 centred below, like a standard telephone rather than the numeric keypad on most computers.

Table 11.1 Some recommended switch parameters. The first row is taken from STANAG 3869 A1[1], a standard for military aircraft. The others are from MIL-STD-1472B[2], which applies to military equipment generally

	Size of main dimension (mm)	Operating force (N)	Displacement (mm)	Separation between adjacent sides (mm)
Pushbutton	10 to 25	2.8 to 5.6	2 to 13	> 13
Finger-tip pushbutton	10 to 19, but >19 if operated by thumb or heel of hand	2.8 to 11, but 1.4 to 5.6 for little finger operation	3 to 38	>13, but 50 preferred for single finger >8, but 25 preferred for single finger sequential use 13 preferred if operated by several fingers
Keyboard	10 to 19, but 13 preferred 19 preferred when used with arctic mittens	1 to 4 (numeric) 0.25 to 1.5 (alphanumeric) 0.25 to 1.5 (dual)	0.8 to 4.8 (numeric) 1.8 to 6.3 (alphanumeric) 0.8 to 4.8 (dual)	6.4 preferred
Soft key	19 to 38	2.8 to 11	3 to 6	3 to 6 vertical 5 to 6 horizontal

The recommendations for keyboard characteristics were largely derived from investigations involving skilled telephonists and typists working in an office, where the main requirements are high keying speed, typically about five presses per second, and the ability to correct mistakes easily. These conditions are pertinent to civil flying and perhaps to some military aircraft, for instance the operators in the rear of maritime patrol aircraft. However combat aircrew are at the opposite end of the speed-accuracy trade-off. They usually have only the left hand available and want to input relatively short sequences without error when working in gloves and exposed to vibration. They tend not to hold their hands poised over a keyboard and press keys with their finger-tips, using almost all of their fingers. Instead they reach out with the left hand, rest the fingers alongside the keypad to stabilise the hand and press the sequence of keys with the thumb.

To meet these needs within the constraints of the combat aircraft cockpit, the designers have mainly chosen to site the keyboard and associated display beneath the left glare shield and put other keys in a panel on the protruding HUD, calling the latter the up-front control panel. The size, inter-key separation and movement recommendations are largely met, but to avoid erroneous activation they have made the individual keys so that they need a considerably heavier, and therefore a more deliberate, activation force. A survey by Taylor and Berman[3] found that in a variety of UK military helicopters and fast jets the switching resistance of keyboard keys ranged from 4.9N to 9.8N, about three times greater than recommended for use in a benign environment. Pushbuttons in grip-tops typically required 17N, again considerably greater than the standard recommendations.

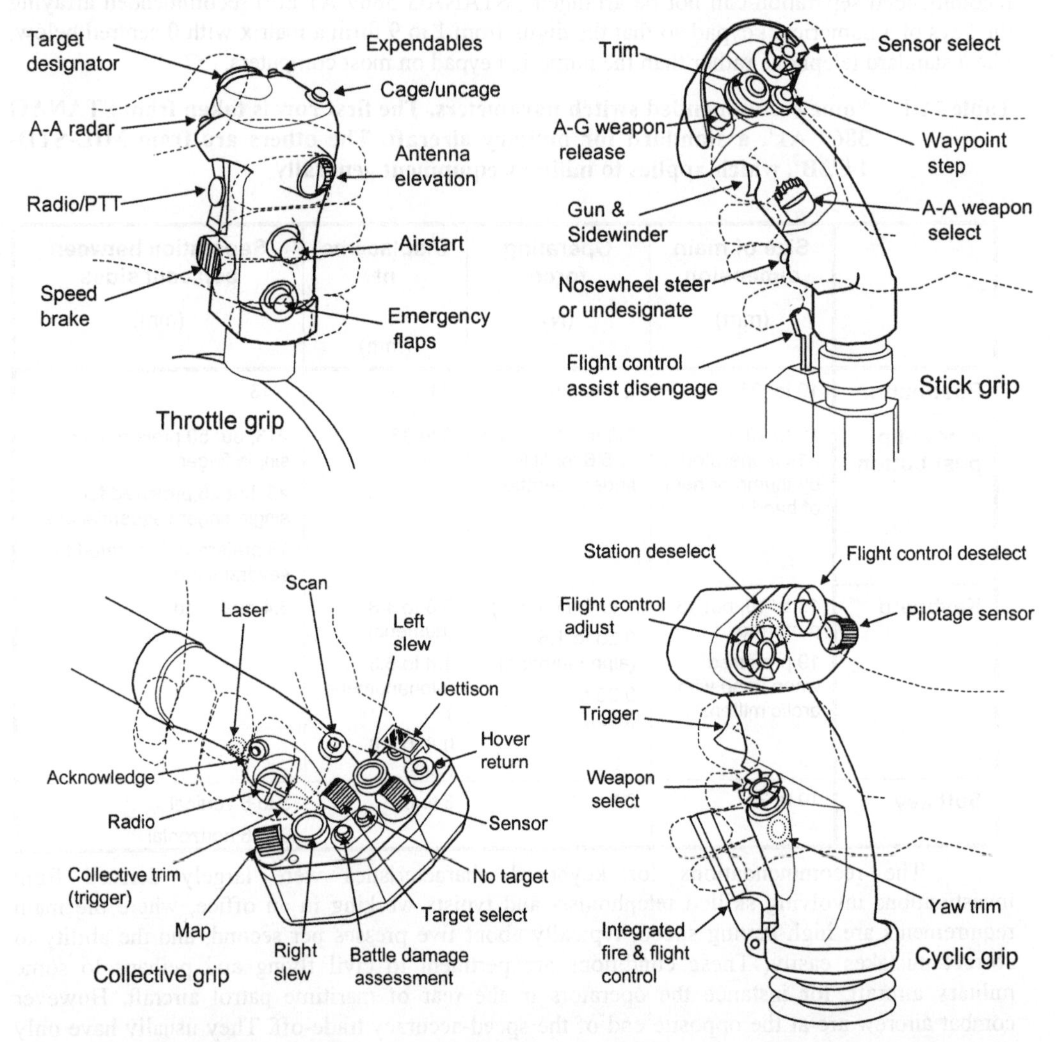

Figure 11.3 Illustrating (above) the HOTAS controls in the AV-8B Harrier II+ and (below) the HOCAC controls in the RAH-66 Comanche Attack Helicopter (adapted from *Flight vehicle integration panel working group 21 on glass cockpit operational effectiveness*, AGARD-AR-349, 1996)

The approach has been applied to the large, prominent pushbutton for emergency jettisoning of the under-wing stores, and has been taken to the limit in the pull of about 250N needed to operate the ejection seat firing handle. The alternative way to prevent accidental activation has been to fit mechanical guards. These have taken many forms. For instance a sprung flap can be used to cover a key, while a toggle or a rotary knob can be sprung so that it must be pulled outwards before it can be shifted, and a trigger switch can be locked out by a safety catch. All are effective, but need to be big enough to be manipulated by aircrew wearing gloves and strong enough to resist damage. They also introduce an extra delay, and during the operation the pilot invariably needs to look at the guard. The only switches that are intended to be found and operated by touch are those embedded in the grip-tops of the primary flying controls of combat aircraft. These are used to make frequent, or time-critical, selections without the crewman moving his hands from the flying controls. With the increasing complexity of on-board systems, the sheer number of embedded switches has resulted in grips that resemble half a saxophone. The facility they provide is known as the "hands-on-throttle-and-stick" concept, or HOTAS.

As illustrated in Figure 11.3, the intention behind the design of a HOTAS grip is to leave sufficient surface for it to be grasped between the heel of the hand and the fingers without inadvertent activation of any of the switches, yet arrange for individual switches to be reached by a specific finger or thumb. The number of switches, their function, position, alternative states and force-displacement characteristics are peculiar to the aircraft type. As shown, most are sited in a protuberance on the end of the grip where they can be pressed by the thumb. There are a few consistencies, for instance weapon release is invariably sanctioned by the forefinger on the right hand grip, and the press-to-talk (PTT) button is always on the left grip where it can be pressed to connect the microphone to the radio and put the radio in transmit mode. In multi-seat aircraft the PTT is usually arranged as a two-way switch that can be rocked to the second position to enable inter-crew communications. If the aircraft is fitted with a speech recognition system, the PTT becomes a three-position switch that it can be rocked into the third position to activate voice command input.

One considerable advantage of the HOTAS principle arises because the act of grasping the handgrip steadies the whole hand, and it is possible to exercise accurate control under vibration and high-G. As well as push-on/release-off buttons, a variety of devices are mounted in the grips, for instance a miniature thumb-operated two-axis inceptor that can be used to slew a cursor across a HDD for pointing a thermal sensor, and a cruciform switch such as used for roll and pitch trimming. Their adoption presumes that aircrew will become almost reflexive familiar with the switches, and will locate the appropriate switch rapidly and not actuate it unintentionally. The main problems arise for aircrew whose hands are markedly smaller or larger than the median group for whom the grip was designed. Those with smaller hands cannot reach some of the switches without sliding their palms along the grip, while large hands tend to make inadvertent contact and the occasional mis-selection. As hand dimensions differ greatly, for instance the hand length of a sample of male military personnel was found to range from 159mm to 228mm with a mean of 192mm, the poor fit is likely to affect about 30% of aircrew[4]. A similar sample of female hand lengths ranged from 152mm to 205mm with a mean of 176mm, so a considerably larger proportion of female aircrew are likely to experience difficulty in operating HOTAS switches arranged for male aircrew. It is also notable that switching delays and mis-selections are exacerbated by the clumsiness and tactile isolation induced when neoprene rubber gloves are worn inside the normal cape leather flying gloves for NBC protection.

One solution to the problem of switch crowding is to employ a second grip, as in the Tornado, where the additional hand controller is placed behind the throttles to house three push-button switches and a slewing inceptor for operation by the thumb. In the AH-64D Apache the collective lever has two separate grips, one held for routine flying and another at the end of the lever to which the pilot transfers his grasp to operate the fire control radar. In a combat aircraft such grips are more useable than other manual control devices, such as a mouse, tracker-ball, lightpen, miniature joystick, touchpad, graphics tablet or touch-sensitive panel. However, the tracker-ball and a variety of touch-sensitive panels that allow direct interaction with an underlying display have been adopted successfully in the relatively benign environment available a number of civil types.

11.3 Voice command

The history of automatic speech recognition (ASR) is relatively short, spanning only the last thirty years, but in that time in parallel with the exponential increase in speed and memory of the enabling computers, numerous systems have become mature. There have been two main drivers; to enable the substitution of dictation typists, language translators and telephone operators by real-time automatic machines, and to give individuals unable or impeded from using their hands a way of interacting with computers. As noted, aircrew fall into the latter group.

The essential function is to transform the electrical signal transduced by a microphone from an audible sequence of utterances into a stream of computer-interpretable codes. This is done in two stages. The first creates templates to represent each word, and the second assesses in real-time how well the templates match the features in the electrical signal. To date all systems that have been devised to perform this function with a useable chance of obtaining a good match do so only for a limited number of words, their recognition vocabulary. This can be quite a small number, less than 200, or a large number, say 1000 to 5000 words. However, it should be noted that to enable the system to keep pace with the speaker only a sub-set of the templates within the total vocabulary is examined at any juncture. The rules used to select the sub-set constitute the grammar or syntax that prescribes what words can legally follow each other. For speech-based control over aircraft systems the rules can be expressed simply as a branching network that is similar in structure to the sequence of keys that must be pressed to input data using a keyboard or navigate around the HDD menus. An example is shown in Figure 11.4. The average number of words that can follow each word is known as the perplexity of the grammar.

Systems also differ in two other respects. Those that are intended to classify the utterances of a particular person are known as speaker-dependent because the templates are set up using word sounds produced by that individual. Speaker-independent systems require no such training. Early isolated word recognisers made it necessary for the speaker to introduce silent gaps of more than about 200msec between words. Later connected word systems reduced this to a requirement for much shorter pauses, and current continuous speech recognisers will accept fluent speech without inter-word pauses. It should however be noted that it is necessary for the user to press an input key, usually a HOTAS key operated by the left thumb, to activate the recognition process. This primes the recogniser and acts as the primary safeguard against an unintended command to one of the airborne systems.

When installed in aircraft, voice command facilities are commonly called Direct Voice Input (DVI) or Voice Input Control (VIC). Although a wide vocabulary would be attractive, and a high perplexity could for instance allow the speaker to input a command using alternative words, the main requirement for the aviator is that at least 98% of command utterances should be recognised correctly. A greater probability of error and the consequent need for correction or backtracking compromise the speed, convenience and lack of attention diversion offered by voice interaction. It should be noted that this is the mean error rate for a whole command string; the compounded effects of errors in recognising individual words in the string. These encompass the deletions when a word code is omitted, insertions when an extra word code is included and substitutions when a word code is output that does not correspond with the spoken word. To gain this accuracy it is at present necessary to constrain the vocabulary to a small number of words, typically about 200, and operate with an inflexible syntax and minimal perplexity. The primary engineering difficulty is to provide this level of performance in a noisy cockpit and cater for the variation in the character of words spoken by the crewman over the course of a mission, particularly the variation induced by stress under high-G.

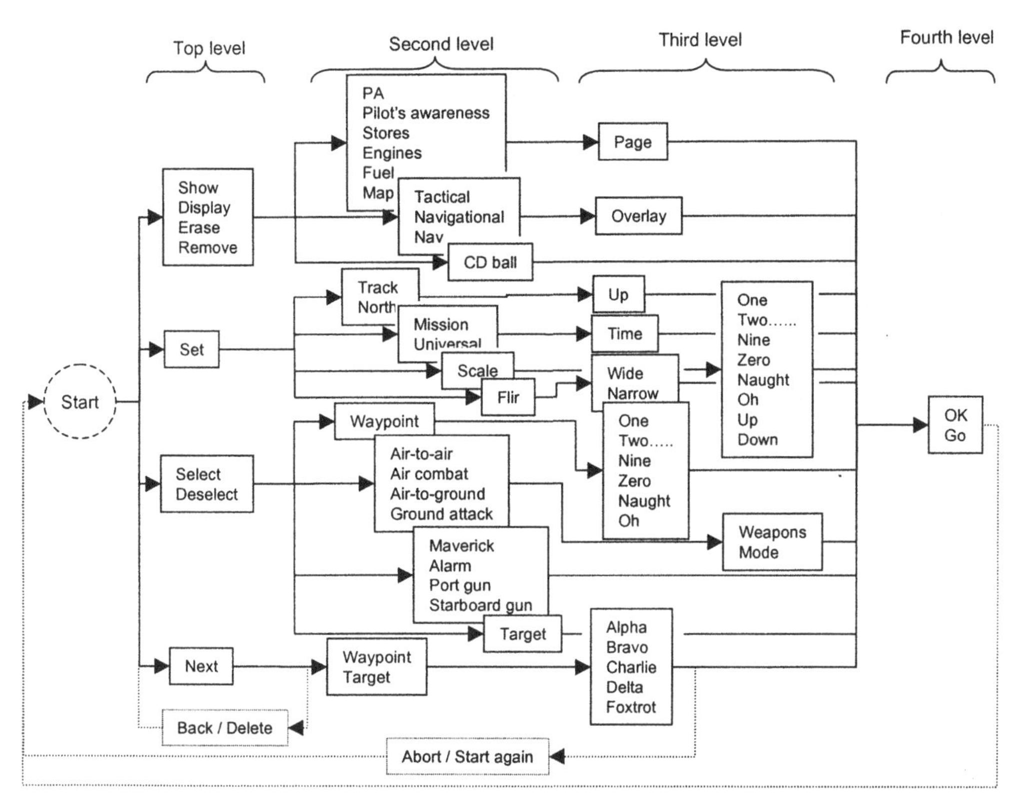

Figure 11.4 Syntax tree for a simple aircraft voice command system. Note that the "Back" or "Delete" commands, which erase the last level uttered, and the "Abort" or "Start again" commands, which erase the complete string, are included in all segments. The vocabulary is 64 words and the perplexity is 7 words (Reprinted with permission from QinetiQ)

11.3.1 Speech recognition techniques

The essential function of automatic speech recognition is analogous to the extraction of syntactical information by the series of acoustic-to-physiological-to-phonological transformations involved in human speech understanding, as outlined in Chapter 2. All speech recognition systems:-

a) use a microphone to transduce the acoustic signal into an electrical signal
b) apply some form of processing to identify features within the signal
c) apply algorithms to obtain the best match between these features and a sequence of templates
d) output a string of text codes for the sequence of matching templates.

In most civil aircraft the microphone is mounted on a short boom from the head-set so that it can be placed close to the lips, and as described in Chapter 10, it is usually noise-cancelling in the sense that it responds mainly to the directional acoustic waves emitted from the lips rather than the ambient sound field. Similar devices are used in military transport aircraft and helicopters. The microphone has a reasonably faithful response over the speech frequency band between about 300Hz to 6kHz, and the main problem of electronic saturation, when the microphone is placed too close to the lips, is overcome by balancing the feedback to the earphones. However in fast jets, or whenever under NBC threat, the speech signal is transduced by a microphone embedded in an oxygen mask or an AR-5 head cowl. The signal is therefore far from ideal; much of the ambient field is picked up, the speech acoustics are modified by resonance within the breathing cavity, and the mask valves introduce clicks and hissing sounds. The signal is modified on entry to the recogniser by an amplifier having automatic gain control (AGC) to match the dynamic range of the downstream analogue-to-digital converter (ADC). It is not uncommon for the AGC to produce transient signal saturation.

The main form of processing applied to reduce the signal to component features is to transform it into a time sequence of short-term power spectra by chopping it into frames of about 10msec. It is necessary to apply a filter such as a Hamming window to remove the spurious high frequency components introduced by the abrupt changes at the start and end of each sample, and the samples usually overlap to avoid excluding information at the frame boundaries.

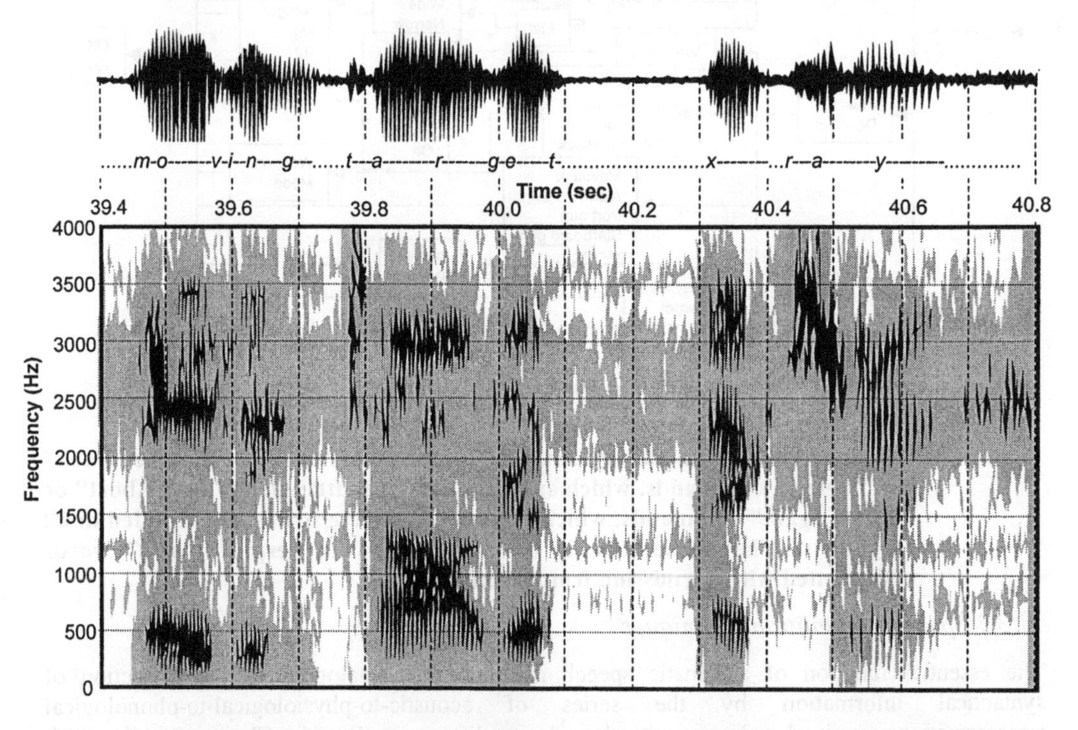

Figure 11.5 Example of the moving window power spectrum taken from airborne recordings made in a Tornado test aircraft at DERA. In this illustration the power in each frequency-time element is represented for simplicity by only three shades; white for negligible power and grey and black for progressively increased levels. The words spoken were "moving target x-ray". The high level of noise at around 2.5 kHz was due to the frequency response of the mask microphone (Reprinted with permission from QinetiQ)

An example of a speech signal represented by a moving window power spectrum computed using Fourier analysis is shown in Figure 11.5. A similar result would be produced by passing the signal through a bank of twenty filters that cover the band from 200Hz to 4kHz in even intervals, then rectifying the output from each filter and passing the rectified waveform through a low-pass smoothing filter. Each of the twenty smoothed signals could then be compressed through a logarithmic filter before sampling and conversion into digital form. The resulting representation of the input signal in the computer memory is a two-dimensional (20xN) array of coefficients describing the power in each frequency-time element. Such an ordered array of numbers can also be thought of as a time-sequence of vectors, each made up from twenty elements.

This process, of reducing the signal to a two-dimensional pattern, followed by an assessment of how well the received pattern matched stored template patterns obtained from training, was the basis of early speech recognition systems. However the process was highly sensitive to the normal variations in the way a word could be uttered and transduced. Although the recognition reliability was poor, the idea of windowing to create two-dimensional patterns that could be matched against templates has remained. Considerable inventive effort has been expended to determine what features, other than the basic power spectrum, could be extracted from the signal to enable words to be distinguished reliably despite normal variations of stress, intonation and speed.

A variety of approaches have been developed that manipulate the sampled spectra. One, known as Linear Predictive (LP) encoding, represents each signal sample by the parameters of an all-pole filter that mimics the vocal tract. Another, called Perceptual Linear Prediction (PLP) uses psychophysical data describing human auditory sensitivity (the equal loudness curve and the intensity-loudness power law, described briefly in Chapter 2) to adjust the signal spectrum before it is approximated by the parameters of an all-pole filter. Others use different characteristics of the human auditory system, for instance using filters with bandwidths that increase with frequency above 1kHz, called the mel scale, coupled with a logarithmic response to signal amplitude. Still others take the logarithm of the power spectrum and apply a reverse Fourier transform so that both axes of the pattern have units of time. In this case the vertical axis is called a cepstrum, and the cepstrel coefficients are considered to represent the shape of the vocal tract. A well-developed technique uses both forms of warping and generates mel-frequency-cepstrel coefficients (MFCC), of which the first ten or so are considered to capture the essential features of the spectral envelope. Another way to optimise feature extraction during the development of the system is to find the combination of features that have the least inter-correlation in the actual conditions in which the speech signal is generated. This is done by Linear Discriminant Analysis. A well-known system has applied this technique to produce the Integrated Mel-Scaled Linear Discriminant Analysis (IMELDA) transform. Yet other system developers have noted that spectral analysis using Fourier transforms is not suited to describing unvoiced sounds, particularly short duration plosives, and they use other forms of signal decomposition such as wavelet analysis.

The multiplicity of ways to extract the features from an acoustic signal has been matched by an equally wide range of techniques for comparing the unknown incoming vector pattern with the stored pattern templates. The earliest of these assessed the similarity between one vector and another as the sum of the squares of the differences between corresponding elements, the Euclidean distance, which was usually modified to give weight to elements that differed consistently between vectors. If each utterance contained words with well-defined start and end points, and each word had consistent duration, the overall similarity between input and template could be quantified by summing the vector differences over the word duration. This could be done for each template in the syntactic subset, and the template with the least difference from the utterance could then be selected for output. Unfortunately, this like-for-like comparison, analogous to the game of snap, failed to accommodate the irremovable variation in the rate at which the component sounds appeared in repetitions of the same word. The answer to this was dynamic time warping (DTW). Here, some vectors in the template were repeated and others were omitted so that requisite sections of the template, usually the vowel sounds, could be stretched or shrunk in time to produce the least overall distance measure. This process was optimised by a linear programming technique called the Viterbi algorithm.

Providing the word boundaries were well defined it was possible to use dynamic time warping to achieve an excellent match. However, concurrent developments in mathematical techniques began to open up the possibility of operating in real-time on a continuous speech signal. Some approaches used Neural Networks (NN), so called because they imitated the properties of the connections between the layers of neurons in the cortex. However, the most productive techniques have used Hidden Markov Modelling (HMM) in which the component sounds, the phonemes, are represented abstractly as a chain of nodes. Each node is characterised by two sets of probabilities calculated from the training utterances. One set describes the probability that individual features, for instance the MFCC values, are present. The other is the set of probabilities that the node is succeeded by each other variety of node, including itself. This statistical approach has the advantages of accommodating natural variations, even between different individuals, of allowing phonetic nodes to be combined in sequence to form representation of words and for these to be combined to form representations of syntactically lawful commands. The pattern matching process uses the Viterbi algorithm to optimise the fit between the transformed utterance and the HMM for each possible node sequence. This is only practical if the number of sequences is first pruned to the most probable by a process such as beam search.

Improvements to the basic mathematical techniques and their efficient implementation in computer code are current topics for research. This is driven mainly by civil applications in which the speech signal is comparatively easily separated from noise and the main requirement is the reliable recognition of unconstrained speech produced by any speaker.

11.3.2 Integration of voice control into the aircraft operating system

Early simulation experiments provided a useful understanding of the ways that DVI could help aircrew. A typical study used a line of text in the HUD field of view and injected synthesised voice into the earphones to give both visual and auditory feedback of the output of the DVI system. The recognition accuracy of the early forms of isolated word recogniser was frustratingly poor, typically about 80%, and this necessitated breaking a transaction into short utterances that could be checked and corrected on the fly. The surprising finding was that despite this stilted technique most pilots preferred to use DVI because, unlike manual alternatives, it was possible to perform the interaction while attending to the concurrent flight control task.

These early studies were very encouraging. It was realised that the pilot's careful way of talking to the aircraft could be transformed if the DVI system had substantially better accuracy and speed of response, and was able to accept continuous speech. With such qualities the pilot's strategy would change. He would for instance be able to input long strings of data by reading the list aloud, then glance immediately at the DVI read-out to check for errors and speak the final confirmatory word to input the whole command into the aircraft system. Without constant self-interruption each transaction could be completed rapidly and assuredly, with least diversion of attention or need to remember a sequence of numbers.

The technique was considered ideal for controlling HDDs, for instance to change the scale of the map or change the display format, where the system response was itself adequate feedback about the correctness of the DVI output. However, for entering more consequential data, for instance waypoint co-ordinates, it has been necessary to trap incorrect responses and have the syntactic tree include a confirmatory utterance. Even with this safeguard, the expected 2% chance of error is too great for safety-critical selections, and DVI will never be appropriate for setting the flight control mode or releasing a weapon. It is currently scheduled for integration into the French Rafale, the four-nation Eurofighter and the US Joint Strike Fighter, but only for routine display selections and for interacting with non safety-critical systems.

Recognition accuracy varies between aircrew. Some individuals articulate clearly and consistently, and others tend sometimes to mumble. For the latter the most practical solution requires modifying their speaking habits, perhaps by receiving appropriate training in elocution. It has also been found essential to train the DVI system in the cockpit environment so that the templates capture the articulation that the individual uses when exposed to the noise and he hears himself speaking via the feedback to his earphones. These factors aside, the chief limitations to the reliability of automatic speech recognition in combat aircraft are, as anticipated, issues associated with ambient noise and the intrinsic variability of speech induced by high-G and stress.

Experiments carried out using recordings made in a Mirage IIIB test aircraft for the French Delegation Generale pour l'Armement (DGA) have shown that enhancing the signal as part of the feature extraction stage can improve recognition reliability significantly[5]. Although dependent upon the individual speaker, the oxygen mask, the noise level and G-level, a Sextant Avionique TopVoice connected word DTW recognition system that would ordinarily recognise less than 90% of command strings could be improved to recognise more than 97%. Much of the benefit was gained by augmenting the acceptance signal from the HOTAS switch by eliminating the noisy periods at the start and end of each utterance using an algorithm that detected when speech was actually present in the microphone signal. Additional benefits were gained by introducing an active noise reduction stage that adaptively modelled the noise spectrum. These additional processes are shown in the upper part of Figure 11.6.

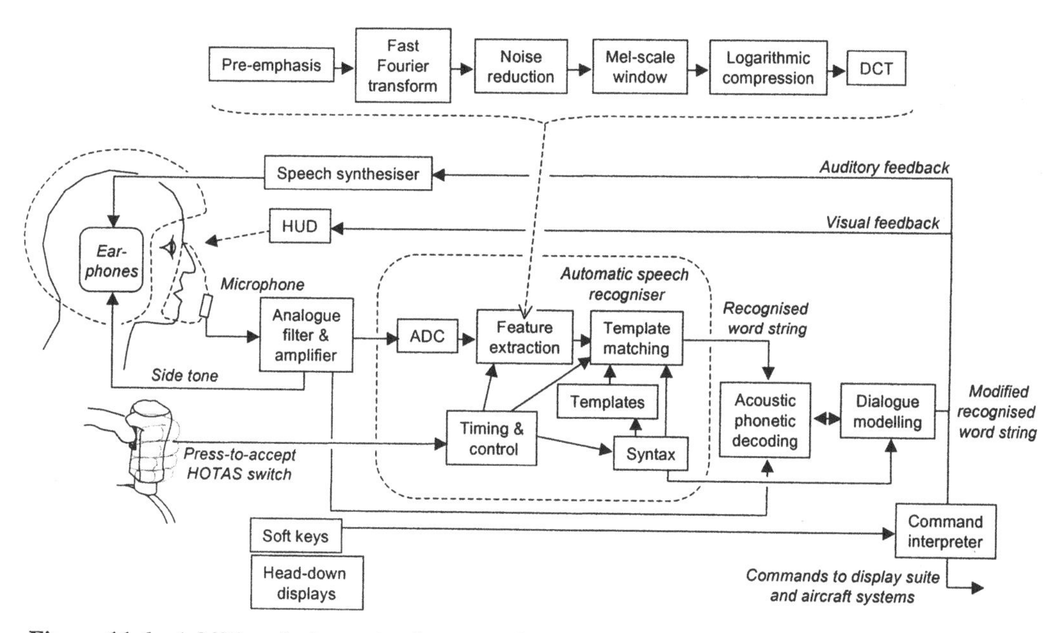

Figure 11.6 Additional elements that may be needed to integrate a Direct Voice Control system into a combat aircraft

Another possible enhancement shown in Figure 11.6 is the incorporation of an extra stage that can check and eliminate obvious errors output by the recogniser. This employs a relatively simple acoustic phonetic decoder that operates like a speech recogniser to perform a high-level classification, for instance categorising the utterance into a sequence of voiced or unvoiced speech, plosives and fricatives. Such an analysis can be used by the downstream dialogue modelling stage to distinguish between the voiced and unvoiced sounds and could help to trap the most common confusions, such as between "five" and "nine", or "fix" and "six". In the Eurofighter the pilot can either make a vocal command or press appropriate keys. As shown in Figure 11.6, this flexibility requires a control interpreter, which is a crucial piece of software that adjudicates between the alternative actions of the pilot to send an unambiguous command to the system under control.

Other auxiliary sensing and processing systems that can supply independent information about speech articulation have been investigated. The most obvious candidates are those that attempt to mimic the skill of lip-reading employed by hearing-impaired individuals and by those with normal hearing in a crowded, noisy environment. A video camera is used to capture images of the speaker's lower face, and the images are analysed by an image-processing computer that, much like the acoustic recogniser, has been programmed to extract salient features in real-time and compare these with pre-formed templates[6]. The characteristic features are the mouth shape, which is sometimes picked out by placing reflective dots around the mouth at specific locations, the positions of the teeth and tongue, and the motions of these features detected by comparing subsequent video frames. The range of template matching techniques that have been used is very similar to those investigated for acoustic feature matching, including DTW, neural networks and HMM. The technique shows great promise, particularly as a complement to an acoustic recogniser, because many of the phonemes that are difficult to distinguish aurally, such as “lay” and “pay”, have markedly different "visemes". However, unless the video camera can be installed with suitable illumination inside an oxygen mask and NBC mask, the approach will not be practical in combat aircraft. An alternative that makes use of the contact area between the mask and the face as a convenient setting for an array of electrodes has been investigated. In this case the electrodes pick up the small myographic signals produced by the nerves that activate the muscles of the jaw, lips and lower face. As with the other approaches it is necessary to amplify and process the signals to extract spatio-temporal patterns that characterise the articulation of specific phonemes. Results at present are promising. The discrimination complements a purely acoustic recognition system, and it offers a high degree of immunity to acoustic noise.

The speech changes that arise due to stress and high-G have proved to be less tractable. The primary effect of increasing psychological stress is a general rise in formant frequency, but there is also a tendency to omit words and use inappropriate phrasing. The most promising approach, that would accommodate a comparatively slow change over time, would be to allow the templates to adapt. Such a process would presume that, if the pilot continues to use the DVI system without making corrections, the templates can be modified gradually so that they better resemble parsed sections of the received utterance. Several speaker-independent HMM systems are already configured in this way; they have no explicit training period and their recognition reliability improves as they adapt to, or learn, the characteristics of the individual speaker. However such systems cannot tolerate a rapid change in speech characteristics, and under a sudden onset of stress their recognition reliability would cope no better than a non-adaptive system.

Most aircrew suffer a gradual inability to speak at all when subjected to a progressively increasing G-level, mainly because breathing is interrupted by the muscular straining used to maintain consciousness. The positive pressure breathing and inflatable jerkin employed in Eurofighter to help maintain cerebral blood oxygenation at more than 6G, may lessen the need for muscular straining but it is unlikely that the overall effect will be anything less than a considerable disturbance to normal speech. Attempts to counter the effects of high-G involve the use of templates recorded at high-G, or the introduction of the effects of high-G to a model of speech articulation, combined with input of the instantaneous G-level to the recogniser. The myographic electrode technique mentioned above may also be less sensitive to breathing disturbances under high-G. In the meantime current systems do not incorporate these sophistications, and it is debatable whether it is worth attempting to make the DVI system respond meaningfully to the limited vocabulary of grunts that aircrew can utter at very high G levels.

11.4 Head pointing and helmet tracking systems

The natural tendency to turn the head toward an object of attention has been harnessed in current combat aircraft to provide the pilot with his only hands-free way of pointing. Currently, this is exploited in visually-coupled systems (VCS) and helmet-mounted sights (HMS), introduced in Chapters 5 and 9 respectively, both of which involve pointing sensors in the same direction as the helmet. The crucial ergonomic issues are aspects of the human ability to control head posture, and the crucial technological issues are those associated with the rapid, accurate measurement of helmet orientation.

The chief requirements of a helmet tracking system are to measure the orientation, and sometimes the position, of the helmet over a wide range of angles and positions without adding a burden to the helmet or restricting head movement. The measurements must be accurate, frequent and undelayed. The system must be highly reliable, tolerate the cockpit environment, easy to install and require little setting up and maintenance. In attempting to devise satisfactory solutions, engineers have exploited electro-mechanical, electro-magnetic, electro-optical, electro-interferometric, electro-inertial and electro-acoustic forms of coupling between elements attached to the helmet and elements fixed within the cockpit. Numerous systems have also combined these phenomena, and each system has unique merits and practical detractions. The following sections give brief overviews of the human ability to control head orientation and the current state of the art of helmet tracking systems in aircraft.

11.4.1 Head mobility in aircraft

The human head rests on the cervical vertebrae and is controlled by a number of large muscle pairs. Although there are wide variations between individuals, the head of the average adult male can be moved over a range of about ± 60° in pitch, about ± 40° degrees in roll towards either shoulder, and about ± 80° in yaw about the spinal column axis. Figure 5.1 in Chapter 5 illustrates the head pointing range for a sample of aircrew wearing full equipment and tight seat harness.

The typical reaction to the appearance of an unpredictable object is a rapid eye movement followed, after an additional delay of about 40msec, by head rotation towards the object. If however the object's behaviour is predictable, the head generally precedes the eye. Voluntary head reactions[8] can be quite rapid and typically reach a peak velocity of about 600°/sec in yaw and about 300°/sec in pitch, but in general the manner in which the head is moved reflects the instantaneous task and conditions. For instance, searching the sky for adversaries needs wide sweeps, and as noted in Chapter 5, when using a HMD or a NVG, the user tends to compensate for the restricted field of view by making larger, more frequent and more rapid head rotations.

It is natural to turn the head toward a target, but it is not natural to adjust head posture finely or maintain a steady posture for an extended period. With training, aircrew can point an HMS within about 2.5° of a target in 0.8 to 1.5 seconds, depending upon the initial angular offset and the angular speed of the target sight-line, but it takes about 2 to 4 seconds to come within 0.3° and the subsequent rms aiming error is unlikely to fall below this value[7]. The tracking error is largely unaffected by the weight of the helmet, or the size and shape of the aiming reticule, or the angle of the target from boresight (the straight ahead direction, and in a HUD-equipped aircraft, the LFD symbol). However it is disturbed by cockpit vibration and high-G.

Vibration, as noted in Chapter 3, is most often an up and down heave force produced in fast jets by changes in lift when flying through turbulent air and in helicopters by the rotor lift pulses. The vibrations travel from the seat through the spine and trunk to the head where the main disturbance is a nodding motion, with some rolling and yawing. These can be controlled by voluntary muscular action if the excitation is below 0.5Hz, and since vibration above 10Hz tends to be damped by the trunk, most crew are affected by vibration between these frequencies[8]. The worst effects, caused by vibrations between 3Hz and 6Hz, can induce head aiming errors in the order of 4°rms.

The head centre of gravity is cantilevered above and forward of the pivot, as shown in Figure 13.3. Wearing a helmet can add about 50% to the tension the neck muscles must exert on the back of the skull to balance the pitch down moment. During manoeuvring flight at about 2G or 3G, the head becomes noticeably heavy. Above this level a stable posture requires progressively strenuous muscular effort, and above 8G most aircrew cannot lift or control their head motion for more than a few seconds. The posture is most easily disturbed if G is applied rapidly, and this can lead to chin contact with the chest. Aircrew are particularly aware that they must avoid placing their head in a posture that they cannot hold and that the strong muscle forces they must exert are inherently fatiguing and can damage muscles, ligatures and bone support structures. Their usual strategy is to use the head-rest on the ejection seat for support and rely on exaggerated eye movements to compensate for the lack of head mobility. In multi-seat combat aircraft, before pulling the stick hard backwards, the pilot usually warns other aircrew and gives them time to adopt a braced posture. As noted in Chapter 3, aircrew are also aware that coriolis forces disturb their sense of balance, making a yawing movement of the head induce a rolling sensation and vice versa. The avoidance of disorientation is further reason to hold still at high-G.

11.4.2 Mechanical helmet tracking systems

Mechanical head trackers couple the helmet to the airframe through a set of linkages connected by flexible joints, and the articulation of each joint is measured by a transducer, typically an optical encoder, a potentiometer, an a.c. resolver or a strain gauge. It is possible to calculate the full six degrees of freedom (i.e. the helmet orientation and position) by sensing all modes of articulations of all of the joints. However, some implementations just measure a sub-set, for instance, the azimuth and elevation angles. Many trackers have been built for use in research and simulation laboratories, and several of these have been adapted from commercial scribing tools that are normally used to trace and digitize the shape of 3D surfaces.

The best-known mechanically-coupled helmet tracker, employed in versions of the US Cobra AH-1 helicopter, used a rigid rod to join the pilot's helmet to the inside of the roof of the helicopter. A universal joint at the helmet enabled the pilot to rotate his head. A second universal joint attached to a runner travelling along an overhead slider allowed the head to be displaced over a reasonable "head motion box". The joint angles were measured with a.c. resolvers and transformed into the helmet azimuth and elevation angles. The pilot could disconnect his helmet rapidly by moving to the limit of the slider travel and jerking his head away from the tethering rod, an action that decoupled a self-aligning magnetic catch attached to the universal joint on the end of the rod. The arrangement could be set to give errors less than 1° over most of the forward hemisphere, which was more than adequate for pointing the turret-mounted gun and wire-guided missile. The added mass to the helmet was in the order of a few hundred grams, the joint friction was negligible and the slider drag tolerable. The main detraction was the added inertia of the rod, slider and complex joints, which was most noticeable when making rapid head movements. Pilots also experienced the occasional inadvertent detachment and had to live with the anxiety that in an emergency, such as a heavy landing, the mechanism could impede movement or jolt their head.

It is difficult to imagine a configuration of linkages between a pilot's headgear and the airframe that would allow unimpeded head movement and not take up valuable cockpit space or occlude vision in some way. These detractions, combined with a probable incompatibility with the ejection seat, virtually eliminate this approach from consideration in fast jets. Despite relatively low cost, good performance and reliability it is also unlikely that systems based on mechanical linkages will find much use in other aircraft types, mainly because alternative approaches have become available that offer adequate performance without the safety risk. Mechanical trackers are therefore most likely to be employed as research and development tools.

11.4.3 Inertial helmet tracking systems

Inertial helmet tracking systems assess the position and attitude of the helmet in exactly the same way that an Inertial Navigation System (INS) determines the position and attitude of an aircraft. The system is set to a known initial state and from then on calculates new positions and attitudes by mathematically manipulating the output from sensors that respond to the rapidity and forcefulness of helmet movement. The system firstly estimates the orientation with respect to the vertical, subtracts the gravitational component from the specific force to get the net acceleration, integrates this to get the velocity and finally integrates the velocity to get the displacement. Essentially, the system performs a six-dimensional dead reckoning.

A set of three angle sensors is used to measure the angular velocity of the helmet about three orthogonal axes. Each sensor operates like a gyroscope in which a spring restrains the spinning gyro wheel from rotating about an axis perpendicular to the wheel spin axis. Rotation about the third axis induces a precession that either extends or compresses the spring in proportion to the rate and direction of rotation. A strain gauge, or a similar transducer that responds to the spring deflection, provides a signal proportional to the rate of rotation. The three linear sensors are simpler. They measure the three orthogonal components of specific force; the vector sum of the Earth's gravitational attraction and any acceleration relative to a stable reference frame. Each of these sensors usually contains a weight constrained so that it can move only in one direction within a housing, and the weight is anchored in the middle of its travel by a spring and prevented from oscillating by a damper. When the housing is accelerated, the weight moves against the spring force a distance proportional to the component of the acceleration in the constrained direction, and the spring deflection is measured by a strain gauge or a similar transducer. Although the concepts have been refined considerably and implemented currently using a combination of micro-miniature tuning forks, spinning wheels, laser ring gyros and piezo-electric elements, the sensors measure the same quantities.

It is of note that the main applications for miniature inertial sensing devices are in missiles and similar robotic devices, rather than for tracking helmets. The measurement of the rate of change of orientation typically has an error between 0.1°/sec to 1.0°/s, the resultant acceleration error is between 0.002m/sec^2 and 0.2m/sec^2 and measurements can be output at about 500Hz with less than 5msec latency. Even when the sensing errors are at the low end of the range, the resulting estimate of orientation and position can drift by several degrees and centimetres per minute. These are satisfactory for controlling a vehicle with a short flight time or where errors can be corrected using a GPS reference, but the accumulated errors soon become problematical for tracking a helmet.

For this application it is necessary to package the set of sensors as a small module attached to the helmet and send the signals to an electronic computing unit mounted in or near the cockpit. To calculate the helmet position and orientation relative to the aircraft axes the orientation and position of the aircraft must be known either from the INS or from a module similar to the unit mounted on the helmet but contained within the electronic unit. INS data is more accurate but requires some compensation for transport delay, while a module in the electronic unit would have similar rate of drift. The chief detractions of inertial helmet sensing are the need to mount a cube of about 35mm with a mass of about 100g on the helmet, combined with the progressive shift in the datum. The latter deficiency is exaggerated by the differences in drift of the helmet-mounted and airframe-mounted sensing modules. On the other hand, inertial trackers provide an unconstrained measurement range and they are unique in supplying information about the rate of helmet rotation at a high enough frequency to counter the "head lag" problem, discussed in Chapter 9, that can de-stabilise imagery presented by a HMD. The most likely use for the technique, given further miniaturisation of the sensing components, is in conjunction with another system having the complementary quality of providing a dependable steady state reference over a narrow range of angles and positions.

11.4.4 Ultrasonic helmet tracking systems

Ultrasonic tracking systems make use of a small, usually piezo-electric, transducer stimulated by a signal to emit short pulses of sound at about 30kHz, a frequency which is well above the range of human hearing but well matched to the sensitivity of receiving microphones. The distance between an emitter and a receiver is calculated by measuring the time-of-flight of the sound pulse and estimating the local speed of sound propagation. The use of more than three receivers enables three simultaneous range measurements, and if the positions of the receivers are known accurately the location of the emitter can be calculated unambiguously.

This triangulation principle was exploited in an experimental helmet tracking system developed by AEG in Germany in the mid 1980s for use in a military helicopter. The system calculated the position and orientation of the helmet from time-of-flight data sensed by an array of receivers, in accurately measured locations within the cockpit, from ultrasonic pulses produced in sequence by emitters arrayed in surveyed locations on the outer helmet shell. The ultrasonic approach was considered a possible way to avoid the sensitivity to sunlight that affected the contemporary optical approaches and the metal-induced field distortions which were the bane of contemporary magnetic systems.

Ultrasonics did however suffer from a number of intrinsic disadvantages, mainly the need for line-of-sight between emitter and receiver, the intrinsically slow propagation of sound and the rapid reduction of detectivity with range produced by the inverse square law. The receiver microphones also had a limited angular sensitivity, typically only a cone of about 45°. As the system was intended to track with reasonable accuracy over a wide range of helmet orientations, and cater for about half a cubic metre of head translation, these factors conspired to limit the update rate to less than 30Hz. Other problems became apparent when the system was installed for testing in a helicopter cockpit, in particular the reflections from the multiplicity of hard surfaces that were roughly perpendicular to the helmet, the occasionally high level of ambient acoustic noise and the large variation of the speed of sound with air temperature. Although much work was done to compensate for the latter by installing a fixed emitter within the cockpit so that the speed of sound could be measured along several paths, in practice the air movement and strong thermal gradients defied correction. Eventually the complexity of the installation outweighed the benefits. It is likely that the main use of an ultrasonic system would be in a much simplified form employing only a few emitters and receivers. This could be used, as suggested above, to complement an inertial helmet tracker.

11.4.5 Optical helmet tracking systems

Over the past 35 years engineers have developed a variety of optical helmet tracking systems in an attempt to attain a satisfactory balance between measurement accuracy and reliability in the cockpit environment. Several have exploited phenomena such as interferometry and pattern recognition, but the most successful have been based upon triangulation and used near IR light to measure a set of angles between cockpit-mounted and helmet-mounted devices. The devices differ, and in some the emitters are fixed in the cockpit while in others they are on the helmet. Their sensitivity to artefacts, particularly those due to incident sunlight, has also depended strongly on the chosen sensor. An outline of the principles of operation of two airborne systems provides a nice illustration of the developer's technical ingenuity.

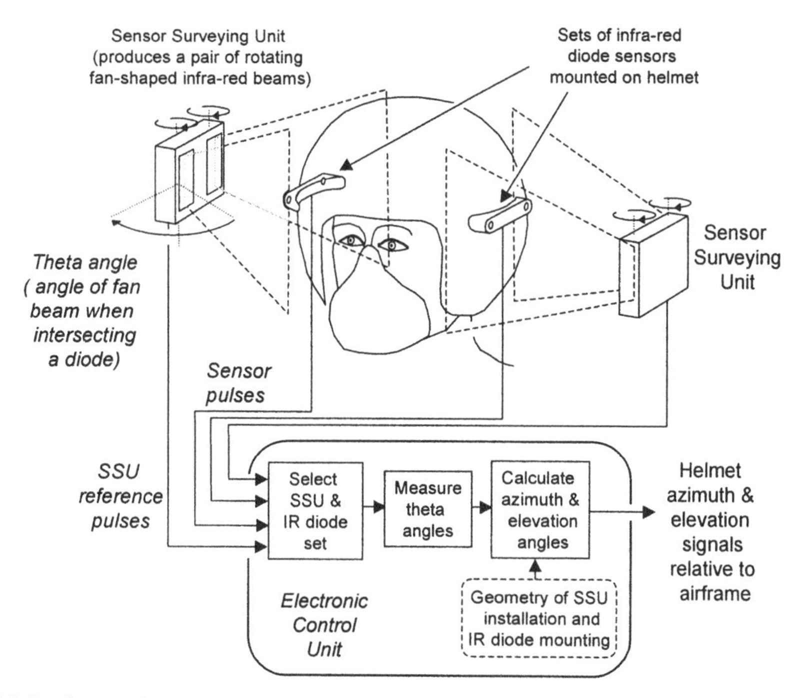

Figure 11.7 Operating principle of the Honeywell MOVTAS helmet tracker

The Honeywell Modified Visual Target Acquisition Set (MOVTAS), shown schematically in Figure 11.7, was devised in the late 1960s and has been installed in a variety of aircraft. It is best known as the helmet tracker employed in the IHADSS for the AH-64 Apache helicopter. In a manner analogous to a sailor observing the flash of a lighthouse, a helmet-mounted IR sensing diode produces a short electrical pulse when swept by a fan-shaped beam from a sensor surveying unit (SSU) mounted in the cockpit. The beam rotates rapidly at a constant angular rate, so the interval between a pulse produced by the helmet-mounted diode and a reference pulse produced by the beam rotating mechanism is proportional to the beam angle to the diode. The diodes are paired, and a pair is surveyed by both beams in a SSU to give four quasi-simultaneous beam angle measurements. Given knowledge of the installation dimensions, the electronic unit solves the trigonometric equations to calculate the helmet pointing direction, which is output each computational cycle as the helmet azimuth and an elevation angle. Several sets of diodes and SSUs are normally used to extend the range of angles measured and the extent of the head motion box. With pairs of SSUs and diode sets, MOVTAS measures the two primary helmet-pointing angles over a range of ±180° azimuth and ±70° elevation to an rms accuracy of about 0.5°. Measurements are updated at 30 Hz, and the latency is about 30msec.

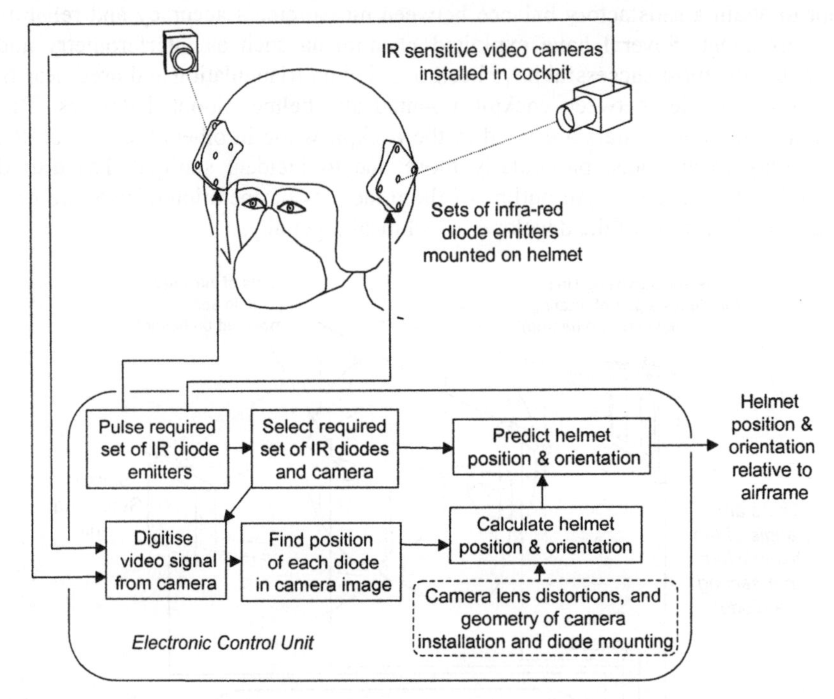

Figure 11.8 Simplified schematic of a modern optical helmet tracker

A more modern approach is illustrated in Figure 11.8. Here, a cluster of LED emitters on the helmet is imaged by a cockpit-mounted camera. An electronic unit, based on digital signal processing (DSP) chips, finds the position of each diode in the 2-dimensional camera image and, knowing the installation geometry and the distortion introduced by the camera optics, calculates both the position and the orientation of the helmet. The update rate of systems employing video cameras as imaging sensors is usually limited by the frame rate of the video signal to either 50 or 60 Hz, although fast frame cameras can be employed to increase the measurement frequency. Sensitivity to sunlight can be reduced significantly by opening the camera electronic shutter only during the brief fraction of the frame period when the diodes are pulsed. Measurement delay can be virtually eliminated by motion prediction algorithms, although rapid changes of direction must be handled carefully to prevent large oscillatory errors.

Some systems use lateral effect photo-sensitive detectors (LEPSD) instead of video sensors. These can increase the measurement update rate and enable sequential pulsing of individual diodes to improve signal detectability and remove any uncertainty in their identity. It is, however, essential to filter the incident light to exclude all but the IR source waveband to prevent sunlight from saturating the detector, but it is possible to compensate for the in-band sunlight by sampling the LEPSD output when all the diodes are momentarily inactive. As with the MOVTAS system, the range of measurements and the allowable head box are invariably extended using several clusters of emitters and several cameras.

The configuration of sensors and emitters must be designed for each aircraft type, but once the components have been installed, residual alignment errors can be corrected within the electronic control unit and routine in-service calibration is unnecessary. All six degrees of motion are measured. The range of measured positions and angles varies with the installation, but ±180° in azimuth, ±60° in elevation and ±40° in roll combined with ±150 mm motion in all three directions is practical. Static errors can be less than about 0.2° in angle and about 1mm in position.

There are however some problems. The camera units must be installed in positions where they can receive light from the emitters but where they do not intruding on the pilot's view and are shielded from direct and reflected sunlight. Reflections from emitters must also be considered. The usual configuration is to site a pair of cameras as high and wide as possible behind the seat, angled downwards and inwards to point slightly below the normal helmet centre. Here they are least susceptible to reflections, and the chance of obstruction by the pilot's arm or hand is negligible. The diodes may be attached to the rear and sides of the helmet, and the emission spread to illuminate the sensors throughout the head motion range. However, as it is necessary to measure and maintain their relative positions to an accuracy of about 0.1 mm to achieve the advertised performance, the useful area of diode sites depends on headgear flexure. At night the mixture of emitted, reflected and scattered IR can make the equipment incompatible with the use of night vision goggles. There is also some concern that IR emission from the cockpit can make military aircraft more readily detected by external surveillance systems.

11.4.6 Magnetic helmet tracking systems

Magnetic helmet trackers create magnetic fields of known orientation and measure the current induced in a sensor fixed to the helmet. The original system, outlined in Figure 11.9 was devised in the early 1970s with support from the US Navy and the USAF[9]. The inventor, W L Polhemus, called the system a Spatial Synchro (or Spacyn) because it was a three-dimensional analogue of a conventional a.c. servomechanical synchro that measured the rotation of a shaft using the relative strengths of currents induced in the pair of sine and cosine windings. The sensor consisted of a set of orthogonal coils wound on formers around an isotropic core to form a cube with sides about an inch long that weighed less than an ounce. A larger but similarly-constructed radiating antenna, fastened in the cockpit close to the pilot's head, was excited with alternating currents at about 30kHz to create a dynamically oriented magnetic field. This was the heart of the invention. Each coil produced a time-varying elementary magnetic dipole, and the resultant field produced by the set of three orthogonal coils was also a time-varying magnetic dipole. However, this nutated, i.e. each cycle of excitation caused the axis of symmetry to rotate around a narrow cone. The axis of this alternating, nutating magnetic vector could be pointed by manipulating the relative phases and strengths of the excitation currents fed to the three coils.

The system operated by measuring the strength and phase of the currents induced in each sensor coil and manipulating the excitations to the radiator to obtain a null. In this condition the nutation axis pointed precisely at the sensor, which enabled the sensor orientation to be computed unambiguously from the residual currents induced by the field nutation. The induced currents were converted into digital form by an ADC, the matrix transformations were performed digitally and the excitation currents were generated by a digital to analogue converter (DAC) and amplified. Initial testing indicated that helmet azimuth, elevation and roll angles could be measured within a constrained a motion box over most of their range with errors between 0.1° and 0.25°.

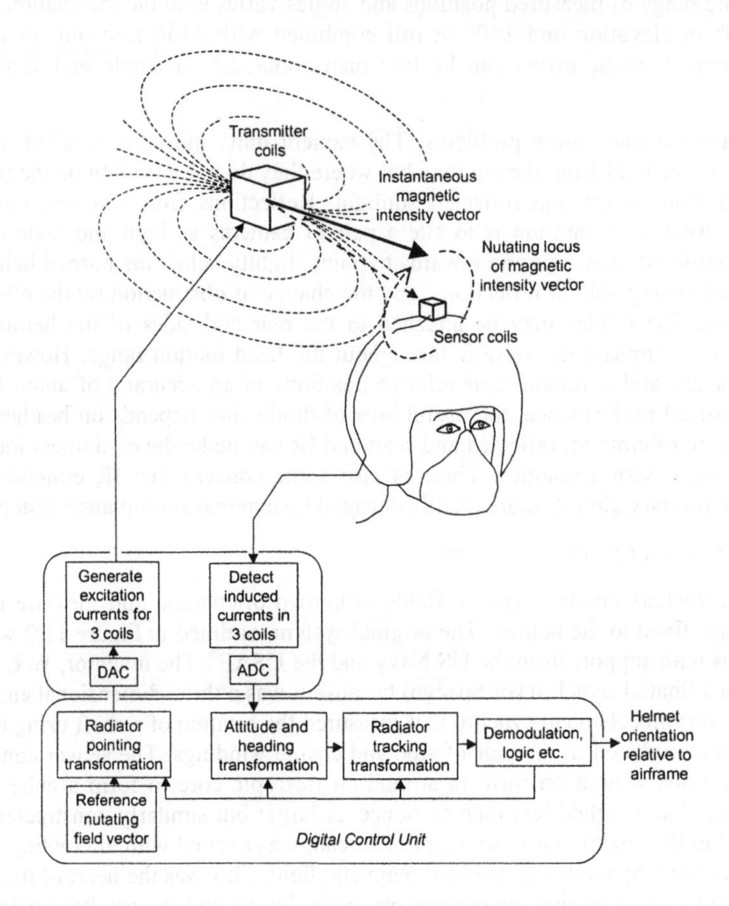

Figure 11.9 Simplified schematic for the Polhemus magnetic head tracker

The inventor anticipated correctly that the basic idea could be used widely, for instance to measure the orientation of limbs. However his prognosis that it could be developed to enable measurement of the lines of sight of multiple operators sited several thousands of feet from the antenna has not come about. The main practical problems have been caused by reactive magnetic fields produced by eddy currents induced in conductive metal structures near the transmitter and sensor. These magnetic fields had the same frequency as the excitation and created errors proportional to the conductivity, size, shape and orientation of the interposed material. A large lump, for instance an ejection seat head box made from aluminium alloy sheet, could introduce errors in the region of 20°.

The advantages of the magnetic approach, mainly its ease of installation, measurement range, robustness, reliability and minimal helmet burden, spurred a number of inventive efforts to remove the field distortion problem. Two classes of system have resulted, called AC and DC magnetic helmet trackers.

The AC types retain coils of much the same size and shape as the Polhemus originals, and the transmitter coils are excited with sinusoidally varying excitation currents to eliminate the effect of static fields in the environment. They differ from the original in two ways. Firstly, they assess all six degrees of freedom by treating the magnitude and phase of the currents induced in the sensor coils as a set of six independent measurements, solving the six equations to derive the three angular and the three translational states of the sensor. Secondly, they then compensate for field distortions using a large look-up table (LUT) containing the necessary corrections. The latter are derived from a mapping process, similar in principle to the use of a pelorus to plot the deviation of a magnetic compass, but in this case the differences between measured and known values are recorded when the sensor is moved, usually by an automated jig, over the full range of orientations and positions in the cockpit. The Honeywell Advanced Metal Tolerant Tracker (AMTT) uses this technique to characterise each installation, and can achieve an accuracy comparable with an uncorrected AC tracker in a field free of perturbations.

DC magnetic tracking systems achieve insensitivity to transient field distortions in quite a different manner. To obtain measurements at a rate of 100Hz, each transmitter coil is excited in turn with a short steady current pulse that lasts about 2msec. This allows sufficient time for eddy currents produced by field changes in nearby metal structures to die out before the strength of the currents induced in the three orthogonal sensor coils are sampled. The sensor acts like a sensitive three-axis magnetometer that produces three current values for each of the three excitation pulses - nine values in all - from which the sensor orientation and position are calculated unambiguously. The currents induced in the sensor coils are also sampled when the transmitter is dormant to enable subtraction of the effect of the Earth's magnetic field and any residual steady field produced by the airframe. Although it is unnecessary to carry out the expensive distortion mapping, to generate adequate field intensity the transmitter coils of DC systems must be larger than those for comparable AC systems. Finding a site for a 70mm cube close to the helmet but out of the pilot's sight is a problem, particularly in a fast jet with an enveloping canopy. Also, the update rate is intrinsically limited by the time needed to wait for eddy currents to die away.

Magnetic tracking devices of both AC and DC types are readily available in both commercial and militarized forms. In a benign environment commercial systems offer accuracies ranging from 0.75mm to 2.5mm in translation and 0.15° to 0.5° in orientation. Accuracy is usually best when sensor and transmitter are very close, and errors become appreciably worse as they separate beyond about 0.5m. The update rate ranges from 60Hz to 120Hz, and latency from 4msec to 150msec with a typical value of about 40msec depending on the type of system and the amount of filtering used. Filtering depends greatly on the noise in the environment, and as the properties of the filters are related to head motion rates, the overall dynamic characteristics of the system and the effective measurement latency can be complex.

The problem posed by metal objects attached to the head-gear, which induces a variable distortion to the field from the radiating coils, is either ignored or reduced to a manageable level by incorporating miniature compensating circuitry at the magnetic sensor. Electromagnetic emissions from other equipment can often be reduced by properly synchronizing the magnetic system with the offending electro-magnetic source. Errors introduced into AC systems by small geometric variations between individual aircraft of the same type, and by flexure of the canopy when the cockpit is pressurised, can in principle be annulled. However the gain is unlikely to warrant the extra expense of mapping individual aircraft and the introduction of further entries in the compensatory LUT.

Table 11.2 summarises the characteristics of the main techniques used to track helmets. It suggests that magnetic tracking is most mature and currently offers the best overall performance. Future developments in inertial and optical techniques may produce systems that can act in combination to give accuracy comparable with magnetic systems together with the higher update rate needed for good stabilisation of HMD images.

Table 11.2 Summary of the main helmet tracking techniques

Method	Typical performance	Main advantages	Main detractions
Mechanical	Accuracy: ~5mm, ~0.2° Speed: ~500Hz	Good accuracy and sample rate Low cost	Imposes inertial & drag forces Difficult installation Accident & egress hazard
Inertial	Accuracy: 0.002 – 0.2m.sec^{-2} 0.1 – 1.0°/sec Speed: >500Hz	High sample rate	Poor static accuracy Appreciable added headgear mass
Acoustic	Accuracy: ~5mm, ~0.5° Speed: ~30Hz	Moderate accuracy and sample rate	Echo & noise interference Affected by air temperature and flow
Optical	Accuracy: ~1mm, ~0.2° Speed: ~60Hz	Good accuracy & moderate sample rate	Susceptible to sunlight Constrained installation IR interactions with other systems
AC magnetic	Accuracy: ~5mm, <0.2° Speed: ~200Hz	Good accuracy & sample rate Large motion box Easy installation	Expensive error mapping
DC magnetic	Accuracy: ~5mm, <0.2° Speed: ~100Hz	Good accuracy & moderate sample rate Large motion box	Large transmitter

11.5 Novel modes of control

A number of tools that were originally devised for psychological and physological research have been extracted from the laboratory and investigated as novel ways for aircrew to interact with the computers that control the aircraft and on-board systems. Eye trackers offer an easy, speedy look-and-shoot capability that is insensitive to vibration and high-G. Gesture recognition systems may enable a more flexible dialogue between man and machine, and there are many bio-potential sensing systems that pick up small electrical potentials from the brain and from muscles. Although the latter are unlikely to be sufficiently reliable for direct control, they may usefully reinforce other evidence about the pilot's intentions, particularly because they precede overt action. The following sections give short accounts of the rationale, the main techniques and the likely uses of these novel control systems.

11.5.1 Eye tracking systems

As outlined briefly in Chapter 2, the orientations of the eyeballs are normally controlled by their extra-ocular muscles so that their visual axes are either tracking a slowly moving object, fixating for between 150msec and 600msec on a stationary object, or jumping rapidly between fixations. Although the eye pointing direction drifts and jitters by about 0.5° when fixating and tracking, measurement of the mean direction could enable easy and speedy designation of a fixated or tracked target and give combat aircrew several important benefits. Firstly, the strain needed to exercise precise control of head posture for a HMS would be alleviated. This and the reduced chance of disorientation with rapid head motion would be particularly beneficial during high-G manoeuvres. Secondly, as shown in Figure 5.1, eye excursions could extend the off-boresight designation range by about 50°. Thirdly, because the eyes are stabilised by the balance sensors in the inner ear, the mean line of sight is largely unaffected by turbulence-induced head motion.

In order to pass the direction of an eye-designated target to the weapon aiming system, the potential benefits must be considered against the increased static errors that result from adding the eye direction, measured relative to the helmet, to the helmet orientation measured relative to the airframe. Although it is only necessary to track one eye, the measurement system must add negligible (<20g) mass to the headgear, be accurate (<0.5° rms error over a range of ±25°), frequent (> 100Hz sampling rate), fast (<5msec latency) and very reliable.

A wide range of phenomena has been exploited to measure eye pointing behaviour in laboratory conditions. Some, devised mainly to study the small movements during fixation, involve attaching fine structures to the eyeball. Others, called electro-oculography (EOG), pick up electrical signals using surface electrodes around the eyes and can be regarded as a serendipitous development from the use of electro-encephalographic (EEG) scalp electrodes in to pick up brain activity. The broadest variety employ some form of optics. The first category is exemplified by the scleral coil, a fine coil embedded in a silicone rubber ring that adheres by suction to the edge of the corneal bulge. This produces a small AC current in proportion to the angle between the coil and an a.c. magnetic field produced by a pair of Helmholtz coils surrounding the head. Although responsive and accurate, this approach is not practical in a combat cockpit where it would be necessary to embed large field coils in the headgear and the aircrew would have to tolerate the scleral ring. Each blink would tend to rub the sensitive eyelid tissue against a loop of delicate wires joining the coil to an anchorage on the side of the nose.

On the other hand the EOG technique, which requires attaching sets of small silver/silver chloride electrodes above and below each eye, and near the outside corners of the two eyes, has been shown to be practical[10] in the combat cockpit. The small voltages, about 20μV/°, that are produced by eye rotation can be amplified and filtered to remove most of the electrical interference. However, the source of the phenomenon, a voltage difference between the front and back of the eye (attributed to different metabolic processes in the cornea and the retina) is almost capriciously unreliable. The strength of the signals differs between individuals, and with the time of day and the ambient light level. Also, it is difficult to eliminate electrode polarisation and associated signal drift when seeking small changes in d.c. surface potential. In consequence, despite frequent re-zeroing with the crewmember looking at a boresight marker, it has not proved possible to use this method to infer the eye pointing direction with errors less than about 5°rms.

The most practical techniques are optical. Systems developed for use in the laboratory exploit different features of the eye and use different combinations of illumination, signal detection and signal processing. The main ocular features, illustrated in Figure 11.10, are either real entities, such as the outer boundaries of the pupil and iris, or virtual entities produced by reflection from the cornea and lens. The circular boundary where the iris meets the white sclera is called the limbus. The virtual features produced as bright spot images of the illuminator from reflections at the outer and inner surfaces of the cornea and the lens are called the 1st, 2nd, 3rd and 4th Purkinje Images (PI) respectively. The most prominent virtual image is produced by reflection from the outer surface of the cornea, the 1st PI, and this is usually called the Corneal Reflex (CR).

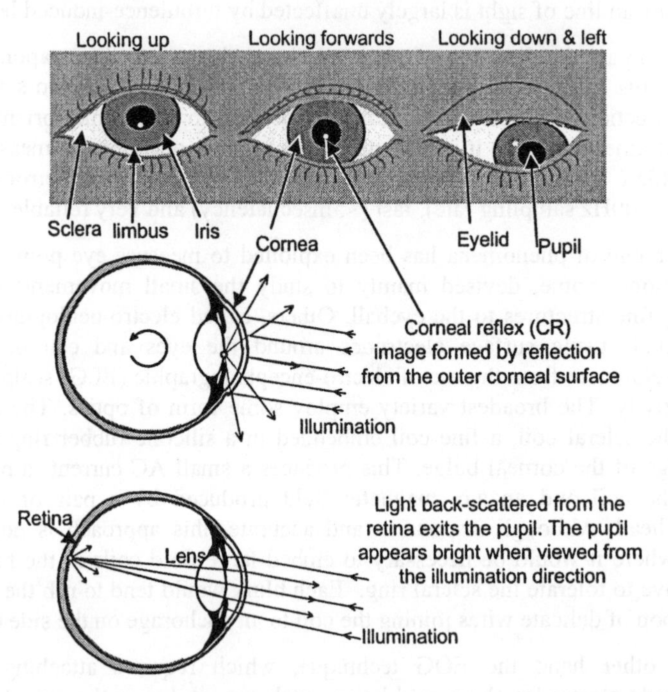

Figure 11.10 Eye features exploited by optical eye tracking systems (Reprinted with permission from QinetiQ)

Figure 11.10 also indicates how illumination entering the eye can be back-scattered by the retina and exit the eye along a reciprocal path. This phenomenon produces the unwanted "red eye" in photographs taken with a flashgun mounted alongside the camera lens, and in some optical eye trackers the illumination is deliberately placed on the same axis as the sensor to exploit this highlighting, mainly so that the pupil can be identified readily. The resulting brightening of the pupil is proportional to the retinal illumination, and hence to the pupil size, so the contrast of the back-illuminated pupil against the front-illuminated iris is only useable with a pupil larger than about 4.5mm. The strengths of the 3rd PI and the 4th PI, which are likewise dependent upon the collection of light reflected back through the pupil, are also reduced when the pupil shrinks to about 2.5mm in normal daylight. It is of note that laboratory refractometers that assess the focal state of the lens by detecting the axial position of the 4th PI can only be used in very dim conditions, and a few drops of a mydriatic drug, such as atropine, are commonly administered to dilate the pupil.

All optical eye direction measurement systems depend upon the transverse movement of the eye features with eye rotation, some assessing the shift of one feature and others the differential shift of two features. The latter have the advantage that they can, in principle, distinguish a shift of the eye features from slippage of the optics relative to the head. Slippage is particularly important because the features move so little with eye rotation, typically only 1mm per 10°, and many of the laboratory systems that depend upon single feature tracking include means, such as a chin cup, brow rest and a bite-bar, to hold the head still.

The simplest such system, the limbus tracker, usually takes the form of an infra-red diode illuminator and several photodiodes that pick up the reflection from the front surface of the eye. Horizontal receiver diodes are positioned out of the user's direct forward view and are set to maximise the differing amounts of light redirected from the sclera and the cornea-iris as the limbus is moved laterally. The vertical receiver diodes are positioned to sense the change brought about by movement of the upper eyelid. Several commercial systems are mounted on spectacles and use modulation of the brightness of the IR diodes combined with phase-sensitive detection to distinguish the IR from the in-band ambient light. Some put a vertical illuminator-detector module in front of one eye and a horizontal illuminator-detector module in front of the other, and assume that the two eyes move together. The technique gives measurements over a wide horizontal excursion range with good resolution and update rate. However, vertical measurements are at best rough estimates and slip induces large errors. A version, developed in Poland specifically for use by aircrew, measures horizontal eye velocity using components integrated into the oxygen-mask[11]. Here, fibre-optic light-guides couple the outgoing and incoming IR from the upper edges of the rubber moulding that forms the nasal seal of the mask to the active elements on a compact printed circuit within a slightly enlarged valve body .

Numerous eye tracking systems have been developed that detect changes in the position of the CR using either a lateral effect photodiode, an array of photodiodes or a video camera focussed on the eye. Unless a special camera is employed, the last approach limits the measurement frequency to the 60Hz or 50Hz field rate of conventional broadcast video systems. The CR can usually be found quite easily because it is the brightest feature in the image and readily induces a signal at the peak video level. Other systems are based upon finding the bright or dark pupil by applying a similar thresholding algorithm to a video signal, and in these cases the bright CR constitutes an anomaly that must be accommodated to avoid increased errors. However, as noted in Chapter 13, any system mounted on a conventional helmet must tolerate slippage of at least 5mm, and as this would create errors between about 20° and 50°, it is essential to use a differential approach. The majority of commercial head-mounted systems therefore find the differential positions of the two most prominent features, the pupil and the CR.

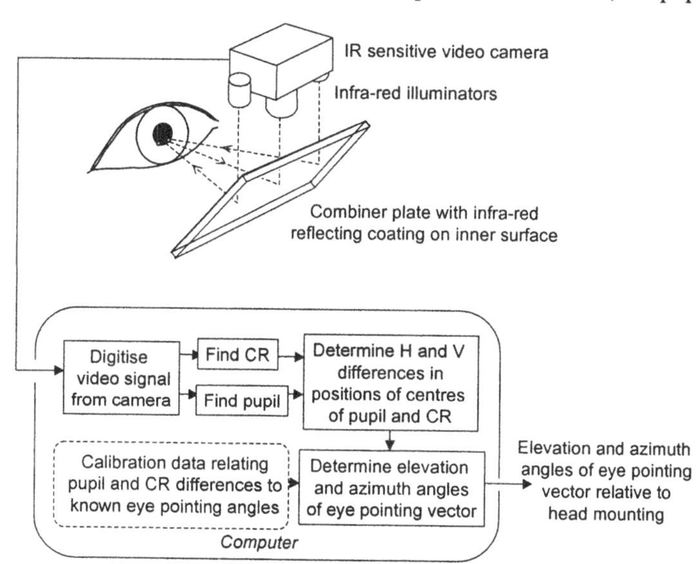

Figure 11.11 A head-mounted eye tracking system that uses the differential motion of the pupil and the corneal reflection

A simplified schematic for a typical head-mounted differential pupil-CR tracker is shown in Figure 11.11. The illuminator and camera are positioned out of the forward view and capture the image of the eye on reflection from a thin flat glass or plastic combiner plate coated to transmit visible and reflect infra-red light. Because the upper eyelid moves down appreciably when glancing downwards, the combiner is usually angled to view the eye from about 10° below the notional forward direction. The video signal from the camera is invariably analysed by a specially configured computer, and after conversion by a fast ADC the digitised pixel data can either be processed on the fly or placed in a section of RAM. A variety of proprietary algorithms are used to find the pixel co-ordinates of the centre of the bright spot and the co-ordinates of the centre of the pupil. If multiple off-axis illuminators are used to produce a dark pupil, the co-ordinates of the centre of the constellation of spots are determined. The vertical and horizontal differences between the spot and pupil co-ordinates are then used to calculate the eye-pointing direction by interpolation from an LUT or by using a 2-dimensional polynomial regression equation. The values of the parameters in the LUT or the polynomial coefficients are calculated from a calibration process in which the user is asked to stare at a succession of markers in known eye-pointing directions and the pupil-to-spot co-ordinate differences are measured during each staring fixation.

Several commercially available systems function on the same basis, but have the camera-illuminator module mounted remotely. These systems tolerate a range of eye positions by either capturing a small image of the eye within a larger frame, usually about 100mm at the working distance, or make use of a head position tracking system to point a narrow field camera towards the eye. These approaches seem viable, especially as a helmet tracker would be available in most future combat aircraft, but a dozen camera-illuminator modules would be needed to cover the likely range of off-boresight eye directions. A single helmet-mounted module is far more practical.

The available head-mounted eye tracking system based on the differential pupil-CR technique perform adequately in optimum laboratory conditions. A variety of algorithms are used to locate the centres of the CR and pupil, and many allow for the ellipticity of the latter when looking away from the camera axis and the distortion produced by drooping lids and lashes. Contact lenses can be worn providing they do not float around on the cornea. So too can some corrective spectacles, although it is usually necessary to view the eye through the lens from a direction where bright reflections from convex lens surfaces do not overlay the eye features. Measurement range is limited by the deterioration of the CR as it approaches the boundary of the cornea, or moves away from the central part of the contact lens. The useful range, of about 25° to 30° from the camera axis, depends on the individual eye and the camera-illuminator geometry.

Versions employing axial illumination and a bright pupil are preferred because the illumination is easily collimated. The drawback, for an airborne system that must function satisfactorily in all ambient light levels, is that the contrast declines rapidly when the pupil shrinks on adapting to higher light levels, and for this reason alone it must be concluded that the bright pupil arrangement is intrinsically unsatisfactory. However, the use of off-axis illumination and detection of a dark pupil has the problem of siting one or more IR emitters, each with collimating optics of about 20mm diameter, at least 15° away from the camera axis. Most arrangements ignore this requirement and instead place uncollimated IR emitters away from the camera axis at about the same distance from the eye as the camera lens. In consequence the CR produced by each emitter moves erroneously in concert with the slip-induced change in direction of illumination, and the measurements suffer a significant proportion of the slippage errors that the differential technique is intended to circumvent.

The development of systems suitable for combat aircrew must address a number of practical problems. In particular, the system design must remove – or at least significantly reduce – sensitivity to slippage error, yet fit within the geometrical constraints imposed by the headgear and display optics. The essential features of the eye in the captured image must retain sufficient contrast in strong sunlight to be found reliably by the image analyser. Safety is also a major concern. No sharp or delicate elements can be placed close to the eye, and the level of IR illumination must be well within the recommended limits. As discussed in Chapter 9, the latter vary between of different national and international Standards organisations. As a guide[12], the American Conference of Governmental Industrial Hygienists (ACGIH) recommends that the cornea should not be exposed to non-laser IR (of wavelength between 770nm and 3000nm) with an irradiance greater than $10mW.m^{-2}$. To protect the retina the radiance should be less than $0.6/\alpha$ $W.m^{-2}.sr^{-1}$ where α is the angular size of the source in steradians as viewed by the user.

The safe illumination levels are unlikely to limit the quality of the eye image, which can be maintained at high ambient illumination in several ways. For instance, new complementary metal oxide on silicon (CMOS) camera chips have sufficient dynamic range to avoid saturation, or the eye could be shielded using an IR absorbing coating on the visor. Alternatively, the opening of the camera electronic shutter could be synchronised to coincide with bursts of illumination. This would reduce the effective sunlight brightness by a factor equal to the shutter opening time as a fraction of the frame period. The chief obstacle is that the introduction of an eye tracker into an aircrew helmet will exacerbate the already difficult integration problem, as discussed in Chapter 13, and the optical packaging must be done very carefully.

One further technique is worth mentioning; the retinal tracker. Any display device transfers an image to the retina of the eye, and given the reversibility of the light paths, the same optical arrangement can act as a camera. It is however essential that such a retinoscope collects the light that passes out through the pupil and rejects external light reflected from the outer parts of the eye. Although a retinoscope is normally used to examine inside the eye, any shift in the position of features relative to the camera axis can be used to measure the eye pointing direction. Processing of the retinal image to identify features, such as the fundus or a pattern of blood vessels, is performed in much the same way as the more conventional system track the pupil or corneal reflex. Figure 9.17 suggests that the optical arrangement of the VRD is particularly appropriate. As well as scanning the laser beam through a convergence point in the centre of the pupil, the light back-scattered by the retina returns along a reciprocal path. A video signal modulated by the retinal reflectivity can be obtained by directing the returned light onto a fast, sensitive detector from a beamsplitter placed between the compensating lens and the three lasers. Although the arrangement seems like a nice way to gain an eye tracker at very little cost, any significant rotation of the eye or any shift of the optics would displace the pupil from the convergence point. It would therefore be necessary to track the pupil and dynamically adjust the position of the convergence point.

11.5.2 Gesture tracking systems

The main idea behind the development of gesture tracking systems is to exploit well-learned expressions and communicative movements, in particular those made either deliberately or incidentally by the fingers, hands, arms and face. Potential applications include the exploitation of the gestures that accompany speech in order to improve the reliability of DVI systems and the automatic recognition of "signing" gestures made by speech-impaired individuals. Gesture recognition would be a major element in a multi-modal dialogue to enable a flexible interaction with computer-controlled equipment that does not require the user to know, or have the physical ability to use, spoken and written language.

Specific gestures vary between the individuals who make them, and they change over time with repetition and fatigue in much the same way as spoken words. There are also similar problems of defining start and end points that cause problems in parsing continuous speech. In an aviation context gestures would also be affected by high-G and vibration, and by the peculiar headgear and clothing. Also, in the same way that a HOTAS press-to-talk button is used to allow the speech signal to enter a DVI system, in many cases it would be necessary to supply a confirmatory signal to indicate that an action is intended. System performance can be described by the same parameters used for speech recognition, such as the size and particulars of the vocabulary, and also the recognition reliability, probability of confusions between individual gestures and the latency before output. Gesture and speech also share similar needs for feedback and correction facilities.

Many investigators consider that some input systems, such as the touchscreen, miniature joystick, lightpen, mouse, trackerball and similar two-dimensional pointing devices function using gestures. However the unique gesture recognition systems are those that assess the dynamic orientation of fingers and hand, and the configuration of facial features. The range of technological approaches that has been investigated[13] includes variants of the mechanical, electromagnetic, ultrasonic and optical systems described previously for tracking the position and orientation of a helmet and assessing eye pointing direction.

The orientation and position of the hand can be measured electro-magnetically by attaching a sensor cube to the back of a tight-fitting glove, and several ways have been devised to measure the flexural state of the fingers and thumb. The Dataglove had a fibre-optic light guide sewn into the outer part of each finger to form a loop between a light source and a photodiode, and the flexure of the digit was assessed from the reduction in the photodiode signal produced by the increased light leakage with flexural strain. The total bend angle could be measured to an accuracy of about 5°, if calibrated for the user, so the glove could readily distinguish a fist from a relaxed hand or a pointed forefinger. A later version called the Cyberglove used strain gauges instead of the too-delicate fibres, and the makers claimed greater accuracy and an update rate of 100Hz. The most ubiquitous device, the Powerglove, was devised by Nintendo for interacting with computer games, and this used simple strain gauges printed with carbon ink to measure digit flexure, in conjunction with a simple acoustic system to track the position and orientation of the back of the hand.

Other devices attempt to remove the fitting and slippage problems associated with gloves that are not tailored exactly to suit the individual. For instance the Dexterous Hand Master has a set of articulating links along the back of each phalangeal segment, rather like an exo-skeleton, and uses three or four miniature Hall effect sensors to measure the change in orientations along the segments of each digit. The SensorGlove is a similarly adjustable mechanism but instead uses an array of miniature accelerometers. Both devices are claimed to allow normal hand movement, but at more than 300g they are heavy. It has proved difficult to build mechanisms that apply forces to the joints as feedback from remote manipulators. The more satisfactory feedback has been incorporated into gloves in the form of miniature piezoelectric vibrators and inflatable pads that can stimulate the fingertips with tactile sensations.

The sensed measurements are interpreted and classified into gesture types by recognition algorithms, most of which, as in speech recognition, involve the extraction of salient features and a comparison between the real-time samples and stored feature templates. Static postures have proved to be more easily identified than dynamic movements, which have greater variation, and the range of sophistications attempted include dynamic time warping and hidden Markov modelling. At present the algorithms handle isolated gestures. The problem of defining the start and end points has been addressed by requiring the user to adopt a standard hand posture between gestures that corresponds to silence in speech.

Facial gesture recognition is invariably done by analysing imagery from a single video camera. This avoids attaching the user to the system, but the head must be placed within the field of view and oriented towards the camera. Analysis of the shape of the mouth, equivalent to lip reading, has been discussed in 11.3.2. Work is also underway to distinguish facial expressions, for instance of anger, surprise, disgust and happiness, with the intention of modifying the computer-mediated interaction so that it is responsive to an individual's reactions.

Several simulation experiments have been conducted to assess the usefulness of gesture tracking devices in an aviation context. These include the use of a video-based finger tracker to guide hand movements when operating a virtual keypad[14] and the use of an electromagnetic hand tracker for pointing at a distant map display[15]. The pilots participating in the first study disliked having to remove their hands from the HOTAS controls and they had to make slower and more deliberate hand movements to compensate for the latency of the finger tracker. They considered that a visual 3-D cursor showing the measured finger position and tactile feedback of switch actuation (simulated by using real but unseen keys) were essential. In the second experiments, the WPAFB participants found that they could keep their hand near the side-stick, point the electromagnetic receiver of the tracker attached to their hand, see the cursor move around the HDD and then depress a left HOTAS switch to activate the selection. The finger-to-cursor movement scaling was best at about 1:1, and "proximity cueing" which automatically highlighted the object nearest the cursor speeded the process. However, it must be borne in mind that in neither simulation were the participants exposed to realistic disturbances.

11.5.3 Biopotential sensing systems

The biopotentials that are relevant here are the electro-myographic (EMG) signals from the activation of skeletal muscles and electroencephalographic (EEG) signals from brain activity, as mentioned in Chapter 2. The use of signals produced by electro-oculographic (EOG) electrodes placed around the eyes has been mentioned in the previous section dealing with eye pointing. Both EMG and EEG signals are sensed using surface electrodes attached to the skin, and do not involve implanting or inserting electrodes within body tissue. Their causal electro-chemical mechanisms are well understood, as is their main use for medical diagnosis. More recent studies have explored ways by which an individual can monitor a biopotential signal, for instance from a set of EEG electrodes, and learn to exert a remedial influence over a bodily process, for instance the rate of beating of the heart or the rate of sweating; a surprising ability because such responses have been regarded as entirely autonomic. Biopotentials are also under investigation to enable direct conscious control over an external system, for augmenting another control modality by reducing selection uncertainties, and to indicate the alertness and mental workload of the individual. Much of the EMG work is motivated by the desire to provide disabled individuals with direct voluntary control over artificial limbs and orthotic devices.

The aspects of particular interest in aerospace applications are that EEG signals indicate preparation for, and EMG signals precede, muscular action. The signals may therefore be an early, accurate indication of an intention when the muscular action itself is impeded by protective clothing or disturbed by vibration and high-G. The immediate objective of the investigators is to use the signals to augment other control mechanisms and provide a reliable, speedy, flexible interface that leaves the hands free to perform parallel activities. In the long-term it is possible that a combination of control modalities that includes biopotential sensing could lead to the intelligent and intuitive interface discussed in Chapter 14. The investigators are also aware that an open-loop monitoring system will not need to be as accurate and reliable as a means of exerting control, and if their ambitious aims prove unattainable their efforts will nevertheless be useful for gauging the physiological state and alertness of the crewmember. They tend to regard the fictional idea of operating a device merely by thinking about the desired result as naively dangerous. Given the gap between what can now be done and what is required for systems that are mature enough for integration into an aircraft cockpit, unless aircrew can be persuaded to accommodate implanted electrodes, it is also exceedingly fanciful. This can be made clear by considering the current state of the art.

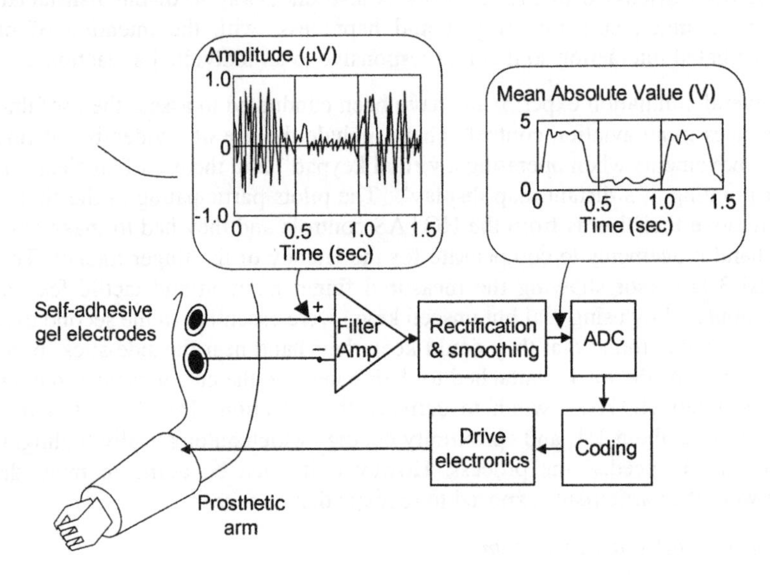

Figure 11.12 Simplified schematic for a single channel EMG sensing system controlling a single degree of freedom prosthesis

The simplest form of EMG system for controlling an artificial limb with a single degree of freedom (DoF) is illustrated schematically in Figure 11.12. When the user voluntarily contracts a muscle, the exchange of ions across the membranes in the constituent muscle fibres produces a collective electrical effect that can be detected as a voltage on the skin surface. It is this myoelectrical activity, a random noise between about 20Hz to 200Hz in the order of 0.1 to 10mV, that is picked up by suitably placed surface electrodes. These must make low resistance contact with the muscle mass through the surface tissue and skin, for which it is usually necessary to clean away the outer layer of dead skin and apply a film of conductive salt gel. The essential phenomenon is the amplitude of the sensed noise, which is related to the level of muscular effort. As shown in Figure 11.12, the electronics amplify the noise signal, assess the amplitude as the mean absolute value (MAV), compare this with some pre-defined set of coding criteria that define the response options and then trigger the prosthesis to make the chosen response. With practice and concentrated attention, the user can for instance induce a prosthetic hand to grasp or release reliably.

This basic scheme admits many subtle variations. For instance, the smoothing applied by the rectification stage can be used to trade accuracy of the MAV against responsiveness to changes of MAV, and this has considerable impact the on the choice of response options. These can be based on the MAV level, the rate of change of the MAV or the use of MAV pulses, for instance a single pulse can be used for grasping and a double pulse for releasing. The last option tends to be slow. Although a proportional control law is possible with the other forms of coding, it has proved most practical to segment the dynamic range into three levels, or three rates, and have these signify "off", "grasp" and "release". More subtle schemes can be used to give better dexterity if other muscle groups provide independent signals, for instance one channel can control the direction of the response and the other the speed.

Complex coding systems using multiple DoF sensing and actuation, combined with electrical and force feedback are current research topics. One strong incentive is to reduce the training and concentration needed for the user to produce deliberate muscle contractions and instead allow spontaneous muscle activation patterns associated with the required manipulation. Most approaches use a larger number of muscle groups and EMG channels, each of which produces an unambiguous on/off MAV output, and the problem has become that of recognising characteristic spatio-temporal features in the activation patterns. As the mathematical techniques are similar to those employed in speech recognition, training is still necessary but this time it is the machine that is doing the learning.

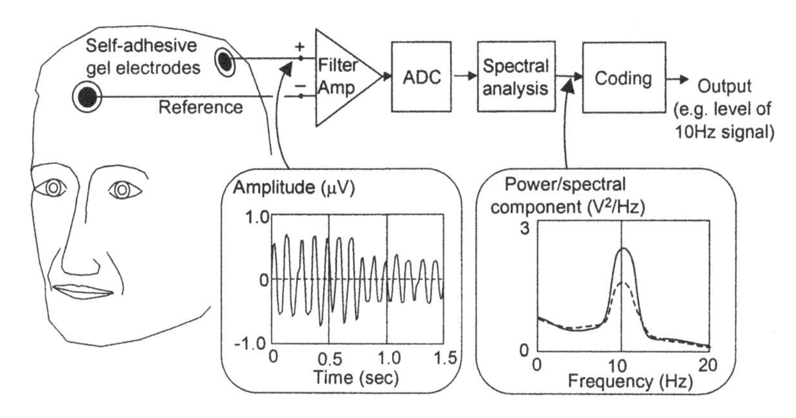

Figure 11.13 The simplest form of processing needed to extract a control signal from a single EEG channel

The exploitation of EEG signals for control is similar to the use of EMG signals, but the means of inducing EEG signals, the nature of the signals and the consequent processing are markedly different. Figure 11.13 shows a simplified schematic for a single channel where the EEG electrodes are placed on the surface of the skull and the signals are the summated effect of the electrical activity of the underlying cranial neurones. The signal strength is weak, in the region of 1 to 50 microvolts, making it necessary to use an amplifier with high impedance (>10Mohm) and high gain (>10^6). The two electrodes are connected to differential inputs to reduce interference from stray electromagnetic fields. The relative weakness of EEG also makes the signals susceptible to masking by nearby muscle activity, particularly from the jaw and eyes, and the electrodes must be located on the scalp in positions that minimise such detrimental artefacts while maximising genuine pick up from the underlying activity of the brain. The crucial difference between EEG and EMG signals is that, although noisy, EEG signals reflect a collective neuronal synchrony, at about 10Hz, and it is the strength of this cyclic variation that is used for control. As indicated in Figure 11.13, the simplest processing uses this "rhythm level encoding". The raw EEG signal is filtered to assess the magnitude of the energy between say 8Hz and 12Hz, and this magnitude is tested to determine whether a threshold value is exceeded in order to select one of two alternative control states for output.

In practice many parameters must be adjusted to obtain reliable voluntary control, mainly because the magnitude of the effect varies between individuals and with small differences in the placement of the electrodes. The techniques tried with EMG sensing have also been explored to enhance the number of degrees of freedom, and the repeatability and speed. There is a similar endeavour to use spontaneous signals rather than contrived, self-regulated responses. Unlike the thinking needed to trigger a muscle contraction, the user has no notion of what he must think about to alter the EEG signal and relies initially on visual feedback of the EEG output. Some individuals develop a way of organising their thoughts, for instance by imagining a pointer moving up and down a vertical scale, but they have no way of knowing whether this causes the change or is merely a correlated irrelevance. Although practice improves performance, this mystery remains.

It is also of note that spontaneous EEG signals can be produced in response to an external trigger, for instance a flickering light. This has the advantage of enabling the repeatedly evoked signals to be discriminated using boxcar integration techniques in which summation synchronised to the triggering events enhances the signal and reduces the random noise. The wide literature describing investigations using event-related potentials (ERPs) and visually-evoked potentials (VEPs) for basic research has stimulated this approach. As an example, it has proved possible to determine the user's direction of fixation within an 8 by 8 matrix on a computer screen by sensing the VEP at the occipital scalp near the visual lobes[16]. In this case the 64 regions of the screen were modulated with different pseudo-random time sequences and the VEP with a matching time sequence could be determined, usually in less than 1.5 seconds. The modulation was however below the critical fusion frequency and the user was aware of the flashing elements on the screen. Visually evoked EEG techniques are unlikely to be tolerated as ways of exercising control.

As yet few simulation experiments have been conducted to assess the practicality of biopotential sensing for aircrew. A considerable amount of work is needed on basic issues, such as dry electrodes that can produce stable signals without skin preparation and hair removal, as well as the optimum mathematical manipulation of the sensed voltages and the incorporation of self-calibration facilities for the user. The most challenging problem is posed by the need to show that aircrew can voluntarily produce and maintain graded EEG signals in real conditions when attending to competing task demands.

11.5.4 Integration of novel control systems

The mixture of human, operational and technological issues involved in introducing new ways of operating an aircraft has been discussed in Chapter 4. The integration of novel controls generates particular engineering challenges, not least the obvious difficulty of packaging the mixture of helmet-sensing, eye-sensing and biopotential sensing devices into headgear, which is considered in Chapter 13. Other issues that must be considered are the overall system architecture and the need for control over the controls.

A way of interconnecting the major elements in the cockpit of a single-seat aircraft fitted with novel control systems, and possibly the other pilot state sensors and intelligent aids discussed in Chapter 14, is shown in Figure 11.14. The important element is a command interpreter that adjudicates between signals produced by both the conventional controls and the novel controls so that a clear, unambiguous command can be sent to the aircraft systems. This function is already supplied by a software module in Eurofighter and Rafale to enable data entry and HDD moding through a flexible combination of vocal commands and key presses. The functions would need to be extended in a future aircraft so that the pilot could for instance select an external object either by fixating with the eye, pointing the helmet sight or indicating with a hand-pointing gesture, at the same time specifying what he wants by saying "target", "lock radar" or "range?" As suggested in the figure, the command interpreter would also arrange feedback to keep the pilot aware of the control system output, and it could relay information between the different control systems to improve performance and supply additional information about the state of the pilot state to an intelligent decision-making aid. Although the performance of the novel controls would be determined by their individual characteristics, the programming of the command interpreter would dictate the usability of the ensemble.

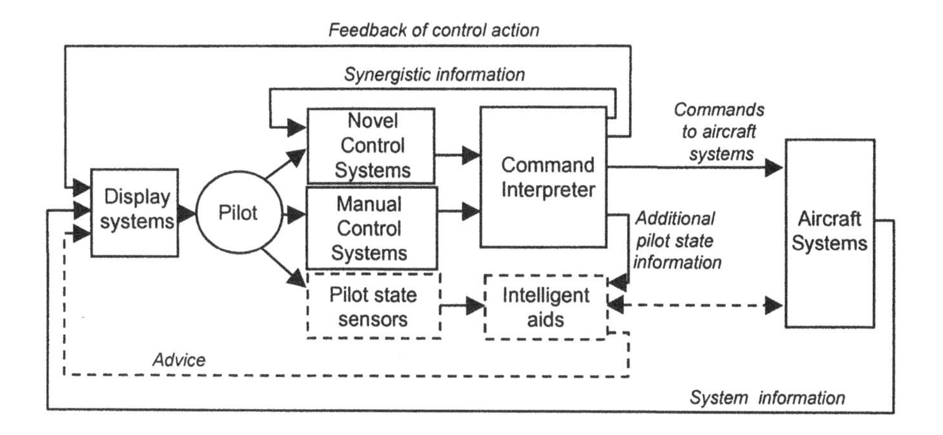

Figure 11.14 Overview of the possible interconnections between cockpit systems in an aircraft fitted with novel control systems (Reprinted with permission from QinetiQ)

At a more pragmatic level, it must be recognised that, unlike conventional manual controls, all novel controls need special provisions so they can be set up to suit the crewmember, and switched off if considered burdensome. A comprehensive training simulator will be essential to enable aircrew to develop confidence in the novel control facilities and complete any time-consuming calibration such as fixating on an array of targets and speaking a list of commands, perhaps also learning whatever mental discipline is needed for biopotential control. As noted, the noise environment must be reproduced authentically so that templates for the DVI system capture the articulation used when the individual is exposed to the noise and he hears himself speaking via the feedback to his earphones. This data can then be transferred to the aircraft by a data-link or portable operational data store to simplify the pre-flight setting up. It is suggested that a high-level means of exercising control over the controls could be engineered as a dedicated panel housing a short row of on/off switches (marked Head tracker, DVI, Eye tracker etc.) that are also connected to the command interpreter. On the other hand, in the absence of understanding how reliably and quickly aircrew recognise the malfunction of a control facility, it may be better to programme the control interpreter to perform this function.

11.6 Challenges

In future, guarded switches, such as those used to jettison underwing stores in an emergency and to release weapons, will continue to be used to avoid the possibility of inadvertent operation. Similarly, the conventional throttle, stick, pedals, and associated HOTAS switches are likely to persist as the fundamental means of steering the aircraft and performing rapid selections. The main challenge is to develop grip-tops and embedded switches which can be operated by a broader range of hand sizes, and avoid increasing the number of such switches in response to the need for confirmation of an input from a novel control device. Other manually operated controls have distinct disadvantages in a combat cockpit; they remove one hand from the primary controls, they are affected by vibration and high-G, and they invariably require the pilots to glance down to locate the switch so that the hand can be guided to reach it. Keys, particularly soft keys whose function depends upon the mode of an adjoining head-down display, and other hand-operated devices such as touch-sensing face-plates on a head-down display, and small joysticks and rollerballs, are more suited to transport aircraft. Here, the environment is relatively benign and head-out attention is unnecessary. The main challenge for the developers of speech recognition systems for combat aircraft is to give DVI sufficient accuracy, vocabulary and insensitivity to operating conditions so that keys can be relegated to a minor reversionary role.

Although helmet position and orientation sensing systems are mature, they currently lack the sub-milliradian accuracy and low latency for a helmet-mounted display to be used in a visually-coupled system to replace a head-up display. Eye trackers are untried, but they are likely to give substantial benefits if integrated properly into aircrew headgear, and measure eye direction imperturbably, accurately and rapidly. Gesture sensing systems have limited usefulness in combat aircraft. The challenge for other airborne applications is to develop recognition systems that are sufficiently robust and accurate for the naturalness of gesture to be exploited as a component of a multi-modal dialogue. Biopotential sensing systems, which are still really in the laboratory, face a particularly daunting challenge. It is necessary to show that in the cockpit environment the sensed information can be related to the operator's intentions with sufficient reliability to exert direct control over airborne systems. If they cannot be accorded this authority, they may still be useful in conjunction with the other novel control technologies to create an intelligent interface that actively assists the crewmember by inferring his intentions. They may also be exploited as a means of assessing the operator's physiological and psychological state.

Introducing novel control technologies into an aircraft cockpit requires the resolution of many human and engineering issues, in particular the detailed design of the control task and the facilities for calibration, feedback and error correction. The main challenge will be to exploit fully the opportunity to create a genuinely natural interface that can be tailored to suit individuals of either sex with a wide variation in size, shape, ability and preference.

References

1. *Aircrew stations control panels*, STANAG 3869 A, NATO Military Agency for Standardisation, Brussels
2. *Human engineering design criteria for military systems, equipment and facilities*, MIL-STD-1472B Department of Defense, Washington D C, 1983
3. Taylor R M and Berman J V F, *Human factors in aircraft keyboard design: standards, issues and further evidence relating to gloves and key characteristics*, in *Advanced avionics and the military aircraft man/machine interface*, AGARD-CP-329, 1982
4. *Operational needs and opportunities for alternative controls*, in *Alternative control technologies*, RTO-TR-7, 1998
5. Pastor D and Gulli C, *DIVA 5 dialogue vocal pour aeronef: performance in simulated aircraft cockpit environments,* Joint ESCA-NATO/RSG10 Tutorial and workshop: applications of speech technology, Lautrach, 1993
6. Finn K E and Montgomery A A, *Automatic optically-based recognition of speech*, Pattern Recognition Letters, 8(3) pp159-164, 1988
7. Barnes G R and Sommerville G P, *Visual target acquisition and tracking performance using a helmet-mounted sight*, Aviation, Space and Environmental Medicine, April 1978
8. Rowlands G F, *The transmission of vertical vibration to the head and shoulders of seated men*, Unpublished MOD report

9. Polhemus W L, *Micro-tracker*, in Proceedings of the 2nd Advanced Aircrew Display Symposium, NATC, Warminster, 1975

10. Vivash J P and Belyavin A J, *Eye movements under operational conditions*, in Waters M and Stott J R R (eds), Journal of Defence Science 1(2), 1996

11. Ober J, *White Box mask-mounted eye tracker*, http://www.ibib.waw.pl

12. Sliney D H and Wolbarsht M, *Safety with lasers and other optical sources: a comprehensive handbook*, Plenum Press, New York, 1980

13. Buxton B, *A directory of sources for input technologies*, available from http://www.billbuxton.com/InputSources.html

14. Ineson J and Parker C C, *The accuracy of virtual touch*, Unpublished MOD report

15. Reising J, Liggett KK and Hartsock D C, *Exploring techniques for target designation using 3-D stereo map displays*, International Journal of Aviation Psychology 3 (3), pp. 169-187, 1996

16. Sutter E E, *The brain response interface: communication through visually-induced electrical brain responses*, Journal of Microcomputer Applications 15, pp. 31-45, 1992

Chapter 12

Emergency Escape

12.1 Rationale

The requirement seems simple and self-evident; to enable the crew to escape from a stricken aircraft and reach ground level uninjured. This was satisfied in early aircraft by supplying a parachute and relying on the crewman's ability to unbuckle the seat harness, clamber from the cockpit, jump, wait till clear and pull the parachute release handle. However, the motion of the aircraft and the force of the windblast, compounded by the likelihood that the crewman would be as incapacitated as the aircraft, made this "bailing out" at best slow. A successful outcome was lucky rather than likely. With the development of enclosed crewstations the crew had also to negotiate their normal means of access and egress, such as a sliding canopy or a series of doors, so the probability of survival declined further. It was then accepted that the crew needed assistance. The requirement has come to mean the provision of facilities that, no matter what the state of the crew or the aircraft, when the need arises the facilities can be triggered to automatically extract the crew from the aircraft, deliver them to ground level, keep them alive and co-operate to ensure their rescue.

12.2 Physical and physiological requirements

Escape from a terminally-damaged aircraft can expose the human body is to a sequence of violent forces; the gyratory tumbling of the stricken aircraft, the thumping heave of ejection, the blast of the airstream, the snatch through the harness when the parachute inflates and the impact with the ground. As noted in Chapter 3, at high altitude there is a dearth of oxygen and the air temperature is very low, for instance at 36,000 ft pressure is about ¼ of that at sea level and the temperature is about –55°C. It is necessary to understand how the human body behaves when subjected to these stresses. The large number of difficult aeromedical experiments that have been conducted, mainly to establish quantitative relationships between the magnitude of the stress and the probability of injury and death, have underpinned the development of escape systems.

Consider firstly some of the forces and accelerations, for which the nomenclature is illustrated in Figure 3.9. Early experiments showed that it was unlikely for a man to be capable of sliding a canopy and clambering out of his seat if the aircraft were generating forces greater than about 2.5G in heave. If he were to slip he would incur considerable delay, which was very likely if the direction of the force changed constantly as the aircraft gyrated. Such experiments provided convincing evidence of the need for assisted escape.

Early experiments also established that during forcible extraction the human body could survive an acceleration of about 20G in the +Z direction if this was imposed for less than about 200msec and the jolt (the rate of onset of the force) was kept below about 200G.sec^{-1}. These dynamics were just within the capabilities of early cartridge-powered telescopic catapults systems. The experiments also established that the spine was more vulnerable than the heart or brain. Over the course of time an understanding has been built up concerning biodynamic stress transmission and its effect on posture, for instance how the head can be thrown forward onto the chest and the torso collapse towards the knees under a $+G_z$ jolt. It was also found that the seat cushion could exaggerate forces by first compressing and then rebounding to release energy into the pelvis. Amongst other lessons, these experiments emphasised the need to support the spine and head, and the value of a self-tightening harness.

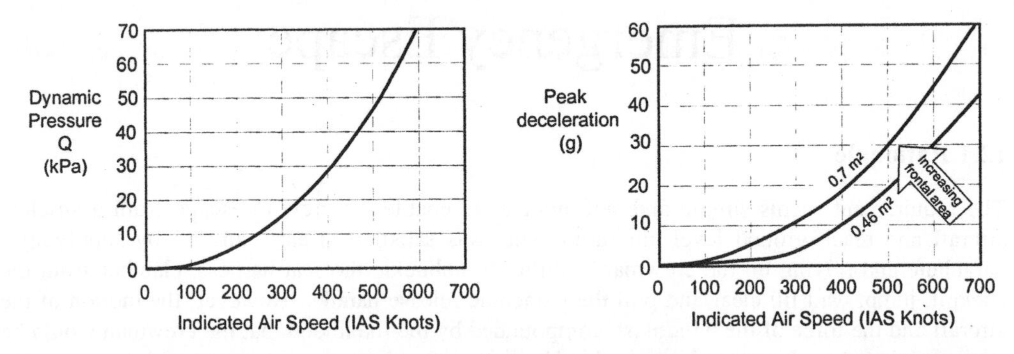

Figure 12.1 (Left) the variation of the peak dynamic pressure with the airspeed of the aircraft. (Right) The peak deceleration induced by the windblast for two values of the frontal area and a combined seat and occupant mass of 145kg (Data are from Mohrlock[1])

Complex effects occur when the man and seat are suddenly released into the airstream. The dramatic $-G_X$ eyeballs-out deceleration caused by the aerodynamic drag of so blunt a body as a seated man depends mainly on shape, mass, effective cross-sectional area and airspeed. As shown by the right graph in Figure 12.1, for airspeeds above 500KIAS this deceleration exceeds the $+G_Z$ extraction acceleration[1]. However, the windblast also pushes the man backwards from the slumped posture produced by the ejection jolt, so that his torso is pinned to the seat back and his head and helmet collide with the head-rest. The magnitude of the blast force can be estimated from the graph relating the dynamic pressure to the windspeed, as shown on the left of figure 12.1. At 500KIAS the dynamic pressure of about 45kPa delivers a force of about 20kN to the helmeted head alone. The blast also tends to spin and pitch the seat because the centre of blast pressure rarely passes through the seat c-of-g, and it is necessary to deploy a stabilising drogue as soon as possible. If unprotected, the soft tissue of the face would distend at about 100KIAS and tear at about 300KIAS, so enveloping head protection is vital. The rapid onset of the blast also makes the head, arms and legs flail and flutter. Above about 450KIAS the forces exceed the most strenuous muscular effort, so without the positive location and support of the head box and firm harness and restraints, limbs would dislocate and probably rupture. It is this vulnerability that has prompted the development of enclosed seats and capsules.

Surviving escape at altitude above about 25,000 ft needs heat-retaining clothing and a supply of breathing oxygen. It is thus necessary to descend to about 8,000ft quickly, and only then deploy the parachute. At this altitude the shock load imposed on the man through the harness when the parachute inflates, arresting the free-fall from about $160 ft.s^{-1}$ to a safe $25 ft.s^{-1}$, is a tolerable 15G to 20G. Inflating the parachute at high altitude where the speed change is considerably greater in the thinner air, or at lower level while still decelerating in the windblast, can add considerably to the shock load on the parachute and the man, and either can be overstressed. The best time to deploy the parachute depends on the circumstances.

The final hurdle, of coming safely to earth, is least likely to injure a parachutist performing a standard rolling fall. If the man still occupies the seat, a descent at $25 ft.s^{-1}$ onto an unyielding surface could induce excessive forces, so some form of impact cushioning is essential. Coming down into water is less of a shock, but unless the man can inflate a life jacket and remove the oxygen mask, drowning is an obvious hazard. Depending on the wind and sea states, it may also be vital to detach the parachute to avoid being dragged under water. At low temperatures, exhaustion from heat loss can be delayed by entering an inflated dinghy, and on land the man must seek shelter. Timely use of radio location aids, and smoke, lights and flares, makes rescue more likely and significantly easier. It is of note that all of these activities must be done by the ejectee, and unless the designers of the system have arranged for the whole process to be fully automated, the ejectee must be delivered the to ground level in a conscious state for him to be capable of managing his subsequent survival.

12.3 An overview of ejection seat development

The story of the development of the engineering solutions to this requirement is interwoven by three threads; the gradual expansion of the conditions in which flight has become possible, the accumulation of knowledge about human tolerance to mechanical stresses, and the realisation that aircrew represented a considerable investment in training and are worth saving on economic grounds alone. In retrospect it is surprising that the invention of the practical parachute in 1919 by Leslie Irvin was regarded by the Royal Flying Corps as a slur to the daring of the aviator. However the attitude soon changed and most aircrew had recourse to a parachute, albeit not normally attached by harness or stowed conveniently.

In the 1930s, the development and the flight testing of the mono-plane with an enclosed cockpit, high performance and high ceiling set aviators and engineers thinking about ways to remove aircrew from a crippled aircraft. The German Luftwaffe considered several approaches, including a powerful pre-compressed coil spring beneath the seat, or a sprung lever pivoted behind the cockpit that could be released to throw the pilot upwards by yanking on a becket attached to the seat harness. Although experiments were conducted using dummies, nothing is known about the results, but it can be imagined that the engineers would have encountered problems, not least with the mass of the spring, the strength of the container and the reliability of the release mechanism. A similar concept, in which the seat was lifted clear of the cockpit by springs contained in telescopic tubes, was devised by a RAF officer but again the idea was not taken up.

By the start of the Second World War the RAF had equipped combat aircraft crew with parachutes and harness that were worn throughout the flight. At the same time Luftwaffe aero-medical experiments suggested that the seated human could tolerate an ejection force of about $20G_z$ if the force was directed along the braced spine and endured for less than about 200msec. The Heinkel Aircraft Company, among several German manufacturers developed seats that could throw the crew clear of the aircraft within these impulse limits using a variety of compressed springs and telescopic cylinders, the latter energised either using pre-compressed gas or gas generated by detonating an explosive. In-service experience confirmed that the explosive propulsion was lightest, most reliable and easiest to maintain in the field. This was confirmed by the Swedish SAAB Aircraft Company, where an ejection system was developed for the J21aircraft which had a pusher propeller behind the cockpit. The ending of the war brought the German and Swedish expertise in emergency escape systems to light, and this, together with the emergence of jet-powered aircraft, caused the military aviation authorities in France, Britain and the USA to instigate indigenous research and development. Most of the companies that manufactured combat aircraft began to think of including such systems in their new projects. In Britain the Martin-Baker Aircraft Company took up the challenge at the request of the RAF Air Staff.

12.3.1 First generation ejection seats

By the end of the 1940s the diverse developers had largely adopted the Heinkel and SAAB model, with all escape systems configured as a seat propelled by gas cartridges acting on a telescopic gun, but there was considerable variation in the details. For instance, researchers in the

US Air Force at Wright Field considered that the best way to induce the ejectee to hold his head erect, straighten his spine and clench his torso muscles, a posture that was least vulnerable to the upward jolt, was for the ejectee to pull upwards then squeeze together a pair of actuating handles in the arm-rests. The US Navy, along with Martin-Baker, considered that the preferred posture was best induced by reaching up and pulling down on a face-blind actuator, as this had the additional benefit of shielding the head from the subsequent windblast. It is interesting to note that both approaches have been supplanted by the between-the-legs handle that can be grasped more easily and operated more rapidly in a gyrating aircraft.

The Martin-Baker Mk1 was typical of the seats fitted to the first jets in UK service, such as the Venom and Canberra. The structure was an aluminium ladder frame with a head-box, seat-back and a seat-pan with thigh-restraining seat extensions and sprung foot-rests. The seat pan could be adjusted vertically using a retracting handle. When the crewman initiated ejection by pulling down the handle of a face-curtain at the top of the head-box, the primary cartridge of the catapult was detonated by a firing unit linked directly to the face-curtain.

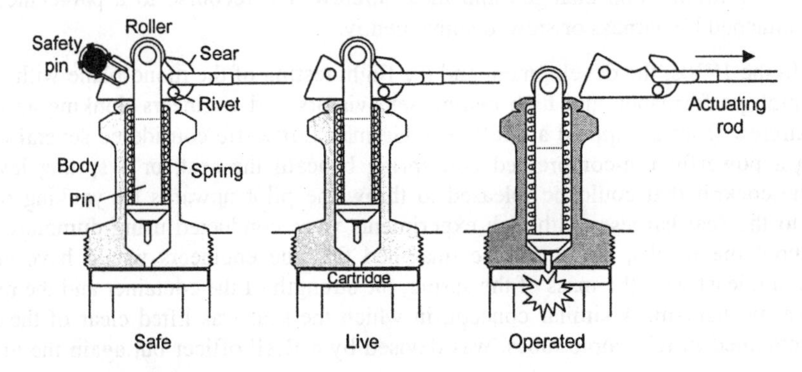

Figure 12.2 Essential elements of a firing unit

The characteristics of a firing unit were central to the reliability and safe management of the seat. As illustrated in Figure 12.2, the unit was a simple, reliable mechanism for setting off a detonator that consisted of a sprung firing pin with a roller, a sear and a locking pin. The firing pin was kept cocked by the sear and this could not be pulled out accidentally if the prominent, red-tagged safety pin was in place. Aircrew checked before each flight that all safety pins were removed and stowed in a designated holder, and they inserted the safety pins as part of the post-flight checks. Ground crew repeated the latter check to ensure that the seat was safe for maintenance. With the safety pin removed, a sideways pull on the sear lifted the firing pin against the heavy spring force of the roller against the ramp on the sear. When the roller reached the ramp apex the sear was released and the spring propelled the firing pin into the detonator.

The catapult gun that sat between the vertical seat frames was a pair of telescopic steel tubes about 100mm diameter and 1m long. The outer was attached to the cockpit floor and bulkhead and had longitudinal groves to guide sliders on the seat frame. The inner tube, which acted as the moving piston, was latched to the top of the seat frame. The primary cartridge was inserted in the upper end of the piston, and a secondary (flat cylindrical) cartridge was mounted about half way up the outer tube. When the primary gun cartridge was detonated, the explosive hot gas forced the seat upwards. As the piston skirt exposed the second cartridge this was detonated by the flame front to maintain the upward impetus. With the seat and occupant clear of the aircraft, a small (about 24 inches diameter) drogue parachute was automatically deployed on a 24 ft static line attached to the top of the seat frame. This stabilised the seat in a feet-into-wind attitude. The drogue was extracted forcibly from its compressed pack in the head-box by a stout nylon line attached to a weighty steel bullet that was fired from a short-barrelled drogue gun mounted alongside the head-box. The sear in the firing unit of the drogue gun was connected to the outer tube of the catapult by a rod, which withdrew the sear as the seat moved up the guide rails. A clockwork escapement mechanism in the firing unit slowed the movement of the firing pin to introduce a delay of one second for the seat to clear the aircraft. However, the crewman had to perform all the other actions. Prior to ejection he had to remove the canopy and as soon as possible after the seat was clear of the aircraft and the drogue set, he had to pull a separation handle that levered a series of rods and cranks to unlatch the catches retaining the seat harness. This allowed the seat to fall away. He used another handle to deploy the parachute that was originally packed into the seat-pan, and when safely under the canopy he had to unclip the survival pack (originally stowed in the seat-back) so that it dangled beneath him from a lanyard attached to the parachute harness.

12.3.2 The ejection envelope

The fundamental requirement of the ejection seat was to impart sufficient transverse velocity to throw the occupant clear of the aircraft. At high aircraft speed the vertical tail-fin of most jets soon reached the spot above where the cockpit had been, so sufficient ejection velocity was needed to make the seat travel more than the fin height in the time it took the fin to move the fin-to-cockpit distance. The dimensions of the aircraft and the maximum escape airspeed therefore governed the necessary seat acceleration. This was usually most relevant at high altitude. When close to the ground other parameters became crucial; the minimum height and maximum descent rate at ejection for the crewman to have time to deploy the parachute and for it to inflate and arrest his fall, and the lowest airspeed that would carry the descending parachutist clear of the abandoned aircraft. The attitude of the aircraft was also important, mainly to avoid catapulting the seat down towards the ground. This and the time taken to remove the canopy or any other impediment before the cartridge could be detonated also had a dramatic effect on the minimum height for successful ejection. These parameters defined the ejection envelope that characterised the performance of the system.

12.3.3 Further developments

Progressive development continued. To separate the man from the seat and deploy the parachute, Martin-Baker exploited the pulling power of the drogue and they devised a variant of the clockwork drogue gun delay mechanism which they called the Time Release Unit (TRU). This was also triggered by a firing unit attached to the cockpit by a static rod, and a delay of 5sec was introduced by an escapement mechanism, unless the ambient air pressure was below that at 10,000 ft, in which case an aneroid capsule in the unit jammed the clockwork escapement. The escapement remained jammed until the seat fell below 10,000 ft whereupon the rise in the ambient pressure allowed the aneroid capsule to shrink and disengage. When the TRU fired, a plunger pulled a series of rods and levers that withdrew the engagement catches from the shoulder, lap and negative-G straps of the seat harness, and an upward extension of the plunger simultaneously unlocked a scissor shackle. The latter uncoupled the drogue line from the seat and allowed it to drag the main parachute from its packing. As it inflated, the parachute pulled the man away from the seat.

An automatic canopy jettison system was devised that used gas-powered jacks connected by pipes to the top of the seat guide rail where a firing unit and cartridge were sited. Both this unit and the catapult were triggered simultaneously by actuating the face-blind handle, but the firing unit of the ejection catapult was modified to incorporate a clockwork escapement mechanism to give a delay of 1sec. Leg restraints were added to prevent exit injuries and to stop flailing. These consisted of a webbing strap attached at one end to the seat base, threaded through a sliding buckle on a strong garter around the upper calf, brought back through a round hole in the seat base and connected to the cockpit floor by a shear pin. As the seat rose on ejection, the strap pulled the leg back till the buckle hit the hole in the seat. When the attachment pin sheared, the strap was held taught by a ratchet. The list of improvements included the between-the-legs actuating handle, a combined seat-&-parachute harness incorporating a manual shoulder strap unlock, and a G-sensor; effectively a rolling ball that jammed the escapement mechanism of the TRU. The latter stopped the drogue being detached from the seat and deploying the main parachute until the drogue-induced deceleration reached a level that would not damage the main parachute.

Although the RAF and the US Navy have conducted research and development in collaboration with Martin-Baker, and largely adopted their seat systems, other Air Forces, and in particular the USAF, have pursued independent courses. From the outset in the early 1950s the US Navy was particularly concerned with saving aircrew at low level and low speed during carrier landing and launch, while the USAF emphasised the need to provide escape from interceptor aircraft flying at high speed and high altitude. Their requirement was actually to enable escape above Mach 1, where two issues became acute. Firstly, their studies had shown that to throw the seated man clear of the tail an explosive-cartridge-propelled gun would have to produce an injurious jolt. Secondly, they were concerned to protect the crewman against exposure to a potentially supersonic airflow. These problems stimulated considerable engineering ingenuity, but some of the earlier pragmatic solutions were not successful. For instance, as the probability of spinal damage from a more powerful gun was seen as outweighing the probability of needing to escape at low level, the first version of the lightweight Lockheed F-104A Starfighter was designed with a downward-firing cartridge-powered seat. This was installed with an automatic sequencer that first depressurised the cockpit, moved the control stick forward, tensioned the harness and an arm-restraint net, reeled the feet back using cables attached to the boot heels and blew the floor hatch away using explosive bolts. Although the system was effective and reliable, a well-publicised fatality during an attempted low-altitude escape caused the USAF to reconsider. The cockpits of later aircraft were therefore modified to allow the installation of an upward-firing ejection seat.

12.3.4 Second generation ejection seats

The USAF recognised that the better solution lay in lessening the need for a high-impulse acceleration produced by an explosive gun. The gun also produced secondary problems. For instance, as soon as the seat cleared the guide rails, the pitching force rotated the seat around the centre of gravity, making it tumble forward as it arced upwards, subjecting the seat and the occupant to complex inertial and aerodynamic instabilities. These continued until the drogue stabilised the seat in a feet-forward attitude. The solution adopted by the USAF and US aircraft manufacturers such the Weber Corporation, was essentially the same as that used in the Martin-Baker Mk6 seat; the use of solid-propellant rockets. These could apply a longer duration thrust to augment the gun impulse, and the combination could throw the seat from ground level to a height that enabled the parachute to inflate and arrest the crewman's descent before impacting the ground. Figure 12.3 shows the under-seat mounting of a pack of small rocket-motors adopted by Martin-Baker and the behind-the-seat pair of rocket-motors, parallel with the gun tubes, used by most of the US manufacturers. In the latter, the thrust of the rear-mounted rockets was angled backwards to avoid inducing a pitching moment. This had the small disadvantage of propelling the seat forward and exacerbating the windblast.

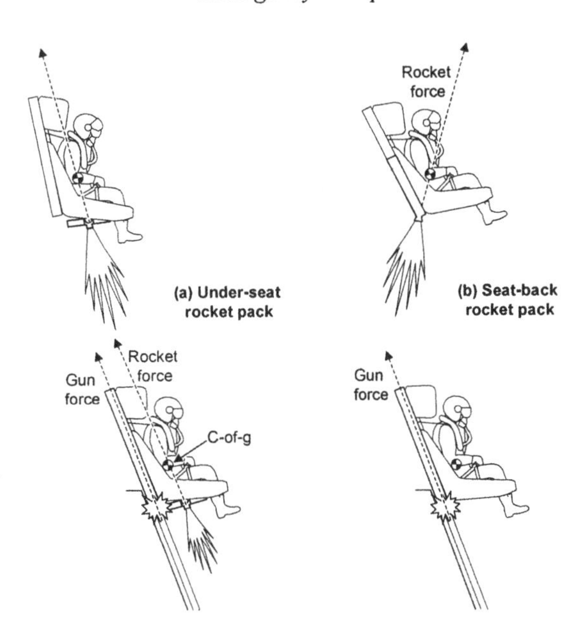

Figure 12.3 Use of a rocket pack to maintain post-escape acceleration, and the direction of the rocket force from under-seat and behind-seat installations

The introduction of rocketry was therefore the solution to the problem of low-level escape and the means of entering the zero/zero corner of the escape envelope. Unsurprisingly, when the USAF and US industry, under the aegis of the Industry Crew Escape System Committee (ICESC), instituted studies in 1957 to develop a system for escape at supersonic speeds from the new series of high-level interceptors, the configurations put forward all featured rocket propulsion. Of these, the Convair-Stanley B-seat, which was built, tested and refined over the early 1960s, was the most technically adventurous. It had a mass of about 260kg – three or four times the mass of a conventional ejection seat – and had three ejection stages. A strong pull on the between-the-legs actuating ring simultaneously jettisoned the canopy, pulled the shoulder harness tight, reeled in the boot heels, raised the foot pans and brought up the knee and thigh guards so that the occupant was trussed in a foetal huddle. A second pull on the actuator fired the seat vertically up a set of rails into the airstream, where it was rotated backwards so that the extended seat base pointed into the airstream to protect the now supine occupant. On latching, cartridges were fired to extend a pair of telescopic stabilisation booms from the rear of the seat, and explosive bolts were triggered to separate the seat from the rotational carriage. The main rocket was then ignited, carrying the man and seat clear of the aircraft. As with the Martin-Baker system, the sequence timer and control unit kept the man in the seat till below 10,000 ft, but at low speed and low altitude seat separation was not delayed and the parachute was deployed after only 3.5sec. The seat was fitted to early versions of the F-106, but the multiple stages proved to be too slow during low-level emergencies, and after several fatalities it was replaced by a more conventional rocket-powered seat made by the Weber Corporation. Again, a system designed primarily to meet the extremely demanding requirements for escape at high speed and high altitudes had, like the downward ejection system in the F-104, failed to satisfy the apparently easier low-speed/low-altitude requirements.

Most US manufacturers, including Vought, Douglas, North American Aviation, Stanley and Stencel, developed rocket systems to augment the gun. When it became available they incorporated the Rocket Assisted Personnel Ejection Catapult (RAPEC), an outcome of an extensive US Navy development and testing programme. In part the history of later developments has paralleled the gradual evolution of the Martin-Baker series, with progressive improvements introduced on successive versions. One notable example was the Douglas Escapac series. These seats were ejected by a RAPEC catapult having 1100 lb.sec impulse and, if fitted to multi-seat aircraft such as the Lockheed S-3A, a small 11lb.sec rocket was added to exert a side thrust for about 20msec to make the trajectories diverge. The seat also employed two other devices. A gimbal-mounted rocket of 250lb.sec impulse, controlled by a gyro, stabilised the seat attitude during the main boost, and then a strong, 100lb.sec thruster was used to ensure that the man separated cleanly from the seat. These systems have evolved into the modern McDonnell-Douglas Advanced Concept Ejection Seat (ACES) series, fitted to US combat aircraft such as the F-16 and F-117. Versions differ between aircraft types but all provide a wide escape envelope, from zero/zero to their respective aircraft ceiling. Most now incorporate electrical sensing and computation, and electrically-initiated gas generators.

There are many other examples of independent developments in the US. One is the series of lightweight systems developed by the Stencel Aero Engineering Corporation, which have been adopted for the US AV8-B and numerous light combat aircraft. Others are the range of encapsulated seats and detachable escape capsules, reviewed briefly in section 12.6. Two other examples illustrate lateral thinking and engineering excellence nicely. In the early 1960s Stanley developed a means of extracting the man from the aircraft that avoided having to catapult him out while strapped to a seat. The principle was simple; clear the escape path, release the pilot's seat harness and simultaneously catapult out a rocket on a nylon pendant attached to the harness. Then ignite the rocket to haul the pilot from the aircraft and launch him upwards. The nylon stretched by about 30%, and this eased the jolt. A second, considerably smaller, rocket was then fired from the back-pack to unfurl the parachute. This "Yankee" system was relatively light and simple and could be used at zero height and zero airspeed, hence it was ideal for lightweight subsonic aircraft including the Douglas A-1D Skyraider. The design was taken up later by Stencel and developed as their Ranger escape system. The second example is the seat developed by North American Aviation in conjunction with the USAF for the X-15 rocket-powered high altitude research aircraft. This was intended to provide safe escape at an altitude of 120,00 ft and a speed of Mach 4, but was also tested to provide satisfactory escape at ground level and 90KIAS. Although the configuration could be regarded retrospectively as that of a conventional open seat with gun and rocket propulsion, it was constructed to withstand the strong blast forces and it included modern refinements such as solid metal restraints for feet and arms, fold-out stabilisation fins and telescopically extending drag booms. The pilot wore a full pressure suit, so the seat also carried a larger-than-usual supply of breathing oxygen, plus a battery to power the pilot's electrically heated visor that could otherwise freeze over.

12.3.5 Third generation ejection seats

Concurrent development of Martin-Baker seat systems continued steadily. For instance, in their Mk6 the rocket pack was mounted on lateral pivots and set at a pitch angle proportional to crewman weight, which minimised tumbling by aligning the thrust with the overall c-of-g of the seat and occupant. The rocket carried the seat upwards about 350 ft, which enabled ejection at 0ft/0KIAS, even from a falling VSTOL Harrier that had the misfortune to suffer a flame-out while hovering. The Mk8, intended for the TSR-2, was a more considerable re-design that eliminated the face-blind handle, largely because the newer helmet and mask systems provided similar face protection and the between-the-legs handle had proven reliable.

Figure 12.4 illustrates the configuration of a typical ejection seat in service in aircraft such as the Jaguar, Tornado and Harrier. The main structural elements are a pair of vertical bearers of rectangular section made from extruded aluminium alloy. On these the other assemblies are mounted; the seat-pan, the seat-back and the head-box, all constructed from riveted plate alloy. The position of the pan can be adjusted vertically over about 140mm by an electrical screw jack to raise the crewman so that his eyes are in the design position when using the HUD. This is the only electrically powered unit on the seat. All the other functions, including ejection propulsion and the gentler forces for control and actuation, operate using the pressure, and sometimes the temperature, of hot gases produced by carefully metered chemical reactions. The sequence of operations is summarised in Figure 12.5.

The reliance on explosive gas generators for control, actuation and propulsion is made clear in Figure 12.6, which shows the essential units of a Martin-Baker Mk10 seat. The catapult gun still sits between the vertical bearers, and is typically a set of three steel telescopic tubes about 100mm diameter and 1m long. The outer is attached to the cockpit floor or bulkhead and has longitudinal groves that guide sliders on the seat bearers. The inner tube acts as the piston and the intermediate tube extends the overall travel and spreads bending loads. The primary cartridge is inserted in the upper end of the piston, which is latched to the top of the seat frame, and two secondary (flat cylindrical) cartridges are mounted on the outer tube.

When the crewman pulls the seat-firing handle a sear is removed from a small firing unit that detonates and sends hot gas through a rigid pipe to the harness retraction unit; a take-up spool powered by a piston and a rack and pinion gearbox. The rigid pipe continues up to an actuator unit at the top of the frame forcing a piston to extract the sear from the firing unit of the primary gun cartridge. This is an elongated version of that illustrated in Figure 12.3, in which the motion of the firing pin is slowed by a clockwork escapement wheel so that detonation is delayed by 300msec to give time for the canopy to be removed or shattered. The latter is initiated through a coupling unit on the primary gas pipe. On detonation, the primary gun cartridge forces the seat to accelerate upwards, and when the piston skirt exposes the second and third cartridges these are detonated in sequence by the flame front. At the limit of its travel the intermediate tube is retained within the outer tube, and the piston continues to push the seat upwards to the limit of the gun stroke at about 2m. The total jolt takes about 500msec and imparts a velocity of about 30m/sec. As well as retracting the leg restraints, the upward motion of the seat pulls in and cleats the arm restraint straps. These are attached to a shear-pin anchored to the cockpit floor, pass through guides in the front of the seat-pan alongside the firing handle and are brought up over the torso to metal rings at the top of the sleeves on the life-preserver stole. The coupling rings can run along braids attached securely to the shoulder and wrist, and are kept at shoulder level by hook-&-loop flaps on the outside of the sleeves that enclose the braids. In normal flight there is ample slack in the straps for normal shoulder movement. On ejection, the upward seat movement pulls the straps tight and the couplings are dragged down the braids, pulling the weakly-locked flaps aside, until they reach the wrists. Like the leg restraints, the straps are tensioned when metal clamps on the straps reach a limit stop. The tension shears the anchor pins, leaving the crewman's wrists tethered by the short lengths of strap between the sleeves of the stole and sprung cleats at the limit stop.

Before separating from the aircraft a trip line uncoils, triggering a firing unit mounted near the top of the seat frame that sends hot gas along an armoured hose to ignite the under-seat rocket pack. This is a set of twelve, parallel, closed-ended tubes screwed into a central manifold on which are mounted a set of downward-pointing efflux nozzles. This gives an impulse of about 1000lb.sec, which is sufficient to throw the seat high enough for zero/zero escape.

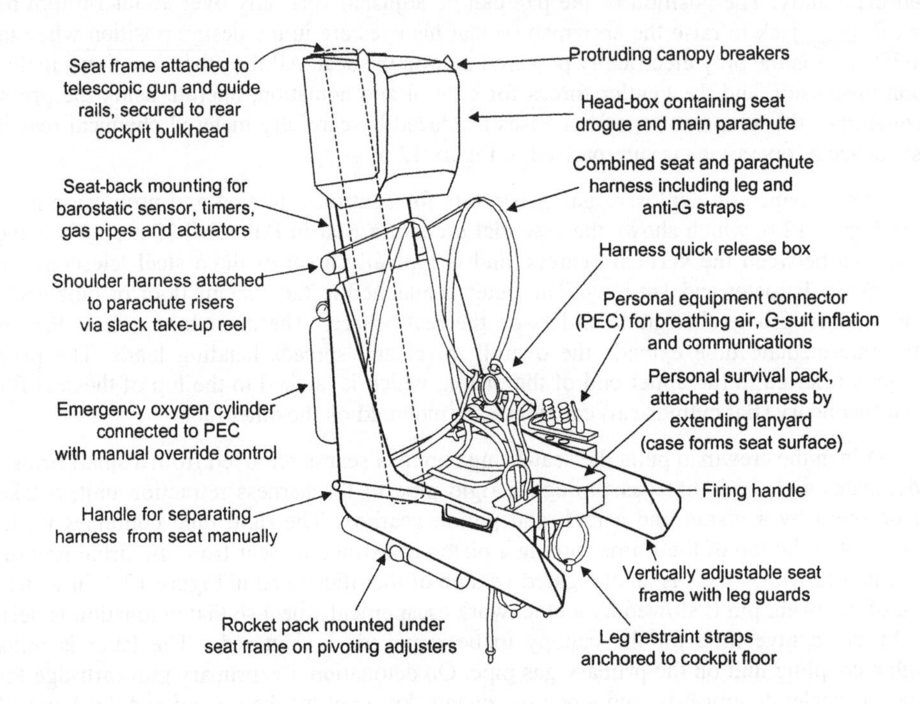

Figure 12.4 Main elements of a typical third generation ejection seat

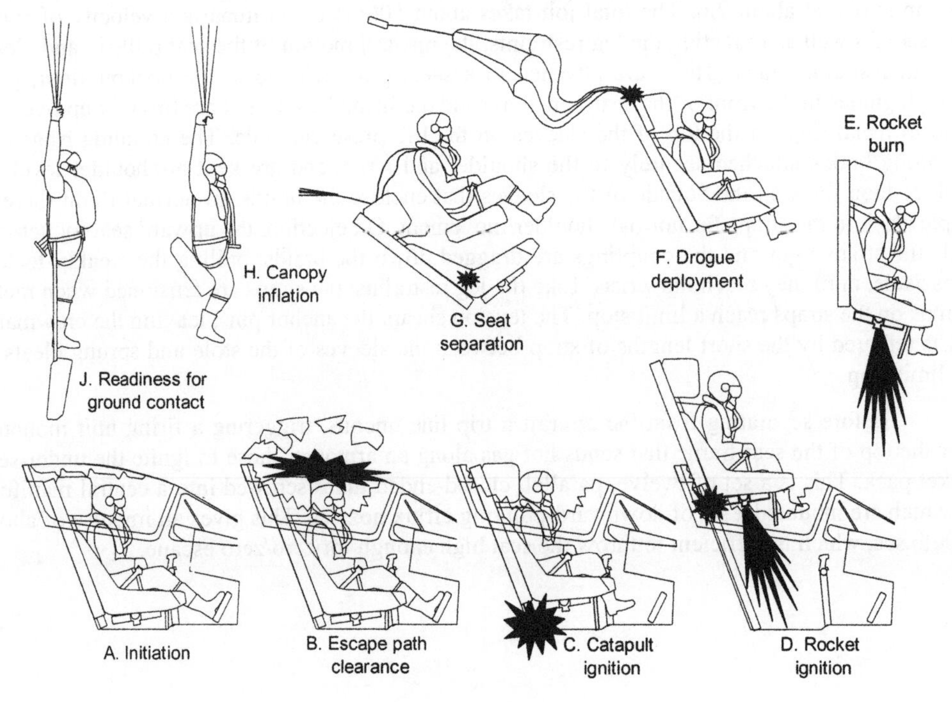

Figure 12.5 Escape sequence for a typical in-service ejection seat

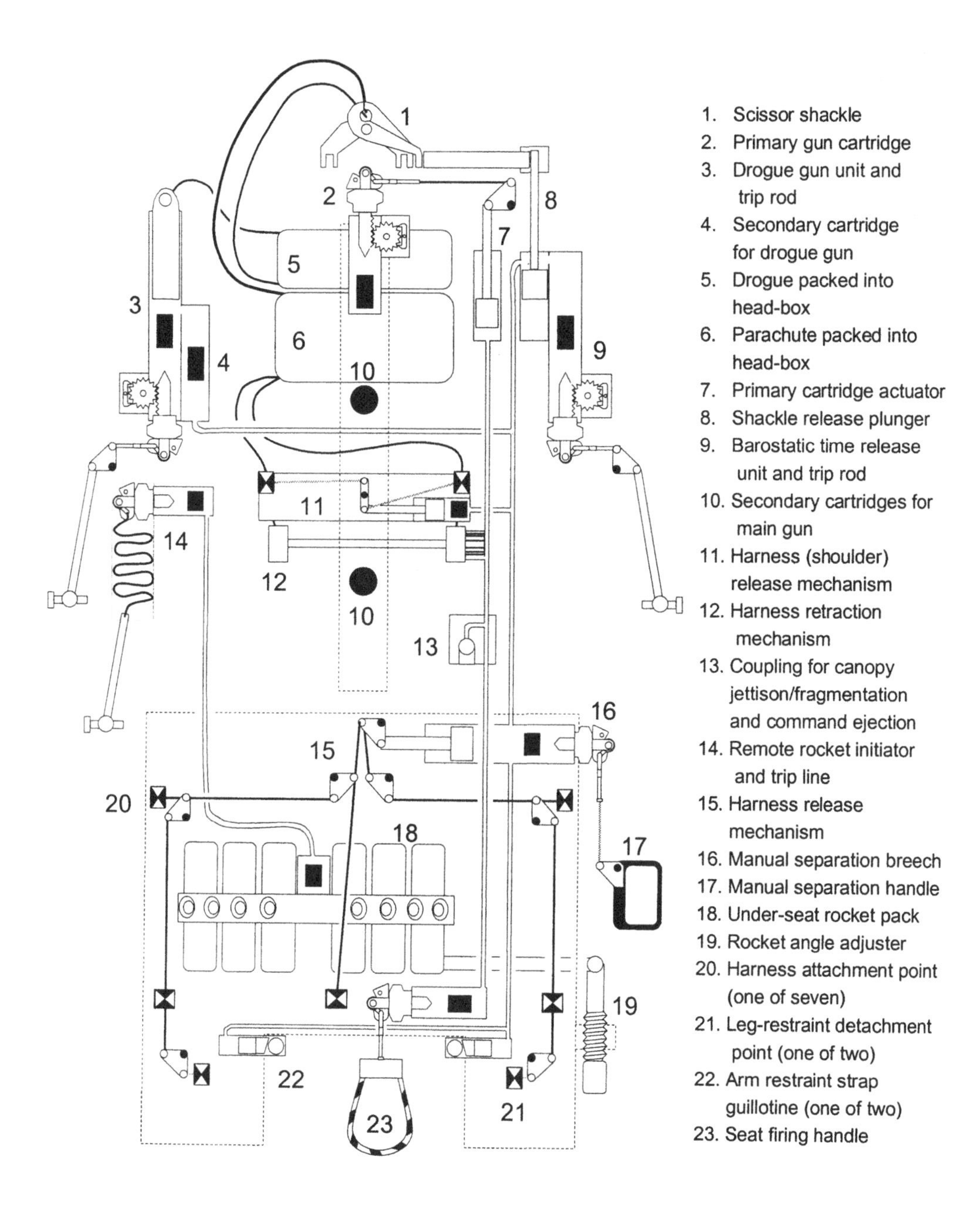

Figure 12.6 Simplified schematic showing the main gas-operated mechanisms of an in-service ejection seat. Explosive gas generators are shown as black rectangles

A Personal Equipment Connector (PEC) is a three-layer coupling for the pipes carrying gas for G-suit inflation and breathing, and the cable for the microphone and earphone signals. The part fixed to the seat is like the filling in a sandwich with holes that allow the pipes and cable connector from the upper man-connected part to mate with the lower aircraft-connected part, both of which are held firmly in alignment by heavily sprung catches. The aircraft portion is only removed for servicing, but the crewman normally connects his helmet, mask and G-suit to the aircraft each flight by engaging the man portion of the PEC to the seat portion. The latter also acts as a manifold for connecting the seat-mounted oxygen tank to the mask supply pipe. This can be activated by pushing a small handle on the side of the seat if the aircraft breathing gas supply fails or is contaminated in flight. The handle operates rods and levers to close off the aircraft supply and open a valve connecting the high-pressure oxygen bottle to a miniature demand regulator mounted near the PEC. During ejection, as the seat begins to travel up the rails, a lanyard releases the aircraft portion of the PEC and pulls the rod to turn on the emergency oxygen supply.

Automatic deployment of the drogue is done by pulling it from the top section of the head-box by a stout nylon line attached to a weighty steel bullet fired from a gun behind the head-box, as in the early seat systems. The barostatic time release unit (BTRU), similar to the original TRU and also triggered by a static rod anchored to the gun outer tube, is used to detach the harness from the seat and deploy the parachute. The delay is 1.5sec, unless the ambient air pressure is below that at 10,000 ft, in which case an aneroid capsule in the unit jams the clockwork escapement until the seat falls below 10,000 ft. The BTRU disengages the harness and the man-portion of the personal equipment connector and unlocks the scissor shackle attaching the drogue bridle to the seat top. This drags the parachute canopy from the lower part of the head box, and when extended the risers pull the man away from the spent seat.

However, with the drogue packed on top of the main parachute, the drogue would balk the deployment of the parachute if for any reason it were to remain in the head-box. Therefore, should the drogue gun or the BTRU fail, the previous technique of severing the drogue-to-parachute link using an explosively actuated guillotine could not be used for manual separation. Instead, as shown in Figure 12.6, an alternative approach was adopted in which the emergency action of the crewman fired a cartridge in the seat strap release mechanism that disengaged the seat and leg straps directly. The gas pressure also cut the arm restraints, released the shoulder catches and the scissor shackle, and fired a secondary cartridge in the drogue gun. The drogue then spilled into the airstream, and immediately dragged out the parachute and pulled the detached man away from the seat.

With this generation of ejection seat, the man must still do most of the remaining actions. He must unlatch the survival pack from the short straps holding it by his buttocks so that it dangles on the 15 ft lanyard, and if over water he must remove his breathing mask and pull the handle to inflate the life jacket. Releasing the weighty pack decreases the landing load on his legs, and allows the inflated life jacket to keep him face-up in the water. The most important action on entering water is to find the buckles on the risers and open the flaps that release the parachute to avoid being dragged under by a surface wind. With this done he can relax slightly. Even in the dark he can find the pack by pulling on the lanyard, and then pull another handle under a cover on the pack that releases the binding and opens the valve on the cartridge containing compressed CO_2 to inflate the one-man dinghy. Speed is however essential in cold water to avoid hypothermia. After pulling himself aboard he must find the hooded cape and fasten it around himself for immediate shelter, and then to get extra insulation he can inflate the cape by blowing into a tube. The next priority is to set up location aids. A small battery-powered light and an emergency position-indicating radio beacon (EPIRB) are extracted from their pockets in the life jacket jerkin and set up and activated. Flares and coloured smoke signals are also prepared in case a vessel or aircraft is sighted. The latter are part of a set of survival equipment, packed and tethered inside the dinghy, that also includes pouches of drinking water, sea-sickness tablets and an assortment of small but useful tools. Among these are usually a short line, a few fish hooks and a 3 inch square metal mirror that can be used as a heliograph for attracting attention.

12.4 Implications for aircraft design

The decision to include ejection seats as the primary means of providing rapid emergency escape from the aircraft brings an interesting assortment of problems, initially for the aircraft design team and later for the operators and aircrew. The provision of effective protection throughout the process, including the parachute descent and ditching into extremely cold water, has considerable effect the requirements for crew clothing and headgear. These are considered in Chapter 13.

In low-level flight in a combat aircraft, with a large enveloping canopy, the speed with which the escape path can be cleared can govern the probability of escaping from the aircraft. At high airspeed, above 250KIAS, unlatching the frame and pushing the front up by about 300mm using explosive jacks creates sufficient blast-induced lift to throw the whole assembly clear in about 300msec. The problem is to get a similar effect at low airspeed. The main approaches are to use rockets as a substitute for the force of the windblast, or to place an explosive cord on the canopy to cut an opening. The choice has depended largely on the size and structure of the canopy, particularly the siting of metal frames, and the material and the thickness of the transparency. The embedded micro-detonating cord (MDC), triggered by an electrical detonator, has proven the most rapid technique and is employed in the majority of UK aircraft designed for the ground-attack role. The explosive cord is cemented on the inner surface of the transparency in a pattern that gives least visually intrusion and creates small fragments on shattering. The cord itself is made as a strip, about 6mm wide, and backed by a lead overlayer that directs the majority of the shock upwards to cut into the polymer. There is however a third approach that relies upon the impact of the top of the rising ejection seat to punch an opening in the canopy. If the canopy is made from acrylic, and is flat and relatively thin – less than about 6mm – this method is surprisingly effective and absorbs little energy from the catapult. It is employed in several types as a secondary system should the primary system fail.

In aircraft with side-by-side or tandem seats, the main requirements are to enable one of the crewman, usually the pilot in a combat aircraft and the instructor in a trainer, to initiate ejection of both crewmembers and to avoid the subsequent possibility of a collision. A commanded ejection is arranged by connecting the gas line from the primary firing unit of one seat to the same gas line of the other. A valve is invariably included to give the crew the facility to disable the link. The escape trajectories are usually deconflicted by delaying one of the extractions by about 250msec and giving the two seats a small sideways thrust in opposite directions, for instance in the Escapac seat using additional low impulse rockets or in Martin-Baker seats by biasing the under-seat rocket.

Although few large aircraft give all of their crew the luxury of assisted escape, those that have, such as the British V-bombers and the US B-52, have provided somewhat unequal facilities. For instance, in the B-52, small hatches above each seat are removed explosively so that the pilot and co-pilot, sitting side by side facing the windscreen, and the EW officer and gunner who sit side by side behind them facing backwards, can be ejected upwards. However, the navigator and radar operator who sit below them on the lower deck are ejected downwards through explosively removed hatches. There is no directly commanded initiation, and each crewmember must pay periodic attention to a small panel in the centre of his console that enables the pilot to signal that all is not well and abandonment is forgivable.

12.5 Modern ejection seats

The more recent versions of the Martin-Baker ejection seats, the Mk 12 used in the Harrier GR7 and the Mk14 (designated NACES by the US Navy and used for instance in the F-14B), differed from the previous types by incorporating micro-processor control with extendable pitot-sensors for airspeed and altitude as well as a gyro attitude sensor. Although actuation still relied upon gas generators, the real-time air data and the electronic computation enabled the timing of the main parachute deployment to be optimised. The advantages were less delay at low height and low speed, and an increased delay at high q-load that minimised the chance of parachute damage.

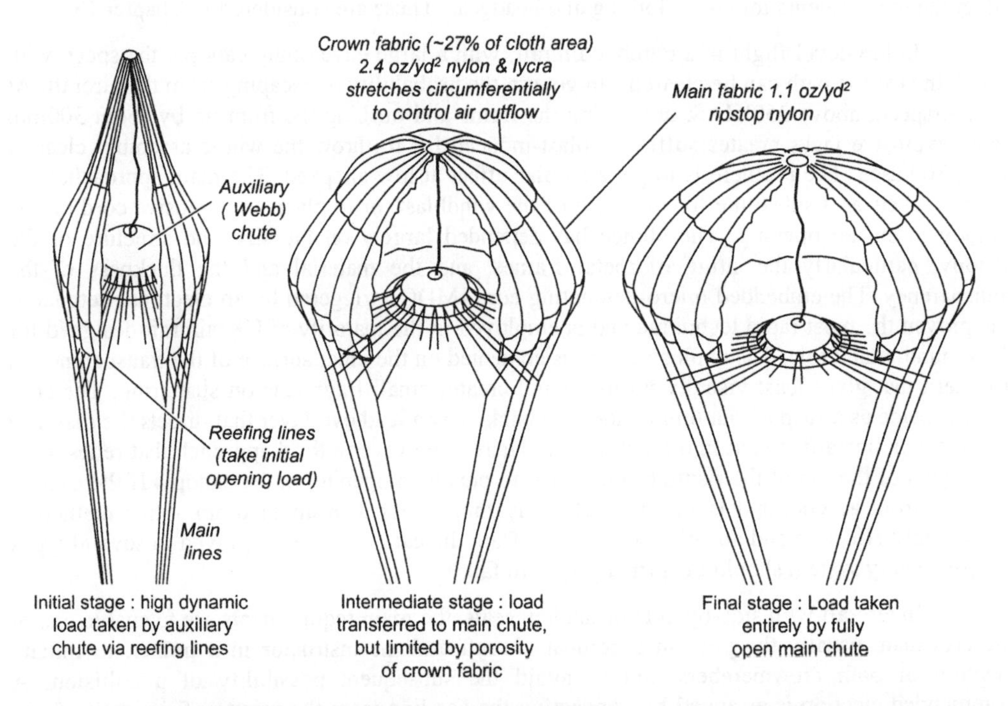

Figure 12.7 Configuration and stages of inflation of an Automatic Inflation Modulation (AIM) parachute developed specifically for use with ejection seats (Reprinted with permission from Irvin Aerospace)

The revision also benefited from the adoption of a bi-porosity conical parachute, as illustrated in Figure 12.7. This 30 ft diameter canopy incorporates an internal auxiliary parachute that inflates as soon as the lines are tensioned by the initial airblast, taking the initial shock and forming a circular opening that feeds air into the main canopy. The latter is restrained from opening suddenly by the reefing lines to the auxiliary chute, and the shock is also controlled by the crown material that stretches and becomes porous under high dynamic pressure. Such a parachute arrests a 300lb load in less than 1sec with a shock of less than 6000lb when deployed at 300KIAS.

In their most modern design, the Mk16, Martin-Baker have sought to extend the escape envelope, improve reliability, improve accessibility and maintenance, reduce the time to parachute inflation, automate some of the human actions needed in the later stages of the escape sequence and accommodate new aircrew-mounted equipment. This will be fitted to Eurofighter, Rafale and JSF to provide an escape envelope that extends from 0ft/0KIAS to 50,000ft/600KIAS. If ejecting at low speed the crewman should be supported on a fully inflated parachute within 1.4sec of initiating the ejection.

The Mk16 configuration seems to be recognisably the same as previous versions, but significant changes are incorporated, the most fundamental being the abandonment of clockwork timing mechanisms and the adoption of digital electronic control and components, such as thermal batteries and electrically-ignited gas generators that are more common in missiles. To get lightness and strength the telescopic catapult gun and ladder frame have been abandoned in favour of an integrated structure in which the main backbone is formed by a pair of multi-cartridge catapults, separated by about 350mm. Similar revisions are apparent elsewhere. The under-seat rocket is a large, single, transverse cylinder with a pair of nozzles at each end, a pair of pitot sensors swing out alongside the head-box and aero-dynamic surfaces are deployed from the head-box and lower flanks. The crewman's arms and legs are all actively restrained. The drogue is anchored to four bridles attached to the back corners of the seat, the head-box is ejected upwards to dispense the parachute, the survival pack is automatically lowered and the dinghy automatically inflated. In addition to the normal PEC coupling, the seat also accommodates connections for a liquid-cooled suit, a helmet-mounted display and a helmet position sensing system. The seat electronics are active and connected to the aircraft power and data-bus in normal flight, mainly to harmonise the seat inertial platform with the more accurate and stable aircraft inertial reference. These are automatically disconnected as the thermal batteries are fired up in the first few milliseconds of the ejection process. The sequence of operations is however the same as that performed by a purely mechanical seat, and the availability of air-data from the pitot heads and inertial data from a triaxial accelerometer is used solely to optimise the timing of the deployment of the main parachute.

Escape systems fitted to Russian combat aircraft have certainly kept pace with developments in the West. For instance, the K-36D ejection seat manufactured by the Zvezda Design Bureau is much like the North American Aviation seat for the X-15 in employing a pair of telescopic booms for controlling the aerodynamic instabilities that are so injurious during escape at high speed. The configuration also includes leg lifters and rigid arm paddles in addition to leg restraints. Additional windblast protection is supplied using a wind deflector about the size of a dinner plate that is propelled into place from between the thighs, plus facings on the head box that help to locate the vented helmet. These give an escape envelope that extends from 0/0 to 56,000ft /Mach 2.5.

Further improvements to the ejection seat are topics for current research. Both the US Navy and the US Air Force have continued to sponsor work under the Advanced Technology Crew Station (ATCS) and Crew Escape Systems Technology (CREST) programmes respectively. Vectoring the rocket thrust to give the seat a vertical trajectory may become possible using a miniature gyro-platform and a pintle rocket motor; a solid propellant rocket with steerable nozzles. The use of an electrical discharge through a buried fuse-wire may give a faster, more reliable way of shattering the heavy, thick canopies used in aircraft such as the Joint Strike Fighter. The wire would certainly be less visually intrusive than detonating cord. Other ideas, such as variable seat geometry, prone rather than supine seating, internal and external air bags, para-wing steerable parachutes, all offer some benefit and some obvious complication. The CREST programme has also evaluated Zvezda systems and collaborated to introduce modifications, mainly to accommodate a larger crew size range and ensure compatibility with US life-support equipment so that the Russian-built seats may be used in Western combat aircraft.

Beyond these ideas, it is difficult to imagine what can improve performance rather than be aimed toward lowering cost or easing maintenance. It is reasonable to conclude that the inherent limitations of the ejection seat have been reached. Other approaches to the requirement have exercised many minds, particularly within the US aerospace research and development community, and it is these that are reviewed, very briefly, in the following section.

12.6 Alternative escape technologies

Table 12.1 Summary of aircraft programmes that investigated or adopted alternative escape systems, in approximately chronological order

Aircraft type	Escape system	Comments
BP8-348	Jettisonable nose portion of airframe	The BACHEM Natter was a German experimental rocket-propelled fighter. Only the prototype was built and tested
Leduc-021	Jettisonable nose cone	The pilot lay almost supine in a "glass ring" that fronted the circular intake nozzle of the large ramjet powering this experimental aircraft
So-9000	Jettisonable nose portion of airframe	The first of the "Trident" rocket-jets had a detachable front fuselage section, also devised by Rene Leduc
XS-2	Jettisonable nose portion of airframe	The pilot of this supersonic research vehicle had to bail out from the separated capsule when this had stabilised. Used once unsuccessfully in 1956
XF-103	Rocket "shoe" with tail stabiliser	Mass 700kg. Propelled downwards to clear aircraft fin
F4D	Ejectable capsule cockpit	Study sponsored by the US Navy. The capsule was stabilised by fins and a drogue
F8U-1	Integrated nose capsule	Study also sponsored by the US Navy
F-102	Seat pivoted up into canopy and sealed by base-plate	Separated by rocket but showed gross aerodynamic instability in yaw & pitch despite use of drogue below 150KIAS
B-58	Enclosable ejection seat	Developed by Stanley. 9 of 13 escapes were successful
XB-70	Enclosable ejection seat	Similar to the Stanley B-58 system
MIG-21	Seat and retained windshield	Purported Russian & Chinese experiments
F-104	Rocket detachment of whole nose	5 modules were tested up to 900KIAS. Stabilised by 3 trailing booms. Not adopted for service
F-111	Whole cockpit section of airframe and leading edge wing stub stabilisers	Much developed and fielded successfully
B-1	Integral part of whole aircraft nose	Tested successfully. Production aircraft reverted to 4 ejection seats

For hang gliders, paragliders and recently some light aircraft, it is possible to carry and deploy a parachute that can bring the whole unflyable aircraft, complete with harnessed crew, safely back to the ground. The viability of this approach depends on the low flying speed and the favourable ratio of the mass and volume of the parachute to the mass and volume of the aircraft. For combat aircraft the mass penalty, including the structural implications, would be too great, and the escape envelope would be severely restricted. Most attention has therefore been given to alternative crew escape systems.

To survive the consequence of using the ejection seat for emergency escape, which for most aircrew is a rare occurrence, they must wear suitably protective headgear and clothing the whole time they are in the air. As described in Chapter 13, equipment with the necessary qualities invariably interferes with the crewman's ability to do the job by affecting his vision, dexterity and head movement, and it imposes a physiological burden that exacerbates fatigue. The limited effectiveness of the headgear and clothing at high dynamic pressure also make successful escape unlikely at airspeeds above 500KIAS. These detractions, and the likelihood of drowning if incapable or unconscious when ditching, have motivated design teams to pursue a wide range of alternative approaches to emergency escape.

As summarised in Table 12.1, it is evident that the topic has stimulated considerable ingenuity, the alternative forms of escape system falling into three broad classes: collapsed canopies, enclosable ejection seats and whole ejectable capsules. Prior to catapult-propelled or rocket-propelled separation from the stricken aircraft, the first folds the seat and floor up into the transparency, the second envelopes the seat in a protective shroud, and the third severs the whole pressurised cabin from the airframe. Only the enclosable seat and the whole ejectable capsule ideas have entered service. The prime example of the former was the Stanley seat for the B-58 Hustler, sketched in Figure 12.8, which had overhead overlapping door segments that pivoted forward and down to form a sealed, pressurised enclosure. The occupant could close and open the doors without ejecting, for instance in case the cabin pressurisation system failed, and his legs were automatically pulled up and inside before the door closed. Although he could see some cockpit instruments through a small frontal window, he had access to few controls. After ejecting the seat/capsule was stabilised by deploying a drogue and a small tail frame, giving minimal gyration and excellent windblast protection. With the man inside, the seat had a mass of about 320kg, so a large parachute was needed to control the descent.

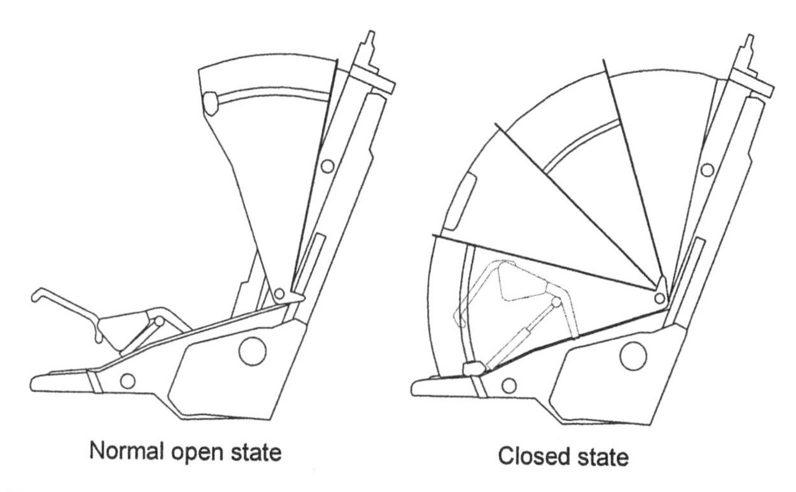

Figure 12.8 Sketch of the encapsulated seat used in the B-58 Hustler

The XB-70 Valkyrie prototype employed an idea similar to the Stanley encapsulated seat and was intended to allow escape from zero feet and 90KIAS to 80,000 ft and Mach 3. In this design the seat was mounted on a set of horizontal rails in front of the open capsule, and on pressing triggers on the armrests the crewman and seat were pitched back by 20° and pulled backwards into the capsule. A pair of clamshell doors then slammed shut, one from above and the other from below, before the enclosure was pressurised. This must have felt like being eaten by a giant fish. The crewman could remain in the pressurised enclosure in the event of cabin depressurisation, and had to make a second trigger action to jettison the surrounding airframe, ignite the rocket thruster and deploy telescopic booms for stability. The mass was about 365kg, including the occupant. The capsule automatically deployed a large parachute, a set of impact cushions and a set of floatation bags.

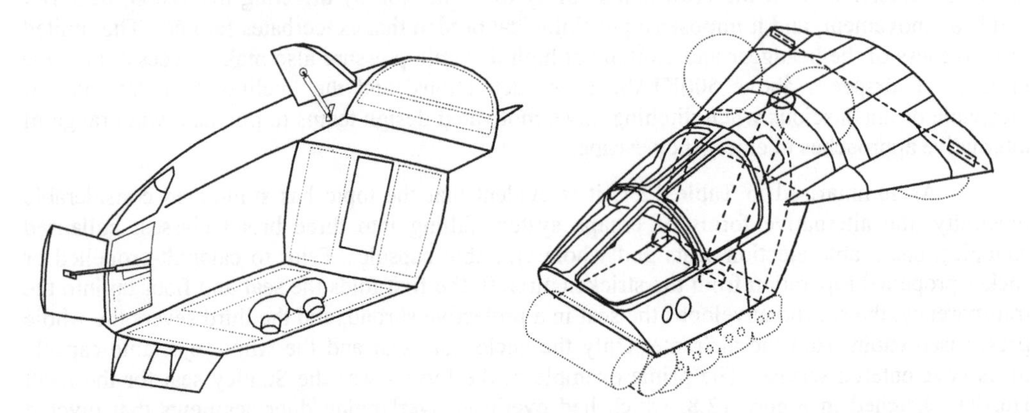

Figure 12.9 Sketches of the large escape capsules as used (left) in the prototype for the B-1 and (right) in the F-111

Development of the F-111 swing-wing bomber started in 1965. The whole cockpit capsule was intended to provide zero/zero emergency escape and comprehensive protection for the two crewmen, including automatic underwater separation from the ditched aircraft. The capsule, sketched on the right of Figure 12.9, was separated from the airframe by a collection of explosive severance systems and carried upwards by a cylindrical solid fuel rocket with two upper nozzles to control pitch attitude and lower nozzles for lift. The mass was about 1420kg including crew, and the rocket impulse was about 10,000kg.sec. Spoilers on the retained stub wing roots and a large drogue stabilised the craft before the main parachute was inflated. Although large external airbags were inflated to buffer the impact with the ground, several cases of spinal injury marred what would otherwise have been completely successful escapes. The injuries were attributed to an increase in the speed of ground contact produced by the swinging of the capsule on the parachute risers.

Development of the B-1 bomber began a decade after the F-111and benefited from the lessons learned. In this case the whole cockpit capsule, as sketched on the left of Figure 12.9, accommodated between four and six crewmembers and was mainly required to for escape above 600KIAS and 50,000 ft. Like the XB-70 and the F-111, it was severed explosively from the adjoining airframe and equipped to act as a survival shelter. It was propelled by two large (20 inch diameter) spherical rockets, one gimballed in pitch, the other in roll. Each operated in two phases, a boost of 44,000lb for 370msec followed by a sustained thrust of 15,000lb for 1100msec. Numerous fins and spoilers and a large drogue were deployed for stability, as were the descent parachutes, landing buffer bags and floatation bags.

Overall, the US experience of developing alternative escape systems suggests that the chief detractions have been due to excess mass, injury during ground impact and cost. None have matched the almost instantaneous zero/zero escape performance of an ejection seat. However, the global adoption of ejection seats in current combat aircraft, and those under development, should not be taken to signify the lasting superiority of the latter. The USAF CREST programme is considering alternative configurations of encapsulated seats and ejectable cockpits for emergency escape from high-altitude, high-speed, trans-atmospheric vehicles, where atmospheric re-entry is a prime requirement. If the capsule could be engineered in a form that gave zero/zero escape, it could provide a better balance of qualities for all combat aircraft. Chapter 14 discusses some possible configurations.

Reference

1. Mohrlock H F, Journal of Aviation Medicine, Vol 28 , 1957

Further reading

Brinkley J W, *Development of aerospace escape systems*, Air University Review, Vol XIX, No 5, pp. 34-49, 1968

Models for Aircrew Safety Assessment: Uses, Limitations, and Requirements, RTO-MP-20, 1998

The SAFE journal and conference proceedings

Chapter 13

Clothing and Headgear

13.1 Introduction

All who take to the air know that what they wear affects both their ability to perform their in-flight role and their chance of surviving accidental exposure to the external environment. A brief review of the natural hazards faced by all aircrew and the additional risks faced those who operate military aircraft is given in Chapter 3.

For the majority of their working life, the physical enclosure of the cabin, the air conditioning and supply of breathing gas provide aircrew with a working environment that protects them from windblast, noise, heat, cold and oxygen starvation. All services also equip aircrew with separate items of clothing and specific devices in the headgear to meet the wider range of hazards such as survivable impact, birdstrike and high-G. For UK forces these form an Aircrew Equipment Assembly (AEA), from which specific combinations are selected to suit the aircraft type, role, season, latitude and whether the flight is across open water. Additional items for protection against NBC agents and lasers are deployed when such threats are anticipated. Most nationalities use a similarly flexible set of equipment, and some include protection against nuclear flash.

It is the need to cater for exceptionally rare but extremely hazardous conditions, such as emergency escape, that has taxed the ingenuity of equipment developers. The chief problem is to ensure that the resulting layers can be worn comfortably in versatile combinations by all the relevant aircrew without compromising their ability to move, reach, grasp and see when strapped into the harness of the ejection seat. This level of functional compatibility has called for very careful design and extensive testing. Unsurprisingly, the development of any new piece usually takes a considerable time, typically about five years. Also, as with any safety-critical element, the items of the AEA must be cleaned, checked, maintained and modified by specialist personnel, and this provision generates a significant logistical problem.

This chapter describes the functions performed by the layers of clothing and headgear, and the approaches taken by engineering teams. The chapter finishes with a discussion of the current difficulties of adding display, control and vision enhancement functions to the headgear to create an "integrated helmet".

13.2 Sizing

Clothing, footwear and headgear can be arranged to suit the dimensions of the wearer using three general approaches; one-size-fits-all; a set of sizes; or bespoke tailoring to suit the individual. The approach taken depends in general on a balance between the need for a precise fit, the dimensional variation in the population and the relative costs of the different approaches, the latter being determined mainly by the number of items to be manufactured. Although the second approach is usually taken when devising items of the AEA, it is assumed that having selected the most appropriate size the garment can be adjusted relatively easily to suit the crewmember, and this individual will have sole use for as long as necessary.

It could be presumed that aircrew vary in size and shape in much the same way as the population from which they are drawn, and that AEA items could be designed using conventional sizing criteria. However, the greater precision of fit needed for comfort and compatibility between the ten or so layers has forced the services and equipment developers to base equipment on more precise knowledge of the dimensions of the aircrew population. Anthropometric surveys of aircrew, amongst other servicemen and women, are therefore carried out periodically at the behest of the interested services. About 60 measurements are typically made on a sample population of between two hundred and ten thousand. A recent standard specifying the postures, the anatomical landmarks and the measurement procedures has been agreed[1]. This should ensure consistency between data-sets.

The service and the equipment supplier are faced with the problem of deciding how many size variants should be made to cover the user population, and the dimensions of each size variant. These determine the criteria for allocating a size variant to an individual and the proportion of items that are likely to be needed in each size.

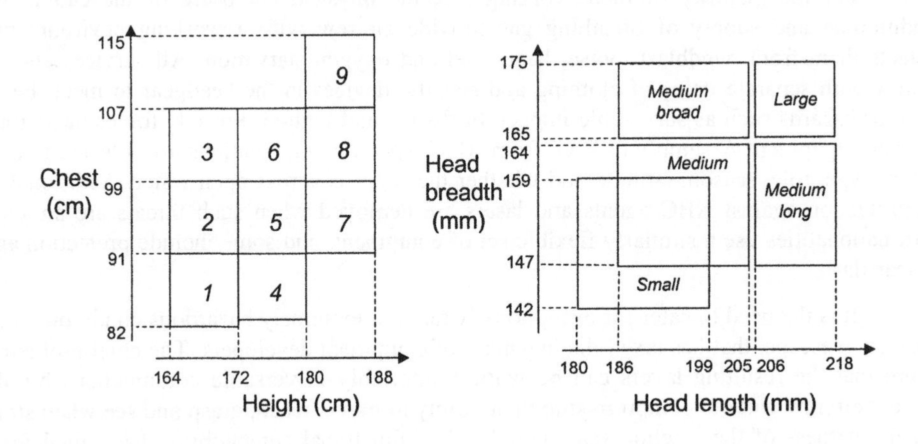

Figure 13.1 The combinations of chest and height measurements used to specify the 9 sizes of RAF aircrew clothing, and the combinations of head breadth and length for appropriate sizes of the Mk10 aircrew protective helmet (Reprinted with permission from QinetiQ)

As illustrated in Figure 13.1, the nine variants in the size roll for most items of RAF clothing are based on a combination of two critical measurements; the standing height and the chest girth. The 8 cm size increments takes account of factors such as the accuracy with which garments can be made and the likely shrinkage and stretching in service. About 40% of aircrew fall within 4cm of the bi-variate median and could be accommodated nicely by size 5 clothing, the other sizes being available for normal outliers. Fitting a helmet can also depend on a pair of measurements[2], as also shown in Figure 13.1. Somewhat surprisingly, in this case the medium/long size has proved most popular.

For those who fall outside the accommodated range the only solution is to manufacture garments by bespoke tailoring. However, with the increasing likelihood of larger proportions of female aircrew, for whom many dimensions are significantly smaller than the males for whom the size rolls were devised, it may be more economical to introduce additional size variants. This may be essential to obtain satisfactorily fitting helmets, for which shells are usually made by laying up composite materials in standard moulds and within-size adjustment is limited.

13.3 Basic clothing

Aircrew underwear is generally a combination of cotton long-sleeved round neck shirt, long-johns and cotton loop-pile socks. In summer a one-piece coverall is worn directly on top of the underwear for overland flights. In winter extra insulating layers are worn beneath the coverall.

Table 13.1 Summary of general flying clothing for RAF aircrew (Data taken from *Catalogue of aircrew clothing and equipment: report for the Commonwealth Committee on Defence*, ML Aviation Co, Maidenhead, UK, 1983, with permission from QinetiQ)

Service Description	Usage	Material	Weight (kg)	Size range
Boots, aircrew, 1965 pattern	General	Leather & rubber	1.4	16 +
Boots, aircrew, lightweight	Hot weather	Leather & rubber	1.2	16
Coverall, aircrew, Mk14A & 15	General, Mk15 accommodates immersion coverall Mk1	Nomex - green	~1	9 +
Drawers and vest, aircrew, long, cotton ribbed	Standard wear	Cotton	0.32	5+
Gloves, sweat resistant	General green or undyed	Cape leather	0.04	7+
Shirt, aircrew, cotton Mk2	General aircrew use	Cotton	0.18	4
Socks, Terry-loop Mk2	Used with boot, 1965 pattern	Terylene/worsted yarn	0.11	6
Suit, aircrew, combat temperate Mk2 (Jacket, trousers & waistcoat)	Cold weather non-fast jet	Cotton-drill, nylon, gauze-backed terylene	3.3	8

The basic clothing used by RAF aircrew is summarised in Table 13.1. The 1965 pattern boots are black leather bonded to rubber soles, and have conventional lacing. They cover the ankles and provide foot protection during emergency escape and good support for walking over mixed terrain. In addition to conventional hip and chest pockets, the coverall has large pockets on the top of the thighs for stowing folded maps and flight reference cards, and small pockets on the outside of the upper sleeves to accommodate pens.

13.4 Thermal protection

It is useful to consider that military aircrew operate in three physical environments; one the air or water outside the aircraft, the second produced by the conditioning system inside the cockpit and a third, a microclimate, inside their clothing. The general conundrum faced by equipment designers is to provide a satisfactory form of the third when routinely immersed in the second, and also provide a tolerable form of the third when exceptionally exposed to the first. The problem can be quantified by considering the thermal insulation capacity of clothing, which operates mainly by trapping a layer of still air in contact with the skin.

The performance of a thermal insulator is assessed by measuring the temperature difference it will maintain for a given rate of loss of heat. In the case of clothing, the *clo* unit of thermal resistance is that needed to provide a comfortable microclimate in the conditions represented in Chapter 3 by Figure 3.5(b). In this case the trapped air layer must maintain about 6°C difference between the external air and the skin surface while allowing about 40 $W.m^{-2}$ outward heat transfer. One *clo* is therefore defined as $0.155°C.m^2.W^{-1}$, and this is provided by about 6mm of still air. Any greater temperature difference needs a correspondingly greater thickness of clothing to maintain core temperature with the $40W.m^{-2}$ resting metabolism. For instance, in 0°C ambience, clothing must give about 4 clo units of insulation and have an effective thickness of about 24mm.

Aircrew who may be ditched into very cold water need clothing that can trap about 24mm of air when immersed in seawater. In the cockpit, the same clothing must dissipate the metabolic load and allow sweat to evaporate from the skin. The current RAF ensemble therefore consists of underclothes and overclothes for the vital insulation against the cold, coupled with an immersion oversuit that can become water-tight when wet to prevent the water saturating the insulation. The heat load imposed by these layers is made tolerable in the cockpit by the ventile fabric of the immersion oversuit, which vents water vapour when dry, and by setting the ECS to produce a cabin temperature of about 15°C. Active cooling garments may also be used to extract the heat retained by the thick insulation.

Table 13.2 Summary of warm clothing available to RAF aircrew (Data taken from *Catalogue of aircrew clothing and equipment: report for the Commonwealth Committee on Defence*, ML Aviation Co, Maidenhead, UK, 1983, with permission from QinetiQ)

Service Description	Usage	Material	Weight (kg)	Size range
Boots, aircrew, mukluk including insoles and socks	Arctic use	Nylon & rubber, plastic, wool	2.3	8
Coverall, inner, Mk3 & 3A	Added warmth in cold	Modacrylic pile & polyester backing	1.1	5 +
Coverall, inner, knitted, Mk1	Inside NBC &/or immersion coverall for warmth	Wool	1	4
Drawers and vest, acrilan pile	Used under cold weather suit Mk3 in extreme cold	Polyester and modacrylic pile	0.8	4
Gloves, silk	Worn inside cape gloves in cold	Double layer silk	0.01	5
Jersey, heavy	For added warmth	Wool	0.68	4
Suit, flying, cold weather, Mk3	For aircrew in cold weather	Gabardine, ventile cotton, nylon net fabric	2.6	8+

The garments supplied in RAF service that can give extra thermal insulation are summarised in Table 13.2. The combination of acrilan pile vest and drawers with a one-piece pile "bunny suit" coverall gives good thermal insulation, and the materials are largely resilient to crushing by the outer layers and the seat harness. The tightly-woven cotton fabric of the outer layer of the cold-weather flying suit vents water vapour at a sufficient rate to keep the microclimate at moderate humidity and prevent sweat saturating the insulation. It also stops most of the cold, rapidly moving air, when for instance the crewman is exposed to an arctic wind, from entering and disturbing the trapped insulating layer of air.

The active forms of thermal protection that have been developed have used warmed air, warmed liquid or direct electrical heating. The former could be used to remove heat by introducing air or liquid from appropriate cooling equipment, albeit with the distinct drawbacks of introducing an extra layer and of extracting heat only when the crewman had started sweating. Such equipment has mainly been employed in aircraft that lack an ECS or as a back-up should the ECS fail, but it could also help aircrew in some types such as the Jaguar and Harrier where the ECS is routinely turned off during take-off and other high thrust manoeuvres. Perhaps the worst cases are encountered when the crew wear the full set of NBC protective clothing in a fast jet that is stood on the flight line under a tropical sun waiting for take-off. It is largely to prevent thermal exhaustion in such conditions that active cooling garments are being developed. These are likely to be introduced in new aircraft, such as the Eurofighter, and retro-fitted later to other types.

Garments that distribute heated or cooled air to the torso, arms and legs have been constructed using a variety of approaches. One design, now obsolete, was used by the RAF. Designated the Mk2 Air Ventilated Suit[3], it consisted of a series of flexible plastic ducts connected to a branching network of fine pipes terminating in 144 diffusing jets. These were sewn into a form-fitting nylon undergarment and were fed at high pressure and a low flow rate. Another arrangement, fielded by the USAF as the Ventilated Suit MA-3, was constructed as two impermeable layers held apart by basket-weave corrugations. This had an array of fine holes punched in the inner layer to direct the temperature-controlled air at the skin, and required a high flow rate but only a low feed pressure. The RAF approach had the benefit of allowing the normal outward sweat evaporation from the area of skin not covered by the pipework and could be tolerated when the air feed was off. On the other hand, the low feed pressure needed for the USAF approach could also be supplied by a portable, battery-powered unit, and this allowed the crewmember to benefit from the suit when transferring between the crewroom and the aircraft. Such portable blower units are also needed temporarily to supply air for breathing and visor ventilation when wearing a head-cowl for NBC protection.

The use of liquid to transfer heat into or away from the body exploits the approximately thirty-fold increase in the rate of heat exchange between a solid and a surrounding liquid in comparison with the exchange between a solid and a surrounding still gas[4]. The vest under development for Eurofighter aircrew contains a set of delivery manifolds and a branching network of flexible pipes that are stitched to form convoluted loops that cover most of the torso with less than about 15mm spacing. The plastic pipework has an external diameter of about 4mm, and the wall thickness is stiff enough to resist pinching. The glycol-water mixture is cooled to a few degrees below the skin temperature on passing through a heat exchanger, and it flows in a closed loop rather like a domestic central heating system. The crewman joins the pair of pipes from the vest to the pipes from the aircraft-mounted unit using a self-sealing two-way connector.

Electrically-heated garments have been favoured as minimally-encumbering underwear inside a partial-pressure suit for use in high altitude aircraft. The suit is made using fine multi-strand copper wires in a thin PVC or PTFE insulation that are knitted with terylene yarn as individual pieces of clothing; socks (~10W each), gloves (~10W each), long-sleeved vest (~35W) and leggings (~20W). The resistive elements receive 28Vd.c and are connected in parallel to prevent open circuit failure affecting the whole suit. The arrangement is flexible, versatile and effective. It has unusual merit of maintaining hands at a comfortable, dextrous temperature, and enables protection against –20°C ambient while wearing passive insulation of only 1.5 clo units. The suit must be cleaned regularly and the electrical connections checked to ensure reliability. To avoid burns through local overheating, which is usually caused by breakdown of the electrical insulation, the whole set is checked periodically using a thermal camera.

13.5 Post-escape protection

As described in Chapter 12, the process of escaping from a fast jet using an ejection seat will certainly expose the crewmember to violent inertial and windblast forces, and possibly to extreme cold. All clothing must be strong enough to withstand the windblast, and this is one aspect of the testing done during development. The man must also be given equipment to enable him to survive immersion in very cold water.

Table 13.3 Summary of RAF aircrew clothing for post-escape survival, (Data taken from *Catalogue of aircrew clothing and equipment: report for the Commonwealth Committee on Defence*, ML Aviation Co, Maidenhead, UK, 1983, with permission from QinetiQ)

Service Description	Usage	Material	Weight (kg)	Size range
Coverall, immersion, Mk10 & 10A	Over-water flying	Ventile cotton, rubber sealed zips	2.0	5 +
Coverall, immersion, quick-don Mk1	Passengers in helicopter flying over water	Polychloroprene-proofed nylon	2.5	1
Coverall, inner, immersion, Mk1	Over-water flying , worn under Mk15 coverall	Ventile cotton, rubber sealed zips and neck and wrist seals	1.4	9 +
Gloves, water resistant	Worn with immersion coveralls, Green or undyed	Treated cape leather, rubber cuff	0.1	7+
Life preserver, Mk27S	Specific to Tornado, includes arm restraints	Nylon/nylon net, butyl bladder	2.45	2
Life preserver, Mk27 series	General fast jet use	Nylon, butyl bladder	2.0	2
Life preserver, Mk25	General non-fast jet use	Nylon/nylon net, butyl bladder	2.4	2

The equipment provided to help RAF aircrew survive after escaping from an aircraft is summarised in Table 13.3. The Mk1 immersion coverall is worn inside the standard Mk15 coverall, and the Mk10 immersion coverall has pockets that make the standard coverall unnecessary. Both are similar in construction to the outermost garments used by sailors and inshore rescue personnel, and are donned through a water-proof zip fastener that cuts diagonally across the front of the torso. The watertightness depends upon this zip, the characteristic expansion of the cloth on immersion in water, the impermeability of the bonded socks and the fit of the seals around the wrist and neck.

The Mk25 and Mk27 life-preservers are also similar in principle to the types used by civilian sailors. They contain a U-shaped bladder that inflates to form a floatation collar extending around the back of the head and two large buoyancy bags over the chest. This arrangement has proved to be best way of holding the head upright and the mouth and nose above water level. To survive the ejection windblast, the cover protecting the bladder is made using heavy nylon fabric and the front closure is a pair of stout interlocking alloy plates. The back and sides form a complete jerkin, the sides having pockets containing a personal radio location device and other survival aids. The bladder is held in its normal tightly stowed state until inflated by release of compressed CO_2 from an attached cartridge, either by a strong pull on a chest-mounted handle or automatically by a mechanism activated by water immersion.

13.6 Head protection

The headgear worn by combat aircrew started in the 1920s as a stout close-fitting leather or cloth outer over a warm liner that, with a pair of flying goggles, gave some protection from the cold, noisy, backwash from the propeller. By the 1930s hearing protectors containing earphones had been added, and by the 1940s attachments to support an oxygen-mask. With the arrival of jet aircraft and ejection seats in the 1950s it was necessary to provide an impact-absorbing shell, and this made it possible to carry a removable, wrap-around, sunlight-attenuating visor instead of goggles. All headgear must adjust to fit a prescribed range of head dimensions and be stable, comfortable, easy to doff and don, have no protuberance that could snag a parachute line, and remain on the head during 500kn frontal and oblique windblasts.

13.6.1 Impact

As noted in Chapter 3, to protect the head and brain from impact damage, the headgear must absorb the impact energy without imposing excessive linear or rotational force. The nature and magnitude of these forces is affected by many factors, in particular the location of the impact on the shell and the angle of the blow as well as the shape, stiffness and friction of the impacted object. The shell absorbs some energy through de-lamination, but most is absorbed by the irreversible compaction of the liner that separates the head from the shell.

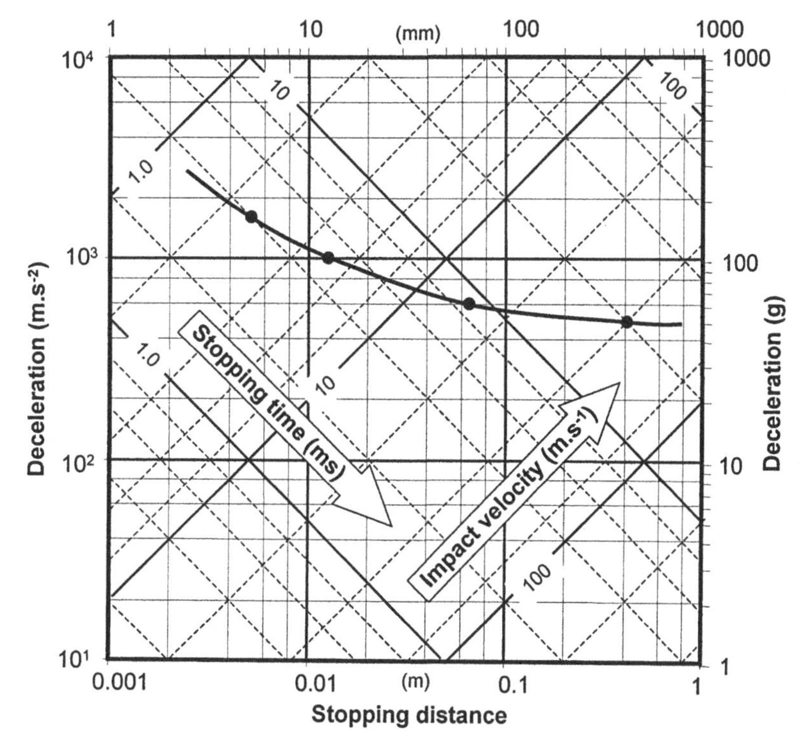

Figure 13.2 Simplified head impact dynamics. The graph represents the relationships between the time taken ($t = v/a$) and the distance travelled ($s = a.t^2/2$) when decelerating an object at a constant rate (a) to produce a change in velocity (v). The four points joined by the curved line mark combinations of deceleration and time above which the human brain is likely to suffer concussion, as assessed by studies conducted at Wayne State University

Figure 13.2 shows the relationships between the dynamic variables for the simplified case where an impact generates a radial blow and the head is brought to a stop at a constant deceleration, any such impact being represented by a point on the graph. The curved line summarises the relationship between the peak value and the duration of the deceleration likely to induce concussion. This data was collated by scientists at Wayne State University from experimental evaluations and epidemiological studies in order to specify head protection for American football players[5]. Testing using dummy heads also suggests that real impacts do not conform to the simple constant deceleration model. Instead the deceleration ramps up to a peak value and descends equally rapidly. The stopping distance is therefore greater than shown on the graph, and the curve should only be taken as an indication of the minimum thickness of energy-absorbing padding needed inside the helmet shell to avoid concussion. For instance, for an impact velocity of $5.0m.s^{-2}$ to impart less than 100G head deceleration, the head should travel at least 11mm within the helmet. In practice, to accommodate the non-linear deceleration, and because foam materials made from polystyrene or polyurethane can only be compacted to about 15% of their volume, the liner must have significantly greater thickness.

In lieu of statistically reliable data relating the nature of head injury to the severity of impact (and the likelihood of a survivable crash or ejection provoking the impact), the physical protection requirements for RAF flying helmets have been based on the UK standards for protecting motorcyclists; BS 2495 and BS 6658[6]. These assume that the flexure of the neck enables the head to be treated as a mass of 5kg separated from the bulk of the body. Interpreted for helicopter aircrew the relevant test requires the deceleration of a standard 5kg instrumented dummy head fitted into the tested headgear to be less than 300G when dropped onto a steel anvil with either a hemispherical or a flat surface. The tests are done so that the rear and the crown come into contact at an impact velocity of $7.5m.s^{-1}$ and are repeated on the same areas at $6.3m.s^{-1}$. The lower impact velocity is also used for assessing impacts to the side of the shell and when testing helmets for fast jet aircrew. These 205J and 135J blows (the kinetic energy is equal to $m.v^2/2$) are quite severe and, as indicated by the Wayne State criteria in Figure 13.2, the maximum head deceleration of 300G may cause concussion. The figure also indicates that about 10mm movement would be produced if the deceleration was constant and at the maximum level. In practice an energy-absorbing liner of about 25mm thickness is needed to cater for repeated blows and the compaction characteristics of the polymer foam.

Similar drop tests are also used to assess the degree of penetration of the shell by a sharp pointed anvil and the minimum snatch force needed to break the chinstrap. It is of note that no testing standard considers frontal, facial impact. Most headgear adopted by US services is required to meet lower impact standards, in which an impact velocity of $5.2m.s^{-1}$ can impart a maximum of 400G deceleration. The energy-absorbing liner is usually about 10mm thick in such helmets. A review of the UK requirements is likely to recommend the retention of the current level of protection, with additional testing in hot and cold environments and after the helmet has been immersed in water.

The likelihood of damage to the brainstem from abrupt rotation of the head depends upon the angular acceleration and the angular velocity[7], and hence the duration of the acceleration. Injury is likely if the peak acceleration exceeds about $5000rad.s^{-2}$ and the velocity exceeds $70rad.s^{-1}$. However, impacts are complex because the rotational force is usually accompanied by some linear force and the resultant depends not only on the velocity and angle of the blow but on friction and mechanical interlock between protrusions on the impacted surface and the shell. Standard tests are done using abrasively treated anvils that impart glancing blows. Rapid head rotation can also be produced during emergency egress if the thin lines connecting the harness to the parachute become trapped in crevices or are wound around protrusions on the headgear. Such snagging hazards are a particular concern to aircrew in fast jets and some transport types. The external surface must be smooth, any gaps must be bridged and any protrusions must break off with little force.

13.6.2 Head-borne mass

Angular acceleration can be reduced by increasing the moment of inertia of the helmet, and in principle the risk of rotational impact injury can be lessened by increasing the headgear mass. As noted above, protection against linear impact requires a thicker liner and this invariably increases the overall helmet mass. On the other hand the head itself can be difficult to support under high-G, and aircrew want their headgear to be as light and well balanced as possible to avoid discomfort, fatigue and neck muscle strain. Although the services of different nations have adopted different criteria for resolving these conflicting requirements, the recent need to mount equipment on the helmet has complicated the trade-off between protection and encumbrance so that the usefulness of these additions is also considered. It has therefore become essential to determine the maximum safe headgear mass and understand how this maximum is influenced by the location of the headgear centre of mass relative to the head.

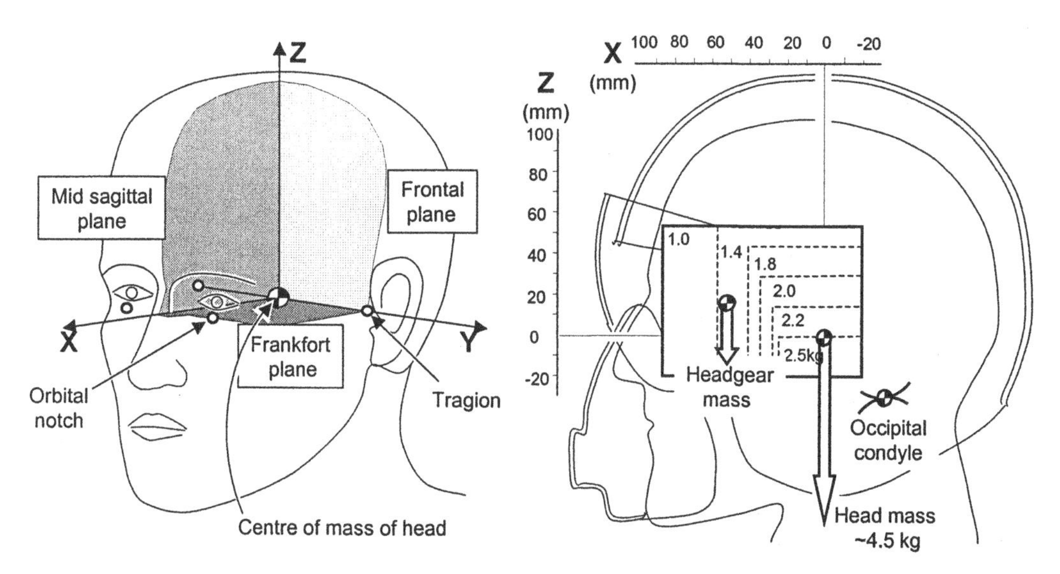

Figure 13.3 (left) Facial landmarks used conventionally to define head co-ordinates. (right) Use of the co-ordinates to specify how the headgear centre of mass can be permitted to move upward and forward if the headgear mass is reduced. In this illustration the maximum mass is 2.5kg and the maximum moment is 1.06 N.m. These values are suggested by the US Army Aeromedical Research Laboratory as the maximum that can be tolerated by helicopter aircrew without risk of severe neck injury

Figure 13.3 shows how the anatomical landmarks on the skull can be used to define a frame of reference for specifying headgear co-ordinates[8]. The tragions (small bony protuberances near the ear canals) define the Y-axis. The point in the centre of the head midway between the left and right tragions is taken to be the head centre of mass. This, together with the lower edges of the eye sockets define the XY "Frankfort plane", and hence the frontal (ZY) and mid-sagittal (ZX) orthogonal planes. The diagram on the right of the figure shows the approximate location of the cartilage at the base of the skull that articulates with top of the vertebral column, the occipital condyle, that is taken to be the pivot for head pitch motion.

The pitch attitude of the head is held by a downward pull on the back of the skull that balances the forward overhang of the head mass, and any headgear that adds to the moment increases the necessary counterbalancing tension of the neck muscles. Experiments conducted by the USAARL for helicopter aircrew[9], who are at most risk of injury from $+G_x$ and $+G_Z$ jolts during heavy landings, indicate that for median head dimensions the headgear mass should be limited to 2.5kg. They also suggest that the maximum overhanging moment should be less than 1.06N.m. The diagram on the right of Figure 13.3 shows how the USAARL permit the centre of mass of lighter headgear to extend forwards and upwards to remain within this limiting moment.

The prospect of routine exposure to $+9G_Z$ during combat manoeuvres and a $+16G_Z$ jolt during ejection suggests that aircrew in fast jets should be asked to bear less head-borne load than deemed safe for rotary-wing pilots. A study undertaken to enable limited flight testing of equipment developed as part of the USAF Interim-Night integrated Goggle and Head Tracking System (I-NIGHTS) recommended that helmet systems should be lighter than 1.58 to 1.81kg (3.5 to 4.0lbs) for general in-flight acceptability[10]. A head-supported mass of 2.04kg was considered safe for ejection if the combined head-&-helmet centre-of-mass shift was less than -20+7mm in X, ±4mm in Y and +38mm in Z. A comparable study carried out for Eurofighter, based mainly on the collation of data from biodynamic modelling, noted that wearing headgear with a mass of 2.5kg would not increase the overall risk of injury during ejection or impact. However, lighter headgear would reduce the incidence of neck pain of sufficient severity to temporarily remove affected aircrew from flying duties, and also lessen the risk of injury due specifically to headgear inertia. Given the known differences in neck muscle strength it was anticipated that these problems would be more likely to affect female aircrew.

13.6.3 In-service headgear

At present the basic helmets supplied for use in most combat aircraft in the UK are the Mk4 and Mk10. These are provided in five sizes and have a mass between 1.4kg and 1.8kg. The Mk10 is lighter and is normally worn by fast jet crew, while the Mk4 is used by helicopter aircrew and by fast jet crew when operating with night vision goggles. Most aircrew flying other aircraft types use versions of these helmets, except Apache and Eurofighter where type-specific headgear is needed.

The open-face design is made by Helmet Integrated Systems Limited using an outer shell of glass-fibre reinforced plastic, about 2.5mm thick, stiffened in places with woven kevlar and carbon fibre. Low density, crushable polystyrene and polyurethane foam mouldings about 25mm thick are bonded inside the crown and brow for impact protection. A pair of ear-shells, described below, are placed over the ears inside the shell for protection against cockpit noise, and a broad stiffened pad is placed at the rear in contact with the nape. In the Mk4 the whole set of components is held together and adjusted to fit the individual by a set of internal nylon webbing straps; one behind the left ear-shell, a second behind the right ear-shell, a third behind the nape pad and a fourth that forms a basket over the crown. Having selected the appropriate size from head length and breadth measurements, as shown in Figure 13.1, the helmet is adjusted by first setting the crown basket and chinstrap so that the shell is as low as possible but the brow is just out of view. The nape strap is then tensioned to produce a firm but comfortable pressure on the brow and nape, and then the straps over the earshells are tensioned until the seals apply firm pressure on the sides of the head. Straps are tensioned through buckles, turned back, and sewn to prevent slippage. To doff the helmet the wearer slackens the earshell tension by turning a pair of over-centre buckles on the outside of the shell, unfastens the chinstrap and, using outwards thumb pressure, flexes the lower edge of the helmet shell by about 20mm so that the earshells can be lifted over the ears. The Mk10 is fitted in a similar way, but a selection of foam pads replace the crown basket.

Electrical cabling for the earphones runs from the standard external connector to the inside of the shell, and to a connector for the microphone in the oxygen-mask. The stiff shell also acts as a support for the pair of swivelling latches that accept the short links to the oxygen-mask or NBC head cowl. The pair of visors is attached at the temples, and a bracket on the brow accepts the NVG mounting. The attachment of a conventional NVG adds about 0.8kg to the 1.8kg helmet and 0.35kg oxygen-mask, which makes the headgear heavy and unbalanced (at present UK aircrew use a NVG as a temporary operational expedient pending the development of lighter, better balanced, alternatives). Helicopter crew can attach the NVG battery case or a counterweight to the back of the shell to shift the c-of-g back towards the head centre, but in fast jets a counterweight would snag the head box of the ejection seat and the aircrew must tolerate the c-of-g shift. In the later case, to avert injurious neck loads during ejection, the goggle is detached automatically using a small electrically-actuated gas cartridge integrated into the goggle mounting, the goggle being propelled beyond the crewman's knees to avoid leg injury.

The basic headgear worn by combat aircrew of other nations is of similar configuration, but details differ considerably, as do the levels of protection. Few have adopted the use of nylon fitting straps, and in most US designs the vertical position of the shell on the head is set by forming a liner that fills the space between the shell and most of the wearer's brow, crown and nape. One approach employs a pad formed from a stack of dimpled sheets made from a proprietary thermoplastic. This is softened by pre-heating and inserted in the shell, then the assembly is placed on the user's head and allowed to cool and set. This material can be re-used. Another uses a tool to form a polyurethane foam moulding of the required volume that is trimmed and faced with a thin layer of compliant foam and leather. Further techniques are under development for integrated helmets. As described in section 13.14.2, one produces individualised liners that conform to the whole of the head.

13.7 Noise protection

The necessity for noise attenuation and hearing protection in combat aircraft has been reviewed in Chapter 3. The requirement is to attenuate noise at both ears to a level that enables reliable speech communicated through earphones and protects the ears from damage. Surveying the characteristic noise in the cockpit of a new aircraft type is a necessary prelude to arranging satisfactory protection, because the character of the noise varies over the flight envelope and protection must be tailored to match the spectrum, level and probability of occurrence of the noise.

Table 13.4 Maximum permissible daily exposure to noise recommended by the Health and Safety Executive in the UK[11]

Exposure (hours)	Maximum level (dB(A))
8	85
4	90
2	95
1	100
0.5	105

To avoid hearing damage, the accumulated daily noise dose should be less than the values given in Table 13.4, and most services strive to provide protection that would enable eight working hours per day. For this the ambient noise must be reduced to a level that, including overlaid speech, is less than 85dB(A). A lower limit of 80dB(A) may be imposed in 2006 by the Health and Safety Executive (HSE) in the UK, in accord with a recent European Union directive.

Attenuation characteristics can be assessed most reliably using a technique that is very similar to the collection of an audiogram. In this case a subject is placed in a sound-isolating chamber and asked to adjust the levels of narrow (1/3 octave width) bands of noise centred on 125, 250, 500, 1000, 2000, 3150, 4000, 6300 and 8000 Hz. The threshold at each centre frequency is established when exposed to the noise and when wearing properly fitted protection. The differences between the thresholds defines the attenuation spectrum. Tests are repeated using at least ten subjects because attenuation effectiveness varies with head and ear shape and the accuracy of fit of the protection. A more rapid approach requires the use of a miniature calibrated microphone to compare the sound pressure level at the entrance to the ear canal when the protection is fitted and when the ear is exposed. Again, tests should be repeated over a representative set of subjects.

13.7.1 Passive techniques

At present the primary way of reducing such distracting and damaging exposure is to cover the ears with a sealed enveloping shell that attenuates the noise level by reflecting and absorbing the sound energy. The circum-aural earshells are flattened hemispheres about 90mm diameter, usually made from polymer composite with glass fibre and carbon fibre stiffening, and lined with acoustically absorbing polymer foam about 6mm thick. The gap between the rim and the side of the head is sealed using a compliant ring, about 10mm wide and 10mm thick, of soft PVC over a compressible attenuating foam core. As noted above, the shells float within the helmet shell so that they can be positioned over the wearer's ears without constraining the fit of the other elements of the helmet. The shell includes means, such as springs, resiliant foam or webbing straps, for pushing the earshells onto the side of the head to compress the seals and avoid gaps that leak noise. A compression force of about 10N is typical. Each earshell has a mass of about 150g, which includes the small acoustic transducer.

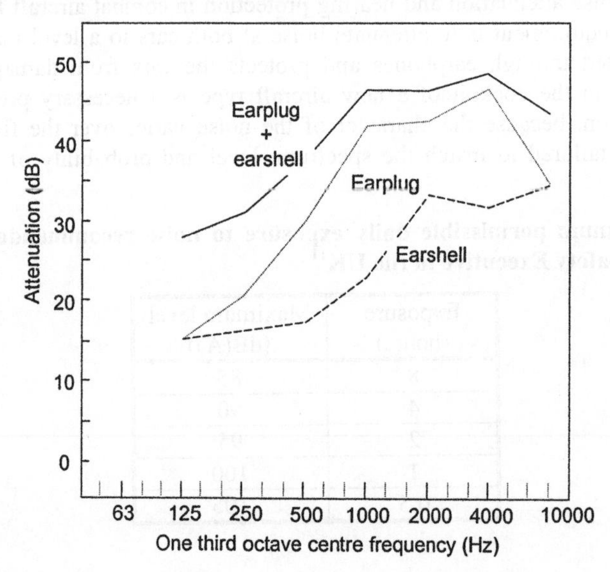

Figure 13.4 Typical earshell and earplug acoustic attenuation characteristics (From Rood G M, *The audio environment in aircraft*, in "Audio effectiveness in aviation", AGARD-CP-596, 1996. Reprinted with permission from QinetiQ)

As shown in Figure 13.4, the earshell provides attenuation of about 25dB above about 1500Hz. This is satisfactory in the cockpits of many aircraft, for instance as shown in Figure 3.6, and the typical noise level in a Tornado can be reduced from an excessive 95dB(A) to about 70dB(A). However, in the many aircraft types that generate a significant proportion of low frequency noise, the performance is not good enough to reduce the accumulated noise dose to the H&SE level of less than the 85dB(A) over eight hours per day. Aircrew in the noisier fast jets, such as Harrier, are advised to use ear plugs for additional protection. As shown in Figure 13.4 the combination can provide nearly 40dB attenuation at 500Hz and reduce the noise level to about 70dB(A).

A variety of plug types are available. The industrial, disposable foam plastic variety and the bespoke silicone rubber are equally effective, but the latter tend to be more comfortable and less likely to dislodge. Pilots of high performance combat aircraft that are prone to changes in cabin pressure during height-changing manoeuvres should always have a vent into the ear canals to prevent an imbalance of pressure across the eardrum. The vent becomes essential during sudden loss of cabin pressurisation. Also, to hear communications, it is necessary to boost the signal at the transducers to compensate for the loss through the ear plug. The Communication Ear Plug (CEP) has been devised to overcome the latter problem by relaying the signals to miniature transducers in the plug. One version, called the Attenuating Custom Communication Earpiece System (ACCES) combines the miniature transducer with the technique of forming a bespoke mould from a cast of the individual ear canal. This can enable a reduction of the volume of air trapped between the plug and the eardrum, which reduces the amount of acoustic energy transmitted to the ear canal through the bones of the skull. The speech signal to noise ratio can then be maintained with less powerful speech input. Both forms of inserted transducer are compatible with the cowls, described below, that are placed over the head and ears to provide NBC protection.

13.7.2 Active techniques

Many aircraft, such as the CH-47 Chinook helicopter which has large, slow rotors, generate low frequency rhythmic sounds that are attenuated inadequately by passive techniques. However, the strength of such ambient sounds within the earshell can be reduced significantly by arranging for the communication transducer to emit sound waves that are exactly 180° out of phase but otherwise match the unwanted sound. The technique relies on the additive cancelling effect of the variations in pressure that constitute the superposed sound fields, and the additional sound is sometimes called "anti-noise".

As shown in the left of Figure 13.5, such active noise reduction (ANR) systems use a set of amplifiers and a miniature microphone. The latter must be placed as close as possible to the entrance to the ear canal, and the amplifiers and transducers must have linear dynamic response and a constant phase response over the cancellation frequency range. As shown in the right of Figure 13.5, the approach is most effective at low frequencies where the acoustic wavelength is larger than the dimensions of the earshell, and this complements the characteristics of passive approaches which supply good attenuation at high frequencies. In practice the additional electronic elements are integrated into the earshell, where the mass penalty is a few grams and the only complication is an additional conductor in the communication cable to carry the low voltage power to the amplifiers.

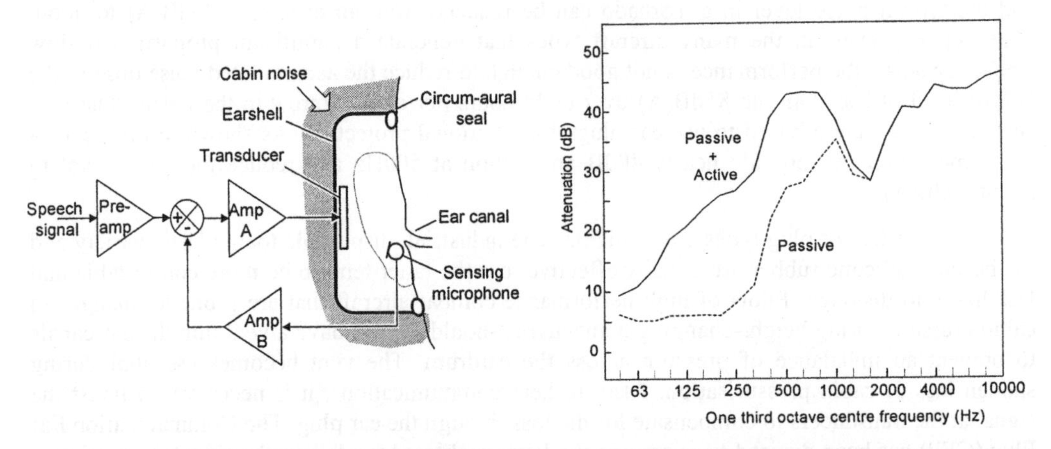

Figure 13.5 (left) Simplified schematic of an active noise reduction (ANR) system. (right) The effective increase of attenuation of low frequency sound gained by combining active noise reduction with passive sound absorption (From Rood G M, *The audio environment in aircraft*, in "Audio effectiveness in aviation", AGARD-CP-596, 1996. Reprinted with permission from QinetiQ)

The earshells incorporated into aircrew helmets account for an appreciable portion of the total head-borne mass. Current developments are centred on reducing the mass and bulk by improving the passive attenuation characteristics of materials without increasing their density, combined with the integration of ANR into the ear insert device and the adoption of digital control over the ANR. The last development seeks to adapt the feedback parameters to the transient character of the noise field.

13.8 Visors and eye protection

Although anything that interrupts, even in a minor way, the aircrew view out into the world and into the cockpit is an annoyance during routine flying conditions, a visor that attenuates the glare of the sun can be essential. A visor is also necessary to protect the eyes from accidental impact by night vision goggles that are now employed in most combat aircraft at night. However, the chief purpose of the visor, or visors, is to protect the eyes and face from a wide range of hazards that are mercifully improbable. A birdstrike or a mechanical failure, or anything that causes the canopy to shatter, can shower the aircrew with debris and turbulent air. In aircraft that employ canopy fragmentation by miniature detonating cord rather than removal of the whole transparency as the first stage of an emergency ejection, it is essential to intercept the mixture of canopy shards and molten lead that spatters the crewmembers. For this, the visor must guard the exposed area of the face and fit against the brow of the helmet and the top of the oxygen-mask. A similarly close, stable overlap must be maintained to deflect the intensely cold and violent windblast that can follow when the head is thrust into the airstream.

At present, radiation from lasers is usually countered by depositing narrow band optical filter coatings on the visor. These are ineffective against the intense and broad spectrum of emissions from nuclear weapons, and the most useful form of eye protection against this hazard depends upon active optical switching.

Table 13.5 Requirements of a typical visor system used by current fast jet aircrew

	Attribute	Typical values
General visor requirements	Binocular field of view	>90% of unrestricted field
	Luminous transmittance (photopic)	< 85%
	Refractive power	See entry A23 of Table 9.1
	Prismatic deviation	See entry A22 of Table 9.1
	Haze	See entry A25 of Table 9.1
	UV transmittance	< 1% over 250 to 320nm range
	Ballistic protection	Not penetrated by a standard 0.22" fragment impacting at 170m.s^{-1}
	Windblast resistance	Withstands 600kn blast for 200 msec without deflecting into the face or breaking away
	Abrasion resistance	Maintain haze value after surface rubbed using standard pumice-loaded eraser
	Cosmetic defects	Pits, scratches, digs & bubbles etc. are specified pragmatically
	Mass	< 280g
Sun visor	Luminous transmittance (photopic)	12%-18%
	Effect on user's colour discrimination	Arbitrarily set level of ability to sort test samples of different standard hues
	Transmission chromaticity	Shift of standard white < 0.03 u,v CIE 1976 chromaticity co-ordinates
Anti-Laser	Transmission in stop band	< 0.01% (4 O.D.) over whole visor field of view
	Effect on colours etc.	As for sun visor
	Closed state transmittance	< 0.02% (3.5 O.D.) over whole visor field of view
Nuclear flash	Open state transmittance	>35%
	Closure delay	< 2 msec
	Recovery delay	< 1.5 sec
	Effect on colours etc.	As for sun visor

The factors that must be considered in order to deflect dangerous forms of energy without impeding the passage of informative light are summarised in Table 13.5. Many of these, such as the various forms of transmittance and chromaticity are assessed in appropriate conditions using a spectro-radiometer to measure the spectral distribution of the incident and transmitted light. The requirements for negligible distortion, spherical power and prismatic deviation are of particular concern in the forward direction. These qualities are assessed from the eye locations, and because differences seen by the right eye and left eye induce misperception and eyestrain, the inter-ocular differences are required to be significantly less than apparent to each eye alone.

At present all visors are made from polycarbonate as this material can be draped, cut, moulded, dyed, and coated, and it has exceptional resistance to brittle fracture especially when subjected to very high strain rate. The last property makes it ideal for deflecting or absorbing a bombardment of debris, but the visor must be about 2.5mm in thickness and convex in shape with well supported edges to survive a 600 knot airblast. The primary clear visor generally has a radius of curvature of about 150mm, so that it conforms to the curvature of the brow of the shell, and it is attached to the shell at the temples using a pair of pivots or guide slots that constrain movement between a raised position above the brow and a deployed position in front of the eyes. As the visor swivels about an axis that passes through the centre of curvature, a second visor of slightly larger radius, which is usually dyed to attenuate sunlight, can be mounted on the same axis and be lowered independently. The pivot mechanism must allow the pilot to stow and deploy both visors with one hand, but when deployed the visor must remain in the down position under windblast. A positive sprung catch with releasable detents is usually employed on both pivots rather than a friction mechanism. In some designs the inner clear visor must always be deployed and the outer visor can be considerably thinner and made from a more brittle material, such as perspex, that accepts a greater concentration of dye.

The anti-laser protection is usually applied as a multi-layer interference filter coating on the inner surface of the outer visor, with the narrow bands of high reflectivity centred on the wavelengths of likely threat lasers tabulated in Chapter 3. The coatings are currently made by vacuum evaporation of materials such as MgF and SiO_2, which are brittle and easily damaged by cleaning as well as rough handling. To extend the useful life the anti-laser visor may be made as a bonded laminate in which the filter is sandwiched between two accurately curved plastic transparencies. Also, IR-absorbing compounds are included in the mixture of dyes dissolved in the material of the outer visor, and these compounds tend to bleach in strong sunlight and UV. Aircrew have to accommodate the small added mass and the increased internal reflections. As the devices are relatively fragile and costly, they are fitted to the helmets only when the crewmember may need protection against laser threats, and are not used for routine flying.

Nuclear flash generates a broad spectrum of emissions and it is not possible to protect the eyes and face by selective filtering. At present several US services provide protection in the form of goggles that mount on the helmet shell to seal against the brow, cheeks and nose to create a light-tight enclosure around the eyes. Light is allowed to impinge on the eyes through a pair of flat optical shutters that can be switched between two states. Each shutter consist of a pair of crossed polaroid sheets separated by a ferro-electric material such as lead zirconate titanate (PLZT) or liquid crystal. In one state the cell passes most of the linearly polarised light, which is attenuated strongly by the second polaroid layer. In the second state, achieved by applying a small voltage to the front and back surfaces of the cell, the material acts like a quarter-wave plate (as described in Appendix A) that rotates the plane of polarisation of the light by 90°. The light is then parallel with the axis of the second polaroid layer and passes with little attenuation.

The device contains a photo-detector and the thresholding and switching circuitry. Power is supplied from a small external battery pack that can be attached to the helmet shell with a fabric hook and loop fastener. Although the device can switch off rapidly and provide the necessary attenuation over the visible spectrum for much of the field of view, most of the other attributes are barely satisfactory. The recovery of PLZT is noticeably slow, and the open-state view is much like looking through a dark visor with additional haze and a masked periphery. Also, should the battery or any of the components fail, the device switches to the optically closed state that totally occludes vision. Few aircrew would consider this to be fail-safe.

13.9 Breathing system

In all modern aircraft that are intended to operate routinely above about 8,000 ft the crewstation and passenger cabin can be pressurised to provide an environment equivalent to that at 8,000 ft. Commercial transport types supply aircrew with an independent supply of oxygen through an oro-nasal mask, and this is considered adequate for the short time needed to descend to a safe altitude should the flight deck pressurisation system fail or become contaminated. However military aircrew routinely breathe an independent source of gas throughout the flight so that they are unaffected by fluctuations of the ECS and they can operate under NBC threat. Breathing through an oro-nasal mask also makes the transfer to pressure breathing automatic should the cockpit pressurisation fail, and there is a similarly automatic transfer to the emergency seat-mounted oxygen supply during ejection.

13.9.1 Mask

As reviewed briefly in Chapter 3, to prevent hypobaric hypoxia when exposed to the external environment at altitudes above the troposphere, about 40,000 ft, the oro-nasal mask must deliver oxygen at a pressure greater than the ambient. This pressure-breathing is also used in newer, highly agile combat aircraft as a way of countering high-G at any altitude. When operating in such a regime the mask must make a good enough seal against the face to maintain the excess pressure, and if using a pressure greater than 30mmHg, it is essential to support the body surfaces, the chest in particular, against the imbalance.

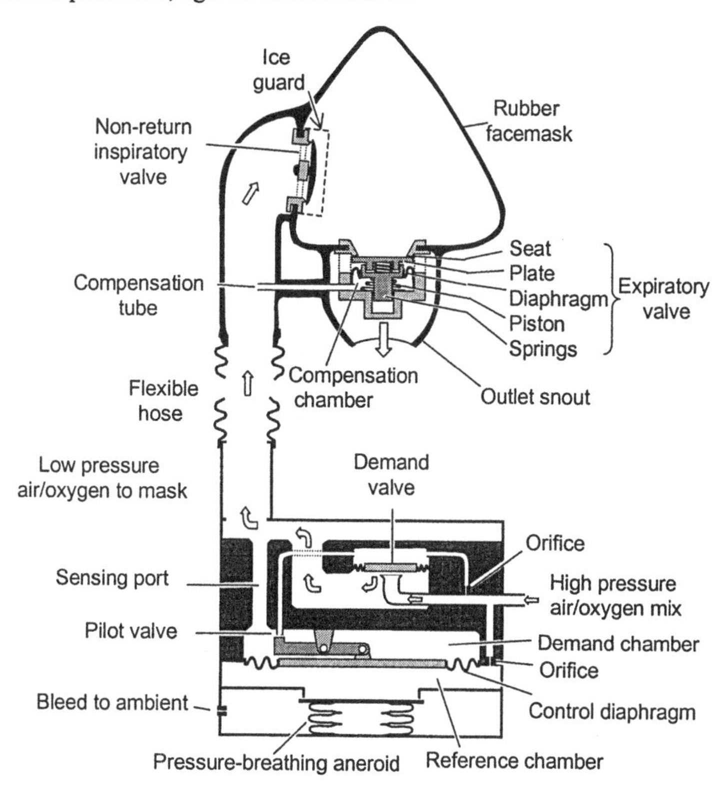

Figure 13.6 Schematic for a typical low-pressure breathing mask and demand pressure regulator (Adapted from Harding RM and Gradwell D P *Oxygen equipment and pressure clothing*, in Ernsting J, Nicholson A N and Rainford D J (eds) "Aviation medicine, 3rd edition", Butterworth Heineman, 1999. Reprinted with permission from QinetiQ))

The ideal qualities of the mask, valves and pressure regulator that make up the cockpit-mounted part of a conventional fast jet breathing system, are to supply the gas reliably without requiring the individual to alter the way that he breathes. For this, the gas must enter the mask at the required pressure and flow with little restriction. The operating principle of the equipment used in most RAF aircraft is shown in Figure 13.6.

The wearer has a choice between two sizes of breathing mask, large and small (called Q and P), that have an internal volume of about 150ml and a mass of about 350g. The inner part is a complex silicon rubber moulding that accommodates the pipes, valves and microphone, and this is held within a stiff, light, external shell. For many years the mask was attached to the helmet by a lateral pair of chains that engaged swivelling hooks on the front edges of the helmet shell, but these have been replaced by a pair of adjustable Mask Quick Release (MQR) catches that incorporate short wire strops that do not become detached when the mask is thrust into the face by the windblast during a high speed ejection. The fastness of the seal between the rim of the rubber moulding and the face depends mainly on the match between the shape of the rim and the shape of the contacted area of the face. The line of contact runs symmetrically from the bridge of the nose, down the side of the nose and then across the cheeks and the front of the chin, which is for the most part against thin bone-supported tissue. The rim itself is fairly unyielding, and it is largely a compliant in-turning lip that beds onto the contours of the face, although a pliable metal strip is incorporated in the nose bridge so that the rubber can be pinched to conform with the width of the nose. Like the helmet, this piece of equipment must be fitted carefully. The tension between mask and helmet must pull the rim against the face to form a leak-tight seal, but the pressure must not cause discomfort or prevent jaw movement for speaking. Although the required mask tension depends greatly on the facial geometry of the wearer it is occasionally found that a small initial leak is sealed by the accumulation of sweat and exhaled water on the contacting surfaces.

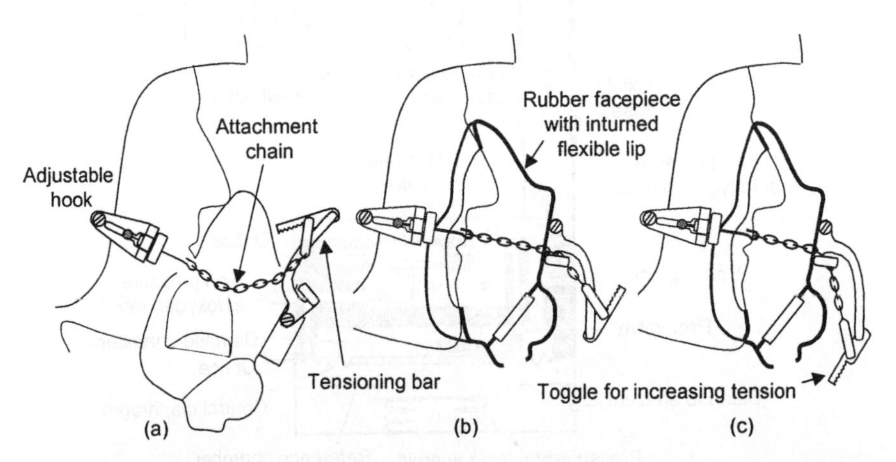

Figure 13.7 The traditional RAF method of attaching the breathing mask to the helmet. (a) The wearer donned the mask by coupling the chains on the raised tensioning bar onto the adjustable hooks on the helmet. (b) The mask was held in position by the down-turned tensioning bar to form a continuous seal against the face. (c) The toggle on the tensioning bar was turned down to pull the mask firmly onto the face to prevent leakage during pressure-breathing. In modern aircraft the chain and hooks have been replaced by quick release catches and wire attachment strops, but the toggle tensioning technique is still employed

Pressure-breathing is unpleasant. Not only must the wearer resist the over-pressure when inhaling, and breathe out forcefully when exhaling, but the mask-to-helmet tension must be increased substantially. This is done in UK fast jets by the wearer, as illustrated in Figure 13.7, who must flip down an over-centre toggle to shorten the couplings by about 8mm. This compensates for the distortion of the mask and helmet, and the compression of the nape pad and tissue, that would allow the mask to push away from the face and break the seal under the force produced by the excess pressure. This force is considerable, for instance an excess breathing pressure of 40mmHg produces about 60 N.

Figure 13.6 shows two valves. The inlet is a simple circular rubber flap that opens inwards when the pressure drops at the start of inhalation and slaps shut against a rigid rim when the mask pressure rises at the start of exhalation. The expiratory valve consists of a sprung metal plate that seats onto a metal rim, the plate being pushed away from the rim to allow the exhaled air to vent into the cockpit. To stop the waste of breathing gas that would otherwise vent through the expiratory valve during inhalation, the plate incorporates a small sealed cylinder and piston that holds the valve shut when the mask pressure is below the pressure in the feed pipe. The latter is linked to the small cylinder by the "compensation tube". Most masks also have a second emergency inspiratory valve near the top of the mask that opens to allow the entry of ambient air when the mask pressure is reduced to about 35mmHg below the ambient pressure. This anti-suffocation valve is intended to allow the crewman to breathe with the mask in place on his face if the end of the supply hose is blocked, for instance by the anti-drowning valve in the personnel equipment connector that closes when the crewman separates from the ejection seat.

13.9.2 Regulator

The pressure regulator shown in Figure 13.6 supplies a flow of breathing gas on demand. The device acts like a pneumatic servo-mechanism in which a sensitive pilot valve controls the opening of a higher capacity demand valve that feeds the gas directly to the mask. When the crewman is exhaling, the pressure in the hose to the mask is raised, so the pressure in the demand chamber exceeds that in the reference chamber by a small amount. The control diaphragm moves towards the reference chamber, and this holds the pilot valve, and hence the demand valve, shut. When the demand chamber pressure is reduced during inhalation, the pilot valve is opened, venting the feed pressure that holds the demand valve shut. The demand valve opens to allow the high pressure oxygen to pass through to the mask, the output pressure being controlled by dynamic feedback through small movements of the control diaphragm. The arrangement holds the pilot valve open when the mask is removed from the face so that breathing gas is not vented wastefully into the cockpit.

When the cockpit pressure falls below that of the gas in the pressure breathing aneroid, which is equivalent to the pressure at 10,000 ft, the aneroid expands to seal off the vent from the reference chamber, thereby automatically raising the pressure in the mask. Figure 13.6 omits the extra components that modulate the expansion of the pressure breathing aneroid to provide a progressive increase in excess pressure above 40,000 ft. Other components are also omitted, for instance the valves for mixing the feed oxygen with cockpit air, and the flow sensor or pressure-sensing switch in the output chamber that are provide the crewmember with a positive indication of flow during each breath. The sensitivity, dynamic response and disturbance by inertial forces, and the effects of variations of the temperature and pressure of the feed gas and the cockpit, depend upon component inertia and fine tuning of the rate of bleed flow through metering orifices. The production of miniature, man-mounted and seat-mounted units depends upon accurate manufacture and knowledge of the effects of numerous design details.

13.10 High-altitude protection

In the 1930s, the American balloonist Mark Ridge used a modified Siebe-Gorman diving suit to assess the viability of surviving the ambient pressure of 17mmHg at an altitude of 84,000ft. He found that an internal oxygen pressure of 150mmHg was just enough for a well acclimatised individual, but a pressure of 280mmHg was necessary for several hours' protection against decompression sickness[12].

To contain such a pressure difference, the first RAF fully enclosing aircrew pressure suit was constructed, much like the diving suit, from heavy-gauge rubberised fabric. The upper and lower parts were joined at the waist by a steel band, the front part of the cylindrical head cover was fitted with a double-glazed visor and a series of external leather straps joined the crutch to the shoulders to limit the distension of the torso. When inflated to the necessary excess pressure the garment took on the appearance of set of butted cylindrical tubes, a humanoid balloon, within which the wearer had to exert considerable force to bend his limbs or hold a sitting posture. The subsequent development of pressure suits, including those worn in the vacuum of space, has therefore been an endeavour to give the wearer the flexibility to move easily as well as providing reliable protection.

Modern forms of pressure suit are made from an inner impermeable layer, such as terylene impregnated with neoprene, butylene or polyurethane, with an outer layer of nylon fabric to restrain the bursting forces. Together they form a single inflatable bag that includes the rigid head covering and transparent visor. When inflated all of the sections take on a circular cross-section, and the helmet needs to be anchored by bands sewn around the torso between shoulders and crutch. Because the force needed at a bend is proportional to the fractional change of volume produced by the bend, the main technique for minimising the articulation force is to make the joints flex in a way that does not reduce the overall suit volume.

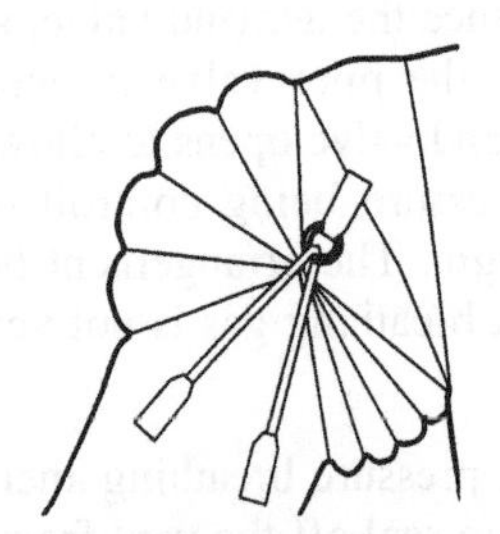

Figure 13.8 Method of tailoring a constant-volume knee or shoulder joint in a pressure suit using a series of collapsible convolutions. To prevent axial extension under pressure, upper and lower elements are linked by nylon cords running through anchored metal loops

As illustrated in Figure 13.8, this can be done using tethered crescent-shaped convolutions that collapse on one side of the hinge line and expand on the other. Judicious use is also made of netted materials to form tubes that increase in girth but shorten in length under pressure, and of "slip-knitted" materials that allow some torsional movement. Other torsional rotations, for instance at the neck and wrists, are usually arranged as a pair of sliding aluminium alloy rings that are sealed to prevent leakage and fitted with ball bearings to minimise friction.

A suit intended for enduring protection, such as that used in space, is constructed very strongly and must provide thermal protection and radiation protection as well as oxygen. The occupant breathes the same gas as that circulated within the suit, and this is maintained by a compact life-support system built into a back-pack. The duration of use depends on the amount of oxygen, the size of the battery for the electronic control system and the capacity of the scrubber that removes carbon dioxide, water vapour and other gases associated with body functions. However, during high-altitude flight aircrew are normally protected by the pressurised cockpit. The purpose of the garment is therefore to provide emergency protection against hypoxia and decompression sickness in the relatively unlikely event of cabin pressurisation system failure. In order to impose the least burden during routine activity, most airforces supply equipment appropriate to the maximum operating altitude of the aircraft. The combination of headgear and clothing provided in RAF aircraft is summarised in Table 13.6.

Table 13.6 Summary of the headgear and garments provided for protection against hypoxia and decompression sickness in RAF aircraft

Headgear and clothing	Altitude (ft)	% O_2 in breathing gas	Pressure-breathing (mmHg)	Protection provided	Comments
A	0 to 10,000	21	-	None needed.	Aircrew can breathe ambient air
B Mask & helmet	10,000 to 40,000	21 to 100	-	Enduring.	O_2 p.p. maintained by increased % of O_2 in breathing gas
C Mask & helmet	40,000 to 50,000	100	< 30	Good, if cabin pressure lost below 45,000ft, but < 1min if cabin pressure lost below 50,000ft.	Must descend below 40,000ft in < 1min
D Mask, trunk pressure jerkin, anti-G trousers & helmet	50,000 to 60,000	100	< 70	Good, if cabin pressure lost below 54,000ft , but< 1min if cabin pressure lost below 60,000ft.	Must descend below 40,000ft in < 2min
E Mask, trunk pressure jerkin, anti-G trousers & partial-pressure helmet	60,000 to 66,000	100	< 110	< 1min if cabin pressure lost at 66,000ft.	Must descend below 40,000ft in < 3min
F Full pressure suit	Above 66,000	100	141 (internal)	Several minutes	Must descend below 40,000ft in < 3min
			226 (internal)	4 to 5 hrs	Above pressure at 30,000ft, hence delays decompression sickness
			282 (internal)	Good	Above pressure at 25,000ft, hence avoids decompression sickness

Combination B is normally supplied in current fast jets, and the automatic application of a controlled excess for emergency pressure breathing enables this to be used up to an altitude of 50,000ft, as in C, providing the pilot makes a rapid descent to an altitude below 40,000 ft. Combinations D and E can best be considered as ways of applying pressure to the chest and the other parts of the body surface to counter the greater difference between the pressure of oxygen in the breathing mask and the ambient pressure to which the airman is exposed. The garments primarily prevent over-extension of the chest and the risk of lung damage, and the transient protection they afford also only suffices to enable the crew to survive a rapid descent to 40,000 ft.

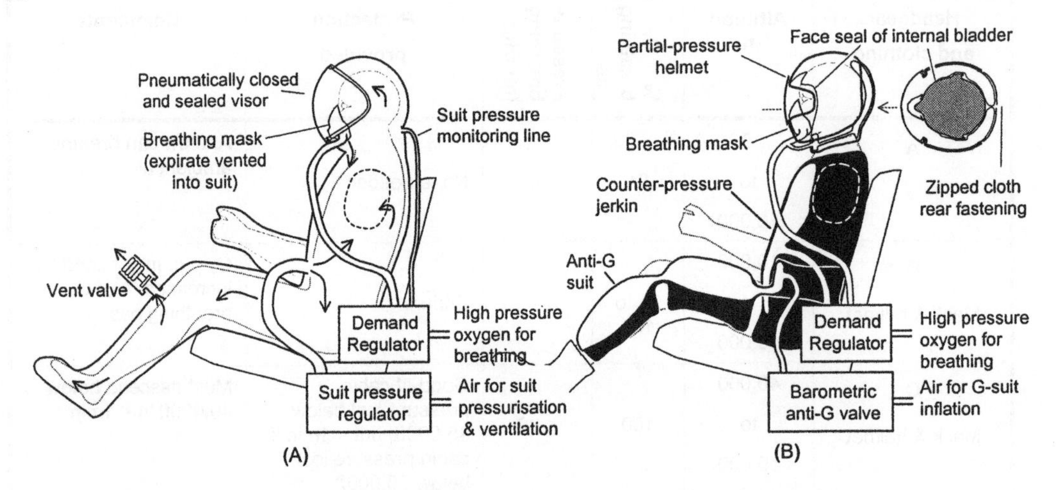

Figure 13.9 (A) A full pressure suit and (B) a partial pressure assembly. These correspond to combinations F and E in Table 13.6

The protection afforded by the full pressure suit depends upon the internal pressure, as summarised in row F of Table 13.6. To prevent the airman breathing air from within the suit, the pressure of the oxygen supplied to the mask must be held slightly above the suit pressure. Figure 13.9 (a) shows one technique that uses an aneroid capsule in the vent valve to control the suit pressure and a pneumatic monitoring line that connects the interior of the helmet to the reference chamber of the breathing demand regulator. When encased inside a full pressure suit the air that ventilates the suit can be heated or cooled to provide a comfortable microclimate, or alternatively the crewman can wear appropriate thermal conditioning layers. However, there is no such choice with the partial pressure assembly, and appropriate layers of thermal protection must be worn. The partial pressure assembly denoted by E in Table 13.6 is illustrated in Figure 13.9(b). The combination consists of a set of G-trousers (described in the next section), a counter-pressure jerkin and a "partial-pressure" helmet.

The partial pressure helmet encloses the whole head and the upper part of the neck and provides a counter-pressure over these surfaces to balance the pressure of the breathing oxygen in the respiratory tract and head cavities. The mask operates conventionally, but unlike the standard items, the helmet shell and visor envelop the mask as well as the head. When activated by the onset of the pressure-breathing regime, the visor is hinged down automatically from its normal raised position and pulled tightly against a seal on the helmet shell by a locking pneumatic mechanism. At the same time a rubber bladder, of somewhat strange shape, is inflated inside the helmet to apply the protective compression to the head and neck surfaces. The bladder can be regarded as a horizontal toroid encircling the head with large holes in the outer and inner surfaces that enable the user to see out through the visor. The outer hole is bonded to the front rim of the helmet shell, so this hole is sealed by closure of the visor. The inner hole is pressed against the brow, cheeks and chin and relies on the continuity of the seal against the face for air-tightness. The rear of the bladder differs between the arrangements for donning and doffing adopted by different helmet designs. The shell of the Taylor partial pressure helmet used by RAF aircrew has a large part of the rear section missing. This is covered by pair of stout cloth curtains, zippered and strapped to form a tight-fitting enclosure, and the bladder extends backwards into the two parts. The Goodrich helmet worn by USAF aircrew is split into front and back halves, joined by a top hinge and a pair of latches at the bottom edges. In this design the front and back halves contain separate bladders. In both arrangements, to prevent the inflated bladder lifting the whole helmet, the bladders do not cover an area over the top of the head equal to the cross-sectional area of the neck.

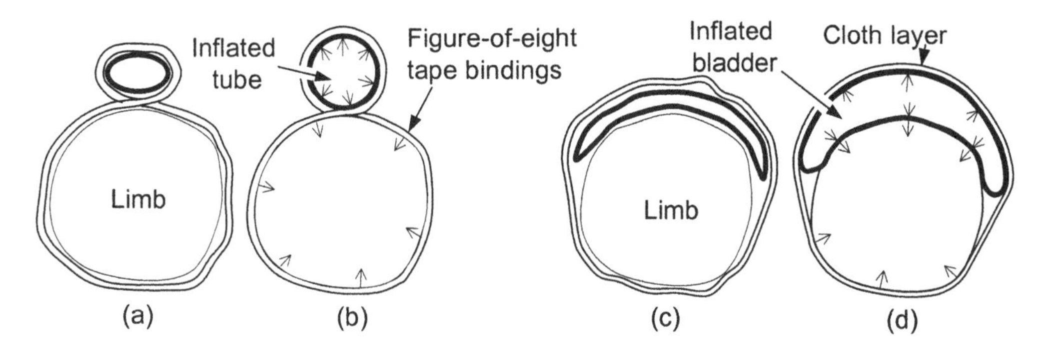

Figure 13.10 The principles of the inflating capstan tube and the inflating bladder for applying pressure. The deflated capstan tube (a) runs through a series of figure-of-eight loops enclosing the limb. On inflation (b), the increased girth of the inflated tube shortens the portion of the loop around the limb, squeezing the limb. If frictionless, the limb: tube pressure ratio would be equal to the tube: limb diameter ratio. A similar effect is produced at 1:1 pressure ratio by a bladder (c) when inflated (d) inside a tight-fitting garment. In both techniques the evenness of the compressive force on the limb surface depends on local friction, on stretching of the materials and on any limb out-of-roundness. The capstan tube allows better ventilation of the limb than the bladder, but it needs greater inflation pressure and becomes quite stiff longitudinally

The USAF originally favoured the technique, illustrated in Figure 13.10, of fitting the crewman with a tight-fitting garment that became even tighter when in-built loops were shrunk by a series of inflatable "capstan tubes". In one design a pair of tubes ran from the outside of the wrists, up the arms, over the shoulders, down the sides of the torso then outside the hips to terminate just above the ankles. The tubes were linked across the back between the shoulders. The capstan cords were threaded through loops stitched to the fabric to form a series of laces that were drawn tight and tied off, and it was necessary to fit the garment very carefully to get uniform compression without local over-constriction. When the high pressure was applied the tubes stiffened and tended to straighten the joints. The more practical approach uses a set of bladders that need to be inflated to considerably lower pressure than capstan tubes. As also illustrated in Figure 13.10, the compression is exerted over most of the area within the tight-fitting non-stretching garment rather than just the area contacted by the bladder. A jerkin covers the whole torso, and this is used in conjunction with a helmet and a set of G-trousers, described in the next section, that stop the blood from pooling in the legs. The combination has greatest effect if the G-trousers are inflated to a higher pressure than the jerkin and helmet using a barometrically controlled G-valve. The level of protection can be enhanced further if tight-fitting sleeves and a lower neck collar also contain inflatable bladders.

13.11 High-G protection

Combat aircraft commonly engage in turning manoeuvres in an attempt to attain a position behind the enemy so that a missile can be launched or a gun can be fired. As described in Chapter 3, the necessarily strong centripedal acceleration in the airframe Z-direction exaggerates the hydrostatic pressure gradient between the top of the pilot's head and the base of his feet. Should the pressure of the blood in the arteries of his brain drop for a few seconds below a critical value of about 20mmHg, the pilot will lapse into unconsciousness. The $+G_Z$ acceleration level that induces loss of consciousness (G-LOC) varies between individuals; if the individual is relaxed, it is between 3G and 5G. This can be raised by about 1G if the seat is reclined backwards, and by about 2G if the airman performs a variety of muscular straining procedures. But even when combined, these measures do not match the 9G or 10G that can be sustained by modern combat aircraft. Special items of clothing and headgear have therefore been devised, originally to raise the threshold by the 3G shortfall and latterly in an attempt to alleviate the effort of muscular straining.

Experiments were started in Germany in the 1930s to assess whether immersion of the body in water would counteract the effects of the increased hydrostatic pressure gradient within the body. These showed that the immersion up to lower rib level raised the G-LOC threshold by about 1.7G. Later, in the 1960s, experiments found that immersion to eye level could have greater benefit, although this varied between individuals; for some the threshold was raised by 3G, and for others by 16G. In the latter cases, the limit was imposed by the difficulty the individuals experienced when attempting to inhale air against the pressure of the water on the chest[13]. As this could be countered by pressure-breathing, it was estimated that full body immersion in water would allow more than 40G acceleration. The limit would be imposed by damaging levels of strain between organs of differing density, and by the imbalance between the top and bottom of the lungs caused by the external pressure gradient.

The early experiments encouraged Franks, during the 1940s, to develop a pair of heavily-constructed, double-layer, water-tight trousers that could be filled with water to about heart level, after the crewman had strapped into the seat. The strange garment was received enthusiastically by some and with reluctance by others. The proponents managed to gain more than 9G benefit, but the detractors only 0.9G. The disadvantages were however very real; a difficulty moving limbs weighted down with a water burden and a tendency to float in the seat with the feet de-coupled from the pedals. The suit also prevented heat loss, and some crew suffered involuntary enuresis when the water was drained from the suit after the flight.

The concurrent experimentation with pressure suits suggested that air pressure, suitably applied, could provide as much protection against high-G as water immersion. Lambert inflated the Franks suit to a pressure that increased with G-level and found that the G-LOC threshold was raised by about 2.2G. This was a greater effect than obtained by filling the suit with water. He therefore devised an air-inflated garment that used the capstan technique, as described in Figure 13.10, to apply the constricting pressure. Initial tests explored the idea of mimicking the action of the water-filled suit by applying less pressure to the abdomen than the calves, but the benefits were insignificant in comparison with the complexity of the pressure regulating valves and multiple connection pipes. Others experimenters suggested that pooling of the blood in the limbs could be prevented during high-G using four pneumatic tourniquets encircling the tops of the arms and the legs. Although this was successful in raising the tolerance by about 3G, the pain in the limbs soon became intolerable. Experiments proceeded using a set of inflatable bladders that covered a larger area at lower pressure. This approach was found to be effective and more acceptable, and is the basis of most current anti-G garments.

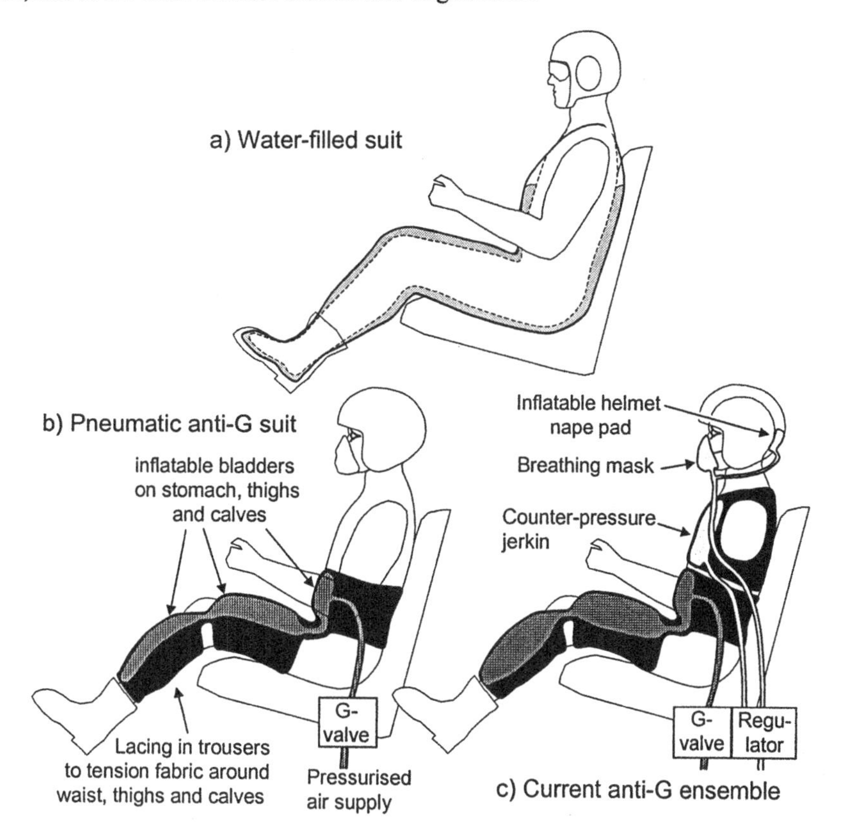

Figure 13.11 Alternative ways of providing high-G protection

The Mk6 anti-G trousers adopted for use in RAF fast jets had five separate, inter-connected butyl bladders inside nylon restrainers and weighed about 1kg. It was worn on top of underclothes so that the bladders were in close contact with the abdomen, the top of the thighs and the front of the calves, as illustrated in Figure 13.11(b). To aid movement, the garment containing the bladders had holes around the crutch and knees, and for ventilation it was tailored from inelastic cotton fabric. Four size variants were available, and the most appropriate size was fitted to the crewman by fastening a set of zips and then carefully pulling the slack out of a set of lacing cords that ran down the outside of the legs and across the small of the back. A slightly heavier Mk4 garment, which has sewn-in thigh pockets, was developed for wearing outside the Mk14 coverall, and this has become the standard. This is inflated to a pressure of about 65 mmHg/G by a mechanical valve that releases air into the bladders through an orifice oriented in the aircraft Z-direction. The flow is regulated by the balance between a closure force exerted by a spring and an opening force exerted by a sliding weight, the latter increasing in direct proportion to the G-level. Other airforces have adopted similar garments and regulators, but with national differences. For instance, US anti-G trousers are inflated to slightly higher pressure, and in the UK the inflation is held off below 2G to avoid transient effects during turbulence. Electronically controlled valves are now available that modulate the pressure in proportion to the rate of onset of G, and these can provide greater protection in aircraft capable of very rapid G onset.

The set of equipment provided in newer combat aircraft, such as Eurofighter, Grippen, F-15 and F-16 is illustrated in Figure 13.11(c). The US equipment is known as the Combined Advanced Technology Enhanced Designed G-Ensemble (COMBAT EDGE). In these aircraft the containment of blood in the upper torso is enhanced by "full coverage anti-G trousers" that have bladders covering most of the abdomen and the legs. Inflating socks are also included. Further protection is provided by raising the partial pressure of oxygen in the lungs using pressure-breathing, and this brings two entailments. The first is the need for a counter-pressure jerkin containing a large circumferential bladder, similar to the partial-pressure garment described above, to ease the difficulty of controlling the chest muscles when inhaling and exhaling. The second is the need to tension the mask to prevent leakage with a pressure schedule that typically starts at 4G and reaches 70mmHg at 9G. In the Gentex COMBAT EDGE helmet this is done automatically by a bladder in the nape of the helmet having an area comparable to the mask rim area. It is inflated to the same pressure as the mask to provide a force on the back of the head that balances the force lifting the mask away from the face.

An alternative form of hydrostatic anti-G protective garment has been developed in Switzerland. Called the "Libelle" (German for dragonfly), it is held to mimic the counter-pressure generated by internal fluids in the dragonfly under the 30G acceleration level inherent in the insect's mode of flight. The suit uses about 1.5 litres of water in a set of sealed tubes sewn into a non-stretchable but tensionable garment. The legs are covered from ankle upwards, the torso to neck height and the arms to the wrist. The mechanism is subtly different from the direct hydrostatic balancing effect of the original Franks suit. Under increased G-level the water tends to swell the tubes around the lower limbs and seat, where it exerts greater tension on the fabric and therefore a greater compression on the underlying tissue. The compression therefore depends upon fabric inelasticity and snugness of fit. The approach obviates the need for external connections and has the additional merits of protecting the arms, allowing good surface ventilation and acting immediately. The developers suggest that respiratory straining can be counter-productive and it is necessary for aircrew using the Libelle suit to learn an alternative AGSM in which they breathe normally and mainly strain the muscles of the lower body. Pilots report less fatigue and easier speech in comparison with the pressure breathing needed for the pneumatic approach. However, preliminary testing in Sweden indicates that, whether or not appropriate forms of AGSM are used, the hydrostatic garment provides protection to only 6.3G in comparison with the 9.0G afforded by an advanced pneumatic assembly[14].

13.12 NBC protection

Protecting aircrew, ground crew and aircraft from the wide variety of lethal nuclear, biological and chemicals agents in gaseous, liquid and particulate forms has, as reviewed briefly in Chapter 3, proved to be difficult. At present it is most effective for aircrew in combat aircraft to wear individual protective equipment (IPE) which acts as an impregnable barrier within a cockpit that cannot itself be regarded as uncontaminated. After exploring the use of over-suits with integral transparent head covers, the UK adopted the aircrew respirator No5 (AR5) for comprehensive above-neck isolation. A combination of garments give comparable below-neck protection.

The AR5 is worn under the standard flying helmet. It has a rigid, polycarbonate face-plate at the front, bonded to an impervious rubber cowl that fits closely over the head and upper part of the neck. The cowl extends down over the shoulders to overlap the suit, with a compliant bellows section at collar level to allow the head to move without dragging the part draped over the shoulders. The face-plate is shaped to fit inside the frontal helmet opening, to which it is attached by the same adjustable MQR linkages as the conventional oxygen mask. The lower half contains the rubber oro-nasal components complete with valves and microphone, and the upper half forms a close-fitting, spherically-curved, transparent visor. Two hoses are attached; one providing the filtered breathing gas, the other the clean, temperate gas for de-misting the inner surface of the visor and ventilating inside the cowl. The exhaled breathing gas is vented into the cockpit through a second expiratory valve that forms a safe air-lock. The spent ventilation gas also vents into the cockpit after passing down the neck and under the shoulder cape. Wearing the AR5 instead of the oxygen mask increases the head-borne load by about 350g.

Below-neck protection relies in the capacity of the material of the coverall to absorb hazardous liquids. This is worn over a one-piece suit made from carbon-impregnated fabric that allows water vapour to pass out but absorbs any lethal gasses. The combination keeps out dust particles.

Dressing under NBC threat requires the crewman firstly to put on his normal undergarments and (if needed) immersion suit, then boots followed by the carbon-impregnated suit. He then dons the AR5, spreads the shoulder cape across his shoulders and puts on the coverall and G-trousers. The two hoses of the AR5 are connected to a portable filter unit that contains a battery-powered fan. This "whistling handbag" pressurises the air slightly to make breathing easier and provide vital ventilation until he is connected to the PEC on the ejection seat in the aircraft. He then pulls on neoprene over-boots, latches himself into the floatation jerkin, puts on leather gloves and neoprene over-gloves. Finally, he puts on the helmet, connects the MQR attachments to the AR5 and plugs in the microphone. Then, so he can communicate with other air and ground crew, he connects the communication pigtail from the helmet to a small battery-powered unit containing a microphone, loudspeaker and amplifier that he clips onto his shoulder.

In addition to the complexities of dressing and undressing, this form of protection causes considerable hindrance to limb and head movement, and hence to vision. Routine operation of the cockpit systems requires more effort and time than normal, and the pilot has to be careful when operating switches because the outer gloves impede dexterity and tactile feedback. The thermal load needs to be managed carefully, particularly by crew operating in a hot climate. They must dress, walk to the aircraft carrying the portable filter and complete the external checks on the aircraft, all before it is possible to climb into the cockpit, strap into the seat harness, detach the blower-filter and transfer to the cooler air from the aircraft ECS. By this time they are sweating profusely with a high enough core body temperature to risk thermal exhaustion.

The NBC-protective helmet produced by Gentex for Tornado aircrew in the German Air Force alleviates some of the headgear problems. As outlined in Figure 13.12, in this design the face-shield and the helmet shell form a sealable enclosure that, together with the outflow of ventilating air and a skirt that drapes over the shoulders, prevent the entry of gas, liquid and dust. The arrangement gives effective NBC protection, and unlike the AR5, aircrew can gain substantial relief and talk normally during the long periods before and after a flight by unlatching the face-shield and hinging it away from the face. The absence of a close-fitting rubber cowl also gives better head ventilation. The design incorporates the components of a standard US MBU-20P oxygen mask for pressure-breathing, and this is strapped to the face-shield but allowed to float so that it sits comfortably on the face. A wound-up tube can be connected to an external bottle so that the wearer can drink with the face-shield closed.

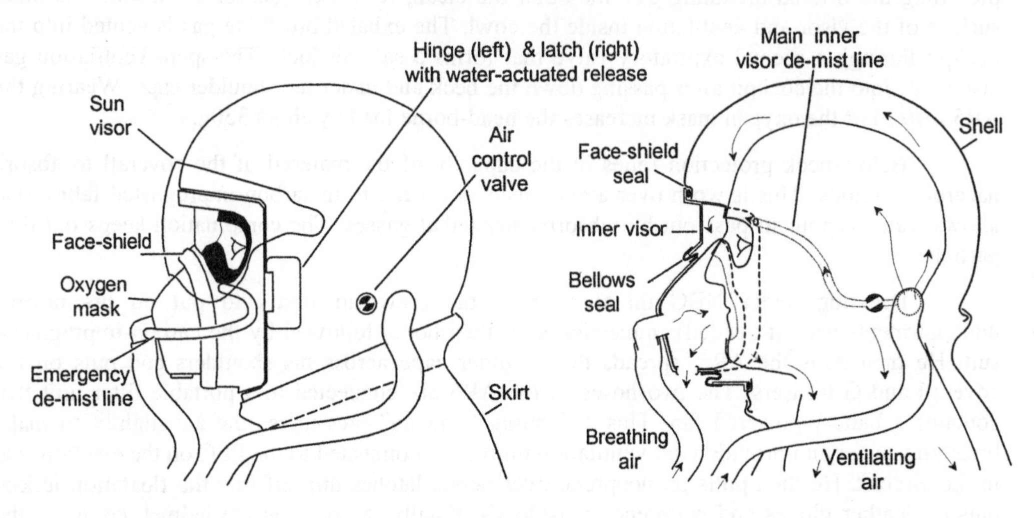

Figure 13.12 Sketch of the enclosed NBC-protective helmet worn by German aircrew in the Tornado

To prevent suffocation or drowning if immersed in water, a water-actuated mechanism automatically unlatches the face-shield and outer visor. The eyes (which are particularly sensitive to chemical agents) are protected by a seal that contacts the face across the brow and runs outside the eyes and over the bridge of the nose. The resultant goggle-like space between the eyes and the visor of the face-shield is fed with clean air through an adjustable bleed from the ventilation supply. Should this supply not be available, or fail, the eye space can also be vented using a bleed from the breathing supply. The possibility of visor fogging is reduced further by a hydrophilic layer on the inner surface. The enclosed helmet is heavier than a conventional combination of flying helmet and breathing mask, but it remains intact and attached to the head during ejection tests at 600 knots.

The problem of providing above-the-neck NBC protection for aircrew flying combat aircraft in the US military services is currently being addressed by the development of a Joint Services Aircrew Mask (JSAM). Like the AR5 in UK service, this item of headgear is intended as a substitute for the conventional oxygen mask when a NBC threat is anticipated. The requirements are stringent. For instance the visor section must meet optical qualities similar to those summarised in Table 13.5, and the mask must provide pressure-breathing for protection against GLOC and hypoxia to an altitude of 60,000 feet. However the stipulated compatibility with clothing, life support and communication equipment in a wide variety of aircraft types is a greater challenge, because the device must be wearable underneath a variety of sizes and types of helmet.

The design is likely to adopt features of the UK AR5 and the German NBC helmet, resembling the former in comprising a face-plate and internal cowl and the latter by arranging for the face-plate to fit into a frame from which it can be easily removed. Such detachment is the most practical way of venting the head and enabling the wearer to vomit cleanly. Automatic detachment is also the most reliable way of preventing suffocation or drowning. The design and the materials employed must not only meet the primary functional requirements but, because most of the procured items are likely to be held in long-term storage and used rarely, the devices must not deteriorate or require frequent checks.

13.13 Avionic helmet

Operators of military aircraft gain considerable operational benefit from display, control and intensified night vision systems that require the attachment of devices to the helmet. As discussed in Chapter 5, night vision goggles are used routinely by aircrew in all types of military aircraft to conduct missions at night in good weather conditions in much the same way as they do during the day. As outlined in Chapter 9, a helmet-mounted sight enables off-boresight weapon aiming, and a helmet-mounted display also provides easier access to vital flight and tactical information. In aircraft equipped with a steerable sensor, such as the Apache attack helicopter, the display can relay the pictures from the sensor as an alternative or complementary way to view the external world.

To gain all of these military advantages, an "avionic" helmet is included in the configuration of the cockpit of new combat aircraft such as the Franco-German Tiger attack helicopter. Fast jet aircraft that are currently under development, such as the Eurofighter and the Joint Strike Fighter also assume such headgear. The problem faced by avionic companies has been to develop systems that incorporate these novel facilities in addition to the already complex systems for protection, life support and communication.

13.13.1 Image combination

It is useful firstly to consider how the companies have approached the problem of configuring binocular electro-optical units that combine intensified night images (the function of the NVG) with sensor and symbolic imagery (the function of the HMD). The essential elements of the latter have been outlined in Chapter 9 and summarised in Figure 9.17. They comprise a miniature display device, eyepiece optics and the supporting electronic units to generate images and drive the display device, together with the helmet-mounted elements of a tracker to measure the helmet position and orientation.

The components that supply intensified binocular images can be added in the two ways shown schematically in Figure 13.13. The pair of matching electro-optical assemblies are shown to employ a CRT image source and a relayed axial eyepiece, but the ideas would apply to any HMD design. The essential element of the "optically combined" arrangement shown in (a) is the partially reflecting mirror placed in front of the CRT. This transmits some of the light emitted by the phosphor screen of the CRT and reflects some of the light emitted from the phosphor of the channel plate image intensifier, so that each of the user's eyes sees the pair of virtual images superimposed on his direct view of the world. In the arrangement shown in (b), the combiner is absent and the weak light from the external scene is instead imaged onto a sensitive helmet-mounted video camera, typically a channel plate intensifier linked to a charge-coupled device (CCD) through a fibre-optic taper. The camera video signal is routed back to the electronic unit, and the CRT can present this intensified image or the image provided by the generator or a mixture of both images. When relaying the intensified images, this "electronically combined" configuration can be regarded as a visually-coupled system, albeit with helmet-mounted sensors.

Both schemes have merits and drawbacks. The main benefit of optical image combination is in preserving the higher resolution of the channel plate image, as shown in Figure 9.11. Subtle differences of image colour, structure and dynamics also help the user to attend selectively to the information in the images, and if powered by a back-up battery the intensified image remains available in the event of display failure. The drawback is the inherent inefficiency of the combiner and the imbalance of luminous output from a channel plate and a CRT, which entail biasing the combiner to reflect about 85% of the intensified light and transmit only 15% of the light from the CRT. This provides a good balance at night, and with suitable mountings it would be possible to swing the combiner out of the optical path to supply a considerably brighter display in daylight, but this has not been implemented in any available design.

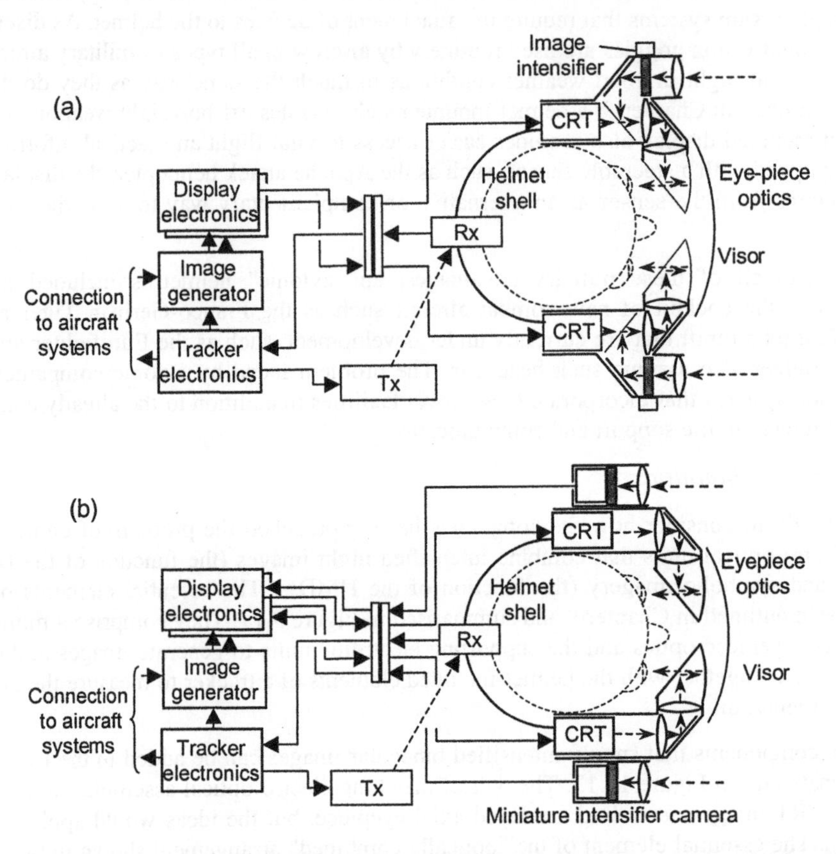

Figure 13.13 Schemes for combining intensified imagery with displayed imagery (a) optically and (b) electronically (Reprinted with permission from QinetiQ)

Optical combination is used in the BAE Systems "Knighthelm" installed in the German Tiger attack helicopter, and in the Kaiser "Strike Eye". Both of these examples have a mass of about 2kg, a field of view of about 35° and a small (~12mm) exit pupil. They use a pair of CRTs driven with a raster scan at night to give images with a maximum luminance of about 200cd.m^{-2}. In daylight the electronic units can be switched to scan symbols cursively to give a peak luminance of about 2000cd.m^{-2}, which is adequate providing the sun visor is lowered to attenuate the competing view of the world. The main advantages of the electronic approach are that the HMD optical design need not be compromised to accommodate a combiner. This gives better packaging and optical efficiency, and the small camera units can be detached and stowed in daylight. In the Thales "Topowl", which is used in the NH-90 and the French Tiger helicopter, a pair of CRTs and a 40° visor-reflection optical scheme are used in this way to project electronically combined imagery from helmet-mounted intensifying cameras.

13.13.2 Modular construction

Topowl is constructed in two parts; an inner helmet and a mating outer carapace. The former provides life-support, communication and impact protection, together with the components that adjust to accommodate the individual head. The outer part contains the electro-optical assemblies, the tracker components and the visors. This is fastened onto the inner using three mounting points that are adjusted to align the optics with the eyes, and because the design allows the user to separate the halves easily without disturbing the optical alignment, the outer can be regarded as a permanent part of the cockpit. This obviates routine electrical disconnection and enables the ground crew to check the system more easily. Also, unburdened by the heavy outer part, the crewman can escape more readily from a ditched or crashed helicopter.

A similar two-part helmet has been adopted by BAE Systems for the Eurofighter. The inner helmet is a new lightweight design and the outer part comprises an additional shell containing the electro-optical assemblies and spherical visor of the 40° by 30° "Viper" optical configuration shown in Figure 9.16(b). The headgear mass is 1.9kg including an oxygen mask, and this is raised to 2.3kg when a pair of intensifier cameras is attached for night operations. To provide NBC protection, a second version of the inner has been developed that effectively incorporates the components of an AR5. When this is used, the resultant masses of the day and night headgear become 2.595kg and 3.0kg respectively. The developers have suggested that the inner helmets could be treated as an anchorage for a range of mission-specific modules that, for instance, provide either a simple day-time monocular display, a binocular display, a night vision goggle or a day/night electronically combined NVG and HMD. These could meet the needs of all combat aircrew, simplify the logistics of in-service support and allow the use of common life-support elements[15]. The approach is also advocated in the US as a way of providing a night module and an interchangeable day module[16]. The night unit would house a Panoramic NVG, described in Chapter 5, together with a unit for injecting symbolic information into the forward channel for one eye. The day unit would be configured as a monocular CRT-based visor-projected display akin to the JHMCS. The use of the PNVG in the US set of modules would undoubtedly give better resolution and a wider field of view at night than the pair of intensifier camera images relayed by the CRTs in Eurofighter. The PNVG would also avoid the wide spacing between the objective lenses of the add-on intensifiers and so avoid giving the user a strange hyper-stereoscopic view of the world.

It is of note that the inner part of a two-part integrated helmet must flex when doffing and donning to allow the wearer to pull the earcups apart and slide them over his ears. However, the outer must be stiff enough to preserve the alignment between the electro-optical assemblies. The design must therefore confer the requisite stiffness to the two parts, and they must be joined "kinematically" at three points by articulating linkages that do not transfer the strain of the inner to the outer. The detailed designs for such secure, adjustable, detachable, articulating mounting points are fine examples of mechanical ingenuity.

13.13.3 Helmet-seat interconnection

The means of coupling the helmet-mounted systems to the supporting equipment in the cockpit presents a detailed electro-mechanical design problem. CRT-based displays need high tension (HT) anode excitation at about 13kV, 3kV and 1.5kV, with auxiliary drive signals for the cathode heaters, grid modulation and the currents for the scan coils. Conductors are also needed to carry the signals for the helmet tracking system, communications and any built-in test functions. About forty conductors are needed for a binocular system, and all of these must pass though a secure connector mounted on the side of the ejection seat. The man-portion must engage and disengage the seat-portion easily, and it must uncouple automatically during rapid egress and ejection. No sparks can be allowed during detachment, and the man-portion of the connector must not flail during the ejection windblast, or snag the harness and parachute lines.

Several suitable connectors have been devised. In one, the HT voltages are quenched at source by a signal from the seat-mounted portion during the initial phase of the separation, and the conductors are short-circuited to the earth conductor. In addition, the HT leads pass through sealed contactors that are opened before the main connecting pins leave their mating sockets so that any stored energy creates an isolated spark in the inert gas[17] In another, adopted by Eurofighter, the man-portion of the connector contains the high-voltage generators as well as video-amplifiers, built-in test and characterisation memory. The protrusion in the man-portion and the asymmetric mating cavity in the seat portion are long enough to ensure individual pins and sockets align before engagement, and a force of about 100N is required to push the connectors home. A toggle-mechanism on the man-portion is connected to the pilot's jerkin by a short lanyard, and a tension of about 30N on the lanyard pulls the toggle which acts on a lever to crank the two portions apart. Both of these connectors are about the size of a paperback book and weigh about 500g. The use of flat panel display devices in the helmet will obviate the need to handle HT complexities, but robust and necessarily weighty connectors will be needed for mechanical integrity.

13.14 Future developments

Although the clothing and headgear worn by current aircrew provides effective protection, the compounded layers undoubtedly hinder mobility, vision and dexterity. A crewman is faced with a choice between accepting reduced performance and protection or suffering accelerated fatigue by exerting increased effort to compensate for the restrictions.

13.14.1 Integrated clothing

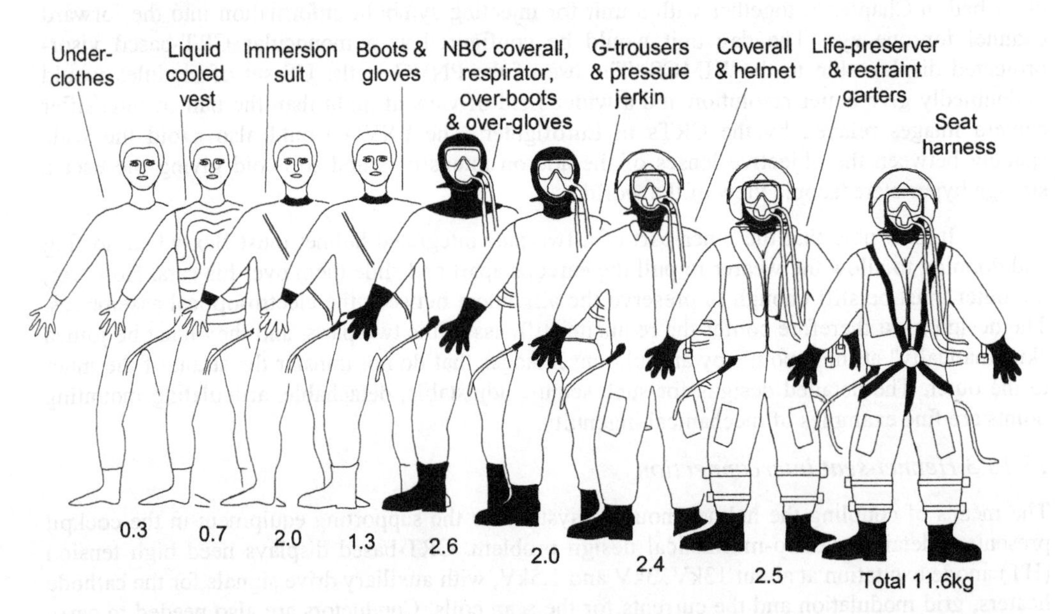

Figure 13.14 AEA likely to be worn by aircrew in RAF fast jets for summer over-water flying

The combination of clothing and headgear that may be worn in summer by RAF aircrew in a typical fast jet, such as the Tornado, is summarised in Figure 13.14. In all there are likely to be ten layers of material around the torso, nine around the legs, four around the feet and two around the hands. The mass burden of about 12kg is most evident to the wearer when walking, and this does not include the hoses, the man-portion of the PEC or the battery-powered filter. Winter operations add extra insulating layers and increase the mass slightly. Although aircrew in helicopters do not usually wear the layers that give anti-G or altitude protection they commonly wear body armour, and this is at least as heavy and cumbersome. Figure 13.14 also omits other man-mounted equipment, such as NVGs and the outer part of an integrated avionic helmet, and the extras such as a pistol, water and food that are supplied to aid post-escape survival.

The most pressing need is to supply the current facilities and levels of protection in a way that reduces the burden on the user and lessens the logistical complexity to the service. The most feasible remedy involves replacing the large set of compatible single-function garments by fewer multi-functional garments. For instance, the integrated flight jacket that forms a major part of the Eurofighter AEA contains the locator beacon and survival aids as well as the inflatable bladders that apply counter-pressure to the chest when positive pressure breathing is invoked at high-G and low ambient pressure. Such a garment is also likely to be adopted as part of the AEA for other types of fast jet. Other engineering synergies are theoretically possible, for instance a full-coverage tight-fitting garment containing a network of pipes filled with temperature-controlled fluid that can provide cooling, heating and anti-G protection.

13.14.2 Integrated headgear

A helmet that can supply all of the requirements without imposing an uncomfortable, encumbering and dangerous head-borne load has become essential. To be satisfactory in a fast jet it is suggested that the mass should be less than 1.5kg, and that for stable balance the centre-of-mass must be located within 20mm of the head centre-of-gravity and on the mid-sagittal plane. Such a genuinely lightweight piece of headgear can only be produced by abandoning the conventional approach, which adds extra layers and devices to a basic helmet. Instead, the helmet must be configured with lightness as the chief aim, with each element performing as many functions as possible[18].

Figure 13.15 illustrates an approach, devised in the UK at QinetiQ, which is based upon two design tenets; firstly that electro-optical elements should be mounted directly onto a stiff enveloping shell, and secondly that the liner should be tailored to conform exactly to the shape of the wearer's head. A practical form of shell can be made using a sandwich structure with a tough outer skin formed from two plies of glass fibre separated from a stiff inner carbon fibre skin by about 10mm of medium density expanded foam. The shell is split into front and back halves for donning and doffing, and these are joined at the crown by a plastic hinge and at the lower edges by bayonet latches. The mating edges form a pair of inverted-Us that interlock and give a gas-tight seal.

Comfort and stability depend upon the conformal liner. This is made in four steps. The wearer's head shape is firstly measured using a 3-D laser scanner. The model head and a model of the helmet are then manipulated using computer-aided design tools to chose the optimum shell size, place the wearer's eyes on the display optical axes and define the cavities around the mouth, nose, ears and eyes. The internal shapes of the two halves of the liner are then machined to an accuracy of about 0.5mm from polyurethane foam mouldings using a computer-controlled milling machine. Finally, the contact area is faced with a thin layer of anti-bacterial leather, the seals are bonded in place and the liner is inserted into the shell.

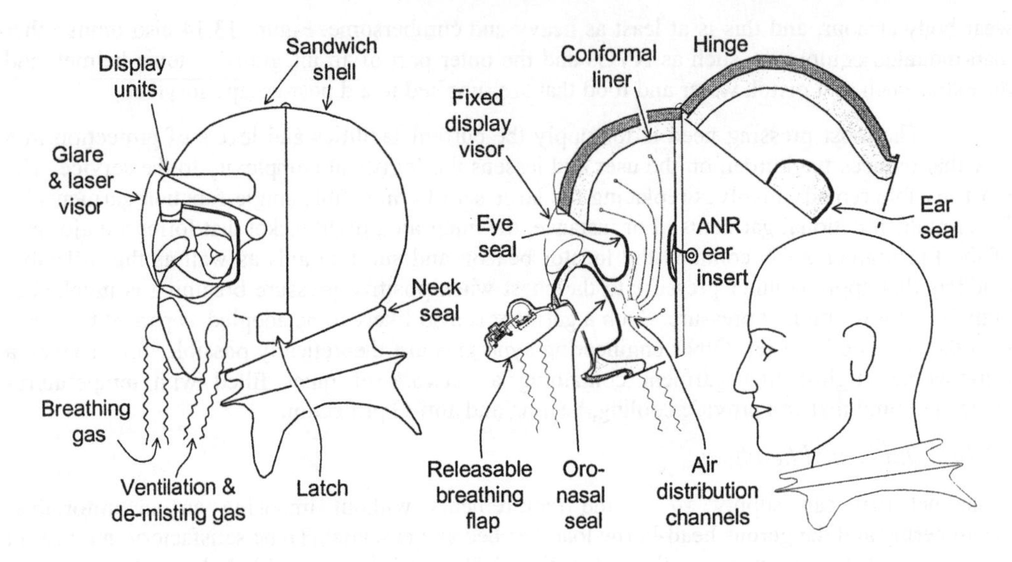

Figure 13.15 Possible configuration for a lightweight integrated helmet (Reprinted with permission from QinetiQ)

As indicated in Figure 13.15, the front part contains most of the active components; the fixed visor and display units, the releasable flap that houses the breathing valves, the outer protective visor and the entry hose for the ventilation and de-misting air. The breathing cavity contains a microphone, and each ear cavity a communication/ANR earplug. The seal at the edge of the breathing cavity has an inverted lip, much like that in the conventional mask, and the seals and the liner for the ear cavities provide acoustic attenuation. Prior to donning the helmet the wearer pulls a rubber bellows over his head and with the helmet closed stretches the upper edge of the bellows so that it seats into a mating flange around the bottom edge of the shell. This guard forms the link between the gas-tight shell and the below-the-neck NBC protective clothing. The choice of electro-optical configuration for the binocular display must give good image qualities with minimal visual obstruction, and the elements must be assembled into neat, low-slung, lightweight packages. Although none is ideal, many existing configurations, such as illustrated in Figure 9.16, could be used, and most synergy would be gained from a configuration that enabled the sealed visor to be employed as an active image-forming element.

Experimental evaluations have shown that the combination of a sandwich shell and a bespoke whole-head liner gives better protection against impact and penetration than a conventional helmet structure, and can readily exceed BS6658 criteria. The heat burden is the same as current in-service helmets. Stability is improved considerably because the stiff conformal liner does not distort or slip more than a few millimetres over the scalp. This lessens the need for tolerant exit pupil dimensions in the display, which therefore reduces the size and mass of the display optics. The liner is comfortable, unlike conventional pads and straps that must apply localised pressure to grip the scalp. A basic helmet that gives comprehensive protection, including protection against NBC agents and windblast at 600kn, can be built with a mass of less than 1.1kg. This would become 1.4kg with display and position-measuring components.

The use of computer-aided design techniques not only helps to optimise the choice of components and their placement to suit the individual, but the shapes of the size variants can be based upon real head shapes rather than conventional anthropometric data describing the distributions of landmarks. The latter were satisfactory for making decisions about a conventional size roll, where it was only necessary to accommodate the range of head widths and lengths of the aircrew population. However, the novel arrangement must take account of natural human asymmetries, the limited adjustment range of the display optics and the limited range of positions that components can take in the vicinity of the mouth, nose and ears. The position and orientation of the head within the shell are therefore critical. In the QinetiQ studies the head shape data were aggregated by shifting individual head forms so that the mid-point between the eye pupils was at the datum position, then orienting the head so that the line joining the eye pupils pointed sideways and the line joining the eye mid-point to the chin cleft pointed down. With these criteria the aggregated data showed least variation in the vicinity of the eyes, and mouths were spread along a vertical line. As the variation was greatest at the top and back of the aggregated heads, the tooling costs for six helmet sizes could be reduced significantly by making the complex front part in only three size variants and the back in two length variants to match each of the three fronts. The alternative, of using a single large size helmet, was rejected because the result would be dangerously unwieldy for small aircrew.

This form of headgear is an intrinsically better platform for accommodating the sensing elements needed for the novel control modalities discussed in Chapter 11. For instance, the enhanced stability would make it easier to attach a set of illuminators and a miniature video camera for an eye sensing system. Also, surface electrodes with active amplifiers for EOG and EEG sensing can be embedded in the stiff, conformal liner. This has the useful advantages of holding the electrodes in light contact with the skin, of obviating the usual need to attach electrodes individually and of simplifying calibration because the electrode sites are unaffected by repeated doffing and donning.

Further work is needed to optimise the internal cooling airflow and to integrate facilities, such as a deployable drinking tube and a nose pinching device that the wearer can use without opening the mouth flap and breaking the integrity of the NBC enclosure. The design will also need further refinement to simplify manufacture and to take account of issues raised during acceptance testing.

References

1. Air Standardisation Co-ordinating Cttee Air Standard 61/83A, Aug 1990
2. Simpson R E, *Specimen size rolls for aircrew headgear based on an analysis of the head measurements of 2000 Royal Air Force aircrew*, Unpublished MOD report
3. Kerslake D M, Nelms J D and Billingham J, *Thermal stress in aviation*, in Gillies J A (ed) *A textbook of aviation physiology*, Pergamon Press, 1965
4. Burton D R and Collier L, *The development of water-conditioned suits*, Unpublished MOD report
5. Glaister D H, *Head injury and protection*, chapter 12 of Ernsting J, Nicholson A N and Rainford D J (eds) *Aviation medicine*: 3rd edition, Butterworth Heinemann 1999
6. *Protective helmets for vehicle users, BS 6658:1985*, British Standards Institution, London, 1985
7. Lowenheim C G P, *Ph.D Thesis : On bridging vein disruption and rotational head injuries due to head impact*, Department of Forensic Medicine and Division of Solid Mechanics, University of Lund, Sweden, 1977

8. Rash C E, Mozo B T, McLean W E, McEntire B J, Haley J L, Licina J R and Richardson L W, *Assessment methodology for integrated helmet and displays in rotary-wing aircraft*, Report No. 96-1, U S Army Aeromedical Research Laboratory, Alabama, 1996

9. McEntire B J and Shanahan D F, *Mass requirements for helicopter aircrew helmets*, in *Impact head trauma: Responses, mechanisms, treatment and countermeasures*, AGARD CP-597 1997

10. Stiffler J A and Wiley L, *I-NIGHTS and beyond*, in Lippert T M (ed) *Helmet-Mounted Displays III*, SPIE Vol 1695, 1992

11. *Reducing noise at work: Guidance on the noise at work regulations*, published by the Health and Safety Executive, London 1998

12. Ernsting J, *The principles of pressure suit design*, in Gillies J A (ed) *A textbook of aviation physiology*, Pergamon Press, 1965

13. Howard P, *The physiology of positive acceleration*, in Gillies J A (ed) *A textbook of aviation physiology*, Pergamon Press, 1965

14. Eiken O, Kölegoård R, Lindborg , Aldman M, Karlmar K-E and Linder J, *A New Hydrostatic Anti-G Suit vs. a Pneumatic Anti-G System: Preliminary Comparison*, Aviation, Space and Environmental Medicine 73, pp. 703-8, 2002

15. Carter S J and Cameron A A, *Eurofighter helmet-mounted display: status update*, in Lewendowski R J, Haworth L A and Girolamo H J (eds), *Helmet-and Head-Mounted Displays V*, SPIE Volume 4021, 2000

16. Frank D L, Geiselman E E and Craig J L, *Panoramic night vision goggle flight test results*, in Lewendowski R J, Haworth L A and Girolamo H J (eds), *Helmet-and Head-Mounted Displays V*, SPIE Volume 4021, 2000

17. Bapu P T, Aulds J M, Fuchs S P and McCormick D, *Quick-disconnect harness system for helmet-mounted displays*, in Lippert T M (ed) *Helmet-mounted displays III*, SPIE 1695, 1992

18. Jarrett D N and Karavis A, *Integrated flying helmets*, Proc. Inst. Mech. Eng. Vol 26 Part G, Journal of Aerospace Engineering, pp. 47-61, 1992

Chapter 14

Future Crewstations

14.1 Introduction

Chapter 1 has outlined the way that aircraft crewstations have been developed to their present form, and the intervening chapters have described the component systems and suggested how these may be developed further. Adopting such improvements would undoubtedly bring worthwhile increments to aircrew safety and effectiveness, but the general arrangement of the cockpit and the way the crew go about their job would remain much as they are at present. The aim of this chapter is to discuss a few ideas that have either not been broached already or offer benefits of kind rather than degree. These ideas could have a substantial influence on the physical form of the crewstation and the job of the crew.

14.2 Physical reformulation

14.2.1 External vision

Aircrew can currently view the world in the four ways described in Chapter 5: directly through the cockpit transparencies; indirectly using a forward-looking imaging sensor and head-up display; indirectly through night vision goggles (NVG); and indirectly using a visually-coupled system (VCS). The NVG can be regarded as a VCS that senses near-IR radiation and has a particularly rigid connection between sensor and display, while the combination of sensor and HUD is equivalent to a VCS that has been bolted to the airframe. The choice is really between a natural, direct view through passive transparencies and some form of VCS giving an indirect view artificially via active sensors and displays.

It is interesting to consider the relative advantages and disadvantages of these two alternatives, and suggest how the cockpit could evolve in the light of such a consideration. Natural vision has, at present, several obvious superiorities. As noted in Figure 2.17 in Chapter 2, human vision encompasses almost the whole forward hemisphere. This contrasts markedly with the tiny field of about 40° provided by current artificial vision systems. The panoramic NVG, with a sideways field of about 100°, extends this considerably but still imposes a noticeable restriction. The other outstanding natural quality is the ability of the eye-brain to resolve detail. A VCS typically supplies an image having only about 40% of the acuity of natural vision. The NVG is better at about 70%, and it also obviates most dynamic oddities that plague a VCS, such as the lag that causes features in the projected image to shift from their intended location in visual space when the head is moved rapidly. Different sensors have their own characteristic limitations, such as the loss of thermal contrast after rain, smearing of moving objects and blooming when overexposed. Current forms of artificial vision therefore provide a colourless, fuzzy, blinkered view of the world that is sometimes wobbly and misleading, all the while imposing a considerable burden to the muscles of the neck.

On the other hand, glare from strong sunlight combined with canopy haze and distortion commonly degrade the clarity and large instantaneous scope of natural vision. The flexibility afforded by suitably-engineered forms of indirect vision also deserves credit. For instance, by sensing electro-magnetic radiation outside the narrow eye-sensitive band, it is possible to see through some clouds, and the enhanced sensitivity can provide clear images at night. The potentially large field of regard, which is limited only by the size of the sensor window and the mobility of the head, can give a view unimpeded by the aircraft structure. The comparatively poor system resolution can be countered by incorporating a facility for zooming the sensor field of view. Perhaps the greatest flexibility is the intrinsic ability to provide overlaid symbolic information from other systems and databases.

The relative merits must also be moderated by the chance of operating in conditions that favour the two forms of vision. Commanders who can deploy aircraft equipped with artificial vision systems feel that they "own the night" and increasingly conduct missions at night to place a lesser-equipped adversary at a tactical disadvantage. This, coupled with the increased use of electro-optically guided and satellite-guided weapons for air-to-ground missions and the prevalence of "beyond visual range" air combat missions, has reduced considerably the need for the pilot to see the target directly. Although colourless, fuzzy, blinkered, wobbly and misleading, an artificial view is bettered by natural vision only in good weather during the day, particularly when manoeuvring close to another aircraft, such as a wingman or a tanker. Given that most fixed-wing and rotary-wing combat aircraft are fitted with systems that give the crew some form of artificial vision, it is possible to suggest the direction in which developments will progress and how these might influence aircraft design.

At present the individual channels of an NVG provide better resolution than a sensor-&-display system because the image intensifier tube provides more pixels than a miniature display panel. However, it is likely that this advantage will be eroded as sensor-and-display devices are developed to operate at higher video standards and the field of view is divided up so that it is a patchwork of butted pictures. Combat aircraft such as the JSF will operate with a combination of head-mounted image-intensifying sensors and aircraft-mounted thermal sensors until the image-intensifier sensors are transferred to the airframe and their images are fused with those from the other sensors.

At some further stage, when the indirect vision system surpasses natural vision in all operational contexts, it would be advantageous to optimise the crewstation for the former rather than the latter. This would be done by enclosing the whole cockpit within an opaque shield[1] to remove the competing view of the outside scene and provide a controllable ambient light level. The latter would ease the difficulties, discussed in Chapter 9, of constructing a high resolution, wide field of view helmet-mounted display, because very bright images would be unnecessary and the focus could be set to about two metres to ease the transfer of attention between head-in and head-out information. Other issues, such as head-lag mis-registration, refractive distortion, transparency scatter, protection against glare and laser weapons, and the transition between day and night operations would cease to be problems. Rather than battle with the structural weakness and airflow perturbation of a protruding canopy, the teams designing new manned military aircraft would enjoy the freedom to devise lighter and stealthier configurations that enclose the aircrew within an aerodynamically smooth envelope. The accommodation would be quieter and cooler, because of the reduced dynamic pressure and solar heating, and because no de-misting would be needed, there would be considerably less demand on the environmental conditioning system. The engineering advantages are particularly relevant to future civil aircraft; for instance supersonic transport aircraft could do away with the complex droop nose mechanism used in Concorde to provide a forward view when landing and taking-off.

It may however be necessary for the pilot to have recourse to a limited, direct, forward view of the world to confirm the veracity of the artificial view and provide a reversionary facility for landing with a failed VCS. A roof-mounted, pivotable periscope could suffice. The topic is however made more complex by a consideration of "synthetic vision systems" (SVS) in which, as in many flight simulators, a computer generates an image of the external scene from a terrain data-base. To distinguish indirect vision systems that employ sensors, it is useful to call them "enhanced vision systems" (EVS). It is slightly surprising that SVS is being investigated for use in manned aircraft because the synthetic representation only corresponds with the real world if the data describing the shape of the terrain and measurement of the aircraft position and orientation are all reliably accurate. However, the idea warrants investigation if these can be made trustworthy. A synthetic view could have the advantage of containing only the relevant, useful information and omitting unnecessary and misleading clutter. The result could be less confusion and less likelihood of human error. Perhaps some combination of EVS and SVS would be most effective. For instance, a synthetic mesh, or in hilly terrain a set of ridgelines, could overlay a sensor image. Alternatively, a small image from a relatively inexpensive and reliable boresighted sensor could be inserted into the broad imagery generated from the terrain database.

In any event, the collation of data from diverse sources to construct pictures containing stimuli that enable the viewer to receive cues and perceive details of maximum relevance to head-out tasks will be an enduring research topic. Such "image fusion" will complement the topic of "data fusion" discussed in section 14.3.2.

14.2.2 The virtual cockpit

Pilots' eyes are busy looking out at the world, looking at a display or looking around the cockpit to find a switch. The need to remain aware of events outside the aircraft has driven the development of devices to display information that complements what can be seen through the windscreen. The design of head-down displays has therefore sought to ensure that information can be discerned and assimilated during the briefest of glances, and head-up and helmet-mounted displays have been devised primarily to facilitate attention transfer by superimposing symbols onto the external view.

The "Virtual Cockpit" (VC) proposed by Tom Furness[2] at Wright-Patterson AFB in the late 1970s was intended to take this evolution a step further. As illustrated schematically in Figure 14.1, the central idea was to extend the VCS and use helmet-mounted visual and auditory displays to meld all information about the outside world with the view of the outside world. This would allow flight parameters, attitude references, navigational directors and the locations and states of co-operating aircraft and any targets or threats to be transformed into this common ego-centric frame of reference and presented conformally over a wide field of regard. The pilot would no longer need periodically to examine the conventional set of head-down display formats, or mentally collate their content using appropriate co-ordinate rotations. Fewer interruption to head-out attention would reduce the chance of disorientation and improve flight path control. The 3-D sound system would reinforce the ego-centric location of tactically relevant entities. For instance, voice communications from a wingman would be heard to originate where the accompanying aircraft could be seen, and an intermittent bleep picked up by the passive warning receiver (PWR) would be heard to come from the direction of the external searching radar.

Information that was geometrically unrelated to the outside scene, such as a tactical overview and fuel and weapon states, would be displayed head-in on virtual 3-D volumes and panels using the two channels of the HMD to project images of appropriate stereoscopic disparity. The virtual imagery would be complemented by compatible head pointing, eye pointing, finger pointing and speech command systems, so that only the pedals, throttle, stick and embedded HOTAS switches need be retained from the conventional control set. It was anticipated that the novel devices would provide more natural forms of control, for instance, to allow the pilot to fixate a threat in the outside scene by eye, say "what's that?", receive an auditory response from the defensive aids suite and then say "jam" to initiate the countermeasure. Interaction with head-in entities could also follow a spatial-designation-with-spoken-instruction dialogue, and perhaps employ the benefits of three-dimensional finger pointing.

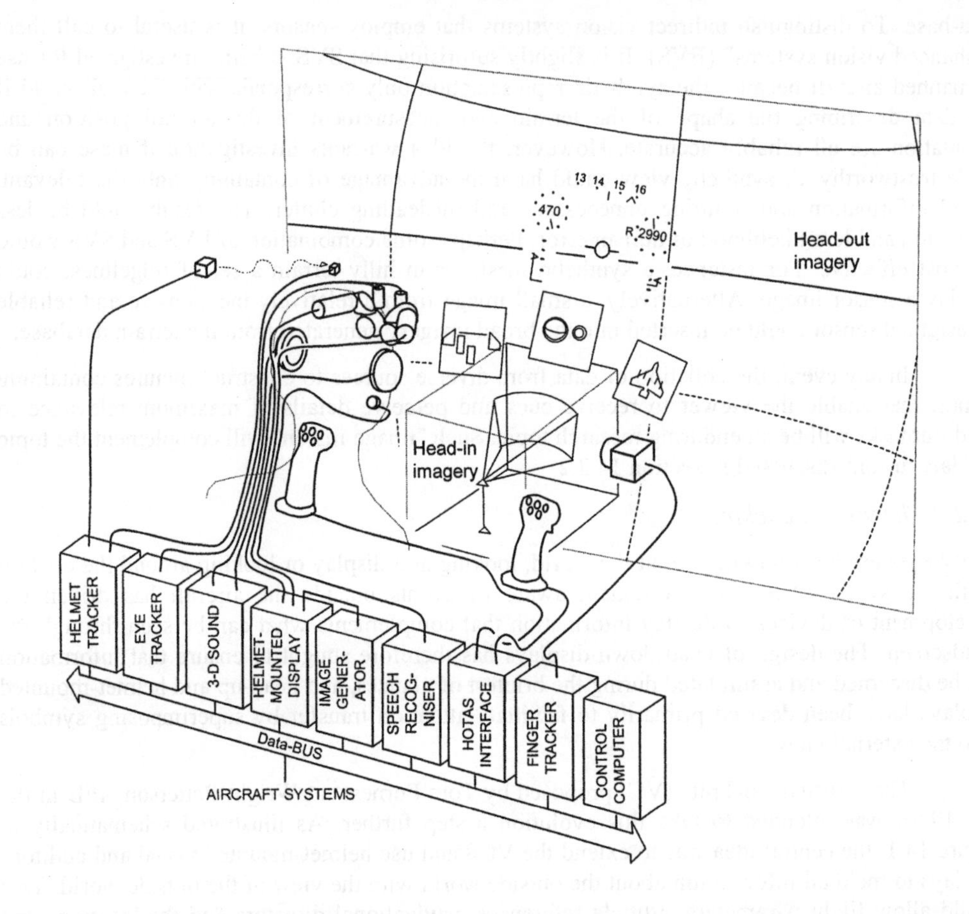

Figure 14.1 The Virtual Cockpit retains HOTAS controls but replaces the remaining conventional controls and displays by a suite of helmet-mounted systems. These provide cockpit-stabilised 3-D images, instead of conventional HDDs, plus a head-directed window into an ego-centric enhancement of the external world. The pilot interacts with both using speech and a mixture of head, eye and finger pointing (Reprinted with permission from QinetiQ)

The originators suggested that in addition to the ergonomic benefits, the VC would have two operational advantages over the conventional interface. Firstly, because the virtual images could be tailored for superimposition on the natural view of the world, or on sensor and computer-generated views, the VC offered a way of working that would be unaffected by variations of ambient light level and meteorological visibility. Secondly, the VC could be used in conjunction with an opaque visor or canopy that could be deployed to protect aircrew against DEW weapons. It would also be possible to enclose the crewstation permanently, as discussed above.

It is worth noting that the phrase "Virtual Cockpit" can mean many things. For instance, NASA have used the term to describe a programme to enhance the safety of business and commercial aircraft in visually occluding weather conditions. In this case, the term signified the replacement of conventional primary flight and navigation formats on colour head-down displays by "tactical" and "strategic" representations of the aircraft state. Head-mounted displays are also used in a wide range of computer-generated Virtual Reality (VR) simulations, for instance to help visualise architectural options and manufactured products or provide a more compelling computer game. There are obvious similarities between the VC and VR ideas. Both use head-mounted displays to immerse the user in ego-centric imagery, both provide the user with compatible control devices and both seem to mask sight and sound of the real world. However, the former is intended to enhance interaction with what is real and consequential, while the latter deals with a fictitious environment. This crucial distinction is easily overlooked or misconstrued, especially when the product being visualised in a VR simulator is an aircraft crewstation.

In attempting to determine whether this novel interface had the advocated advantages over a conventional arrangement and whether it would be acceptable to aircrew, investigators faced two inter-linked problems. Firstly, no customer would accept a novel man-machine interface that had not been tested thoroughly under the physical and psychological conditions of real flight. Secondly, the idea obviously called for equipment with qualities that could not be engineered in a flight-worthy form. The most practical approach adopted by the USAF Supercockpit programme at Wright-Patterson AFB[3] and a programme conducted by DERA at Farnborough in the UK[4] during the 1980s and 1990s split the endeavour into two phases. The first phase used heavy, cumbersome equipment on the ground to explore and optimise facets of the idea and understand what attributes should be specified for flight test equipment. The second was intended to develop the equipment, install it in a test vehicle and conduct the determinative flight tests.

The simulation facility constructed at DERA was based on a binocular HMD, each channel of which received an image from a cabinet-size projector through a flexible, two metre long, coherent fibre-optic bundle. At the time this was the only way to get bright, coloured, high-resolution (1024 by 1280 pixels) images with wide fields of view. Each projector housed a separate red, green and blue CRT, so a semi-automated procedure was needed to apply corrections to the convergence scan coils to bring the component images into pixel-perfect registration. To reduce the effective mass of the head-borne optics and the awkwardness of the fibre-optic bundles a pneumatically-supported mechanical head-tracker was built that allowed almost full head rotation and about a cubic foot of displacement. This assembly was integrated with the other sub-systems shown in Figure 14.1; a control computer, the throttle, stick and pedals with HOTAS grip-tops, a finger tracker, a voice recogniser together with pairs of image generators, 3D-audio modulators and infra-red eye trackers.

As with most exercises arranged to test hypothetical working conditions, the majority of the engineering lessons were learned during the construction, integration and commissioning stages to produce an entity that could be exposed to the critical judgement of test pilots. A number of experiments were also conducted. Some addressed issues inherent to the use of the HMD, as discussed in Chapter 9. These included the trade-off between field of view and resolution, the effects of partially overlapping fields, the spatial accuracy of stereoscopic imagery, the effect of head lag on symbol legibility and ways of stabilising images to counter the head lag phenomenon. Others considered the detailed design of primary flight information and the use of alternative control modalities. For instance, it was concluded that pointing by eye would be useful if the eye direction could be measured reliably to within about 0.5° over a fair range of excursions with no perceptible delay. However, tracking the fingers to operate virtual keys proved much less useful. Visual feedback of the measured finger position, in the form of a simple cursor, was needed to enable the pilot to compensate for tracker errors and delays. Tactile and/or auditory feedback were needed to signify that the finger was in contact with a virtual switch and that the switch had been operated. In addition, the removal of the hand from the grip-top prevented easy use of HOTAS switches for concurrent interaction with the speech recogniser. In flight, the pilot would also have difficulty controlling the finger position accurately under vibration and high-G.

Although the simulation experiments left numerous questions and possibilities unexplored, the idea was considered to be a feasible and beneficial way to add value to a VCS. The most immediate application could be in aircraft such as Eurofighter that will have the essential elements, a VCS and a speech recognition system, albeit the first with a limited, monochromatic field of view and the second with a small vocabulary. In this case the helmet could be fitted with a totally light-tight outer visor that the pilot could deploy, temporarily, for robust protection against laser weapons. Work would be needed to investigate how an essential sub-set of the information normally available from the full-colour HDDs could be presented using the HMD, and how the pilot could reach and operate soft-keys and other switch panels that he could not see. A more powerful dual channel image generator could possibly be used to exploit the binocular HMD to present virtual head-down panels, or a pair of close-focused cameras could be fitted onto the opaque visor in line with the normal eye lines of sight to relay a stereoscopic view of hand space and the HDDs.

In the mid term it is not fanciful to consider that aircraft such as the JSF could, at some stage in their working lives, be retro-fitted to operate in this way. Minimal structural change would be needed to replace the transparent canopy by a similarly-shaped opacity with a small DEW-protected window that would open to provide a reversionary view forward for landing. The existing large head-down display that provides coloured, high quality imagery would be retained. The head-out imagery projected by an improved helmet-mounted display system would be blanked where it overlaps the large HDD, and the benefits of head-pointing and eye-pointing would extend to include interactions with the head-down imagery.

The VC would have most advantage in the long term as the physical display-control interface in the visually closed crewstation of a fast jet where advanced forms of all the component systems could be allied to a comprehensive suite of imaging sensors. The physical configuration of the crewstation could take the form described in section 14.5. Perhaps an additional non-attention demanding display, such as the tactile suit described in Chapter 6, could be introduced to reinforce the pilot's sensation of motion.

14.2.3 Plug-in cockpit

Chapter 12 concluded that the current form of the ejection seat, represented by the Martin-Baker Mk16 and Zvezda K-36D, has almost reached the limit of practical development. Although capsules tested in the past have had poor records of success, it is suggested that the capsule idea can be developed much further and give substantial benefits. As well as providing enduring, unencumbering protection and a temperate, relatively quiet, NBC-clean working environment, aircrew would be spared from wearing the complex assembly of protective clothing needed to survive ejection in an open seat. The services would be spared the considerable logistical difficulties of supplying and maintaining such clothing, and there are other operating advantages, as discussed below. The problem is to devise a form of capsule that can enable escape over the wide envelope achievable with ejection seats.

Figure 14.2 sketches two possible forms of box-shaped capsule, one with a transparent top to provide a direct view of the world and another that is fully enclosed. Both forms would be inserted into an accommodating cavity within the airframe and have air, data and electrical power transferred through a mating latch mechanism. The pilot would enter and exit by raising the transparency of the visually open box or raising a hatch in the top of the visually closed box. For emergency escape, the explosive bolts and airframe disruption used in previous forms of capsule would be unnecessary. Instead, the capsule would be ejected upwards from the cavity, and this would not incur the delay needed to clear the path for an ejection seat. Lightweight composite materials would be used to achieve a target mass of about 300kg, including the occupant, and a pair of guns combined with a solid-propellant rocket motor of about 6,000kg.sec impulse would be sufficient to give zero/zero to high/high escape. Pintle-nozzle vectoring, aerodynamic surfaces and a drogue would be used to stabilise the box in a feet-forward attitude immediately after ejection. The torso of the occupant would be restrained by a normal five-point harness, and the head by an airbag. No limb restraints would be needed. A small battery would power the control system and provide internal lighting, and an internal oxygen supply would be connected automatically when the aircraft supply is disconnected, as done by the personal equipment connector on an ejection seat. A single parachute would arrest descent, and although air bags and floatation cells would be deployed, the sandwich structure would be arranged to crush progressively to absorb a considerable proportion of the ground impact energy in much the same way as the shell and liner of a helmet.

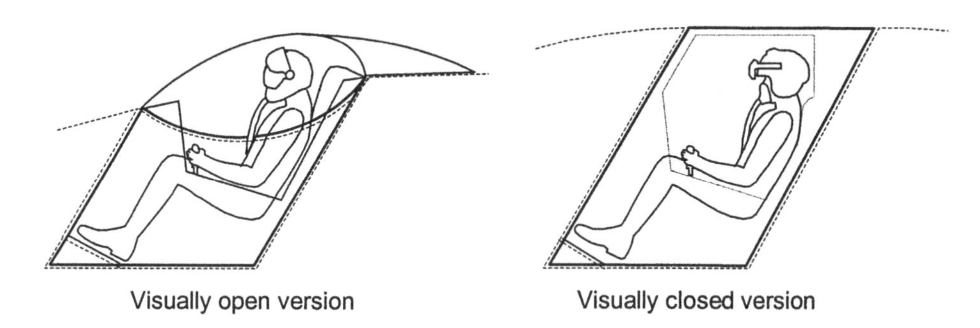

Figure 14.2 Suggested forms of compact capsule (Reprinted with permission from QinetiQ)

To achieve the necessary low mass, the display-control interface would be based on the virtual cockpit idea, described in section 14.3, and employ helmet-mounted systems in conjunction with HOTAS manual side inceptors. The front and side internal panels would house the minimum number of dedicated switches, for instance for the landing gear and emergency jettisoning of externally carried stores. Minimal display and control panels would be needed to provide get-you-home information in the event of HMD system failure and for operating the post-escape facilities of the capsule. The closed version would rely on synthetic external vision, as discussed in section 14.2, with a periscope that could be deployed at low airspeed to give a direct view forward.

An aircraft designed around the idea of the plug-in cockpit could be very versatile. For instance, when operating under NBC threat, the capsule interior could be maintained as a clean environment, and the crew could be transferred between the aircraft and NBC clean room by detaching the capsule from the airframe. It could also be possible to change between open and closed versions of the capsule to suit the mission, and detach the capsule so that it could be coupled to a simulator for training or mission rehearsal. With a suitable substitute capsule containing the decision-making electronics, it would also be possible to convert between manned and unmanned modes of operation. With this provision an aircraft with multiple cavities could be converted on-station to operate with one, two or zero crewmen. The practical feasibility of this idea has not been investigated.

14.3 Re-formulating the job

14.3.1 Intelligent interfaces; the structural coupling paradigm

Scientists at Wright Patterson AFB suggest that the way operators interact with controls, by issuing commands to which the machine responds deterministically, is a legacy from the time when the operator's muscular effort actually steered a vehicle. Now that the machine has become self-powered and is controlled directly by a system of computers, they suggest that the interface between the human and the machine can be as flexible as that between a human and a computer. They propose a revolutionary change, a control metamorphasis, whereby the dumb one-way servo interface is replaced by a "structural coupling paradigm" to create an actively helpful interface[5]. The aim is to end the master-slave relationship and enable the operator and machine to adjust to each other through a two-way dialogue to reach implicitly shared goals. A similar ideal has also been put forward as the ecological interface[6] , which is defined as "constructed so that the operator need not work at a higher level of control than required by the situation". Both ideas have the laudable, human-centred aim of unloading the operator so that he no longer has to make nitty-gritty machine-compliant actions but can instead exert control flexibly as though interacting with another trained co-operating human.

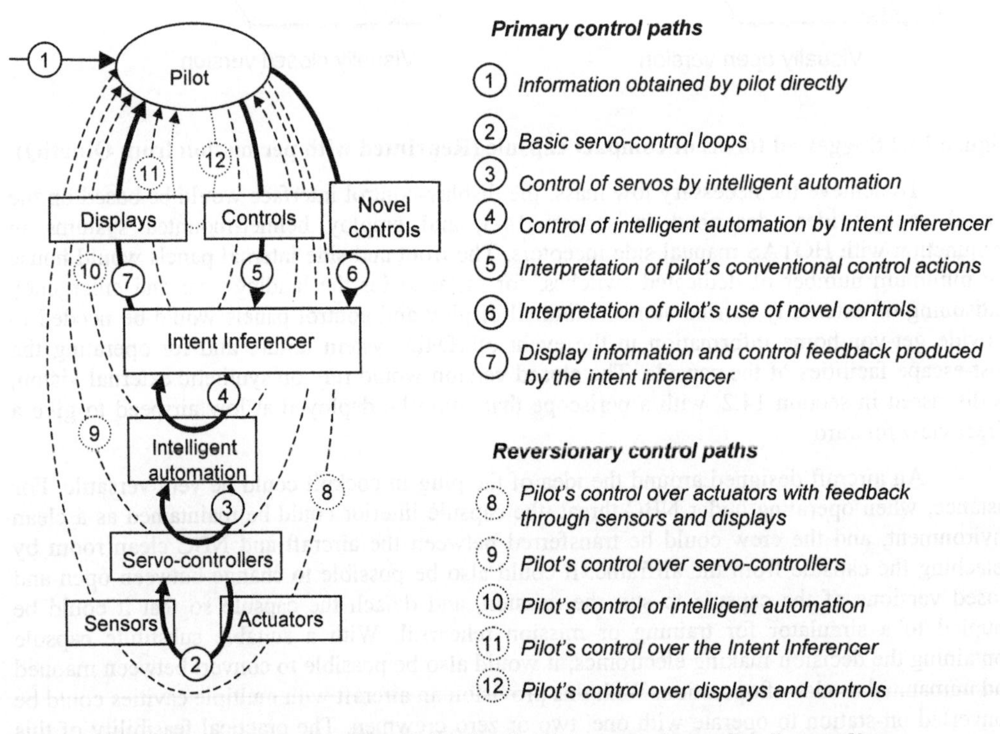

Figure 14.3 Classes of man-machine links for the "structural coupling" paradigm

Figure 14.3 uses the scheme from Figure 4.2 to show how the extra elements of the structural coupling paradigm would fit into the automation strata of an advanced avionic system. Here, the outputs from both the conventional and novel control devices would be received by an "intent inferencer" that uses knowledge of the mission plan and the on-going state of all the on-board systems to infer what the operator wants to have done. This intelligent mechanism would then exert control over the aircraft and the systems, provides timely feedback for the pilot to know that he has operated the controls successfully and inform him about the commands it has passed on to be executed. It therefore subsumes the function of the deterministic "command interpreter" shown in Figure 11.15.

The introduction into the aircraft cockpit of automatic speech recognition systems and helmet trackers is a step towards this operating philosophy. The introduction of eye trackers, and perhaps gesture-based and bio-potential controls, described briefly in Chapter 11, would certainly add to the gamut of inter-human communicative behaviour that can be assessed by the machine. However, the viability of the approach rests fundamentally on the reliability of the inference about what it is that the operator wants the aircraft and the on-board systems to do. The most applicable software techniques have been developed under the banner of Artificial Intelligence (AI), which is based on five inter-linked processes:-

a) Collation of the available data about the state of the system (in this case the human operator)
b) Predictions about how this state can change in the near future
c) Delineation of the alternative courses of action
d) Assessment of the consequences of choosing each action given each predicted situation
e) Choice of the optimum action

What AI brings to the programmer's armoury, which makes the approach feasible in real-time applications, is the use of heuristic rules – meta-knowledge available to the programmer – to eliminate the large number of combination of actions and predicted situations that are not worth considering. Unlike conventional software, in which the relationship between input and output can in principle be tested exhaustively, such software is non-deterministic and cannot be guaranteed to produce a consistent output. Although it would be feasible to limit the authority of the actively helpful interface to pass only non-safety-critical commands, the allowable probability of inferential error would still need to be high, in the order of 10^{-5}, to preserve the operator's trust. Such reliability must be demonstrated before this form of interface can receive serious consideration.

14.3.2 Comprehensive automation

As noted in Chapter 4, avionic systems for civil transport aircraft have been developed to the state that all of the tasks performed by the flight crew are assisted by automatic aids. The crew can put the electronic systems in full control of the flight unless they need to depart from the flight plan. As summarised in Table 14.1, computer-based systems that will give combat pilots comprehensive electronic assistance are either already available or are under development.

There are also plenty of ideas for the more effective use of existing sensing systems. One nice illustration is analogous to the subconscious linkage between ambient vision and locomotor control that enables rapidly moving animals to avoid collisions. Vection cues (illustrated in Figure 2.14) could be extracted from the successive pictures obtained by an imaging sensor, such as a forward-looking thermal imager, and used to compute the direction of travel and the instantaneous time to contact anything in the scene[7]. This passively sensed information may not be sufficiently reliable for use as the sole input to the flight control system, but it would be valuable as independent evidence, in conjunction with a data-base estimates, to lower the reliance on an active obstacle sensor such as a radar or lidar.

To bring these component systems together, avionic systems developers must deal with a fair number of complexities. Firstly, in contrast to existing systems that carry out skill-based and rule-based control, many of the systems under development employ non-deterministic AI and neural network software techniques. These may have to be partitioned from safety-critical functions. The second complexity arises from the functional interdependence of the decision-making systems, and in particular their need to share access to sensing systems. For instance the radar of the future will need to jump from a track-while-scan mode to find and identify airborne threats, to a synthetic-aperture mode for finding potential targets on the ground, and a moving target mode to discriminate between tracked and wheeled vehicles. Similarly, an infra-red imaging sensor will need to operate in a wide-field mode as a VCS, a narrow search-and-track mode to identify airborne objects and in a further mode to steer a laser to illuminate a target.

Table 14.1 A summary of on-board automation for combat aircraft

Task	Available automation	Automation under development
Aircraft control	Flight Control System Terrain Following System	Ground Collision Avoidance System Obstacle Avoidance System
Navigating	Navigation computer	Navigation system manager
Communicating	Identification Friend or Foe	Communication system manager
Aircraft system management	Fuel manager Aircraft systems monitor	Utility systems manager
Targeting	Radar system computer	(Sensor-specific) automatic target recognisers
Weapon management	Weapon aiming computer	Stores management computer
Countering threats	Defensive aids suite	Countermeasures manager Emissions control manager
Tactical decision making		Sensor data fuser Threat prioritiser Tactical situation advisor
Mission Management	Mission planning aid Mission systems computer	Tactical re-router Mission management aid

The third complexity arises because combat missions will only rarely call upon an aircraft to act autonomously. Most will be planned as a co-operation between a number of role-specific combat, reconnaissance and anti-radar aircraft, under the co-ordination of command and control aircraft such as the Airborne Warning and Control System (AWACS). Information will be communicated via secure systems such as the Joint Tactical Information Distribution System (JTIDS), which will contribute greatly to the pilot's appreciation of the tactical picture. A fourth complexity arises from the need to minimise the chance of detection by enemy surveillance systems. This will entail timely use of radio and radar transmission and the avoidance of manoeuvres that present a radar-reflective aspect of the airframe to known defensive radar sites.

In addition to the profusion of sensed and communicated data, future combat aircraft will carry large data-sets describing the enduring characteristics of terrain, obstacles, threats and weapons. To help aircrew assimilate all of this information, so that they can understand the dynamically changing tactical circumstances and decide how to respond, the developers of electronic aids treat the problem as an exercise in "data fusion". This proceeds through the stages shown in Table 14.2 in conformance with formulations known as the Joint Directors of Laboratories (JDL) model and the Observe-Orient-Decide-Act (OODA) model.

Table 14.2 Stages of Data Fusion in the Joint Directors of Laboratories and the Observe-Orient-Decide-Act formulations

Stage	JDL	OODA
Receipt of sensor data		
Association of plots or tracks	1	Observe
Fusion to form entities		
Classify, de-clutter, identify		
Picture formation	2	Orient
Threat Prioritisation	3	
Situation awareness		Decide
System-aided pilot decision-making	4	
Tactical manoeuvre		Act
Countermeasure allocation		
Mission replanning		
Pilot displays		
Situation reporting		
Weapons control		

In the JDL formulation, the earliest stage, Level 1, takes data from different sensors and checks whether plots and tracks from one sensor, for instance plot N°.P from the IR imager, can be associated with that from another sensor, for instance the radar-tracked object N°.Q. If so, a better estimate of the object position can be obtained by taking range from the radar and angular position from the IR sensor. The tactical overview, produced in Level 2, places the identified, tracked entities into groups and infers their intentions, for instance a set of fighters flying in the same direction as a bomber would be treated as escorts. Levels 3 and 4 assess the priority for dealing with the entities and decide what to do about them, for instance to ignore the entity, initiate a tactical manoeuvre or use an active countermeasure.

The USAF Pilot Associate[8], the UK Mission Management Aid[9] and the French Co-pilote Electronique[10] are examples of programmes that have demonstrated the possibility of linking software modules that can plan a mission, deal with emergencies, formulate the tactical picture, consider options and suggest short-term flight-path changes. If these systems are augmented by developments to control the vehicle trajectory at low level safely, and by systems that find and identify targets from data supplied by specific sensors, all of the tasks that make up the job of the combat pilot will be supported by automation. Such comprehensive automation, effectively an electronic crewmember, will alter the relationship between the pilot and the aircraft. For this to be satisfactory, it will be necessary to implement a form of interaction that exploits the complementary qualities of human and machine intelligence while acknowledging the lessons concerning the pitfalls of inappropriately integrated automation discussed in section 4.3.2 of Chapter 4. The fundamental requirement is to have a logical basis for the whole system that enables the pilot to understand what each part does, together with a simple, unambiguous control and display that enables him to put the system into a chosen mode of operation.

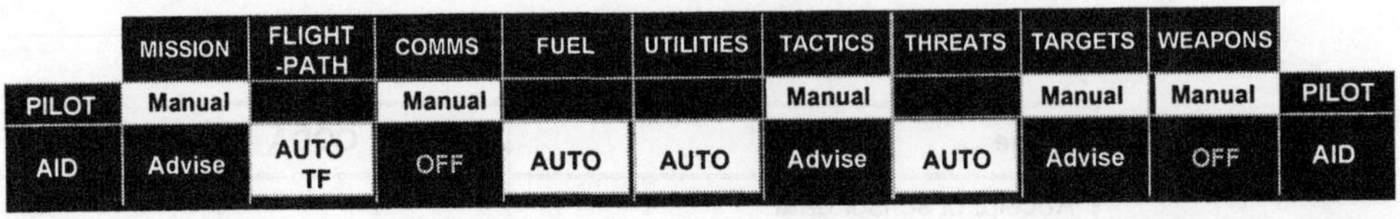

	MISSION	FLIGHT -PATH	COMMS	FUEL	UTILITIES	TACTICS	THREATS	TARGETS	WEAPONS	
PILOT	Manual		Manual			Manual		Manual	Manual	PILOT
AID	Advise	AUTO TF	OFF	AUTO	AUTO	Advise	AUTO	Advise	OFF	AID

Figure 14.4 Display/control panel showing the operating states of the suite of automatic aids that may become available in future combat aircraft. The upper row of captions delineates the separate tasks. The middle row shows what tasks are performed by the pilot and the bottom row shows the modes of the electronic aids. The pilot could change a mode by repeatedly pressing the lower caption light or use voice commands. Differences in colour could be used to emphasise the selections

One way to organise this degree of supervisory control would be to arrange that the intelligent agencies performing specific tasks could be put into one of six states of authority over appropriate on-board effector systems:-

Level 5	AUTOMATIC	Full authority
Level 4	DIRECT SUPPORT	Full authority, keeping the pilot informed
Level 3	IN SUPPORT	Providing advice and exercising limited authority
Level 2	ADVISORY	Providing advice
Level 1	ON CALL	Providing advice only if requested
Level 0	MANUAL	Inactive

Figure 14.4 illustrates the idea of an authority allocation panel that uses these levels of authority in conjunction with the task definitions described in Chapter 2 for the case of a fast jet performing a ground attack role. The panel allows the pilot to shed and resume authority easily, and reminds him about the elements of the job requiring his direct attention. In this illustration the pilot has delegated flight-path control, management of fuel and utility systems and deployment of countermeasures to silent automatic aids. This allows him to manage the communication and weapon systems without advice, and concentrate on target identification, tactical decision-making and mission management with appropriate advice. He could however put all of the aids in "direct support" mode and intervene if he considers a decision to be inappropriate. The form of the advice is a research topic. It can range from the display of visual cues and director symbols, as used for instance in a helmet-mounted display to point out the location of a waypoint, to an audio-visual question and answer dialogue with the electronic aid, for instance about the intentions of an unidentified aircraft and the options for dealing with it.

In addition to meeting certification criteria by performing safety-critical functions reliably and making any failure states obvious, the electronic crewmember, like any member of a team, must gain the trust of the pilot. This will depend partly upon the quality of the decisions it makes, and therefore on the rigour of the software engineering, and partly on the pilot's familiarity with its capabilities. Exploring the limits of the system will require good ground simulation facilities and testing of the advisory modes in the air. The latter may have to make use of "in-flight simulation" as a way of playing out interactions with large numbers of friendly and enemy aircraft in order to exercise the advisory systems in authentically complex, dynamic circumstances. In this case, the entities would be purely virtual and reside only in the memory of the on-board computers that model their behaviour. If implemented, the activation of this extra mode must not allow weapon release, and it must be signalled very clearly. An additional SIMULATION mode should, perhaps, be added to the captions on the authority allocation panel.

14.3.3 Pilot state monitoring

Incapacitation of the pilot through GLOC, hypoxia and spatial disorientation has been a major cause of fatal accidents in combat aircraft during peacetime. Several research programmes[11] have been set up to develop systems that can provide a safety net in aircraft, with appropriate flight control systems, to detect pilot incapacitation and automatically command a manoeuvre to bring the aircraft to a safe flying state.

Although combat pilots would be the main beneficiaries, they have been ambivalent about the practicality and usefulness of such systems. Any implementation must be very reliable, have a very low false alarm rate, not present a burden and not require setting up. To prevent unnecessary intervention or interference to an intended manoeuvre, most pilots would prefer to have the system detect incapacitation, ideally incipient incapacitation, and only initiate a recovery manoeuvre if the pilot is in immediate danger and does not respond to some form of interrogation.

As the main causal factor of hypoxia and GLOC is a deficiency of oxygen in the brain, several systems that measure qualities of the blood in the head have been investigated. One used multi-wavelength near-IR spectroscopy to measure the ratio of oxy-haemoglobin to haemoglobin, and others have used ultrasonic doppler pulses to measure blood velocity. To be practical the clinical equipment would have to be miniaturised and integrated into the headgear. Less burdensome approaches have been based on the interpretation of signals obtained from EEG electrodes attached to the scalp to pick up cranial activity and from EOG electrodes around the eyes, as described briefly in Chapter 11. Some of the EEG studies[12] have been directed towards the measurement of cognitive workload from the shifts of d.c. potentials, the topic of the next section of this chapter. Others[13] suggest that a.c. EEG signals in the 0.5Hz to 32Hz band, particularly the ratio of the theta (4 to 7.75Hz) to the alpha (8 to 13.75Hz) signal strength, are related to the subject's attention to the task, and may be useful for detecting performance degradation. The EOG signals have mainly been used to detect blinks. Electrodes have also been attached to the chest to pick up electro-cardiographic signals and measure the mean heart rate and the variation in the inter-beat interval.

A promising form of prototype Loss of Consciousness Monitor[14] (LOCOM) has pooled the data from a collection of sources. These include a head-lolling sensor, an eye blink sensor attached to the oxygen-mask, a blood pressure sensor, signals from the breathing regulator and data about altitude, G-level and throttle and stick movement available from the aircraft data-bus. Earlier concerns about the speed and capacity of the computer needed to host the data collation software and assess the subject's alertness have been allayed by ongoing processor development. The performance, the need for calibration and whether the system can detect the onset of incapacitation are not known. It would however be reasonable to suppose that the suite of speech recognition, helmet tracking, eye tracking and biopotential controls needed to implement the actively helpful interface, described in 14.3.1 above, would supply as much relevant data as the proposed suite of sensors. The helpful interface could therefore incorporate the LOCOM decision-making.

The practical problems associated with the development of flightworthy physiological sensing systems have largely been overcome by alternative approaches to the problem. The Automatic Ground Collision Avoidance System (AutoGCAS) has been developed by the USAF on the AFTI F-16 and can be regarded as a watchdog that, if activated by the pilot, provides a safeguard against inattention and misjudgement as well as incapacitation. This system continuously computes the ground clearance that can be achieved using a 5G pull-up manoeuvre, and if this is less than a settable value – say 100ft – the system intervenes and commands a wings-level roll followed immediately by the 5G pitch-up. Control is handed back to the pilot as soon as the predicted ground clearance exceeds the set value. The control is much like that exercised by a terrain-following system in which the aircraft position is measured by a GPS receiver and a radio altimeter. A data-base of the terrain height is used to assess the shape of the ground along the aircraft track rather than a laser or radar. This facility is scheduled to be introduced into later C and D versions of the F-16.

Another approach, the Disorientation Recovery Capability (DORC) provided in Eurofighter, gives the pilot a HOTAS switch that he can press to initiate a recovery manoeuvre that will automatically bring the aircraft to a wings-level attitude at a safe altitude. This is intended for use in incidents of type-1 spatial disorientation where the pilot realises he is in danger and needs help.

14.3.4 Adaptive automation

The possibility of comprehensive automation in future combat aircraft has switched the debate from a consideration of what can be automated to a consideration of what should be automated. It has also stimulated research into alternative approaches to the interface between man and machine that can exploit their complementary intellectual abilities.

In dealing with this issue it is important to distinguish *authority*, the ability to exercise control over behaviour, from *responsibility*, the liability for the consequences of behaviour. As soon as a pilot takes command of an aircraft he assumes responsibility for the behaviour of the aircraft. Thus when he engages an automated mode of an on-board system, he knows that he is still accountable for the consequences of actions committed by that system. He can transfer authority but not responsibility. This formal and legal requirement, which is obviously underpinned by a normal concern for human life - including his own - is why the pilot needs to understand the workings of the automation and have facilities that enable him to monitor what the automation is doing and allow him to shed and resume authority easily.

Table 14.3 Belling's Principles of Human-Centred Systems

Premise	Humans are responsible for outcomes in human-machine systems
Axiom	Humans must be in command of human-machine systems
Corollary 1	Humans must be actively involved in the processes undertaken by these systems
Corollary 2	Humans must be adequately informed of human-machine processes
Corollary 3	Humans must be able to monitor the machine components of the system
Corollary 4	The activities of the machines must therefore be predictable
Corollary 5	The machines must also be able to monitor the performance of the humans
Corollary 6	Each intelligent agent in a human-machine system must have knowledge of the intent of the other agents

A lucid view of the way that knowledge-based automation should co-operate to produce a human-centred system has been put forward by Charles Billings[15]. As summarised in Table 4.3, the premise, the axiom and the first four corollaries are cogent rules for ensuring that an operator can supervise automation. They re-state the need for the operator to understand how the automation functions and for the automation to behave predictably. The fifth and sixth corollaries are however a little surprising. These recognise the operator's fallibility and are based on the notion that operator and electronics constitute a team, in which the latter act collectively as an intelligent assistant that can ease the cognitive burden on the human and pick up *his* errors of omission and commission.

A way for automation to adapt to the state of the pilot in a combat aircraft, by monitoring his physiological state and inferring his workload and level of consciousness, was put forward in 1985 as the Symbionic Cockpit[16]. More recently the idea has been investigated in the UK at QinetiQ as the Cognitive Cockpit[17], which can be regarded as an ambitious attempt to link together elements from three ideas described above; the structural coupling paradigm, the safety net and comprehensive automation.

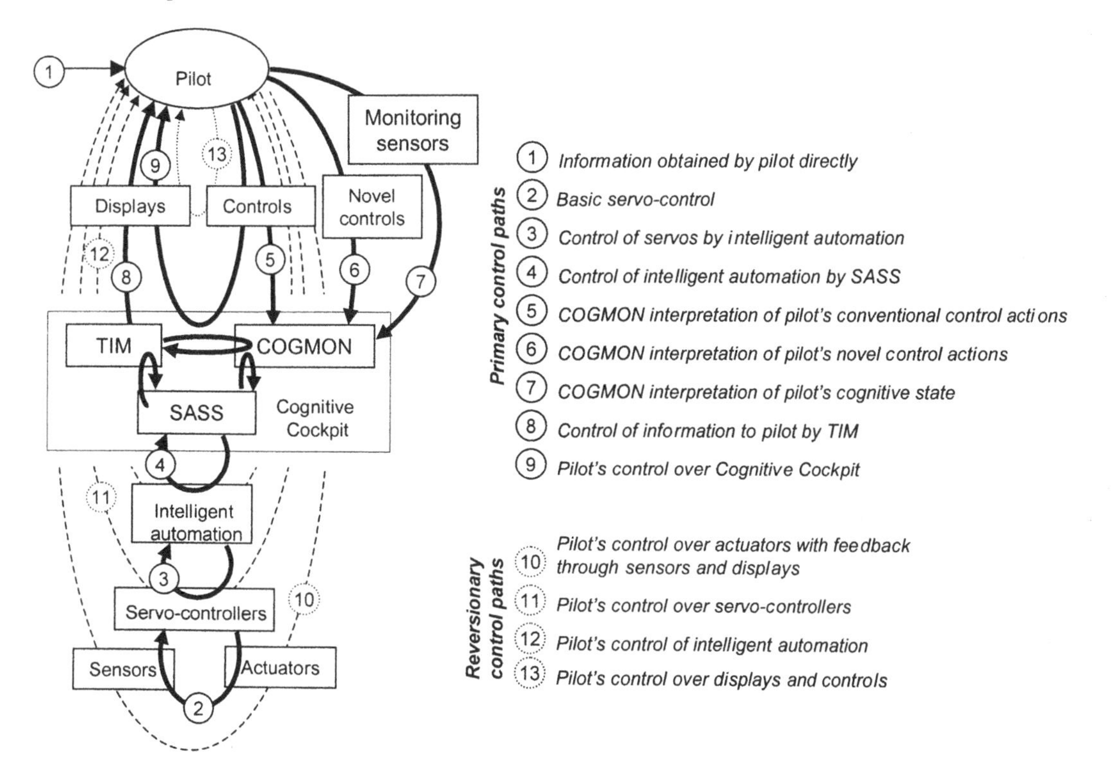

Figure 14.5 Classes of man-machine links for the Cognitive Cockpit. The idea links three elements, a Cognitive Monotor (COGMON), a Task Interface Manager (TIM) and a Situation Assessor Support System (SASS)

Figure 14.5 uses the scheme introduced in Figure 4.2 to show how the three elements of the Cognitive Cockpit would fit into the automation strata of an advanced avionic system. The Situation Assessor Support System (SASS) could either be the whole suite of automated functions or it could be the element that carries out the top level function of managing the mission and supervising the other functions. The Cognitive Monitor (COGMON) normally passes on the explicit commands made by the pilot straight on to the executive systems, but it also infers the intentions of the pilot, as in the structural coupling paradigm, and assesses the pilot's current cognitive state, i.e. what the pilot is attending to. The third element, the Task Interface Manager (TIM) takes the information about the pilot's state from the COGMON and compares it with information from the SASS about what the pilot should be doing. The TIM can then either alert the pilot to take notice of any task he is neglecting or, if the pilot is judged to be too busy, give the relevant automatic system the authority to assume control. In the latter case, perhaps the only thing the pilot would notice out of the corner of his eye would be a change to "AUTO" on the Authority Allocation Panel shown in Figure14.4, if such a panel is presented.

All three elements are technically ambitious. The COGMON is likely to be based on a model of human perceptual, cognitive and neuro-muscular processes, such as Wickens' Multiple Resources Theory, as outlined in Chapter 2. The software must assess the degree of utilisation of the pilot's processing resources from data supplied by the collection of control devices and electro-physiological sensors. It is necessary to show that the simplified model captures the salient human attributes, that the computations can be done speedily and reliably, and that due account is taken of the characteristics of the individual pilot. With comparable speed and reliability, the SASS must assess whether the mission is proceeding to plan while keeping abreast of changes to the plan. It is notable that the SASS must be able to update the TIM about the proficient performance of all tasks, and be able to execute any of them should the TIM decide it is necessary. The TIM must make a judgement about the tasks that are, and are not, receiving the pilot's attention from the evidence supplied by the COGMON and the SASS. This must be done with sufficient reliability for the pilot to trust that it will intervene appropriately.

Aside from these practical difficulties, it is also worth examining the assumptions of the scheme. As a start, it must be acknowledged that Charles Billings' fifth and sixth corollaries of Table 4.3 are contrary to the spirit of the first four, simply because an adaptive system must bring greater complexity. The Cognitive Cockpit will certainly be much less understandable and predictable than, for instance, an autopilot, and the pilot will have to accommodate a more intricate internal model for the machine. Secondly, the developers of such an involving form of automated assistance must take on board the lessons learned from the use of the earlier, simpler systems. As discussed in Chapter 4, automation is fine when functioning properly, but the developers must consider how the combination of pilot and electronic assistance can fail, and anticipate the consequences of failure. The classical "what's it doing now?" reaction[18] of the pilot when an unexpected form of control is exerted by an autopilot, could well become the start of an argument, especially when the electronics makes a false inference concerning the intentions of the pilot. The idea is to have an assistant, not an arguing partner.

The acceptability of the Cognitive Cockpit will depend on the pilot's perception of the gain in usefulness that it brings and whether the system is judged by the regulatory bodies to be safe. Before adaptive automation can be granted the same supervisory authority over fully autonomous modes as the pilot, the regulators must be assured by a rigorous and comprehensive analysis of failure cases that this will reduce significantly the probability of the loss of the aircraft. Perhaps a simpler scheme could be implemented more readily by omitting the COGMON element as this is likely to be least acceptable to aircrew and least reliable. In this case the TIM could call in autonomous assistance whenever the SASS indicates a significant or urgent shortfall in the pilot's performance in much the same way that the AutoGCAS in the AFTI F-16 can instigate a recovery manoeuvre.

It should however be noted that the development of the suite of intelligent aids themselves does not depend upon the acceptability of the adaptive means of control. Developments will undoubtedly continue, and increasingly sophisticated automatic systems will be engineered for use in manned aircraft. However, their real pay-off will be in unmanned aircraft.

14.4 Unmanned air vehicles

The argument for developing unmanned air vehicles (UAV) is straightforward: with no crew there is no crewstation. The large, heavy windows, the pressure cabin, the control and display facilities and all of the life support and protective systems are unnecessary. This reduces the size, mass and complexity of the aircraft and the on-board systems, and allows the designers to use airframe structures with better aerodynamic performance and lower radar reflectivity. It is unnecessary to recruit, train and maintain the aircrew, the aircraft can be manoeuvred with no concern for human physiological limitations, and it can be operated without risking the lives of the crew. Crew-less aircraft therefore offer better performance, in terms of range and flight duration, and they may be considerably cheaper. The problem is to provide a substitute for the crew. This requires a careful delineation of the functions the crew perform and a consideration of how these can be split between on-board electronic systems and a remote operator.

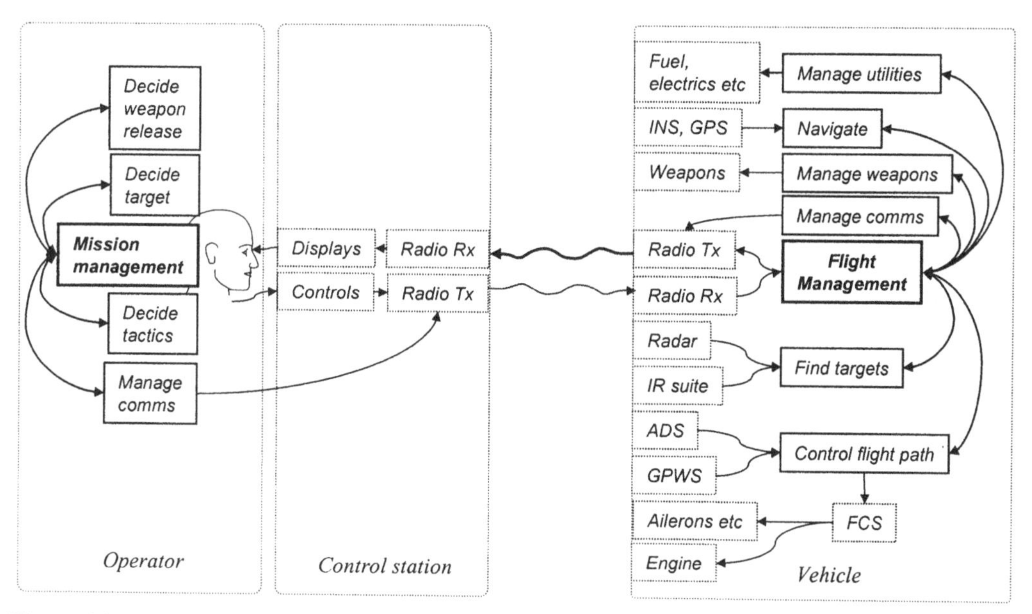

Figure 14.6 Hypothetical split of functions between the systems on an unmanned air vehicle and a remote operator

An example helps to illustrate the issues. Figure 14.6 shows the simplified avionic architecture of a hypothetical combat UMA that can be regarded as a semi-autonomous robot capable of performing a limited repertoire of programmed flight manoeuvres, such as take-off, transit, loiter and landing under the direct supervision of on-board *flight management* electronics. These are similar to the mission task elements devised for assessing rotorcraft handling qualities, as discussed in Chapter 4. By requesting an appropriate sequence of manoeuvres with appropriately set airspeed and clearance height, the operator can get the vehicle to a specified location where it can search for a target and assign - but not release - a weapon to the target. In this illustration the vehicle avoids collisions autonomously, but no defences or countermeasures are carried. The operator receives radar information, IR images and status information through a high bandwidth communication channel, and sends brief instructions through a low bandwidth channel to the on-board flight management electronics to specify the required manoeuvre.

Before take-off the operator formulates the mission plan and transfers the mission information to the flight manager. After take-off, he monitors the progress of the vehicle, decides whether the marked target is appropriate and if so he can release the weapon. He has little to do after instructing the vehicle to return to base and land, which it can do autonomously. The control station has no decision-making aids and could consist of a single large display showing a map of the area annotated with the vehicle position relative to the planned route and tactical information, such as the known defences and real-time information received from JTIDS. One inserted window could show a HUD-like depiction of vehicle attitude, speed, height and heading, and others could relay the IR sensor picture, the radar picture and a summary of the state of the fuel and the vehicle systems. Control may be exercised using a conventional keyboard and roller-ball to move a cursor and call pull-down menus. An alternative control station, utilising the idea of the virtual cockpit, could improve the operator's tactical awareness by giving the sensation of remote presençe in the vehicle.

This example gives the operator and the automation much the same authority over the individual tasks as the pilot could choose in a manned aircraft fitted with a comprehensive suite of automatic aids, as show in Figure 14.4. The vehicle has the authority to perform skill-based and rule-based functions, while the tasks requiring knowledge-based judgements, for instance about the way a target can be concealed or the political significance of other features in the imagery, remain the remit of the operator.

It is necessary however to acknowledge some inherent operational limitations. The split of authority is fixed by the architecture of the system, so the operator cannot intervene and take over tasks with the flexibility available to the pilot of a manned aircraft. The limited choice of manoeuvres available to the operator may produce inappropriate vehicle behaviour or behaviour that is predictable to an intelligently informed adversary. The operator will be keenly aware of the vulnerability of the vehicle, and will use tactics such as terrain screening and low flying, but will have to be careful not to interrupt radio links that depend upon a line of sight. An enduring break in communications would put the whole mission in jeopardy. The latency and infrequent update of imagery due to bandwidth limits will also be a concern, particularly during the crucial seconds when the operator must identify the target.

The benefits of remote operation must be offset against the cost of providing secure, reliable communication channels and competent and reliable on-board automation, but the latter can be addressed in several ways. Communications can be made more reliable using satellite channels or existing aircraft such as AWACS. Alternatively, it would be possible to deploy additional unmanned air vehicles that act purely as radio relays. The unmanned vehicle could also be operated within line of sight of an accompanying manned aircraft, with the former carrying the target-finding sensors and the latter the weapons. The crew would then have the additional tasks of managing the UMA in much the same way as a ground-based operator. A long term solution would involve developing a fully autonomous, long range, high speed robotic vehicle that needs no en-route communication and can be trusted to find the target and apply the rules of engagement with the reliability of the best human operator. Perhaps the ultimate vehicle would be a supersonic cruise missile, made cheaply relative to a re-usable vehicle, and able to pick out the intended target with such a high degree of reliability that damage assessment would be unnecessary.

14.5 Prognosis

Three of the ideas outlined in this chapter could have a substantial effect on the physical form of the cockpit; a reliance on synthetic external vision, the virtual cockpit as a man-machine interface and the versatile plug-in compact capsule. The first two could be used in any future aircraft and could be retro-fitted with varying ease to some existing aircraft types. The last would be useful only in a new fast jet designed to perform a combat role. As the ideas are complementary, in the sense that the adoption of one would make the others more practical, the greatest advantage would be gained by adopting all three. However, most services are currently more concerned to reduce equipment costs than enhance their capability, and there is no imminent programme to build a totally new combat aircraft. Development of the plug-in capsule seems therefore to be stuck in limbo awaiting a tangible need. Development of the two other ideas depends on persuading the customer's accountants of their merits. If military customers maintain their policy of exploiting commercial off-the-shelf systems and devices, they will wait until the sensing, display and control components are almost mature before funding the crucial work to integrate and test the systems in military aircraft.

On the other hand, changes that confer greater capability and autonomy on computer-based systems can be introduced progressively. Existing types of fixed-wing and rotary-wing manned combat aircraft will therefore be modified to incorporate increasingly comprehensive automation, but the displays and controls will be modified only to the extent needed for easy, reliable interaction with the automation. The complications needed to adapt the automation mode in response to the pilot's dynamic physiological and cognitive states are unlikely to warrant the expense and the effort needed to gain certification. The main factor driving the adoption of automated systems in manned aircraft will be the need to operate in concert with unmanned types in an environment that is rich with data but sensitive to surreptitious spoofing. The increased complexity is barely imaginable, like the internet breeding viruses and worms, and will result in new problems. The enemy will attempt, for instance, to disable unmanned aircraft and turn them against their owners with the knowledge that an automaton is like a brave, ingenuous child; seeing what it has been programmed to see, and doing what it has been programmed to do. One solution will make the relationship between manned and unmanned vehicles almost parental. As well as supervising aerial robots during flight, the versatility and operational flexibility of manned aircraft could supply a safe, supervised context for the development of software that gives the robot a measure of tactical astuteness.

The long term prognosis for fast combat aircraft is clear. It is unlikely that new types will be needed, and those remaining in service will be withdrawn gradually from front line roles. Although some current developments will be brought to maturity, the history of the combat aircraft cockpit is coming to an end.

References

1. Jarrett D N, *Alternative crewstation configurations for future combat aircraft*, in *Countering the directed energy threat: Are closed cockpits the ultimate answer?*, RTO-MP-30, 1999

2. Furness T A and Kocian D F, *Putting humans in virtual space*, The Society for Computer Simulation, Simulation Series 16 (2), San Diego, CA, pp214-230, 1986

3. Martin W L, *Developing virtual cockpits*, in *Advanced aircraft interfaces: The machine side of the man-machine interface*, AGARD-CP-541, 1992

4. Ineson J, *The DRA virtual cockpit research programme*, in *Virtual interfaces: Research and applications*, AGARD-CP-541, 1993

5. McMillan G R, Eggelston R G and Anderson T R, *Nonconventional controls*, in Salvendi G (ed), *Handbook of human factors and ergonomics*, 2nd edition, Wiley, New York, 1997

6. Rasmussen, J. and K. Vincente, *Coping with Human Errors through System Design: Implications for Ecological Interface Design*. International Journal of Man-Machine Studies 31, pp. 517-534. 1989

7. Zinner H, Schmidt R and Wolf D, *Navigation of autonomous air vehicles by passive imaging sensors*, in *Guidance and control of unmanned air vehicles*, AGARD-CP-436, 1988

8. Banks S B and Lizza C S, *Pilot's Associate: a cooperative, knowledge-based system application*, IEEE Expert Vol 6 (3) pp. 18-29,1991

9. Pipe H J, *A new class of mission support for combat aircrew*, in *Combat Automation for airborne weapon systems: Man/machine interface trends and technologies*, AGARD-CP-520, 1993

10. Le Fort N, Abou Kahled O, Ramamonjisoa D, *A Copilot Architecture Based on a Mutli Expert System and Real Time Environment*. in IEEE International Conference on Systems, Man and Cybernetics, Le Touquet, France 1993

11. *Safety net to detect performance degradation and pilot incapacitation*, AMP symposium, AGARD CP-490, 1990

12. Hicks M R, *EEG indicators of mental workload: Conceptual and practical issues in the development of a measurement tool*, in *Safety net to detect performance degradation and pilot incapacitation*, AMP symposium AGARD CP-490, 1990

13. Offenloch K and Zahner G, *Computer aided physiological assessment of the functional state of pilots during simulated flight*, in *Safety net to detect performance degradation and pilot incapacitation*, AGARD CP-490, 1990

14. Albery W P and Van Patten R E, *Current status of an artificial intelligence-based loss of consciousness monitoring system for advanced fighter aircraft*, in *Safety net to detect performance degradation and pilot incapacitation*, AGARD CP-490, 1990

15. Billings C E, *Principles of human-centred aviation automation*, Lawrence Earlbaum Associates, 1996

16. Reising J M, *The Symbionic Cockpit*, IEEE National Aerospace and Electronics Conference Proceedings 85CH2189-9, pp. 1051-1054, 1985

17. Taylor R M, Bonner M C, Dickson B, Howells H, Milton N, Pleydell-Pearce K, Shadbolt N, Tennison J and Whitecross S. *Cognitive cockpit engineering: Coupling functional state assessment, task knowledge management and decision support for context sensitive aiding*. Human Systems IAC Gateway Vol XII (1) 2001

18. Abbott K, Slotte S M and Stimson D K, *The interfaces between flightcrews and modern flight deck systems*, FAA Report 1996

Appendix A

Light and Polarisation Phenomena

A.1 Electromagnetic waves

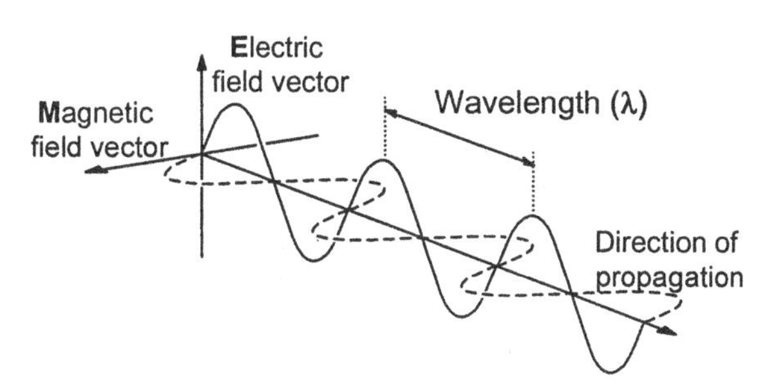

Figure A.1 Transverse electromagnetic wave

Electro-magnetic disturbances travel through space with a characteristic speed "c" (2.998 x 10^8 m.s^{-1}), and are slowed in any transparent medium by a factor called the refractive index "n" (= $\varepsilon^{1/2}$, where ε is the relative dielectric constant of the material). As shown in Figure A.1, the electrical and magnetic fields propagate in phase and are mutually perpendicular to the direction of energy flow.

A.2 The spectrum

An idealised sinusoidal propagating disturbance has a wavelength "λ" and a frequency "f" (= c/λ). The frequency range is theoretically infinite. In practice, waves have been classified into the broad regions shown in Figure A.2.

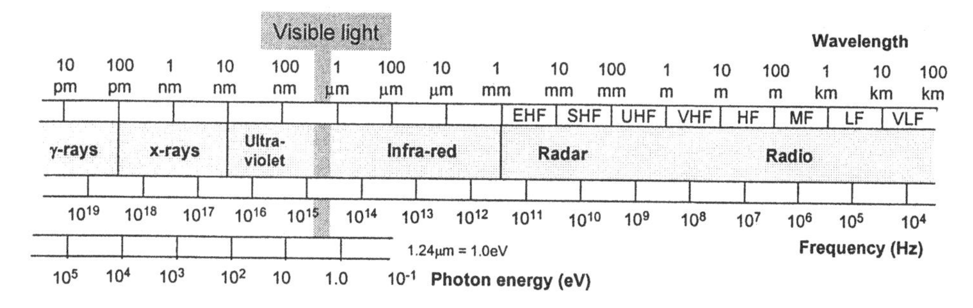

Figure A.2 The electro-magnetic spectrum

The source of the disturbance is generally the acceleration of an electrical charge. Microscopically this is the change of state of a charge carrier such as an electron, and macroscopically the change to the multiplicity of charge carriers that constitute an electrical current. E-M waves are absorbed by matter in a reciprocal fashion, causing a change in the state of a charge carrier or inducing a change to an electrical current. Light is E-M radiation in the wavelength range between about 0.1μm and 20μm. The human eye senses the narrow octave from 0.35μm to 0.7μm.

A.3 Quanta and thermal emission

Energy is exchanged as tiny but finite quantities of radiation called "quanta", which are related to the wave frequency f by $e = h.f = h.c/n.\lambda$, where e is the quantum energy expressed in electron-volts, f is in Hz, λ is in metres, c is the velocity of light, n is the material refractive index and h is Planck's Constant, 6.626×10^{-34} J.sec.

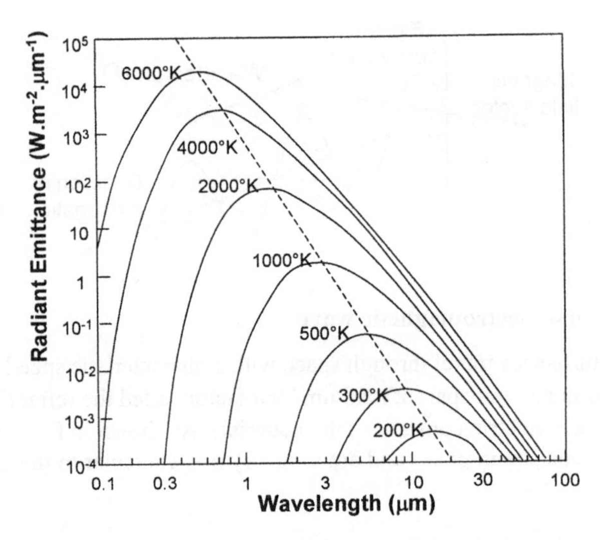

Figure A.3 Characteristic radiation emitted from a perfect black body

Heat energy, the random constrained motions of atoms and molecules in solids, liquids and gases, produces electro-magnetic emissions. Figure A.3 shows the characteristic emission from a "black body", where no other form of energy exchange occurs. The amount and the spectral range of the radiation depend only on the temperature of the emitter.

A.4 The Sun and the atmosphere

The thermal emission from the Sun is characteristic of a black body at about 5800°K. The total thermal energy incident on the Earth, the "Solar Constant", is 1.37 kW.m^{-2}. Only a proportion of this reaches the Earth surface through spectral windows in the UV, visible, near IR, 3-5μm IR and 8-13μm IR bands where atmospheric gases, mainly water vapour and carbon dioxide, have little absorption.

A.5 Polarisation

The direction of the E-vector is the "plane of polarisation" of linearly polarised electro-magnetic radiation. Light from the Sun and most man-made sources such as incandescent lamps, fluorescent strip lights and CRT screens, is a jumbled collection of waves polarised in all orientations. Such "unpolarised" light can however be considered as composed of two mutually perpendicular but arbitrarily oriented, waves of linearly polarised light. In "circularly polarised" light the E-vector rotates about the direction of propagation once every wavelength. It is equivalent to two identical orthogonal linearly polarised waves displaced by a quarter wavelength

(90° phase difference) as illustrated in Figure A.4. Both "clockwise" and "anti-clockwise" waves occur.

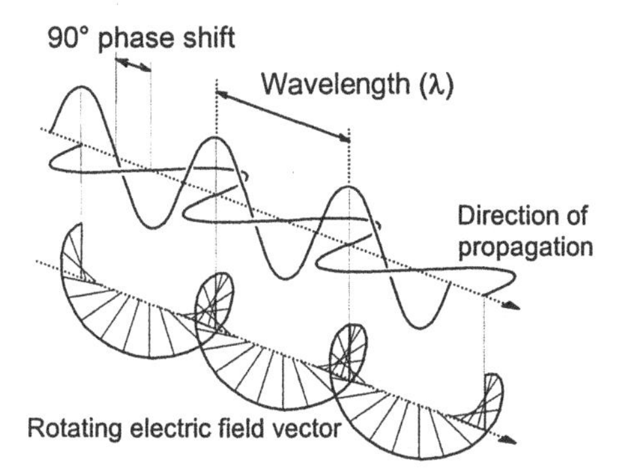

Figure A.4 Circularly polarised light represented by two orthogonal waves of linearly polarised light

In "elliptically polarised" light the two 90° shifted waves have unequal E-vectors, so elliptically polarised light can be considered to be a mixture of circularly and linearly polarised light. Linearly polarised light can also be considered as two in-phase waves of counter-rotating circularly polarised light.

A.6 Anisotropic materials and polarised light

Isotropic materials, such as most liquids and glasses, have no 'grain', but many materials, particularly crystalline solids and long-chain polymers, are anisotropic; their optical characteristics vary with the orientation of the E-vector, and hence the plane of polarisation of light travelling through the material. Although many materials exhibit differences in three orthogonal directions (triaxial anisotropy), in many instances the grain is aligned along a single direction, and it is the uni-axial differences in light absorption and refractive index that are most relevant.

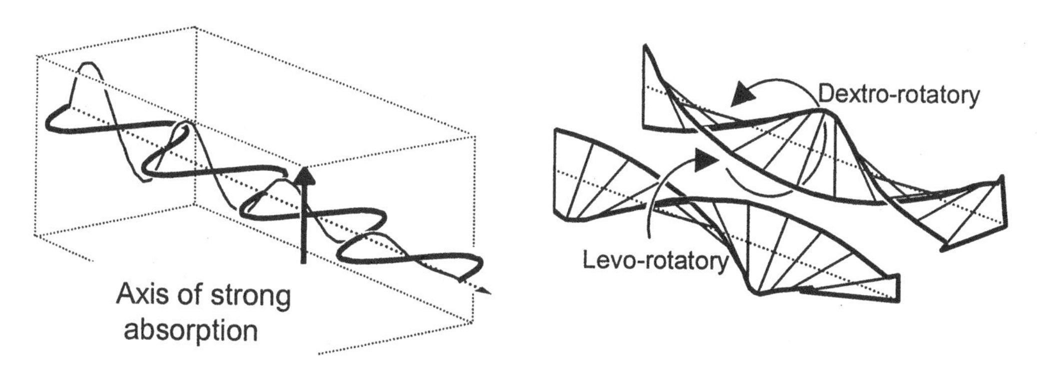

Figure A.5 Dichroic absorption of light

Figure A.6 Optical activity

Dichroism, occurs in anisotropic materials where the absorption of light depends upon the alignment of the E-vector to the optical axes of the material. Figure A.5 illustrates the case of dichroic materials, such as tourmaline or iodine, which attenuate light with a plane of polarisation that is oriented along the axis of high absorption, but pass orthogonally polarised light with little absorption. Such materials are invaluable for transforming randomly polarised light into linearly polarised light. The ubiquitous "polaroid" sheet exploits the phenomenon. It consists of a plastic sandwich encapsulating a thin layer of aligned, sub-microscopic iodine crystals. The name "pleochroism" is usually applied to the more general case in which the optical absorption also depends upon the wavelength of the light. This phenomenon is exploited to manufacture polarisation-sensitive colour filters by aligning dye molecules in much the same way as the crystals are manipulated in polaroid sheet.

Some materials exhibit optical activity, where the plane of polarisation of linearly polarised light is rotated about the direction of propagation. Both senses of rotation occur; "dextro" being clockwise and "levo" anti-clockwise in the direction of propagation. One explanation considers that the two counter-rotating circularly polarised component waves propagate at slightly different velocities.

In anisotropic materials the wave velocity depends upon the alignment of the E-vector with the optical axes of the material, and the refractive index is a complex wavelength-dependent tensor. Birefringence, or double refraction, occurs in uni-axial materials (the simplest case) as a result of the phase difference imparted to the components of linearly polarised light aligned with "fast" and "slow" directions of the material.

As illustrated in Figure A.7, these "Extraordinary" and "Ordinary" component rays are exploited in the "quarter wave plate" that can convert linearly polarised light into circularly polarised light. A phase shift of $\pi/2$ (a quarter of a wavelength) is imparted to the equal components if linearly polarised light is incident with the plane of polarisation at 45° to the birefringence axis. The effect is reversible; circularly polarised light incident on a quarter wave plate is converted into linearly polarised light.

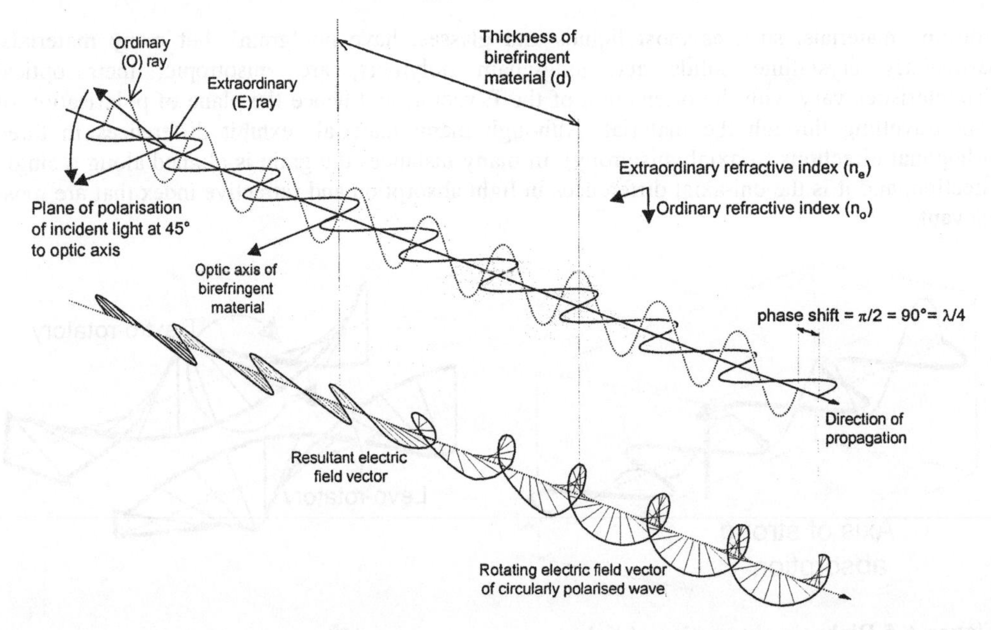

Figure A.7 The quarter wave plate

The "half wave plate", which introduces a π phase difference between the extraordinary and ordinary waves, is also extremely useful because it can be used to turn the plane of polarisation of linearly polarised light through any angle, as shown in Figure A.8. For instance, when the incident plane of polarisation is at 45° to the axis of birefringence, the induced rotation is 90°.

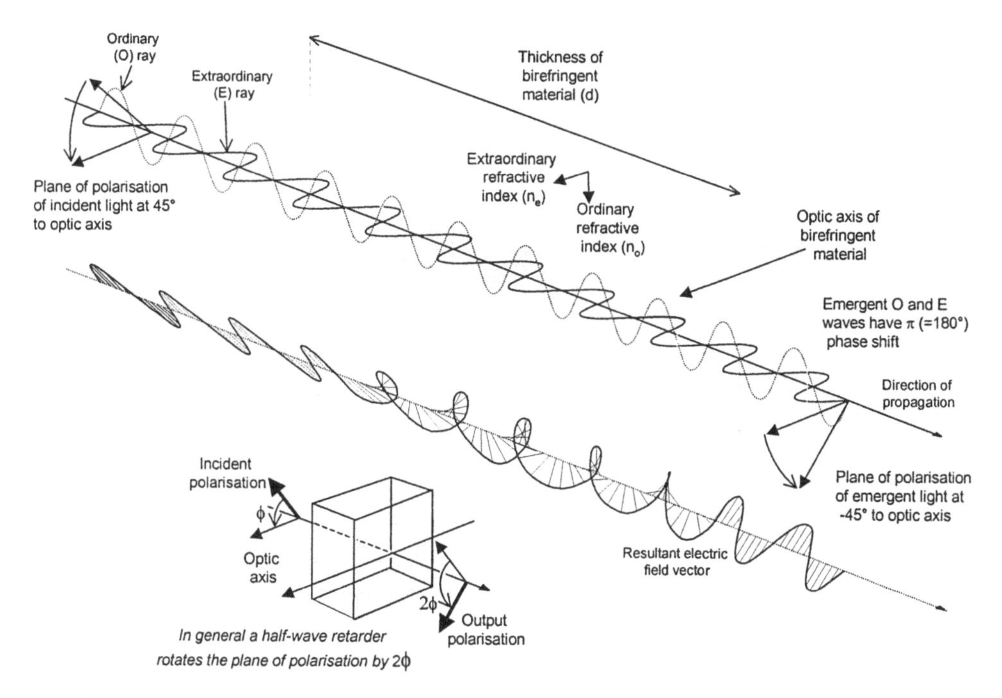

Figure A.8 The half wave plate

A.7 Reflection and refraction

Reflection and refraction occur when light is incident on a sharp boundary between materials of different refractive index. The reflected angle ϕ_r is always equal to the incident angle ϕ_i. The transmitted light travels on at an angle given by Snell's Law, $\phi_t = \sin^{-1} (\sin \phi_i / n)$, whether or not the light meets the discontinuity from the less-to-denser refractive index direction or vice versa. To meet this requirement, no light can be transmitted when incident from within the greater index material at any angle above the critical angle $\phi_c = \sin^{-1} (1/n)$. This "total internal reflection" is exploited widely, for instance in reflective prisms and optical fibres.

As summarised in Figure A.9, the proportions of the intensity reflected and transmitted depend not only on the angle of incidence and the refractive index, but also on the state of polarisation of the light. The figure shows the convention used to describe the directions of polarisation as "p-polarised" light and "s-polarised" light, which are parallel and perpendicular to the "plane of incidence" respectively.

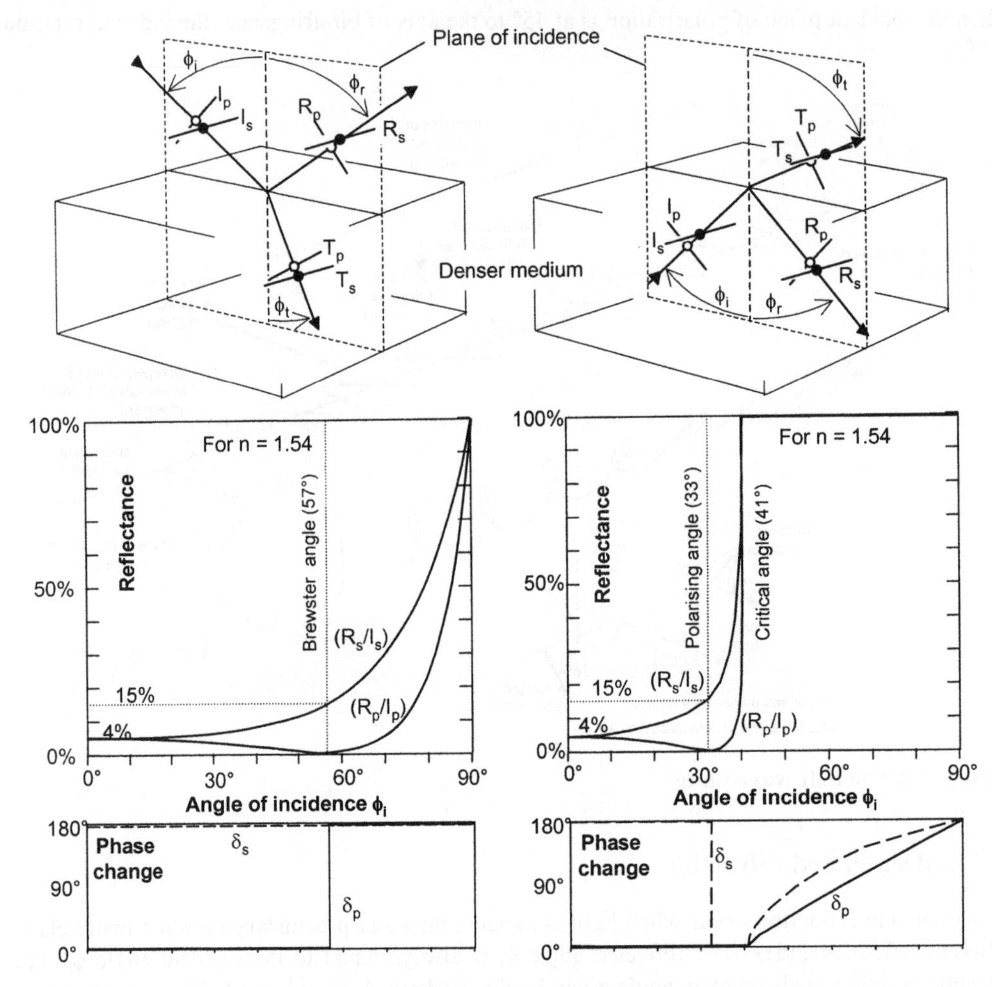

Figure A.9 Reflection at a dielectric interface. Graphs show the behaviour of light at an interface between materials having an index difference of 1.54

Figure A.10 shows that an interface is slightly reflective at normal incidence, completely reflective at grazing incidence, and that in between the two planes of polarisation are reflected quite differently. At the Brewster angle only s-polarised light is reflected and all p-polarised light is transmitted. Thus a fisherman can see beneath the water best by looking down at the Brewster angle and by wearing spectacles containing a polaroid material aligned to absorb the strong s-polarised reflection of sunlight and skylight.

The lower graphs show that the light undergoes a change of phase on reflection, and again the two planes of polarisation behave differently. The 180° phase change of s-polarised light is notable because it induces a change of the direction of rotation to circularly polarised light. Right-handed circularly polarised light is reflected as left-handed circularly polarised light, and vice versa. Thus unwanted reflections can be reduced dramatically, for instance to reduce the effect of glare on display devices, by covering the display surface with a filter containing a linear polariser and a quarter wave plate. The reflected light is polarised with the opposite sense of rotation and is attenuated strongly by the crossed polariser.

Appendix B

Image Quality

B.1 Optical resolving power

The resolving power of an optical system is a measure of the ability of the system to form separate images of objects that are very close together.

Light waves (see Appendix A) are diffracted by an aperture such as a focusing lens so that the image of a point source with a narrow wavelength range consists of a small bright disc surrounded by concentric rings of rapidly diminishing intensity. Rayleigh suggested that the images produced by neighbouring points could just be separated if the central peak of one image coincided with the first dark ring of the other. This "diffraction limit" occurs for a circular lens when the angle subtended between two point objects is equal to 1.22 λ/D, where λ is the wavelength of the light and D is the lens diameter. The resolving power of the human eye is limited by diffraction when the pupil diameter is less than about 2.5mm.

B.2 Resolution

The resolution of an image-forming device is usually tested subjectively by forming an image containing progressively finer features of known dimensions.

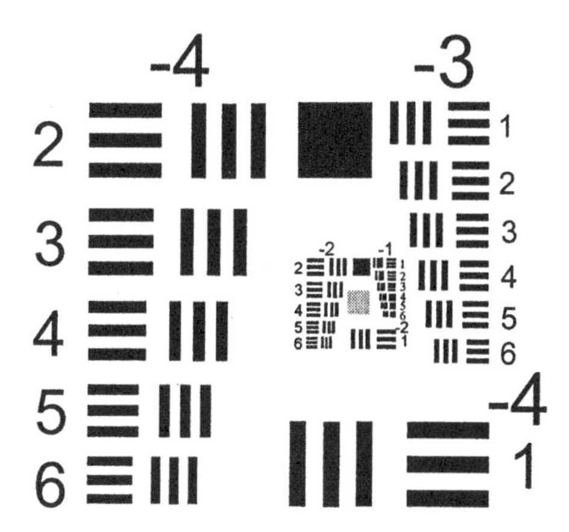

Figure B.1 The 1951USAF tri-bar test pattern (N.B. this should have a contrast of 0.7)

Figure B.1 shows an example of a typical test pattern that is either generated electronically or obtained from a high quality camera so that it can be presented on a display screen in a variety of locations. In this case the high contrast black bars have a length to width ratio of 5:1 and the three bars are spaced one bar width apart to form a square element. Successive elements with horizontal and vertical alignment are scaled down progressively to form a spiral of gradually diminishing dimensions (the finest cannot be reproduced in the figure). In formal tests a panel of observers judge where an element can be seen to be composed of separable lines and any finer elements blur into uniformly grey squares. Separate judgements are made for vertical and horizontal bars. The width of the finest distinguishable bars is quoted as the resolution of the device.

B.3 Image contrast

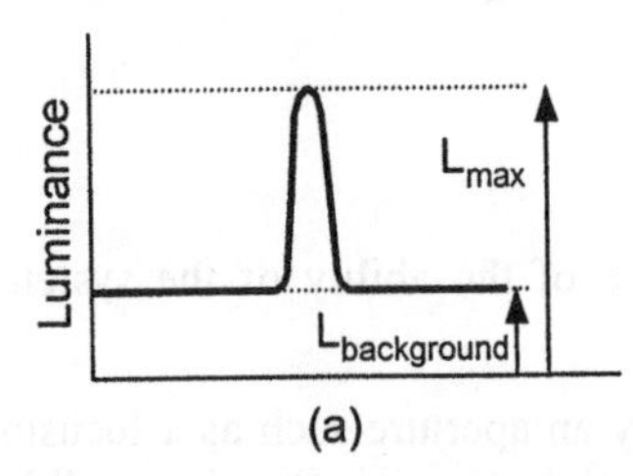

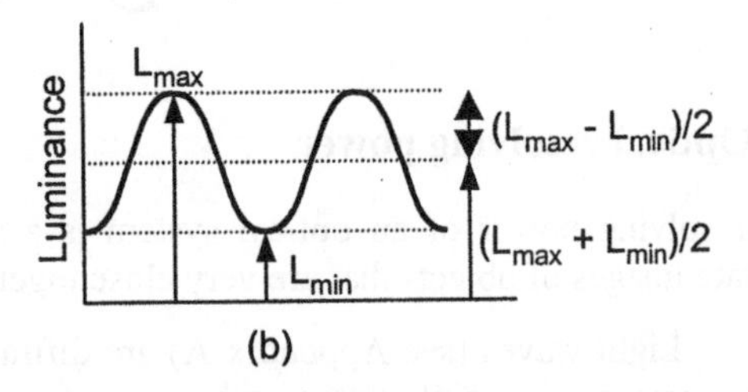

Figure B.2 Contrast definitions of (a) line images and (b) tonal images

As illustrated in Figure B.2(a), the contrast of a narrow line against a broad background is usually quantified as the ratio of the excess luminance of the line (L_{max} -$L_{background}$) to the luminance of the background ($L_{background}$). This has a minimum value of –1 for a totally black line, and there is no upper positive limit for a bright line.

The contrast of tonal imagery, as in Figure B.2(b), is the ratio of half the luminance variation (L_{max} -L_{min})/2 to the mean (L_{max} + $L_{background}$)/2. Tonal contrast therefore spans a scale from zero to unity.

B.4 Modulation transfer function

The dynamics of linear systems are usually described in terms of the response of the system to oscillatory disturbances over a range of excitation frequencies. The approach, called Fourier Analysis, represents the response as a gain transform, which shows the magnitude of the output in relation to the input, combined with a phase transform, which shows how the response lags behind the input. As noted in Appendix C, the relationship between the input video signal level and the output luminance of most display devices is non-linear. It is therefore formally inappropriate to characterise displays using Fourier spectral analysis because the measured gain depends upon the excitation level and the output contains harmonics of the excitation frequency.

Figure B.3 Illustrating (a) the appearance of a perfect display, and (b) the visible output of a typical display that exhibits decreasing contrast with increasing spatial frequency

The "modulation transfer function" (MTF) has nevertheless become the standard way to describe the quality of a tonal image. The display is tested using a sinusoidally varying video signal that produces a grating pattern of parallel lines running either across or down the screen. For consistency, testing should be done with an input that modulates a small, central part of the total dynamic range. With real images the spatial frequency is the number of sinusoidal cycles per millimetre, and for virtual images it is the number of cycles per degree.

Figure B.3(a) shows the high tonal contrast of the sinusoidal grating pattern of increasing spatial frequency that would be apparent with a perfect display system, and (b) shows the typical diminution of contrast with increasing spatial frequency that is reproduced by most practical display systems. For each spatial frequency, the output contrast is measured using a microphotometer, and this is divided by the maximum contrast, or the contrast expected from a perfect display, to give a normalised modulation transfer value between zero and unity.

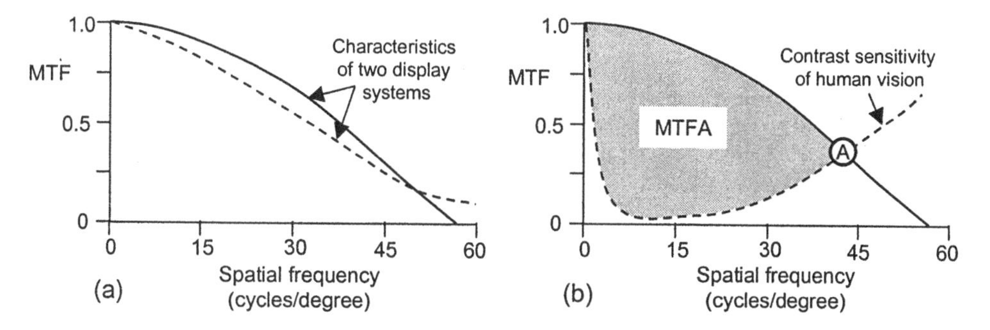

Figure B.4 Illustrating (a) typical MTF characteristics for display devices and (b) the definition of the Modulation Transfer Function Area (MTFA)

As illustrated in Figure B.4(a), the "modulation transfer function" (MTF) is a plot of the output modulation transfer against spatial frequency. The approach is objective and it enables a system designer to treat the display as an element in a picture relaying chain. For instance, the effect of viewing the display screen through imperfect optics can be predicted by multiplying the MTF by an analogous "optical transfer function" (OTF) for the optics.

B.5 Image quality metrics

As shown in Figure B.4(b), the human limit can be included as a curve describing the threshold grating contrast (the minimum contrast of the grating pattern discernible on 50% of test trials) over the same range of spatial frequencies. For most spatial frequencies the human demand is easily satisfied by the contrast supplied by the device. The two curves cross at point A in the figure, and this indicates the highest spatial frequency the device can supply. The inverse of this value of spatial frequency correlates well with judgements of the device resolution obtained subjectively using a test chart, as described above.

Also shown in Figure B4(b) is the “Modulation Transfer Function Area” (MTFA), the area between the curves for the minimum contrast required by the eye and the maximum contrast available from the display. This metric has also been examined as an indicator of the quality of displayed and photographic images[1]. If the image content exploits the available contrast and available range of spatial frequencies, and the display has low noise, the metric correlates well with a subjective ranking of goodness.

Alternative formulations have been examined. One, in which the area is based on the logarithms of the MTF and the logarithm of the spatial frequency, has been shown to correlate well with the ability of trained operators to recognise targets present in the displayed images. Another suggests that spurious image content in the form of random time-varying noise, or any fixed overlaying structure from raster lines or AMLCD pixel edges, could be accommodated in the MTFA as factors that lift the threshold sensitivity curve[2]. This formulation necessitates carrying out psychophysical experiments to characterise the shift of the contrast threshold curve due to these forms of noise.

References

1. Pincus A R and Task H L, *Display system image quality*, in *Visual effects in the high performance aircraft cockpit*, AGARD-LS-156, 1988

2. Snyder H L, *Image quality and observer performance*, in Bieberman L M (ed) *Perception of displayed information*, Plenum Press, 1973

Appendix C

Video Signals

C.1 Pictorial information

A video signal represents a sequence of pictures as a time-varying voltage that enables the picture sequence to be stored, transmitted, modified and displayed. A standard defines the temporal characteristics of the signal, the method of breaking the pictures into discrete elements and the method of representing the brightness and colour of each element.

All video pictures are rectangular and are decomposed into horizontal lines of elements, scanned from left to right, forming a series that starts at the top of the picture and finishes at the bottom. This form of scan is called a "raster" (the Latin word for a rake). Standardised pulse signals are inserted between the lines to mark the start and end of each line, and also between the successive pictures to mark the start and end of each picture. Successive pictures are known as "frames". The video signal voltage can either be a direct analogue representation of the variation in luminance and chrominance of the elements along the lines, or a sequence of digital codes that represents the luminance and chrominance.

C.2 Raster scan standards

A variety of raster scan standards have been developed, originally for television broadcasting and latterly for computer terminals. The former are in general use in aviation, and the latter are becoming increasingly prevalent with the general trend toward computer-controlled automation. Figure C.1 summarises the form of the raster structure defined by the Consultative Committee on International Radio (CCIR) for the video signals broadcast in the UK where a frame is refreshed at 25Hz and composed of 625 horizontal lines to form a rectangle with 4:3 horizontal to vertical aspect ratio.

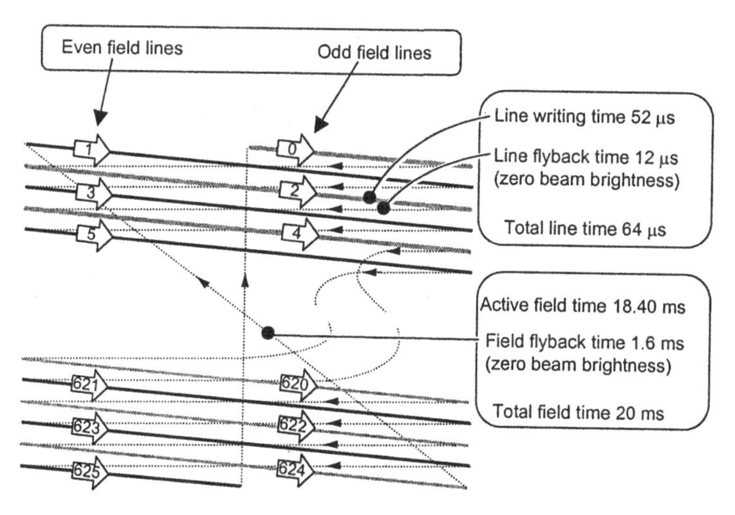

Figure C.1 CCIR interlaced raster scan pattern for broadcast TV in the UK

The figure shows that although each frame has 625 lines it is scanned as two interlaced fields, each of 312.5 lines. This was done to remove visible flicker by effectively doubling the apparent refresh rate of the picture to 50Hz while keeping the modulation signal that conveys good picture detail within the available 5.5MHz broadcast bandwidth.

The originators of the standard also acknowledged the imperfections of the vacuum valve technology employed in sawtooth generators during the 1950s. A period of 12μs was allowed for the flying spot to skip from the end of one line to the beginning of the next, and 1.6msec (25 lines) were blanked during the vertical flyback period between fields. Although the picture information occupied only 575 active lines, the inactive periods are used for transmitting information such as Teletext, and the 12 μs line flyback periods are used to encode synchronising pulses and to establish the black level for the tonal picture information. The field flyback periods have also been exploited, particularly in CRT displays for aviation, to create a hybrid raster/cursive drive mode for drawing a few clear, bright symbols on top of a raster image.

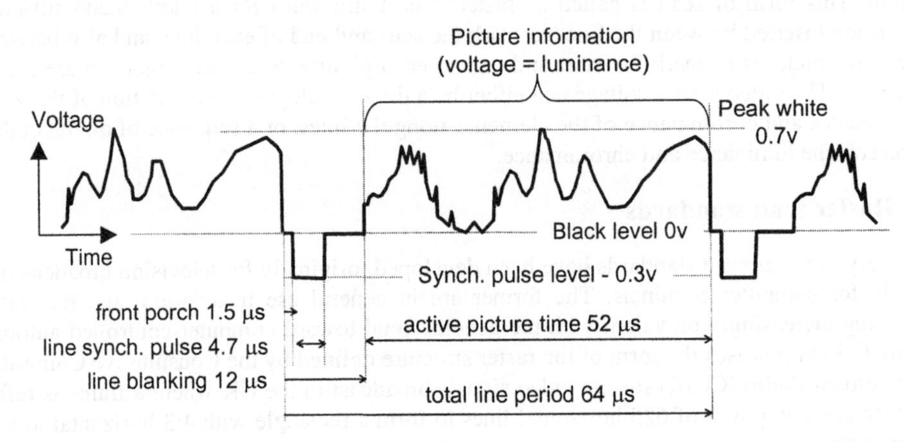

Figure C.2 Standard CCIR 625 line 25Hz video signal

The form of the CCIR interlaced video signal is shown in Figure C.2. The line synchronisation pulse precedes the picture information at the start of each line. Frame start pulses have the form of an elongated and modulated line synchronisation pulse. The video picture information is an analogue signal of 700mV range, the short period between the end of the line synch pulse and the start of the video serving as a reference to establish the black level. This is particularly important because most signalling is AC-coupled and the mean signal level therefore floats up and down with the net brightness of the scene represented by the video waveform.

C.3 Dynamic range and gamma

When used to present a shaded picture in this way, the relationship between the instantaneous signal voltage (V) and the luminance (L_d) of the spot on the display screen is approximately logarithmic:-

$$\log L_d = \gamma_d \cdot \log V + \operatorname{Log} L_o$$

where γ_d is the gamma of the display and L_o is a scaling constant.

A plot of the voltage-to-luminance transfer characteristic using logarithmic scales is illustrated in Figure C.3 for a typical monochromatic CRT display. This display can show pictures at a mid-scale luminance of about 70cd.m^{-2}. The dynamic range runs from the minimum, set by the noise level, to the maximum saturation level, in this case from about $4cd.m^{-2}$ to $700cd.m^{-2}$.

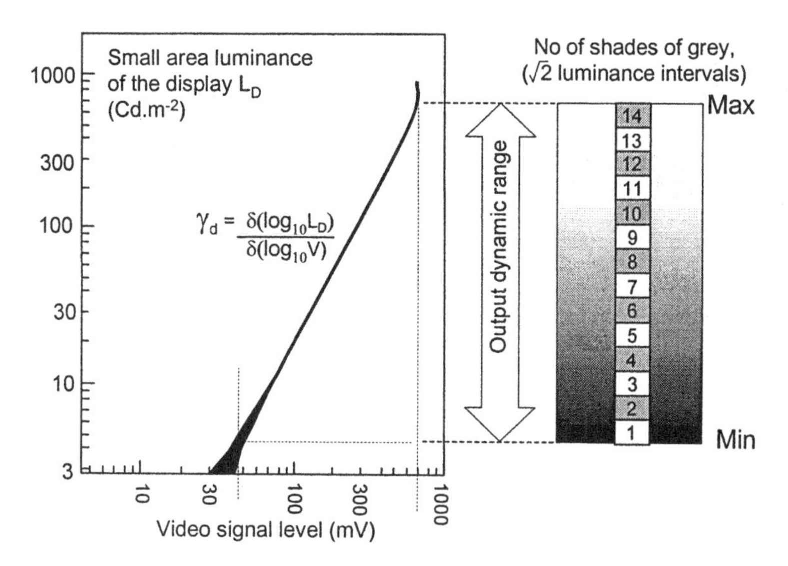

Figure C.3 Transfer characteristic for a typical avionic CRT display

The human visual system adapts to the ambient light level in such a way that the appearance of a surface is not affected by the level of light reaching that surface, and this is why white, grey and black areas remain white, grey and black whether they are seen in daylight or twilight. Because the subjective sensation of a difference between areas is related to their luminance ratio, "brightness" is defined as a unit-less, subjective, quantity that is proportional to the logarithm of the ratio of surface luminances. To handle the proportionality it has become a convention to regard a "shade of grey" as an increase or decrease in the luminance of a small area by a factor of $2^{1/2}$, i.e. a surface is regarded as a shade of grey brighter or dimmer than another if their luminances differ by about 40%. Defined in this way, a display with a characteristic illustrated in Figure C.3 would reproduce about 14 discernible shades of grey.

It is of note that the value of the γ_d (gamma) parameter that defines the non-linearity of the display transfer characteristic is important, mainly because it must complement the non-linearity of the signal produced by a TV camera. As shown in figure C.4, L_d is proportional to L_s if the gamma of the sensor and the gamma of the display are reciprocals ($\gamma_d \cdot \gamma_c = 1$). Thus, although the display can present imagery at a mean luminance that is markedly different from the original scene, such a combination of camera and display preserves the relative luminance of features present in the original scene.

Standards vary between national bodies. To conform with the European CCIR Standard, camera and display systems should have γ_d and γ_c values of 2.8 and 0.36 respectively. To comply with the US Standard, the values should be 2.2 and 0.45. Most cameras have, in practice, a gamma of about 0.45, which gives a slightly exaggerated contrast to dim features presented on a CCIR display. It is of note that many video standards for computer monitors specify a display gamma of unity ($\gamma_d = 1$) and video signals imported into computers invariably need appropriate gamma correction.

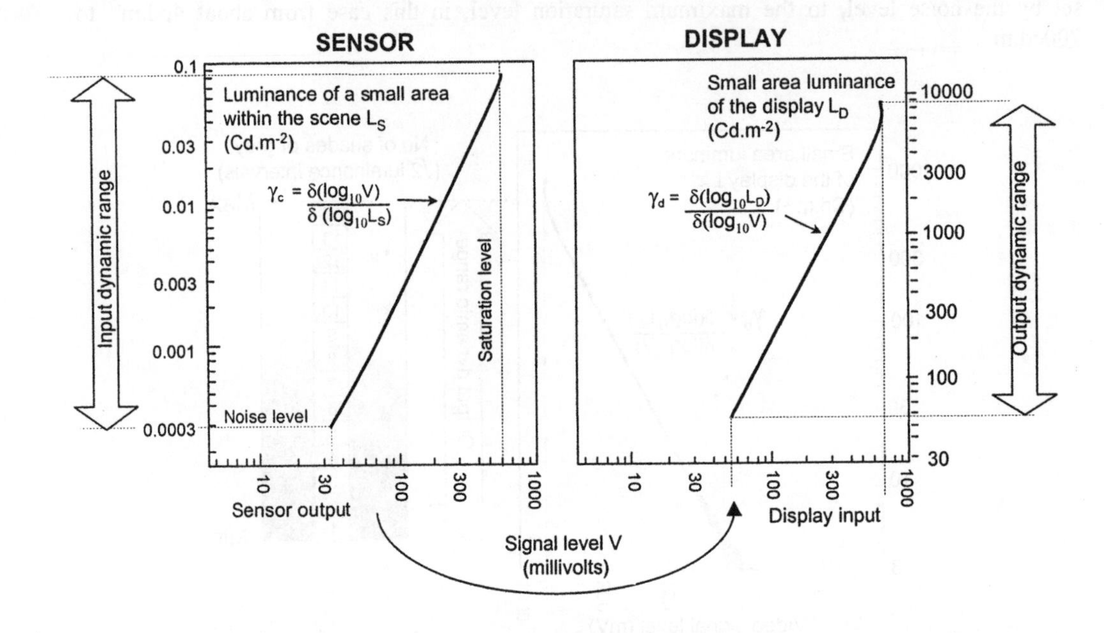

Figure C.4 Illustrating complementary transfer characteristics for sensor and display devices

Appendix D

A Hierarchical Control Model of Human Behaviour

D.1 Background

Consciousness, the dynamic interaction between thought and emotion of ordinary human experience, has been treated by psychologists as a scientific shibboleth. In reaction to the introspective techniques used by psychologists in the early part of the 20th century, the Behaviourist school of psychology advocated the observation of overt, measurable activity. In consequence, behaviour was considered only as an elaborate set of habits, much like the conditioned responses of Pavlov's dog, and the internal experiences of the human animal were treated as incidental epi-phenomena. In the 1960s, a philosophical shift was induced by insights from cognitive psychology, neurology and electro-physiology. This was accelerated in the 1980s by computer-based modelling and by techniques that enabled detailed observation of the activity of the live brain. The result has been an explosion of ideas[1] about emotion, awareness, attention, intention, volition and their evolutionary mechanisms[2].

This appendix outlines a qualitative model for human behaviour that treats perception as the generation of internal representations and consciousness as a useful idea. The conceptual framework is intended to underpin the discussion of attention and multiple task performance in Chapter 2 and the mathematical models of the human operator in Chapter 4. The idea of perception as the erection of a representation may help the discussion in Chapter 9 concerning the visual rivalries experienced by aircrew using HUDs and HMDs.

D.2 Perceptual integration and internal representations

During rapid saccadic eye movements, which take about 100msec, the image is smeared across the retina and almost all processing of visual stimuli is suppressed[3]. The extraction of information is only resumed when the eyeball is relatively stable. Thus what is experienced as a veridical view of the environment with no gaps in time or evident spatial discontinuities is assembled from the stimulation accumulated over a series of transient probing foveal examinations into selected areas of visual space. This is pasted onto the poorly resolved background provided by ambient vision, augmented by cues extracted from tactile, auditory, olfactory and kinaesthetic stimuli.

One explanation for the sensation of visual continuity is that the forebrain, the place that puts the "what" and the "where" information together, does so by inventing a model of the environment that is perpetually refined and updated by such piecemeal input. This explanation suggests that the brain does not receive stimulation passively but creates internal representations that correspond with the objects in the environment that give rise to the stimuli. For most of the time these mental inventions are good enough to allow successful interaction with real entities.

This idea[4] of perception as an inferential mechanism, analogous to the scientific process of erecting hypotheses that are tested against evidence, was advanced originally by Helmholtz in the late 19th Century. A simplified schematic for the cycle of processes is given in Figure D.1. The mechanism accounts for numerous visual phenomena. For instance, it suggests why humans tend to see whole entities when presented with sketchy outlines, and why, despite having two eyes with individual views, the world is composed of single entities that are viewed from a single point. It is of note that in most individuals, this "ego-centre" is shifted from the mid-point between the eyes slightly towards the sighting-dominant eye[5].

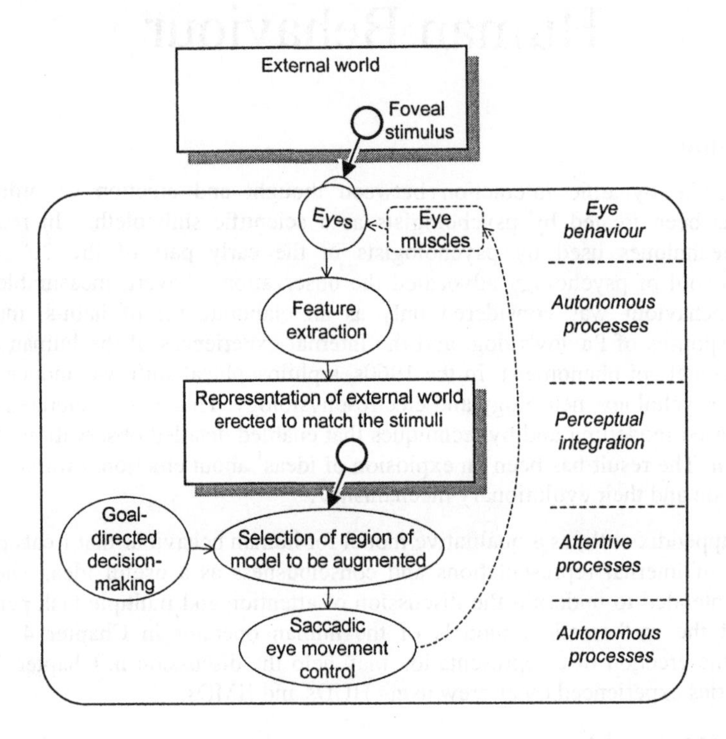

Figure D.1 The cycle of processes that erect an internal representation of the environment to match cues extracted from stimuli received by the eyes. The eyes are directed to jump and fixate on the region of space where the representation needs relevant detail for high-level decision-making

This "singleness of vision" may also account for binocular rivalry, retinal rivalry and the attentional suppression of momentarily irrelevant components in overlaid images. Illusions can be explained as inappropriate representations induced by bias. The converse failure to erect an appropriate representation explains why entities in the world frequently pass unnoticed, particularly when stimuli are unexpected, weak, brief or fall outside the fovea. The explanation is also consistent with the idea that high level intentions direct the spotlight of attention. For instance, as described in Figure 2.18 when reading text the eyes are directed to jump and fixate on the region of space where the relevant detail is the set of symbols making up the next word.

D.3 A hierarchical control model

At the beginning of Chapter 2, Figure 2.1 showed a simplified overview of the main human systems and their principal linkages. These are rearranged in Figure D.2 to show the hierarchy of processes that both compete and co-operate to determine human behaviour.

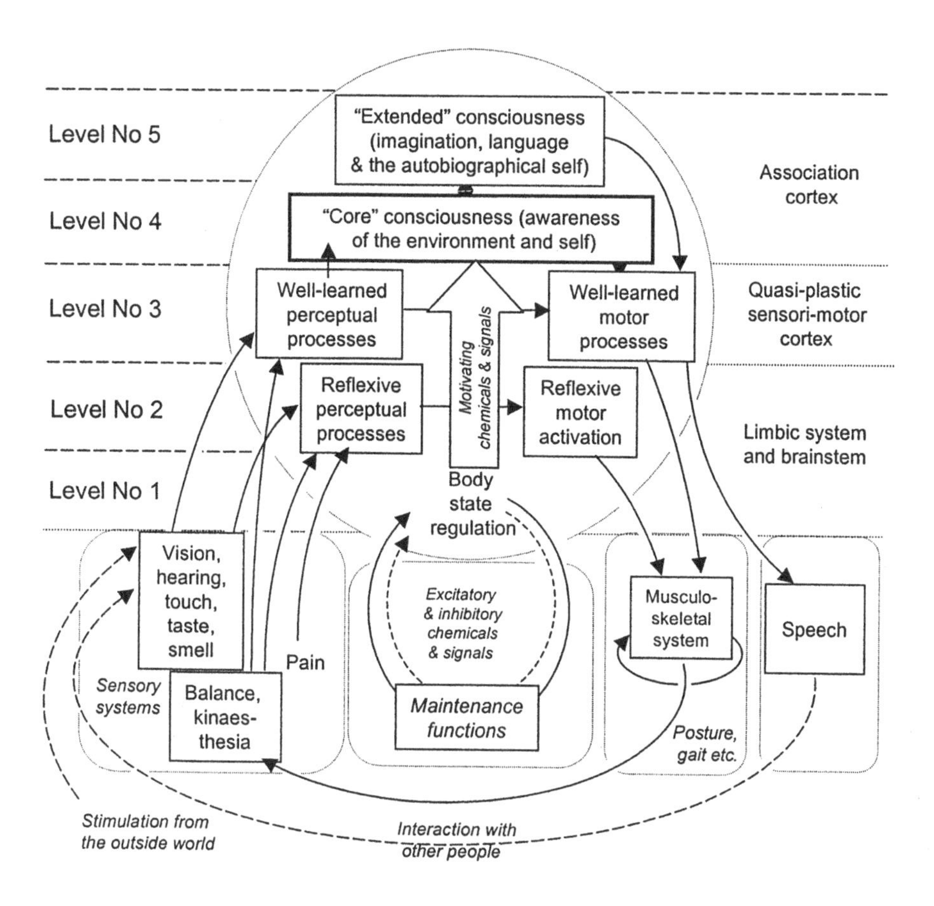

Figure D.2 The brain as a five-level hierarchical control mechanism

For descriptive convenience these processes are placed in five levels, the boundaries between levels having a rough correspondence with neuro-anatomy. Those in the first level, which control basic body maintenance functions manifest as behaviour such as breathing and sweating are performed by neurone clusters in the brainstem in conjunction with the sympathetic and parasympathetic pathways. Most of the neural interconnections are formed in the womb and remain immutable.

Processes performed in the second level are largely in the limbic system, where they receive kinaesthetic, somatosensory and proprioceptive stimulation and produce behaviour such as maintaining an alert posture. Those in the third level process information from stimuli originating outside the body that are received by the special sensing systems. Like second-level processes, they occupy areas of the limbic system, cortex and cerebellum that, by virtue of the proximity to input and output pathways, are quasi-plastic in the sense that they are primed to process information in particular ways. Given appropriately repeated stimulation these regions adapt to extract consistent features from the barrage of sensory stimulation, and generate complex excitatory patterns to activate precise control over muscles. Level-3 processes include ambient vision, and they control behaviour requiring precise dynamic interaction with the world, such as walking and running.

The fourth level occurs in the cortex, largely in the frontal lobes, where associative tracts deliver a barrage of highly processed level-3 information derived from visual, auditory and chemical stimuli originating in the external world. This, together with level-2 information originating from touch, kineasthetic sensors and the internal organs, is integrated to form a self-consistent representation of the source of the stimuli for controlling behaviour appropriate to detailed entities in the environment and the state of the individual. Damasio[6] suggests that the dynamic maintenance of this representation of the body's relationship with the immediate environment, a multi-sensory elaboration of the scheme shown in Figure D.1, is experienced as "core consciousness".

This fourth level of control is common to most animal species. Damasio delineates the "extended consciousness" in level-5 of Figure D.2, as distinctly human. Essentially, this fifth level of control can be considered as extending the ability of the forebrain to form representations to also allow their manipulation. These acts of the imagination enable selected models to run ahead of natural time so that the consequences of alternative actions can be anticipated. The immediate response of level-4 processing can then be over-ridden, leading to a better chance of surviving and reproducing in the long term. This faculty of imaginative forethought can be regarded as the basis of intelligence, and the experience of making a longer-term overriding decision corresponds to the exercise of free will. The fears and desires attributed to the heart that contrast with the rationality of the head are the fictional expressions of the intrinsic conflict between the level-4 and level-5 control mechanisms.

Other uniquely human characteristics result largely from the progressive abstraction and manipulation of conscious representations. As outlined in Chapter 2, section 2.3.2, inter-human communication has arisen from the incorporation of arbitrary but consistent symbolic stimuli, recognised as originating from another human, into the internal representations. The highest level of human information processing, such as the knowledge gained through schooling, deals solely with these abstracted, communicable representations. One crucial model formed by the individual is that of the individual himself; an amalgam of experiential history and self-knowledge.

D.4 Emotional states

As the forebrain can be excited or inhibited by neuro-chemicals secreted by the brainstem in response to the stimuli, and also by the endocrine system via the hypothalamus, the speed and selectivity of forebrain processing depends indirectly on the content of the constructed representation. These reactions act as feedback mechanisms with differing dynamics that tend to exaggerate the automatic reaction to a stimulus and result in the cortex taking on a wide range of excitatory or inhibitory states. Short duration states are experienced as emotions, and those that persist long after the causal stimuli have abated are experienced as moods.

Although emotional states can have an almost overwhelming influence on actions and behaviour, there is only a broad understanding of the mechanisms that trigger the release of specific chemicals and the regions in the brain that respond, and also of the relationships between concentration and effect and the dynamics of dissipation. The effects of a wide range of exogenous chemicals that can be ingested or injected and thence diffuse into the cerebro-spinal fluid from the bloodstream are better known. The gamut of emotional states is unmapped, in part because the phenomena are definitively subjective, and in part because of their complexity.

The third consequence of the secretion of neuro-chemicals is to include, in the cluster of attributes forming the template for a represented entity, efferent signals from the somesthetic and other body systems affected by their excitatory or inhibitory reaction. In Edelman's terminology[7], the qualia (qualities) that form the schema (internal representation) of an object or situation include a pre-disposed emotional judgement. It is suggested that these associated qualities not only form the basis of aesthetics – the liking or disliking of specific visual, auditory, chemical and tactile sensations – but by acting as rewards and punishments they constitute the essential motivating mechanism. In particular, the chemically-mediated rewards induce the urge to repeat sense-decide-act sequences that become useful habits and skills.

D.5 Control of behaviour

This scheme is an attempt to attribute human behaviour to a hierarchy of control processes that correspond to the general notions of being alive, being responsive, being alert, being aware, being attentive and being purposeful.

The pilot's ability to do his job is obviously dependent on blood circulation, vaso-dilation, homeostasis, digestion, sweating and the other body maintenance functions summarised in Figure 2.1. His responsiveness relies on processes such as breathing, pupil dilation and holding an upright sitting posture. His alertness depends upon the extraction of cues from the stimulation received from the external world, and it is from these that he builds a conscious awareness of the surroundings. All of these processes are essentially underpinning. Without a goal the pilot's behaviour, like any passenger's, would default to a set of behavioural habits, and it is the appropriate functioning of the higher levels that transform him into a useful element in the aircraft. The top level holds the mission plan and decides what skill should be brought to bear at any instant, and the goal it should achieve. The skills themselves are complex decision sequences that need relevant information. This is gained by commanding lower level processes to augment the detail within a limited span of the underpinning awareness, which in the case of the pilot, is largely done by directing foveal vision. Other lower level processes are also routinely over-ridden, for instance breathing is modulated when speaking and suspended during high-G muscular straining.

The highest level processes, in particular those that formulate intentions and erect the finely grained short-term goals that direct skilled performance, are the least understood. Much that is considered in the context of attention – the stimuli that receive selective processing and the nature of this processing – ignores the mechanism that selects the goals; any transfer of attention must be triggered by a transfer of intention. An examination of intentions would, however, need an exploration of the wider issues of motivation and include the intrinsic rewards and emotional burdens of the job.

References

1. Posner M I and Rothblat M K, *Attention, self-regulation and consciousness,* Phil. Trans. R. Soc. Lond. B 353, pp1915-1927, 1998
2. Edelman G, *Bright Air, Brilliant Fire,* Penguin, 1992
3. Matin L, *Visual localisation and eye movements*, in Boff K R, Kaufman L and Thomas J P (eds), *Handbook of perception and human performance, Volume 1: Sensory processes and perception*, Wiley, 1986

4. In *Perception as hypothesis*, in Gregory R L (ed) *The Oxford companion to the mind*, Oxford University Press, 1987

5. Barbeito R, *Sighting dominance: an explanation based on the processing of visual direction in tests of sighting dominance*, Vision Research 21 (6), pp855-860, 1981

6. Damasio A R, *Investigating the biology of consciousness*, Phil. Trans. R. Soc. Lond. B 353, pp1879-1882, 1998

7. Edelman G, op.cit.

Acronyms and Abbreviations

AAIB Air Accidents Investigation Branch 36
AAR Air-to-Air Refuelling 72
ACCES Attenuating Custom Communication Earpiece System 329
ACES Advanced Concept Ejection Seat 304
ACGIH American Conference of Governmental Industrial Hygenists 230
ADC Analogue-to-Digital Converter 268
ADC Air Data Computer 129
ADP Adenosine DiPhosphate 49
ADS Aeronautical Design Standard 91
AEA Aircrew Equipment Assembly 317
AEW Airborne Early Warning 72
AGC Automatic Gain Control 268
AGSM Anti-G Straining Manoeuvre 57
AI Air Interdiction 72
AI Attitude Indicator 131
AI Artificial Intelligence 361
AIMIS Advanced Integrated Modular Instrumentation System 183
AMEL Active Matrix Electro-Luminescent 117
AMFLCD Active Matrix Ferro-electric Liquid Crystal Display 176
AMLCD Active Matrix Liquid Crystal Display 117
AMTT Advanced Metal Tolerant Tracker 281
ANR Active Noise Reduction 329
ANVIS Aviator Night Vision 112
AOA Angle Of Attack 131
AR Anti-Reflection 159
AR-5 Aircrew Respirator Mk5 342
ASI Air-Speed Indicator 129
ASR Automatic Speech Recognition 266
ASR Air-Sea Rescue 72
ASRS Aircraft Safety Reporting System 37
ATC Air Traffic Control 131
ATP Adenosine TriPhosphate 49
AutoGCAS Automatic Ground Collision Avoidance System 365
AWACS Airborne Warning And Control System 362
BOFOV Binocular Overlap Field Of View 221
BTRU Barostatic Time Release Unit 308
CAD Computer-Aided Design 86
CALP Caution and Advisory Lights Panel 146
CAP Combat Air Patrol 72
CAS Calibrated Air Speed 130
CAS Close Air Support 72
CAT Clear Air Turbulence 43
CCD Charge-Coupled Device 345
CCIR Committee Consultative Internationale de Radio 383
CDA Climb/Dive Angle 203

CEP	Communication Ear Plug	329
CFIT	Controlled Flight Into Terrain	210
CHIRP	Confidential Human Factors Incident Reporting Programme	37
CIE	Commission Internationale de L'Eclairage	23
CJND	Chrominance Just Noticeable Differences	160
CMOS	Complementary Metal Oxide on Silicon	287
COGMON	COGnitive MONitor	367
COLPRO	COLlective PROtection	63
COMBAT EDGE	COMbined Advanced Technology Enhanced Designed G-Ensemble	342
COMED	COmbined Map and Electronic Display	4
COTS	Commercial Off-The-Shelf	371
CPL	Commercial Pilot Licence	38
CR	Corneal Reflex	286
CREST	CRew Escape Systems Technology	311
CRT	Cathode Ray Tube	163
CUD	Chin-Up Display	159
DAC	Digital to Analogue Converter	279
DAS	Distributed Aperture System	121
DCG	Dichromated Gelatin	199
DEF STAN	DEFence STANdard	71
DERA	Defence Evaluation and Research Agency	243
DEW	Directed-Energy Weapon	65
DG	Directional Gyro	134
DGA	Delegation Generale pour l'Armement	271
DMD	Digital Micro-mirror Display	177
DME	Distance Measuring Equipment	134
DOE	Diffractive Optical Element	199
DoF	Degree of Freedom	290
DORC	DisOrientation Recovery Capability	366
DRAWS	Defence Research Agency Workload Scale	89
DS	Deep Strike	72
DTED	Digital Terrain Elevation Data	143
DTW	Dynamic Time Warping	269
DVE	Degraded Visual Environment	92
EBI	Equivalent Background Illumination	113
ECS	Environmental Conditioning System	44
EEG	Electro-EncephaloGraphic	289
EEMD	Engine and Energy Management Display	141
EMG	Electro-MyoGraphic	289
EMP	Electro-Magnetic Pulse	64
EOG	Electro-OculoGraphic	289
EPIRB	Emergency Position Indicating Radio Beacon	308
ERP	Event-Related Potential	292
EW	Electronic Warfare	72
FAA	Federal Aviation Authority	77
FCS	Flight Control System	75
FET	Field-Effect Transistor	171
FLILO	Flying Infra-red for Low-level Operations	121
FLIR	Forward-Looking Infra-Red	119
fNMRI	functional Nuclear Magnetic Resonance Imaging	15

FOHMED	Fibre-Optic Head-Mounted Electronic Display	239
FoR	Field of Regard	213
FOV	Field Of View	227
FPV	Face Protection Visor	114
GCAS	Ground Collision Avoidance System	362
GLOC	G-induced Loss Of Consciousness	55
GLS	Grid Line Slope	108
GPS	Global Positioning System	130
GVE	Good Visual Environment	92
HD	Hazard – Distilled mustard agent	62
HDD	Head-Down Display	136
HEPA	High Efficiency Particulate Air Filter	63
HFI	Human Factors Integration	70
HL	Hazard – Lewisite	62
HLD	Head-Level Display	144
HMD	Helmet-Mounted Display	213
HMM	Hidden Markov Modelling	270
HMS	Helmet-Mounted Sight	213
HOE	Holographic Optical Element	199
HOM	Human Operator Model	81
HOTAS	Hands On Throttle And Stick	265
HQR	Handling Qualities Rating	89
HRIR	Head-Related Impulse Response	252
HRTF	Head-Related Transfer Functions	252
HSD	Horizontal Situation Display	136
HTL	Hole Transport Layer	181
HUD	Head-Up Display	189
IAS	Indicated Air Speed	129
ICAO	International Civil Aviation Organisation	41
ICESC	Industry Crew Escape System Committee	303
ICNIRP	International Commission of Non-Ionizing Radiation Protection	230
IFR	Instrument Flight Rules	45
IHADSS	Integrated Helmet and Display Sighting System	121
IID	Inter-aural Intensity Difference	251
ILS	Instrument Landing System	45
IMA	Integrated Modular Avionics	8
IMC	Instrument Meteorological Conditions	45
IMELDA	Integrated Mel-Scaled Linear Discriminant Analysis	269
I-NIGHTS	Interim-Night integrated Goggle and Head Tracking System	326
INS	Inertial Navigation System	275
IPE	Individual Protective Equipment	63
IPME	Integrated Performance Modelling Environment	83
IPT	Integrated Product Team	70
ITD	Inter-aural Time Difference	251
ITO	Indium/Tin Oxide	106
JDL	Joint Directors of Laboratories	362
JEFTS	Joint Elementary Flying Training Squadron	38
JHMCS	Joint Helmet-Mounted Cueing System	344
JSF	Joint Strike Fighter	215
JTIDS	Joint Tactical Information Distribution System	362
LANTIRN	Low Altitude Navigation and Targeting Infra-Red for Night	119

LC	Liquid Crystal	169
LCOS	Liquid Crystal on Silicon	176
LDP	Logical Display Plane	222
LED	Light-Emitting Diode	181
LEDD	Light-Emitting Diode Display	180
LEM	Light Emission Layer	181
LEPSD	Lateral Effect Photo-Sensitive Detector	278
LFD	Longitudinal Fuselage Datum	203
LJND	Luminance Just Noticeable Differences	160
LPC	Linear Predictive Coding	269
LUT	Look Up Table	281
MANPRINT	MANpower and PeRsonnel INTegration	70
MAV	Mean Absolute Value	290
MDC	Micro-Detonating Cord	309
MEG	Magneto-EncephaloGraphic	15
MFC	Multi-Function Control	75
MFCC	Mel-Frequency-Cepstrel Coefficient	269
MFD	Multi-Function Display	75
MHP	Model Human Processor	82
MIDAS	Man-Machine Integrated Design and Analysis System	84
MIL-HDBK	MILitary HanDBooK	72
MLS	Microwave Landing System	45
MMD	Master Monitor Display	141
MOVTAS	Modified Visual Target Acquisition Set	277
MPE	Maximum Permissible Exposure	229
MQR	Mask Quick Release	334
MRT	Modified Rhyme Test	247
MTE	Mission Task Element	92
MTF	Modulation Transfer Function	380
MTFA	Modulation Transfer Function Area	381
NBC	Nuclear Biological and Chemical	62
NIRS	Near-Infra-Red Spectroscopy	15
NMRI	Nuclear Magnetic Resonance Imaging	15
NN	Neural Network	270
NVG	Night Vision Goggle	146
NVIS	Night Vision Imaging System	150
OAR	Oto-Acoustic Reflex	17
OBOGS	On-Board Oxygen Generation System	51
OCA	Offensive Counter Air	72
OCU	Operational Conversion Unit	38
OLED	Organic Light Emitting Diode	181
OODA	Observe–Orient–Decide–Act	362
PDP	Programmable Display Processor	142
PEC	Personal Equipment Connector	87
PET	Positron Emission Topography	15
PFD	Primary Flight Display	45
PI	Purkinje Image	283
PIA	Precision Instrument Approach	46
PJND	Perceivable Just Noticeable Difference	160
PLP	Perceptual Linear Prediction	269
POC	Performance Operating Characteristic	34

POP	Prediction of Operator Performance	84
PPL	Private Pilot Licence	38
PROM	Programmable Read-Only Memory	240
PSM	Post-Stall Manoeuvrability	48
PTT	Press-To-Talk	265
PWR	Passive Warning Receiver	355
QFE, QNE, QNH		129
QWI	Qualified Weapons Instructor	38
RAP	Reactive Action Packet	84
RCDH	Rate Command Heading Hold	93
RCHH	Rate Command Height Hold	93
RVR	Runway Visibility Range	46
SA	Situation Awareness	90
SASS	Situation Assessor Support System	367
SAM	Surface-to-Air Missile	72
SE	Specific Energy	47
SEAD	Suppression of Enemy Air Defences	72
SEP	Specific Excess Power	47
SII	Speech Intelligibility Index	247
SNR	Speech-to-Noise Ratio	247
SOP	Standard Operating Procedure	37
SPL	Sound Pressure Level	18
SRAM	Static Random Access Memory	177
SRD	System Requirements Document	72
SSU	Sensor Surveying Unit	277
STANAG	STANdardisation AGreement	71
STI	Speech Transmission Index	247
TAS	True Air-Speed	130
TFEL	Thin Film Electro-Luminescent	179
TFOV	Total Field Of View	220
TIM	Task Interface Manager	367
TLX	Task Load Index	89
TRU	Time Release Unit	301
UAS	University Air Squadron	38
UAV	Unmanned Air Vehicle	368
UCE	Useable Cue Environment	93
UCS	Uniform Colour Space	159
URD	User Requirements Document	72
USAARL	US Army Aeromedical Research Laboratory	325
VC	Virtual Cockpit	355
VCR	Visual Cue Rating	93
VCS	Visually-Coupled System	120
VEP	Visually-Evoked Potential	292
VFR	Visual Flight Rules	45
VHF	Very High Frequency	134
VIC	Voice Input Control	266
VIS	VISibility	46
VMC	Visual Meteorological Conditions	45
VOR	Vestibulo-Ocular Reflex	57
VOR	Voice Omni-directional Radio	59
VOS	Voice-Operated Switch	248

VR	Virtual Reality	356
VRD	Virtual Retinal Display	230
VSD	Vertical Situation Display	136
VSI	Vertical Situation Indicator	135
VSI	Vertical Speed Indicator	130
VSTOL	Vertical/Short Take-Off and Landing	208
VTAS	Visual Target Acquisition System	214
VV	Velocity Vector	203
Windex	Workload index	84
YAG	Yttrium Aluminium Garnet	164

Index

For Product Safety Concerns and Information please contact our EU
representative GPSR@taylorandfrancis.com Taylor & Francis Verlag GmbH,
Kaufingerstraße 24, 80331 München, Germany

Printed and bound by CPI Group (UK) Ltd, Croydon, CR0 4YY
01/05/2025
01858513-0004